The Metal-Driven Biogeochemistry of Gaseous Compounds in the Environment

Metal Ions in Life Sciences

Volume 14

Guest Editors:

Peter M.H. Kroneck and Martha E. Sosa Torres

Series Editors:

Astrid Sigel, Helmut Sigel, and Roland K.O. Sigel

For further volumes:
http://www.springer.com/series/8385 and http://www.mils-series.com

Astrid Sigel • Helmut Sigel • Roland K.O. Sigel
Series Editors

Peter M.H. Kroneck • Martha E. Sosa Torres
Guest Editors

The Metal-Driven Biogeochemistry of Gaseous Compounds in the Environment

 Springer

Guest Editors
Peter M.H. Kroneck
Fachbereich Biologie
Universität Konstanz
Universitätsstrasse 10
D-78457 Konstanz, Germany
peter.kroneck@uni-konstanz.de

Martha E. Sosa Torres
Departamento de Química Inorganica y Nuclear
Facultad de Química
Universidad Autónoma de México
Ciudad Universitaria
México, D.F. 04510, México
mest@unam.mx

Series Editors
Astrid Sigel • Helmut Sigel
Department of Chemistry
Inorganic Chemistry
University of Basel
Spitalstrasse 51
CH-4056 Basel, Switzerland
astrid.sigel@unibas.ch
helmut.sigel@unibas.ch

Roland K.O. Sigel
Department of Chemistry
University of Zürich
Winterthurerstrasse 190
CH-8057 Zürich
Switzerland
roland.sigel@aci.uzh.ch

ISSN 1559-0836
ISBN 978-94-017-9268-4
DOI 10.1007/978-94-017-9269-1
Springer Dordrecht Heidelberg New York London

ISSN 1868-0402 (electronic)
ISBN 978-94-017-9269-1 (eBook)

Library of Congress Control Number: 2014954800

Cover illustration: Cover figure of the MILS series since Volume 11: RNA-protein interface of the Ile-tRNA synthetase complex held together by a string of Mg^{2+} ions, illustrating the importance of metal ions in both the protein and the nucleic acid world as well as connecting the two; hence, representing the role of *Metal Ions in Life Sciences*. tRNA synthetases are not only essential to life, but also serve as a target for novel classes of drugs making such RNA-protein complexes crucial also for the health sciences. The figure was prepared by Joachim Schnabl and Roland K. O. Sigel using the PDB coordinates 1FFY.

Printed on acid-free paper

Springer is part of Springer Science+Business Media (www.springer.com)

Historical Development and Perspectives of the Series

Metal Ions in Life Sciences[*]

It is an old wisdom that metals are indispensable for life. Indeed, several of them, like sodium, potassium, and calcium, are easily discovered in living matter. However, the role of metals and their impact on life remained largely hidden until inorganic chemistry and coordination chemistry experienced a pronounced revival in the 1950s. The experimental and theoretical tools created in this period and their application to biochemical problems led to the development of the field or discipline now known as *Bioinorganic Chemistry*, *Inorganic Biochemistry*, or more recently also often addressed as *Biological Inorganic Chemistry*.

By 1970 *Bioinorganic Chemistry* was established and further promoted by the book series *Metal Ions in Biological Systems* founded in 1973 (edited by H.S., who was soon joined by A.S.) and published by Marcel Dekker, Inc., New York, for more than 30 years. After this company ceased to be a family endeavor and its acquisition by another company, we decided, after having edited 44 volumes of the *MIBS* series (the last two together with R.K.O.S.) to launch a new and broader minded series to cover today's needs in the *Life Sciences*. Therefore, the Sigels new series is entitled

Metal Ions in Life Sciences.

After publication of the first four volumes (2006–2008) with John Wiley & Sons, Ltd., Chichester, UK, and the next five volumes (2009–2011) with the Royal Society of Chemistry, Cambridge, UK, we are happy to join forces now in this still new endeavor with Springer Science & Business Media B.V., Dordrecht, The Netherlands, a most experienced Publisher in the *Sciences*.

---

[*]Reproduced with some alterations by permission of John Wiley & Sons, Ltd., Chichester, UK (copyright 2006) from pages v and vi of Volume 1 of the series *Metal Ions in Life Sciences* (MILS-1).

The development of *Biological Inorganic Chemistry* during the past 40 years was and still is driven by several factors; among these are (i) the attempts to reveal the interplay between metal ions and peptides, nucleotides, hormones or vitamins, etc., (ii) the efforts regarding the understanding of accumulation, transport, metabolism and toxicity of metal ions, (iii) the development and application of metal-based drugs, (iv) biomimetic syntheses with the aim to understand biological processes as well as to create efficient catalysts, (v) the determination of high-resolution structures of proteins, nucleic acids, and other biomolecules, (vi) the utilization of powerful spectroscopic tools allowing studies of structures and dynamics, and (vii), more recently, the widespread use of macromolecular engineering to create new biologically relevant structures at will. All this and more is and will be reflected in the volumes of the series *Metal Ions in Life Sciences*.

The importance of metal ions to the vital functions of living organisms, hence, to their health and well-being, is nowadays well accepted. However, in spite of all the progress made, we are still only at the brink of understanding these processes. Therefore, the series *Metal Ions in Life Sciences* will endeavor to link coordination chemistry and biochemistry in their widest sense. Despite the evident expectation that a great deal of future outstanding discoveries will be made in the interdisciplinary areas of science, there are still "language" barriers between the historically separate spheres of chemistry, biology, medicine, and physics. Thus, it is one of the aims of this series to catalyze mutual "understanding".

It is our hope that *Metal Ions in Life Sciences* proves a stimulus for new activities in the fascinating "field" of *Biological Inorganic Chemistry*. If so, it will well serve its purpose and be a rewarding result for the efforts spent by the authors.

Astrid Sigel and Helmut Sigel
Department of Chemistry, Inorganic Chemistry,
University of Basel, CH-4056 Basel, Switzerland

Roland K.O. Sigel
Department of Chemistry,
University of Zürich, CH-8057 Zürich, Switzerland

October 2005,
October 2008,
and August 2011

Preface to Volume 14

The Metal-Driven Biogeochemistry of Gaseous Compounds in the Environment

In this volume of the *Metal Ions in Life Sciences* series the transition metal-driven activation, transformation, and production of environmentally essential gases are discussed. These include dihydrogen (H_2), methane (CH_4), carbon monoxide (CO), acetylene (HC≡CH), dinitrogen (N_2), nitrous oxide (N_2O), and ammonia (NH_3), as well as the smelly gases hydrogen sulfide (H_2S) and dimethylsulfide (CH_3-S-CH_3). Several of these gases, which are covered in this volume, serve as important fuels worldwide for the increasingly energy-demanding societies. Other gases – most likely since life arose on Earth – have been consumed and are still in use as energy source for a large variety of different microorganisms. To conserve energy by living on these gases, or to synthesize them from starting materials such as sulfate/sulfite (SO_4^{2-}/SO_3^{2-}), nitrate/nitrite (NO_3^-/NO_2^-), or carbon dioxide (CO_2), highly skilled *Chemists* are required which can develop sophisticated metal-dependent catalysts to overcome unusually high activation barriers, and which can cope with extreme and toxic environmental conditions. To achieve the same is a tremendous challenge for "human" chemists.

In Chapter 1 the reader is shortly introduced to the Early Earth atmosphere and Early Life catalysts, with a special focus on clay minerals and their potential role (i) in the synthesis of essential molecules, such as amino acids, and (ii) in creating a chiral template for enantiomeric selection. Chapter 2 deals with a rather exotic and ancient gas on Earth, namely acetylene, and its transformation to acetaldehyde (CH_3CHO) by bacteria. This transformation is mechanistically clearly a non-redox reaction, however, a redox-active tungsten center in close proximity to an iron-sulfur cluster is required for catalysis.

Chapters 3 and 4 are devoted to carbon monoxide, both a toxic gas - *The ol' Detroit perfume* praised in Papa Hobo Lyrics by songwriter Paul Simon - and an important fuel for anaerobic as well as aerobic microorganisms. Again, a sophisticated catalytic machinery hosting either nickel, sulfur and iron, or copper, sulfur, and molybdenum, had to be developed to ensure proper handling of CO. Dihydrogen, in terms of chemical composition, is the simplest gas in the atmosphere, but biological H_2 chemistry could have been key in the origin of life on Earth as discussed in Chapter 5. Consequently, in view of its fundamental importance, a remarkable set of tools exists in Nature to interconvert protons and H_2.

Chapters 6 and 7 focus on the two gases dinitrogen and methane which require top catalysts for transformation and synthesis. To form bioavailable NH_3 from N_2, the stable N,N triple bond has to be cleaved under ambient conditions by the enzyme nitrogenase which carries the largest iron-sulfur cluster in Nature, clearly a masterpiece in transition metal catalysis. At the industrial level, the production of ammonia, by using dinitrogen from the atmosphere as its key ingredient, was invented by Fritz Haber and Carl Bosch to solve a problem that farmers faced across the globe. Churning out ammonia in industrial quantities triggered a green revolution. Several billion people are alive today only because Haber and Bosch found a way to turn atmospheric nitrogen into ammonia fertiliser. *Bread from air*, ran the slogan that advertised this work at the time (R. McKie, The Observer, November 2013). Microorganisms involved in the production and degradation of methane, both a fuel and a greenhouse gas, employ an unusual nickel enzyme, methyl-coenzyme M reductase, equipped with a tetrapyrrole cofactor called coenzyme F_{430}, which can traverse the oxidation states I, II, and III of nickel.

In the following Chapters 8 and 9, the biogeochemistry of two further nitrogen gases, nitrous oxide and ammonia, is addressed, representing key components within the global nitrogen cycle. N_2O, a potent greenhouse gas and kinetically inert similar to N_2, requires a unique tetranuclear copper-sulfur cluster to become split into N_2 and water. Ammonia, the most basic gas in our atmosphere, can be produced from nitrite in a remarkable multi-electron, multi-proton transfer process performed by an array of five to eight c-type cytochromes tightly packed into a protein matrix.

Finally, Chapters 10 and 11 are devoted to the sulfur gases hydrogen sulfide and dimethylsulfide. The conversion of sulfate/sulfite to H_2S (and the reverse reaction) represents an ancient process on Earth, powered by another unique redox center, in this case a green cofactor, called siroheme, which is bridged to an iron-sulfur cluster. Despite its toxicity (higher than that of CO), H_2S must be regarded as an essential molecule for both anaerobic and aerobic organisms. Last but not least, the focus is on dimethylsulfide and its molybdenum-dependent chemistry. Dimethylsulfide is a volatile organic sulfur compound with a characteristic unpleasant odor. It is the largest biogenic source of sulfur to the atmosphere, its sea-to-air transfer represents an important link within the global sulfur cycle that is responsible for sulfur transport from the oceans to the continents. Atmospheric transport of sulfate and methanesulfonic acid (CH_3SO_2OH), produced from CH_3-S-CH_3, and their deposition in terrestrial environments contribute to maintaining sulfur levels in soils which is important for plant productivity.

In summary, this volume offers a wealth of information about important processes in our biosphere. The fundamental role of transition metals (iron, nickel, copper, molybdenum, tungsten) is addressed in the activation, transformation, and synthesis of essential gases and their impact on the environment.

Note, the next volume in this series (MILS-15) is closely related to the present one and entitled *Sustaining Life on Planet Earth: Metalloenzymes Mastering Dioxygen and Other Chewy Gases*. The evolution of dioxygen-producing cyanobacteria was *The Event* in the history of life after the evolution of life itself. The accumulation of O_2 in the atmosphere forever changed the surface chemistry of the Earth. As electron acceptor it is used in the respiration of numerous different organisms that conduct a wide variety of chemically complex metabolisms. In summary, relationships between life, dioxygen, and the surface chemistry of the Earth are evident. However, there exist "impossible" microorganisms which use the oxidative power of nitric oxide (NO) by forging this molecule to ammonium, thereby making hydrazine (N_2H_4). Others can disproportionate NO into N_2 and O_2. This intracellularly produced O_2 enables these bacteria to adopt an aerobic mechanism for methane oxidation, as discussed in the upcoming volume MILS-15.

Peter M.H. Kroneck
Martha E. Sosa Torres

Contents

Historical Development and Perspectives of the Series   v

Preface to Volume 14   vii

Contributors to Volume 14   xv

Titles of Volumes 1–44 in the *Metal Ions in Biological Systems* Series   xvii

Contents of Volumes in the *Metal Ions in Life Sciences* Series   xix

1   The Early Earth Atmosphere and Early Life Catalysts   1
     Sandra Ignacia Ramírez Jiménez

     Abstract   1
     1   The Early Earth Atmosphere and Lithosphere   2
     2   Catalysts in the Early Earth   4
     3   Clays as Possible Catalysts in the Synthesis of Biomolecules   6
     4   General Conclusions   12
     References   13

2   Living on Acetylene. A Primordial Energy Source   15
     Felix ten Brink

     Abstract   15
     1   Introduction   16
     2   Acetylene   16
     3   Bacteria Living on Acetylene   18
     4   Acetylene Hydratase from *Pelobacter acetylenicus*   20
     5   Conclusions   32
     References   34

**3   Carbon Monoxide. Toxic Gas and Fuel for Anaerobes
      and Aerobes: Carbon Monoxide Dehydrogenases**   37
Jae-Hun Jeoung, Jochen Fesseler, Sebastian Goetzl,
and Holger Dobbek

Abstract   38
1   Introduction   38
2   Structure and Function of Carbon Monoxide Dehydrogenases   43
3   Concluding Remarks and Future Directions   64
References   65

**4   Investigations of the Efficient Electrocatalytic Interconversions
      of Carbon Dioxide and Carbon Monoxide by Nickel-Containing
      Carbon Monoxide Dehydrogenases**   71
Vincent C.-C. Wang, Stephen W. Ragsdale, and Fraser A. Armstrong

Abstract   72
1   Direct Carbon Dioxide/Carbon Monoxide Interconversions
    in Biology   72
2   Nickel-Containing Carbon Monoxide Dehydrogenases   75
3   Protein Film Electrochemistry   80
4   Carbon Monoxide Dehydrogenases as Electrocatalysts   82
5   Potential-Dependent Reactions with Inhibitors   85
6   Demonstrations of Technological Significance   91
7   Conclusions   95
References   96

**5   Understanding and Harnessing Hydrogenases, Biological
      Dihydrogen Catalysts**   99
Alison Parkin

Abstract   100
1   Introduction   100
2   Dihydrogen Cycles and Hydrogenases   100
3   [NiFe] Hydrogenases   111
4   Nickel-Free Hydrogenases: [FeFe] and [Fe] Enzymes   115
5   Insights into Hydrogenase Mechanism from Small
    Molecule Mimics   119
6   General Conclusions   121
References   122

6   Biochemistry of Methyl-Coenzyme M Reductase:
The Nickel Metalloenzyme that Catalyzes the Final Step
in Synthesis and the First Step in Anaerobic Oxidation
of the Greenhouse Gas Methane   125
Stephen W. Ragsdale

Abstract   126
1   Introduction   126
2   Structure and Properties of Methyl-Coenzyme M Reductase
    and Its Bound Coenzyme F_{430}   129
3   Redox and Coordination Properties of the Nickel Center
    in Methyl-Coenzyme M Reductase   133
4   The Catalytic Mechanism of Methyl-Coenzyme M Reductase   137
5   Summary and Prospects for Future Science and Technology   140
References   142

7   Cleaving the N,N Triple Bond: The Transformation
of Dinitrogen to Ammonia by Nitrogenases   147
Chi Chung Lee, Markus W. Ribbe, and Yilin Hu

Abstract   148
1   Introduction   148
2   The Structural and Biochemical Properties of Mo-Nitrogenase   150
3   The Catalytic Mechanism of Mo-Nitrogenase   157
4   The Distinct Structural and Catalytic Features of V-Nitrogenase . . .   166
5   Conclusions   172
References   173

8   No Laughing Matter: The Unmaking of the Greenhouse Gas
Dinitrogen Monoxide by Nitrous Oxide Reductase   177
Lisa K. Schneider, Anja Wüst, Anja Pomowski, Lin Zhang,
and Oliver Einsle

Abstract   178
1   Introduction: The Biogeochemical Nitrogen Cycle   179
2   Nitrous Oxide: Environmental Effects and Atmospheric
    Chemistry   181
3   Nitrous Oxide Reductase   184
4   Biogenesis and Assembly of Nitrous Oxide Reductase   200
5   Activation of Nitrous Oxide: The Workings of Nitrous Oxide
    Reductase   203
6   General Conclusions   206
References   208

9   The Production of Ammonia by Multiheme Cytochromes *c* 211
Jörg Simon and Peter M.H. Kroneck

Abstract 211
1   Introduction 212
2   Ammonia and Its Role in the Environment 215
3   Enzymes Involved in Ammonia Turnover 217
4   Cytochrome *c* Nitrite Reductase as Paradigm 217
5   Other Multiheme Cytochromes *c* 231
6   Environmental Issues and Conclusions 231
References 233

**10   Hydrogen Sulfide: A Toxic Gas Produced by Dissimilatory
Sulfate and Sulfur Reduction and Consumed
by Microbial Oxidation** 237
Larry L. Barton, Marie-Laure Fardeau, and Guy D. Fauque

Abstract 238
1   Introduction 238
2   Enzymology of Hydrogen Sulfide Production from Sulfate 250
3   Enzymology of Hydrogen Sulfide Production from Elemental
     Sulfur 262
4   Microbial Oxidation of Hydrogen Sulfide to Sulfate 266
5   Conclusions 270
References 272

11   Transformations of Dimethylsulfide 279
Ulrike Kappler and Hendrik Schäfer

Abstract 280
1   Introduction 280
2   Enzymology of Dimethylsulfide Transformations 285
3   General Conclusions 306
References 308

Index 315

Contributors to Volume 14

Numbers in parentheses indicate the pages on which the authors' contributions begin.

Fraser A. Armstrong Department of Chemistry, Inorganic Chemistry Laboratory, University of Oxford, South Parks Road, Oxford, OX1 3QR, UK, fraser.armstrong@chem.ox.ac.uk (71)

Larry L. Barton Department of Biology, University of New Mexico, MSCO3 2020, Albuquerque, NM, USA, lbarton@unm.edu (237)

Holger Dobbek Institut für Biologie, Strukturbiologie/Biochemie, Humboldt-Universität zu Berlin, Unter den Linden 6, D-10099 Berlin, Germany, holger.dobbek@biologie.hu-berlin.de (37)

Oliver Einsle Institute for Biochemistry, Albert-Ludwigs-Universität Freiburg, D-79104 Freiburg im Breisgau, Germany, einsle@biochemie.uni-freiburg.de (177)

Marie-Laure Fardeau Institut Méditerranéen d'Océanologie (MIO), Aix-Marseille Université, USTV, UMR CNRS 7294/IRD 235, Campus de Luminy, Case 901, F-13288 Marseille Cedex 09, France, marie-laure.fardeau@univ-amu.fr (237)

Guy D. Fauque Institut Méditerranéen d'Océanologie (MIO), Aix-Marseille Université, USTV, UMR CNRS 7294/IRD 235, Campus de Luminy, Case 901, F-13288 Marseille Cedex 09, France, guy.fauque@univ-amu.fr (237)

Jochen Fesseler Institut für Biologie, Strukturbiologie/Biochemie, Humboldt-Universität zu Berlin, Unter den Linden 6, D-10099 Berlin, Germany (37)

Sebastian Goetzl Institut für Biologie, Strukturbiologie/Biochemie, Humboldt-Universität zu Berlin, Unter den Linden 6, D-10099 Berlin, Germany (37)

Yilin Hu Department of Molecular Biology and Biochemistry, 2236 McGaugh Hall, University of California, Irvine, CA 92697-3900, USA, yilinh@uci.edu (147)

Jae-Hun Jeoung Institut für Biologie, Strukturbiologie/Biochemie, Humboldt-Universität zu Berlin, Unter den Linden 6, D-10099 Berlin, Germany (37)

Ulrike Kappler School of Chemistry and Molecular Biosciences, The University of Queensland, 76 Molecular Biosciences Bldg, St. Lucia, Qld 4072, Australia, u.kappler@uq.edu.au (279)

Peter M.H. Kroneck Fachbereich Biologie, Universität Konstanz, Universitätsstrasse 10, D-78457 Konstanz, Germany, peter.kroneck@uni-konstanz.de (211)

Chi Chung Lee Department of Molecular Biology and Biochemistry, 2236 McGaugh Hall, University of California, Irvine, CA 92697-3900, USA, chichul@uci.edu (147)

Alison Parkin Department of Chemistry, University of York, Heslington, York, YO10 5DD, UK, alison.parkin@york.ac.uk (99)

Anja Pomowski Institute for Biochemistry, Albert-Ludwigs-Universität Freiburg, D-79104 Freiburg im Breisgau, Germany (177)

Stephen W. Ragsdale Department of Biological Chemistry, University of Michigan Medical School, Ann Arbor, MI 48109-0606, USA, sragsdal@umich.edu (71, 125)

Sandra Ignacia Ramírez Jiménez Centro de Investigaciones Químicas, Universidad Autónoma del Estado de Morelos, Av. Universidad # 1001, Col. Chamilpa, Cuernavaca 62209, Morelos, México, ramirez_sandra@uaem.mx (1)

Markus W. Ribbe Department of Molecular Biology and Biochemistry, 2236 McGaugh Hall, University of California, Irvine, CA 92697-3900, USA, and Department of Chemistry, University of California, Irvine, CA 92697-2025, USA, mribbe@uci.edu (147)

Hendrik Schäfer School of Life Sciences, Gibbet Hill Campus, University of Warwick, Coventry, CV4 7AL, UK (279)

Lisa K. Schneider Institute for Biochemistry, Albert-Ludwigs-Universität Freiburg, D-79104 Freiburg im Breisgau, Germany (177)

Jörg Simon Microbial Energy Conversion and Biotechnology, Department of Biology, Technische Universität Darmstadt, Schnittspahnstrasse 10, D-64287 Darmstadt, Germany, simon@bio.tu-darmstadt.de (211)

Felix ten Brink Laboratoire de Bioénergetique et Ingenerie des Protéines, Institut de Microbiologie de la Méditerranée, CNRS/Aix-Marseille Université, F-13402 Marseille Cedex 20, France, ftenbrink@imm.cnrs.fr (15)

Vincent C.-C. Wang Department of Chemistry, Inorganic Chemistry Laboratory, University of Oxford, South Parks Road, Oxford, OX1 3QR, UK (71)

Anja Wüst Institute for Biochemistry, Albert-Ludwigs-Universität Freiburg, D-79104 Freiburg im Breisgau, Germany (177)

Lin Zhang Institute for Biochemistry, Albert-Ludwigs-Universität Freiburg, D-79104 Freiburg im Breisgau, Germany (177)

Titles of Volumes 1–44 in the
Metal Ions in Biological Systems Series

edited by the SIGELs
and published by Dekker/Taylor & Francis (1973–2005)

Volume 1:   **Simple Complexes**
Volume 2:   **Mixed-Ligand Complexes**
Volume 3:   **High Molecular Complexes**
Volume 4:   **Metal Ions as Probes**
Volume 5:   **Reactivity of Coordination Compounds**
Volume 6:   **Biological Action of Metal Ions**
Volume 7:   **Iron in Model and Natural Compounds**
Volume 8:   **Nucleotides and Derivatives: Their Ligating Ambivalency**
Volume 9:   **Amino Acids and Derivatives as Ambivalent Ligands**
Volume 10:   **Carcinogenicity and Metal Ions**
Volume 11:   **Metal Complexes as Anticancer Agents**
Volume 12:   **Properties of Copper**
Volume 13:   **Copper Proteins**
Volume 14:   **Inorganic Drugs in Deficiency and Disease**
Volume 15:   **Zinc and Its Role in Biology and Nutrition**
Volume 16:   **Methods Involving Metal Ions and Complexes in Clinical
             Chemistry**
Volume 17:   **Calcium and Its Role in Biology**
Volume 18:   **Circulation of Metals in the Environment**
Volume 19:   **Antibiotics and Their Complexes**
Volume 20:   **Concepts on Metal Ion Toxicity**
Volume 21:   **Applications of Nuclear Magnetic Resonance to Paramagnetic
             Species**
Volume 22:   **ENDOR, EPR, and Electron Spin Echo for Probing
             Coordination Spheres**
Volume 23:   **Nickel and Its Role in Biology**
Volume 24:   **Aluminum and Its Role in Biology**
Volume 25:   **Interrelations Among Metal Ions, Enzymes, and Gene
             Expression**
Volume 26:   **Compendium on Magnesium and Its Role in Biology, Nutrition,
             and Physiology**

Volume 27:   **Electron Transfer Reactions in Metalloproteins**
Volume 28:   **Degradation of Environmental Pollutants by Microorganisms
             and Their Metalloenzymes**
Volume 29:   **Biological Properties of Metal Alkyl Derivatives**
Volume 30:   **Metalloenzymes Involving Amino Acid-Residue and Related
             Radicals**
Volume 31:   **Vanadium and Its Role for Life**
Volume 32:   **Interactions of Metal Ions with Nucleotides, Nucleic Acids,
             and Their Constituents**
Volume 33:   **Probing Nucleic Acids by Metal Ion Complexes of Small
             Molecules**
Volume 34:   **Mercury and Its Effects on Environment and Biology**
Volume 35:   **Iron Transport and Storage in Microorganisms, Plants, and
             Animals**
Volume 36:   **Interrelations Between Free Radicals and Metal Ions in Life
             Processes**
Volume 37:   **Manganese and Its Role in Biological Processes**
Volume 38:   **Probing of Proteins by Metal Ions and Their Low-Molecular-
             Weight Complexes**
Volume 39:   **Molybdenum and Tungsten. Their Roles in Biological Processes**
Volume 40:   **The Lanthanides and Their Interrelations with Biosystems**
Volume 41:   **Metal Ions and Their Complexes in Medication**
Volume 42:   **Metal Complexes in Tumor Diagnosis and as Anticancer Agents**
Volume 43:   **Biogeochemical Cycles of Elements**
Volume 44:   **Biogeochemistry, Availability, and Transport of Metals in the
             Environment**

Contents of Volumes in the
Metal Ions in Life Sciences Series

edited by the SIGELs

Volumes 1–4
published by John Wiley & Sons, Ltd., Chichester, UK (2006–2008)
<http://www.Wiley.com/go/mils>

Volume 5–9
by the Royal Society of Chemistry, Cambridge, UK (2009–2011)
<http://www.rsc.org/shop/metalionsinlifesciences>

and from Volume 10 on
by Springer Science & Business Media BV, Dordrecht, The Netherlands (since 2012)
<http://www.mils-series.com>

Volume 1   Neurodegenerative Diseases and Metal Ions

1   **The Role of Metal Ions in Neurology. An Introduction**
    Dorothea Strozyk and Ashley I. Bush

2   **Protein Folding, Misfolding, and Disease**
    Jennifer C. Lee, Judy E. Kim, Ekaterina V. Pletneva,
    Jasmin Faraone-Mennella, Harry B. Gray, and Jay R. Winkler

3   **Metal Ion Binding Properties of Proteins Related
    to Neurodegeneration**
    Henryk Kozlowski, Marek Luczkowski, Daniela Valensin,
    and Gianni Valensin

4   **Metallic Prions: Mining the Core of Transmissible
    Spongiform Encephalopathies**
    David R. Brown

5   **The Role of Metal Ions in the Amyloid Precursor Protein
    and in Alzheimer's Disease**
    Thomas A. Bayer and Gerd Multhaup

6   **The Role of Iron in the Pathogenesis of Parkinson's Disease**
    Manfred Gerlach, Kay L. Double, Mario E. Götz,
    Moussa B.H. Youdim, and Peter Riederer

7   ***In Vivo* Assessment of Iron in Huntington's Disease
    and Other Age-Related Neurodegenerative Brain Diseases**
    George Bartzokis, Po H. Lu, Todd A. Tishler, and Susan Perlman

8   **Copper-Zinc Superoxide Dismutase and Familial
    Amyotrophic Lateral Sclerosis**
    Lisa J. Whitson and P. John Hart

9   **The Malfunctioning of Copper Transport in Wilson
    and Menkes Diseases**
    Bibudhendra Sarkar

10 **Iron and Its Role in Neurodegenerative Diseases**
    Roberta J. Ward and Robert R. Crichton

11 **The Chemical Interplay between Catecholamines
    and Metal Ions in Neurological Diseases**
    Wolfgang Linert, Guy N.L. Jameson, Reginald F. Jameson,
    and Kurt A. Jellinger

12 **Zinc Metalloneurochemistry: Physiology, Pathology, and Probes**
    Christopher J. Chang and Stephen J. Lippard

13 **The Role of Aluminum in Neurotoxic and Neurodegenerative Processes**
    Tamás Kiss, Krisztina Gajda-Schrantz, and Paolo F. Zatta

14 **Neurotoxicity of Cadmium, Lead, and Mercury**
    Hana R. Pohl, Henry G. Abadin, and John F. Risher

15 **Neurodegerative Diseases and Metal Ions. A Concluding Overview**
    Dorothea Strozyk and Ashley I. Bush

Subject Index

Volume 2   Nickel and Its Surprising Impact in Nature

1   **Biogeochemistry of Nickel and Its Release into the Environment**
    Tiina M. Nieminen, Liisa Ukonmaanaho, Nicole Rausch,
    and William Shotyk

2   **Nickel in the Environment and Its Role in the Metabolism
    of Plants and Cyanobacteria**
    Hendrik Küpper and Peter M.H. Kroneck

3   **Nickel Ion Complexes of Amino Acids and Peptides**
    Teresa Kowalik-Jankowska, Henryk Kozlowski, Etelka Farkas,
    and Imre Sóvágó

4   **Complex Formation of Nickel(II) and Related Metal Ions**
    with Sugar Residues, Nucleobases, Phosphates, Nucleotides,
    and Nucleic Acids
    Roland K.O. Sigel and Helmut Sigel

5   **Synthetic Models for the Active Sites of Nickel-Containing Enzymes**
    Jarl Ivar van der Vlugt and Franc Meyer

6   **Urease: Recent Insights in the Role of Nickel**
    Stefano Ciurli

7   **Nickel Iron Hydrogenases**
    Wolfgang Lubitz, Maurice van Gastel, and Wolfgang Gärtner

8   **Methyl-Coenzyme M Reductase and Its Nickel Corphin**
    Coenzyme F_{430} in Methanogenic Archaea
    Bernhard Jaun and Rudolf K. Thauer

9   **Acetyl-Coenzyme A Synthases and Nickel-Containing**
    Carbon Monoxide Dehydrogenases
    Paul A. Lindahl and David E. Graham

10 **Nickel Superoxide Dismutase**
    Peter A. Bryngelson and Michael J. Maroney

11 **Biochemistry of the Nickel-Dependent Glyoxylase I Enzymes**
    Nicole Sukdeo, Elisabeth Daub, and John F. Honek

12 **Nickel in Acireductone Dioxygenase**
    Thomas C. Pochapsky, Tingting Ju, Marina Dang, Rachel Beaulieu,
    Gina Pagani, and Bo OuYang

13 **The Nickel-Regulated Peptidyl-Prolyl *cis*/*trans* Isomerase SlyD**
    Frank Erdmann and Gunter Fischer

14 **Chaperones of Nickel Metabolism**
    Soledad Quiroz, Jong K. Kim, Scott B. Mulrooney,
    and Robert P. Hausinger

15 **The Role of Nickel in Environmental Adaptation**
    of the Gastric Pathogen *Helicobacter pylori*
    Florian D. Ernst, Arnoud H.M. van Vliet, Manfred Kist,
    Johannes G. Kusters, and Stefan Bereswill

16 **Nickel-Dependent Gene Expression**
    Konstantin Salnikow and Kazimierz S. Kasprzak

17 **Nickel Toxicity and Carcinogenesis**
    Kazimierz S. Kasprzak and Konstantin Salnikow

Subject Index

Volume 3   The Ubiquitous Roles of Cytochrome P450 Proteins

 1   **Diversities and Similarities of P450 Systems: An Introduction**
     Mary A. Schuler and Stephen G. Sligar

 2   **Structural and Functional Mimics of Cytochromes P450**
     Wolf-D. Woggon

 3   **Structures of P450 Proteins and Their Molecular Phylogeny**
     Thomas L. Poulos and Yergalem T. Meharenna

 4   **Aquatic P450 Species**
     Mark J. Snyder

 5   **The Electrochemistry of Cytochrome P450**
     Alan M. Bond, Barry D. Fleming, and Lisandra L. Martin

 6   **P450 Electron Transfer Reactions**
     Andrew K. Udit, Stephen M. Contakes, and Harry B. Gray

 7   **Leakage in Cytochrome P450 Reactions in Relation
     to Protein Structural Properties**
     Christiane Jung

 8   **Cytochromes P450. Structural Basis for Binding and Catalysis**
     Konstanze von König and Ilme Schlichting

 9   **Beyond Heme-Thiolate Interactions: Roles of the Secondary
     Coordination Sphere in P450 Systems**
     Yi Lu and Thomas D. Pfister

10   **Interactions of Cytochrome P450 with Nitric Oxide
     and Related Ligands**
     Andrew W. Munro, Kirsty J. McLean, and Hazel M. Girvan

11   **Cytochrome P450-Catalyzed Hydroxylations and Epoxidations**
     Roshan Perera, Shengxi Jin, Masanori Sono, and John H. Dawson

12   **Cytochrome P450 and Steroid Hormone Biosynthesis**
     Rita Bernhardt and Michael R. Waterman

13   **Carbon-Carbon Bond Cleavage by P450 Systems**
     James J. De Voss and Max J. Cryle

14   **Design and Engineering of Cytochrome P450 Systems**
     Stephen G. Bell, Nicola Hoskins, Christopher J.C. Whitehouse,
     and Luet L. Wong

15   **Chemical Defense and Exploitation. Biotransformation
     of Xenobiotics by Cytochrome P450 Enzymes**
     Elizabeth M.J. Gillam and Dominic J.B. Hunter

16 Drug Metabolism as Catalyzed by Human Cytochrome P450 Systems
F. Peter Guengerich

17 Cytochrome P450 Enzymes: Observations from the Clinic
Peggy L. Carver

Subject Index

Volume 4   Biomineralization. From Nature to Application

 1 Crystals and Life: An Introduction
Arthur Veis

 **2 What Genes and Genomes Tell Us about Calcium Carbonate
Biomineralization**
Fred H. Wilt and Christopher E. Killian

 3 The Role of Enzymes in Biomineralization Processes
Ingrid M. Weiss and Frédéric Marin

 **4 Metal–Bacteria Interactions at Both the Planktonic
Cell and Biofilm Levels**
Ryan C. Hunter and Terry J. Beveridge

 **5 Biomineralization of Calcium Carbonate. The Interplay
with Biosubstrates**
Amir Berman

 6 Sulfate-Containing Biominerals
Fabienne Bosselmann and Matthias Epple

 7 Oxalate Biominerals
Enrique J. Baran and Paula V. Monje

 8 Molecular Processes of Biosilicification in Diatoms
Aubrey K. Davis and Mark Hildebrand

 9 Heavy Metals in the Jaws of Invertebrates
Helga C. Lichtenegger, Henrik Birkedal, and J. Herbert Waite

10 Ferritin. Biomineralization of Iron
Elizabeth C. Theil, Xiaofeng S. Liu, and Manolis Matzapetakis

**11 Magnetism and Molecular Biology of Magnetic Iron
Minerals in Bacteria**
Richard B. Frankel, Sabrina Schübbe, and Dennis A. Bazylinski

12 Biominerals. Recorders of the Past?
Danielle Fortin, Sean R. Langley, and Susan Glasauer

13 Dynamics of Biomineralization and Biodemineralization
Lijun Wang and George H. Nancollas

14 **Mechanism of Mineralization of Collagen-Based Connective Tissues**
    Adele L. Boskey

15 **Mammalian Enamel Formation**
    Janet Moradian-Oldak and Michael L. Paine

16 **Mechanical Design of Biomineralized Tissues. Bone and Other Hierarchical Materials**
    Peter Fratzl

17 **Bioinspired Growth of Mineralized Tissue**
    Darilis Suárez-González and William L. Murphy

18 **Polymer-Controlled Biomimetic Mineralization of Novel Inorganic Materials**
    Helmut Cölfen and Markus Antonietti

Subject Index

Volume 5   Metallothioneins and Related Chelators

1 **Metallothioneins. Historical Development and Overview**
   Monica Nordberg and Gunnar F. Nordberg

2 **Regulation of Metallothionein Gene Expression**
   Kuppusamy Balamurugan and Walter Schaffner

3 **Bacterial Metallothioneins**
   Claudia A. Blindauer

4 **Metallothioneins in Yeast and Fungi**
   Benedikt Dolderer, Hans-Jürgen Hartmann, and Ulrich Weser

5 **Metallothioneins in Plants**
   Eva Freisinger

6 **Metallothioneins in Diptera**
   Silvia Atrian

7 **Earthworm and Nematode Metallothioneins**
   Stephen R. Stürzenbaum

8 **Metallothioneins in Aquatic Organisms: Fish, Crustaceans, Molluscs, and Echinoderms**
   Laura Vergani

9 **Metal Detoxification in Freshwater Animals. Roles of Metallothioneins**
   Peter G.C. Campbell and Landis Hare

10 **Structure and Function of Vertebrate Metallothioneins**
Juan Hidalgo, Roger Chung, Milena Penkowa, and Milan Vašák

11 **Metallothionein-3, Zinc, and Copper in the Central
Nervous System**
Milan Vašák and Gabriele Meloni

12 **Metallothionein Toxicology: Metal Ion Trafficking
and Cellular Protection**
David H. Petering, Susan Krezoski, and Niloofar M. Tabatabai

13 **Metallothionein in Inorganic Carcinogenesis**
Michael P. Waalkes and Jie Liu

14 **Thioredoxins and Glutaredoxins. Functions and Metal
Ion Interactions**
Christopher Horst Lillig and Carsten Berndt

15 **Metal Ion-Binding Properties of Phytochelatins
and Related Ligands**
Aurélie Devez, Eric Achterberg, and Martha Gledhill

Subject Index

Volume 6 Metal-Carbon Bonds in Enzymes and Cofactors

 1 **Organometallic Chemistry of B_{12} Coenzymes**
Bernhard Kräutler

 2 **Cobalamin- and Corrinoid-Dependent Enzymes**
Rowena G. Matthews

 3 **Nickel-Alkyl Bond Formation in the Active Site
of Methyl-Coenzyme M Reductase**
Bernhard Jaun and Rudolf K. Thauer

 4 **Nickel-Carbon Bonds in Acetyl-Coenzyme A Synthases/Carbon
Monoxide Dehydrogenases**
Paul A. Lindahl

 5 **Structure and Function of [NiFe]-Hydrogenases**
Juan C. Fontecilla-Camps

 6 **Carbon Monoxide and Cyanide Ligands in the Active
Site of [FeFe]-Hydrogenases**
John W. Peters

 7 **Carbon Monoxide as Intrinsic Ligand to Iron in the Active
Site of [Fe]-Hydrogenase**
Seigo Shima, Rudolf K. Thauer, and Ulrich Ermler

8   **The Dual Role of Heme as Cofactor and Substrate
in the Biosynthesis of Carbon Monoxide**
Mario Rivera and Juan C. Rodriguez

9   **Copper-Carbon Bonds in Mechanistic and Structural
Probing of Proteins as well as in Situations where Copper
Is a Catalytic or Receptor Site**
Heather R. Lucas and Kenneth D. Karlin

10   **Interaction of Cyanide with Enzymes Containing Vanadium,
Manganese, Non-Heme Iron, and Zinc**
Martha E. Sosa Torres and Peter M.H. Kroneck

11   **The Reaction Mechanism of the Molybdenum Hydroxylase
Xanthine Oxidoreductase: Evidence against the Formation
of Intermediates Having Metal-Carbon Bonds**
Russ Hille

12   **Computational Studies of Bioorganometallic Enzymes and Cofactors**
Matthew D. Liptak, Katherine M. Van Heuvelen, and Thomas C. Brunold

Subject Index

Author Index of *MIBS*-1 to *MIBS*-44 and *MILS*-1 to *MILS*-6

Volume 7   Organometallics in Environment and Toxicology

1   **Roles of Organometal(loid) Compounds in Environmental Cycles**
John S. Thayer

2   **Analysis of Organometal(loid) Compounds in Environmental
and Biological Samples**
Christopher F. Harrington, Daniel S. Vidler, and Richard O. Jenkins

3   **Evidence for Organometallic Intermediates in Bacterial Methane
Formation Involving the Nickel Coenzyme F_{430}**
Mishtu Dey, Xianghui Li, Yuzhen Zhou, and Stephen W. Ragsdale

4   **Organotins. Formation, Use, Speciation, and Toxicology**
Tamas Gajda and Attila Jancsó

5   **Alkyllead Compounds and Their Environmental Toxicology**
Henry G. Abadin and Hana R. Pohl

6   **Organoarsenicals: Distribution and Transformation
in the Environment**
Kenneth J. Reimer, Iris Koch, and William R. Cullen

7   **Organoarsenicals. Uptake, Metabolism, and Toxicity**
Elke Dopp, Andrew D. Kligerman, and Roland A. Diaz-Bone

8   **Alkyl Derivatives of Antimony in the Environment**
    Montserrat Filella

9   **Alkyl Derivatives of Bismuth in Environmental and Biological Media**
    Montserrat Filella

10 **Formation, Occurrence and Significance of Organoselenium
    and Organotellurium Compounds in the Environment**
    Dirk Wallschläger and Jörg Feldmann

11 **Organomercurials. Their Formation and Pathways
    in the Environment**
    Holger Hintelmann

12 **Toxicology of Alkylmercury Compounds**
    Michael Aschner, Natalia Onishchenko, and Sandra Ceccatelli

13 **Environmental Bioindication, Biomonitoring, and Bioremediation
    of Organometal(loid)s**
    John S. Thayer

14 **Methylated Metal(loid) Species in Humans**
    Alfred V. Hirner and Albert W. Rettenmeier

Subject Index

Volume 8   Metal Ions in Toxicology: Effects, Interactions, Interdependencies

1   **Understanding Combined Effects for Metal Co-Exposure
    in Ecotoxicology**
    Rolf Altenburger

2   **Human Risk Assessment of Heavy Metals: Principles and Applications**
    Jean-Lou C.M. Dorne, George E.N. Kass, Luisa R. Bordajandi,
    Billy Amzal, Ulla Bertelsen, Anna F. Castoldi, Claudia Heppner,
    Mari Eskola, Stefan Fabiansson, Pietro Ferrari, Elena Scaravelli,
    Eugenia Dogliotti, Peter Fuerst, Alan R. Boobis, and Philippe Verger

3   **Mixtures and Their Risk Assessment in Toxicology**
    Moiz M. Mumtaz, Hugh Hansen, and Hana R. Pohl

4   **Metal Ions Affecting the Pulmonary and Cardiovascular Systems**
    Massimo Corradi and Antonio Mutti

5   **Metal Ions Affecting the Gastrointestinal System Including the Liver**
    Declan P. Naughton, Tamás Nepusz, and Andrea Petroczi

6   **Metal Ions Affecting the Kidney**
    Bruce A. Fowler

7   **Metal Ions Affecting the Hematological System**
    Nickolette Roney, Henry G. Abadin, Bruce Fowler, and Hana R. Pohl

8   **Metal Ions Affecting the Immune System**
    Irina Lehmann, Ulrich Sack, and Jörg Lehmann

9   **Metal Ions Affecting the Skin and Eyes**
    Alan B.G. Lansdown

10   **Metal Ions Affecting the Neurological System**
    Hana R. Pohl, Nickolette Roney, and Henry G. Abadin

11   **Metal Ions Affecting Reproduction and Development**
    Pietro Apostoli and Simona Catalani

12   **Are Cadmium and Other Heavy Metal Compounds Acting as Endocrine Disrupters?**
    Andreas Kortenkamp

13   **Genotoxicity of Metal Ions: Chemical Insights**
    Woijciech Bal, Anna Maria Protas, and Kazimierz S. Kasprzak

14   **Metal Ions in Human Cancer Development**
    Erik J. Tokar, Lamia Benbrahim-Tallaa, and Michael P. Waalkes

Subject Index

Volume 9   Structural and Catalytic Roles of Metal Ions in RNA

1   **Metal Ion Binding to RNA**
    Pascal Auffinger, Neena Grover, and Eric Westhof

2   **Methods to Detect and Characterize Metal Ion Binding Sites in RNA**
    Michèle C. Erat and Roland K. O. Sigel,

3   **Importance of Diffuse Metal Ion Binding to RNA**
    Zhi-Jie Tan and Shi-Jie Chen

4   **RNA Quadruplexes**
    Kangkan Halder and Jörg S. Hartig

5   **The Roles of Metal Ions in Regulation by Riboswitches**
    Adrian Ferré-D'Amaré and Wade C. Winkler

6   **Metal Ions: Supporting Actors in the Playbook of Small Ribozymes**
    Alexander E. Johnson-Buck, Sarah E. McDowell, and Nils G. Walter

7   **Multiple Roles of Metal Ions in Large Ribozymes**
    Daniela Donghi and Joachim Schnabl

8 **The Spliceosome and Its Metal Ions**
Samuel E. Butcher

9 **The Ribosome: A Molecular Machine Powered by RNA**
Krista Trappl and Norbert Polacek

10 **Metal Ion Requirements in Artificial Ribozymes that Catalyze Aminoacylations and Redox Reactions**
Hiroaki Suga, Kazuki Futai, and Koichiro Jin

11 **Metal Ion Binding and Function in Natural and Artificial Small RNA Enzymes from a Structural Perspective**
Joseph E. Wedekind

12 **Binding of Kinetically Inert Metal Ions to RNA: The Case of Platinum(II)**
Erich G. Chapman, Alethia A. Hostetter, Maire F. Osborn, Amanda L. Miller, and Victoria J. DeRose

Subject Index

Volume 10   Interplay between Metal Ions and Nucleic Acids

1 **Characterization of Metal Ion-Nucleic Acid Interactions in Solution**
Maria Pechlaner and Roland K.O. Sigel

2 **Nucleic Acid-Metal Ion Interactions in the Solid State**
Katsuyuki Aoki and Kazutaka Murayama

3 **Metal Ion-Promoted Conformational Changes of Oligonucleotides**
Bernhard Spingler

4 **G-Quadruplexes and Metal Ions**
Nancy H. Campbell and Stephen Neidle

5 **Metal Ion-Mediated DNA-Protein Interactions**
Barbara Zambelli, Francesco Musiani, and Stefano Ciurli

6 **Spectroscopic Investigations of Lanthanide Ion Binding to Nucleic Acids**
Janet R. Morrow and Christopher M. Andolina

7 **Oxidative DNA Damage Mediated by Transition Metal Ions and Their Complexes**
Geneviève Pratviel

8 **Metal Ion-Dependent DNAzymes and Their Applications as Biosensors**
Tian Lan and Yi Lu

9   **Enantioselective Catalysis at the DNA Scaffold**
    Almudena García-Fernández and Gerard Roelfes

10 **Alternative DNA Base Pairing through Metal Coordination**
    Guido H. Clever and Mitsuhiko Shionoya

11 **Metal-Mediated Base Pairs in Nucleic Acids with
    Purine- and Pyrimidine-Derived Nucleosides**
    Dominik A. Megger, Nicole Megger, and Jens Müller

12 **Metal Complex Derivatives of Peptide Nucleic Acids**
    Roland Krämer and Andrij Mokhir

Subject Index

Volume 11   Cadmium: From Toxicity to Essentiality

1   **The Bioinorganic Chemistry of Cadmium
    in the Context of Its Toxicity**
    Wolfgang Maret and Jean-Marc Moulis

2   **Biogeochemistry of Cadmium and Its Release
    to the Environment**
    Jay T. Cullen and Maria T. Maldonado

3   **Speciation of Cadmium in the Environment**
    Francesco Crea, Claudia Foti, Demetrio Milea,
    and Silvio Sammartano

4   **Determination of Cadmium in Biological Samples**
    Katrin Klotz, Wobbeke Weistenhöfer, and Hans Drexler

5   **Imaging and Sensing of Cadmium in Cells**
    Masayasu Taki

6   **Use of ^{113}Cd NMR to Probe the Native Metal Binding
    Sites in Metalloproteins: An Overview**
    Ian M. Armitage, Torbjörn Drakenberg, and Brian Reilly

7   **Solid State Structures of Cadmium Complexes
    with Relevance for Biological Systems**
    Rosa Carballo, Alfonso Castiñeiras, Alicia Domínguez-Martín,
    Isabel García Santos, and Juan Niclós-Gutierrez

8   **Complex Formation of Cadmium(II) with Sugar
    Residues, Nucleobases, Phosphates, Nucleotides,
    and Nucleic Acids**
    Roland K. O. Sigel, Miriam Skilandat, Astrid Sigel,
    Bert P. Operschall, and Helmut Sigel

9   **Cadmium(II) Complexes of Amino Acids and Peptides**
    Imre Sóvágó and Katalin Várnagy

10   Natural and Artificial Proteins Containing Cadmium
Anna F. Peacock and Vincent L. Pecoraro

11   Cadmium in Metallothioneins
Eva Freisinger and Milan Vašák

12   Cadmium-Accumulating Plants
Hendrik Küpper and Barbara Leitenmaier

13   Cadmium Toxicity in Plants
Elisa Andresen and Hendrik Küpper

14   Toxicology of Cadmium and Its Damage to Mammalian Organs
Frank Thévenod and Wing-Kee Lee

15   Cadmium and Cancer
Andrea Hartwig

16   Cadmium in Marine Phytoplankton
Yan Xu and François M.M. Morel

Subject Index

Volume 12   Metallomics and the Cell
Guest Editor: Lucia Banci

 1   Metallomics and the Cell: Some Definitions and General Comments
Lucia Banci and Ivano Bertini

 2   Technologies for Detecting Metals in Single Cells
James E. Penner-Hahn

 3   Sodium/Potassium Homeostasis in the Cell
Michael J.V. Clausen and Hanna Poulsen

 4   Magnesium Homeostasis in Mammalian Cells
Andrea M.P. Romani

 5   Intracellular Calcium Homeostasis and Signaling
Marisa Brini, Tito Calì, Denis Ottolini, and Ernesto Carafoli

 6   Manganese Homeostasis and Transport
Jerome Roth, Silvia Ponzoni, and Michael Aschner

 7   Control of Iron Metabolism in Bacteria
Simon Andrews, Ian Norton, Arvindkumar S. Salunkhe,
Helen Goodluck, Wafaa S.M. Aly, Hanna Mourad-Agha,
and Pierre Cornelis

 8   The Iron Metallome in Eukaryotic Organisms
Adrienne C. Dlouhy and Caryn E. Outten

9   **Heme Uptake and Metabolism in Bacteria**
    David R. Benson and Mario Rivera

10 **Cobalt and Corrinoid Transport and Biochemistry**
    Valentin Cracan and Ruma Banerjee

11 **Nickel Metallomics: General Themes Guiding Nickel Homeostasis**
    Andrew M. Sydor and Deborah B. Zamble

12 **The Copper Metallome in Prokaryotic Cells**
    Christopher Rensing and Sylvia Franke McDevitt

13 **The Copper Metallome in Eukaryotic Cells**
    Katherine E. Vest, Hayaa F. Hashemi, and Paul A. Cobine

14 **Zinc and the Zinc Proteome**
    Wolfgang Maret

15 **Metabolism of Molybdenum**
    Ralf R. Mendel

16 **Comparative Genomics Analysis of the Metallomes**
    Vadim N. Gladyshev and Yan Zhang

Subject Index

**Volume 13   Interrelations between Essential Metal Ions
            and Human Diseases**

1   **Metal Ions and Infectious Diseases. An Overview from the Clinic**
    Peggy L. Carver

2   **Sodium and Potassium in Health and Disease**
    Hana R. Pohl, John S. Wheeler, and H. Edward Murray

3   **Magnesium in Health and Disease**
    Andrea M.P. Romani

4   **Calcium in Health and Disease**
    Marisa Brini, Denis Ottolini, Tito Calì, and Ernesto Carafoli

5   **Vanadium. Its Role for Humans**
    Dieter Rehder

6   **Chromium. Is It Essential, Pharmacologically Relevant, or Toxic?**
    John B. Vincent

7   **Manganese in Health and Disease**
    Daiana Silva Avila, Robson Luiz Puntel, and Michael Aschner

8   **Iron: Effects of Overload and Deficiency**
    Robert C. Hider and Xiaole Kong

9   **Cobalt: Its Role in Health and Disease**
    Kazuhiro Yamada

10 **Nickel and Human Health**
    Barbara Zambelli and Stefano Ciurli

11 **Copper: Effects of Deficiency and Overload**
    Ivo Scheiber, Ralf Dringen, and Julian F.B. Mercer

12 **Zinc and Human Disease**
    Wolfgang Maret

13 **Molybdenum in Human Health and Disease**
    Guenter Schwarz and Abdel A. Belaidi

14 **Silicon: The Health Benefits of a Metalloid**
    Keith R. Martin

15 **Arsenic. Can this Toxic Metalloid Sustain Life?**
    Dean E. Wilcox

16 **Selenium. Role of the Essential Metalloid in Health**
    Suguru Kurokawa and Marla J. Berry

Subject Index

**Volume 14   The Metal-Driven Biogeochemistry of Gaseous Compounds
             in the Environment** (this book)
    Guest Editors: Peter M.H. Kroneck and Martha E. Sosa Torres

**Volume 15   Sustaining Life on Planet Earth: Metalloenzymes
             Mastering Dioxygen and Other Chewy Gases** (in press)
    Guest Editors: Peter M.H. Kroneck and Martha E. Sosa Torres

1   **The Magic of Dioxygen**
    Martha E. Sosa Torres, Juan P. Saucedo-Vázquez,
    and Peter M.H. Kroneck

2   **Light-Dependent Production of Dioxygen in Photosynthesis**
    Junko Yano, Jan Kern, Vittal K. Yachandra, Håkan Nilsson,
    Sergey Koroidov, and Johannes Messinger

3   **Production of Dioxygen in the Dark: Dismutases of Oxyanions**
    Jennifer DuBois and Sunil Ojha

4   **Respiratory Conservation of Energy with Dioxygen:
    Cytochrome *c* Oxidase**
    Shinya Yoshikawa, Atsuhiro Shimada, and Kyoko Shinzawa-Itoh

5   **Transition Metal Complexes and Activation of Dioxygen**
    Gereon M. Yee and William B. Tolman

6 **Methane Monooxygenase: Functionalizing Methane
   at Iron and Copper**
   Matthew H. Sazinsky and Stephen J. Lippard

7 **Metal Enzymes in "Impossible" Microorganisms Catalyzing
   the Anaerobic Oxidation of Methane**
   Joachim Reimann, Mike S.M. Jetten, and Jan T. Keltjens

Subject Index

Volume 16   The Alkali Metal Ions: Their Role for Life (in preparation)

1 **The Bioinorganic Chemistry of the Alkali Metal Ions**
   Yonghwang Ha, Jeong A. Jeong, and David G. Churchill

2 **Determination of Alkali Ions in Biological and Environmental Samples**
   Peter C. Hauser

3 **Solid State Structures of Alkali Metal Ion Complexes Formed
   by Low-Molecular-Weight Ligands of Biological Relevance**
   Katsuyuki Aoki, Kazutaka Murayama, and Ning-Hai Hu

4 **Discriminating Properties of Alkali Ions towards the Constituents
   of Proteins and Nucleic Acids. Conclusions from Gas-Phase
   and Theoretical Studies**
   Mary T. Rodgers and Peter B. Armentrout

5 **Alkali Metal Ion Complexes with Phosphates, Nucleotides,
   Amino Acids, and Related Ligands of Biological Relevance.
   Their Properties in Solution**
   Silvio Sammartano et al.

6 **Interactions of Sodium and Potassium Ions with Nucleic Acids**
   Pascal Auffinger and Eric Ennifar

7 **The Role of Alkali Metal Ions in G-Quadruplex Nucleic
   Acid Structure and Stability**
   Eric Largy, Jean-Louis Mergny, and Valérie Gabelica

8 **Sodium and Potassium Ions in Proteins and Enzyme Catalysis**
   Milan Vašak

9 **Sodium and Potassium. Their Role in Plants**
   Hervé Sentenac et al.

10 **Sodium-Potassium (Na^+/K^+) and Related ATPases**
    Jack H. Kaplan

11 **Sodium as Coupling Cation in Respiratory Energy Conversion**
    Günter Fritz and Julia Steuber

12   **Sodium-Proton (Na⁺/H⁺) Antiporters: Properties and Roles in Health and Disease**
Etana Padan

13   **Proton-Potassium (H⁺/K⁺) ATPases: Properties and Roles in Health and Disease**
Hideki Sakai and Noriaki Takeguchi

14   **Potassium *versus* Sodium Selectivity in Ion Channels**
Carmay Lim and Todor Dudev

15   **Bioinspired Artificial Sodium and Potassium Channels**
Nuria Vázquez-Rodríguez, Alberto Fuertes, Manuel Amorín, and Juan R. Granja

16   **Lithium in Medicine: Pharmacology and Mechanism of Action**
Duarte Mota de Freitas, Brian D. Leverson, and Jesse L. Goossens

17   **Sodium and Potassium Relating to Parkinson's Disease and Traumatic Brain Injury**
Yonghwang Ha, Jeong A. Jeong, and David G. Churchill

Subject Index

Comments and suggestions with regard to contents, topics, and the like for future volumes of the series are welcome.

Chapter 1
The Early Earth Atmosphere and Early Life Catalysts

Sandra Ignacia Ramírez Jiménez

Contents

ABSTRACT 1
1 THE EARLY EARTH ATMOSPHERE AND LITHOSPHERE 2
   1.1 Earth's Internal Structure 3
2 CATALYSTS IN THE EARLY EARTH 4
3 CLAYS AS POSSIBLE CATALYSTS IN THE SYNTHESIS
   OF BIOMOLECULES 6
4 GENERAL CONCLUSIONS 12
ABBREVIATIONS 13
ACKNOWLEDGMENT 13
REFERENCES 13

Abstract Homochirality is a property of living systems on Earth. The time, the place, and the way in which it appeared are uncertain. In a prebiotic scenario two situations are of interest: either an initial small bias for handedness of some biomolecules arouse and progressed with life, or an initial slight excess led to the actual complete dominance of the known chiral molecules. A definitive answer can probably never be given, neither from the fields of physics and chemistry nor biology. Some arguments can be advanced to understand if homochirality is necessary for the initiation of a prebiotic homochiral polymer chemistry, if this homochirality is suggesting a unique origin of life, or if a chiral template such as a mineral surface is always required to result in an enantiomeric excess. A general description of the early Earth scenario will be presented in this chapter, followed by a general description of some clays, and their role as substrates to allow the concentration and amplification of some of the building blocks of life.

S.I. Ramírez Jiménez (✉)
Centro de Investigaciones Químicas, Universidad Autónoma del Estado de Morelos,
Av. Universidad # 1001, Col. Chamilpa, Cuernavaca 62209, Morelos, Mexico
e-mail: ramirez_sandra@uaem.mx

© Springer Science+Business Media Dordrecht 2014   1
P.M.H. Kroneck, M.E. Sosa Torres (eds.), *The Metal-Driven Biogeochemistry of Gaseous Compounds in the Environment*, Metal Ions in Life Sciences 14,
DOI 10.1007/978-94-017-9269-1_1

Keywords biomolecules • catalysts • clays • early Earth atmosphere • homochirality

Please cite as: *Met. Ions Life Sci.* 14 (2014) 1–14

1   The Early Earth Atmosphere and Lithosphere

The solar system condensed out of an interstellar cloud of gas and dust, known as the primordial solar nebula, about 4.6×10^9 years ago. The formation of the inner planets was the result of innumerable impacts of small, rocky particles in the solar system's protoplanetary disk. Over a period of a few million years, neighboring dust grains and pebbles in the solar nebula coalesced into larger objects called planetesimals. Their mutual gravitational attraction brought a few of them together to form larger objects, known as protoplanets. The accretion of more material continued for another 100 million years and led to the formation of the terrestrial planets: Mercury, Venus, Earth, and Mars [1].

The theory of planet formation and the models that attempt to reproduce the early history of Earth indicate that the actual atmosphere is the third atmosphere to evolve [1]. The earliest Earth's atmosphere was composed of trace remnants of dihydrogen (H_2) and helium (He) left over from the formation of the solar system, but it did not last very long as these gases are too light to be kept. They gained enough energy from the sunlight to overcome the Earth's gravitational attraction and moved away into space [1]. The gases of the second atmosphere came from inside the Earth, vented through volcanoes and cracks in its surface. The evidence indicates that this was a mixture of carbon dioxide (CO_2), dinitrogen (N_2), and water vapor (H_2O_v), with trace amounts of H_2 [2]. There was roughly 100 times as much gas in that second atmosphere as there is in the atmosphere today. Molecules such as CO_2 and H_2O_v stored much more heat than the atmosphere does today, creating a greenhouse effect. The Sun had not reached its full intensity at that time, thus the high concentrations of these gases helped to prevent the young Earth from freezing.

Most of the water vapor that outgassed from the Earth's interior condensed out of the atmosphere to form the oceans. The rain absorbed and carried down lots of CO_2 that was absorbed in the oceans. As life evolved most likely in the primitive oceans, much of the dissolved CO_2 was transformed into carbonates (CO_3^{2-}) that were trapped in the shells of many organisms [1]. It is estimated that for each molecule of CO_2 presently in the atmosphere, there are about 10^5 CO_2 molecules incorporated as carbonates in sedimentary rocks [2]. What remained in the air were mostly N_2 and H_2O_v [1]. Since N_2 is chemically inert, sparcely soluble in water, and noncondensable, most of the outgassed N_2 accumulated in the atmosphere over geologic time to become the atmosphere's most abundant constituent [2]. Some of the oxygen that entered into the early Earth atmosphere came from oxygen-rich molecules, like CO_2, that were broken apart by the Sun's ultraviolet radiation. Then came the dioxygen (O_2) released as a by-product of plant life activity. Early plant life in the oceans removed most of the remaining CO_2 by converting it into dioxygen by photosynthetic processes. As O_2, in the triplet groundstate, is a thermodynamically highly reactive molecule, the early atmospheric O_2 combined quickly with many elements on the Earth's surface, notably metals, to form metallic oxides such as iron oxides or silicon oxides. About 2×10^9 years ago, after all of the surface minerals that could combine with O_2 had done so, the atmosphere began to fill with this gas, and it became the nitrogen-oxygen mixture that we breathe today [1].

The early Earth atmosphere was a mildly reducing chemical mixture, whereas the present atmosphere is strongly oxidizing. It has been estimated that the current level of O_2 in the atmosphere was achieved about 400 millions years ago [3].

1.1 Earth's Internal Structure

Seismic studies reveal that the Earth's interior is divided into three main layers differentiated by its rheological properties: the core, the mantle, and the crust.

The core is made of material of the highest density. It consists primarily of nickel and iron alloys. The Earth's core has two distinct regions, the solid inner core and the molten outer core. The core's density varies from 13.1 to 9.9 g/cm^3. The mantle is formed by rocky material of moderate density, mostly silicate minerals rich in silicon, magnesium, and oxygen. Its density varies from 5.6 to 3.4 g/cm^3. The mantle surrounds the core and makes up most of the Earth's volume. The uppermost layer consist of the lowest-density rocks such as granite and basalt whose density value can vary from 2.9 to 2.2 g/cm^3, and forms the thin crust that is the Earth's outer skin [4].

In terms of rock strength, geologists define the Earth's outer layer as a relatively cool and rigid rock that floats on warmer, softer rock. This layer is called lithosphere, and encompasses the crust and the upper part of the mantle. The crust occupies less than 1 % of the Earth's volume and can be differentiated into the oceanic crust and the continental crust. The oceanic crust is made of high-density igneous rocks such as basalt, diabase (also called dolorite), and gabbro. It is only 5–10 kilometers thick and radiometric dating shows that it is quite young, with an average age of about 70 million years. Even the oldest seafloor crust is less than 200 million years old [5]. The continental crust thickness varies from 30 to 50 kilometers and is mostly composed of less dense rocks such as granite. The oldest continental crustal rocks on Earth have ages in the range of 3.7 to 4.28×10^9 years. The continental crust has an average composition similar to that of andesite. It is enriched in elements such as oxygen, silicon, aluminum, iron, calcium, sodium, potassium, and magnesium, principally found as oxides [5, 6].

2   Catalysts in the Early Earth

From the geological point of view the Earth's surface is continuously changing. An igneous rock is transformed under the action of high pressure or heat into a metamorphic rock, and both igneous and metamorphic rocks can be eroded into a sedimentary rock, which may be carried deep underground to be melted and then it resolidifies as an igneous rock. As all rocks on the Earth's surface are recycled, a rock's type gives information about the way it was made, while its mineralogical composition informs about its chemical nature. Among the naturally occurring inorganic minerals, phyllosilicate and smectite clays with layered structures are the most abundant [7]. Their generic aluminosilicate structure is composed of multiple silicate plates stacked in layers and crystalline defects with divalent metal species (counterions) in the interlayer galleries (Figure 1). In the smectites, the fundamental units are comprised of two tetrahedral sheets sandwiched with an edge-shared octahedral sheet building a 2:1 structure. Smectites have been well characterized with regard to their chemical composition, lamellar structure with high aspect ratio, geometric shape, surface area, and counter-ion exchange capacity [7]. The most representative phyllosilicate clays of the 2:1 type are montmorillonite, bentonite, saponite, and hectorite. They have conventionally been employed as catalysts, adsorbents, metal chelating agents, and polymer nanocomposites [7].

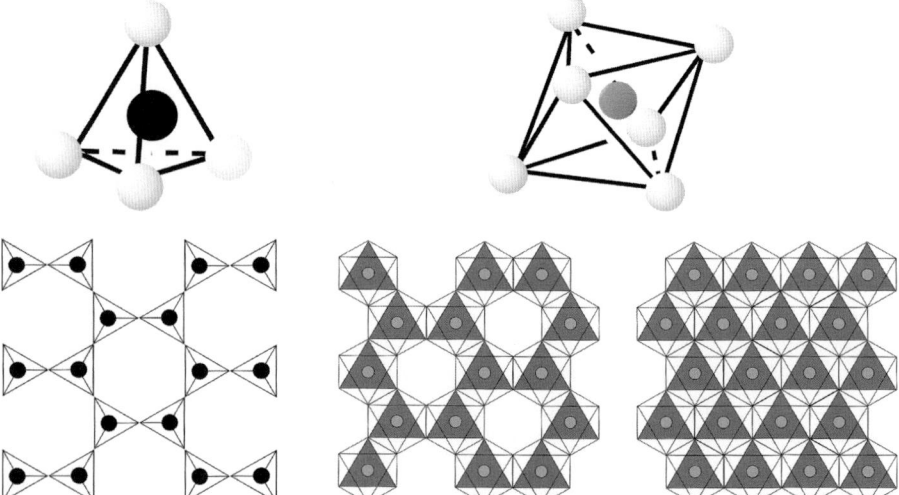

Figure 1 Representation of the tetrahedral and octahedral patterns in phyllosilicate clay minerals. The basic tetrahedron and octahedron units are shown on top. The arrangement for the dioctahedral and trioctahedral sheets is also shown. Black circles represent silicon atoms, gray circles correspond to aluminum, magnesium or iron cations, and white circles correspond to oxygen atoms or hydroxyl groups. Based on [13].

Clay minerals could have formed by weathering of volcanic glass and rocks when getting in contact with water [8]. Investigations about Mars soils indicate the occurrence of clay minerals in the planet's surface older than 3.5×10^9 years, with a chemical composition consistent with Al-Si-O-H and Mg-Si-O-H systems [9]. By analogy, clay minerals could have formed in a similar way on the early Earth [8]. Clay minerals might also have formed as an alteration product of silicate minerals by igneous activity and also during diagenesis of sediments. As soon as liquid water appeared on the surface of the primitive Earth, clay minerals could have probably accumulated on the surface and became suspended in the primitive oceans [10].

Clay minerals are characteristic for near surface hydrous environments including those of weathering, sedimentation, diagenesis, and hydrothermal alteration. Within all of these environments clay minerals may have been newly formed, transformed, or inherited [11, 12]. There are two basic modular components on phyllosilicate clay minerals: (a) sheets of tetrahedrally, and (b) sheets of octahedrally coordinated atoms. Tetrahedral sheets are formed by a cation, usually silicon, surrounded by four oxygen atoms. An octahedral sheet is formed from two planes of closed packed oxygens and/or hydroxyl anions (Figure 1). In the center of such a sheet, and adjacent to every anion, there are three octahedral sites which may be occupied by metal cations, such as aluminum (Al^{3+}), iron ($Fe^{3+/2+}$), and magnesium (Mg^{2+}), each cation being surrounded by six anions. Two kinds of common octahedral sheets are distinguished according to the cation to anion ratio needed for electrical neutrality. If divalent cations fill all three sites, the octahedral sheet is known as trioctahedral (Tri.). Trivalent cations need only two out of every three sites and thus form a dioctahedral (Di.) sheet. The basic crystal units known as layers, consist of tetrahedral (T) and octahedral (O) sheets. The 1:1 unit, also designated as T-O, is formed by linking one tetrahedral sheet with one octahedral sheet. An octahedral sheet between two tetrahedral sheets forms a 2:1 or T-O-T layer (Figure 1). A detailed description of these structures can be found in [13]. Any substitution in 1:1 layers is fully compensated and there is no net layer charge. Substitutions in a 2:1 layer, however, frequently give rise to a net negative charge, which may be neutralized by cations, or hydrated cations. Most clay minerals host two kinds of layers, such as illite-smectite or chlorite-smectite. Interstratification of three kinds of layers are rarely reported. A classification of some of the most representative phyllosilicate clay minerals based on layer type, and the magnitude of any net charge, is presented in Table 1. Cations required to compensate the net negative layer charges, may be exchanged with others. It is this property which defines the cation exchange capacity of the clay mineral.

Table 1 Classification of common phyllosilicate clay minerals.[a]

Layer type	Layer charge (q)	Group	Examples
1:1	0.0	Kaolin-Serpentine	Kaolinite, Berthierine
2:1	0.0	Pyrophyllite-Talc	Pyrophyllite, Talc
	0.2–0.6	Smectite	Montmorillonite, Beidellite, Saponite
	0.6–0.9	Vermiculite	Di. Vermiculite*, Tri. Vermiculite*
	1.0	Mica	Illite, Muscovite, Biotite
	Variable	Chlorite	Sudoite, Chamosite, Clinochlore
	Variable	Sepiolite-Palygorskite	Sepiolite, Palygorskite
Variable	Variable	Mixed-layer	Rectorite, Corrensite

* Di. = dioctahedral sheet. Tri. = trioctahedral sheet.
[a]Based on [13].

Clays have a particular place in studies related to enzyme immobilization, protein fractionation, soil ecosystem safety, genetic engineering, and specifically in the biochemical evolution and origin of life on Earth. In 1949, Bernal suggested that clay minerals played a key role in chemical evolution and the origins of life because of their uptake capability, their ability to protect against ultraviolet radiation, to concentrate, and to catalyze the polymerization of organic molecules [14]. A number of subsequent reports supported this idea. It was claimed that clay minerals such as montmorillonite, might have played a central role in the formation of proteins and nucleic acids serving as primitive templates to concentrate the primordial biomolecules and to catalyze their polymerization. Furthermore, they helped in the preservation of the first biopolymers that eventually initiated the biological evolution on Earth [15–20].

3 Clays as Possible Catalysts in the Synthesis of Biomolecules

A distinctive feature of life's chemistry is its homochirality illustrated by amino acids and sugars, fundamental biomolecules for the construction of proteins and nucleic acids. A chiral molecule is a type of molecule that has a non-superimposable relationship with its mirror image. The feature that is most often the cause of chirality in molecules is the presence of one or several asymmetric carbon atoms. Chiral molecules can exist in two distinguishable mirror-image forms, designated as L- or D-enantiomers (optical isomers) (Figure 2).

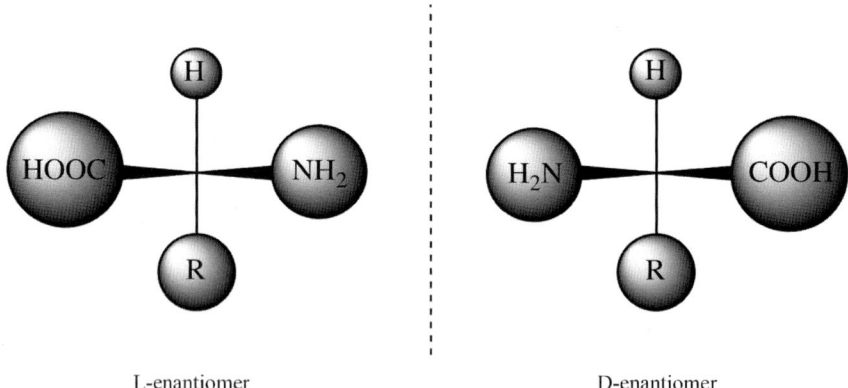

L-enantiomer   D-enantiomer

Figure 2 The two enantiomers of a generic α-amino acid, H_2N-CH(R)-COOH (R = side-chain, see Table 2); L,D assignment is done with reference to glyceraldehyde, HO-CH(CH_2OH)-CHO. The structures are drawn as chemical standard structures, not as the typical zwitterionic forms that usually exist in aqueous solution.

Pairs of enantiomers are often described as right- and left-handed. To name the enantiomers of a compound unambiguously, their names must include the handedness of the molecule according to the R,S nomenclature (Cahn-Ingold-Prelog rules). Enantiomers have, when present in a symmetric environment, identical chemical and physical properties except for their interaction with polarized light (so-called optical activity). For example, an enantiomer will absorb left- and right-circularly polarized light to differing degrees, which is the basis of circular dichroism spectroscopy. To illustrate the three-dimensional structure of a chiral molecule by a two-dimensional drawing, one can use the projection already introduced by Emil Fischer in 1891. It is still in use today, particularly in the case of carbohydrates and amino acids [21].

Proteins are all exclusively made of a series of L-amino acids, whose order dictates the primary structure. This homochirality leads to homochirality in higher-order structures such as the right-handed α-helix found in some secondary structures or the way in which some proteins are folded to originate its tertiary structure (Figure 3). Table 2 presents the side-chain structures of the twenty standard amino acids.

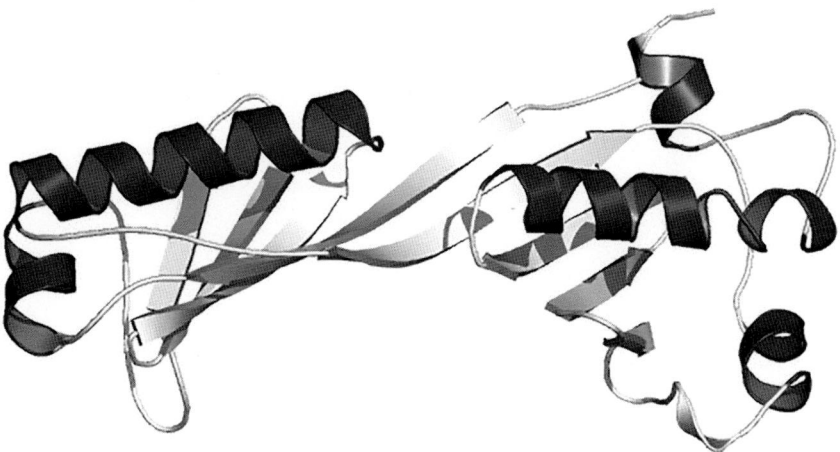

Figure 3 Ribbon diagram for the tertiary structure of the TATA binding protein from *Taenia solium* showing its right-handed α helices (in black). Redrawn from [48] by permission of R. Miranda.

On the other hand, nucleic acids consist of chains of deoxyribonucleosides (for deoxyribonucleic acid, DNA) or ribonucleosides (for ribonucleic acid, RNA), connected by phosphodiester bonds, all based exclusively on the D-deoxyribose or D-ribose sugar ring, respectively [22]. The homochirality in the monomeric sugar building blocks of nucleic acids leads to homochirality in their secondary structures such as the right-handed B-type DNA double helix, as shown in Figure 4.

Table 2 Side chain residues (R) of the twenty standard amino acids, H$_2$N-CH(R)-COOH. [a]

Non polar

H	CH$_3$	CH—CH$_3$ / CH$_3$
Glycine/Gly/G	Alanine/Ala/A	Valine/Val/V

CH$_2$ / SH	CH$_2$—CH$_2$—S—CH$_3$	CH$_2$ / CH—CH$_3$ / CH$_3$
Cysteine/Cys/C	Methionine/Met/M	Leucine/Leu/L

Proline/Pro/P	Phenylalanine/Phe/P	Tryptophan/Trp/W

H$_2$C—CH—H / CH$_3$ / CH$_3$
Isoleucine/Ile/I

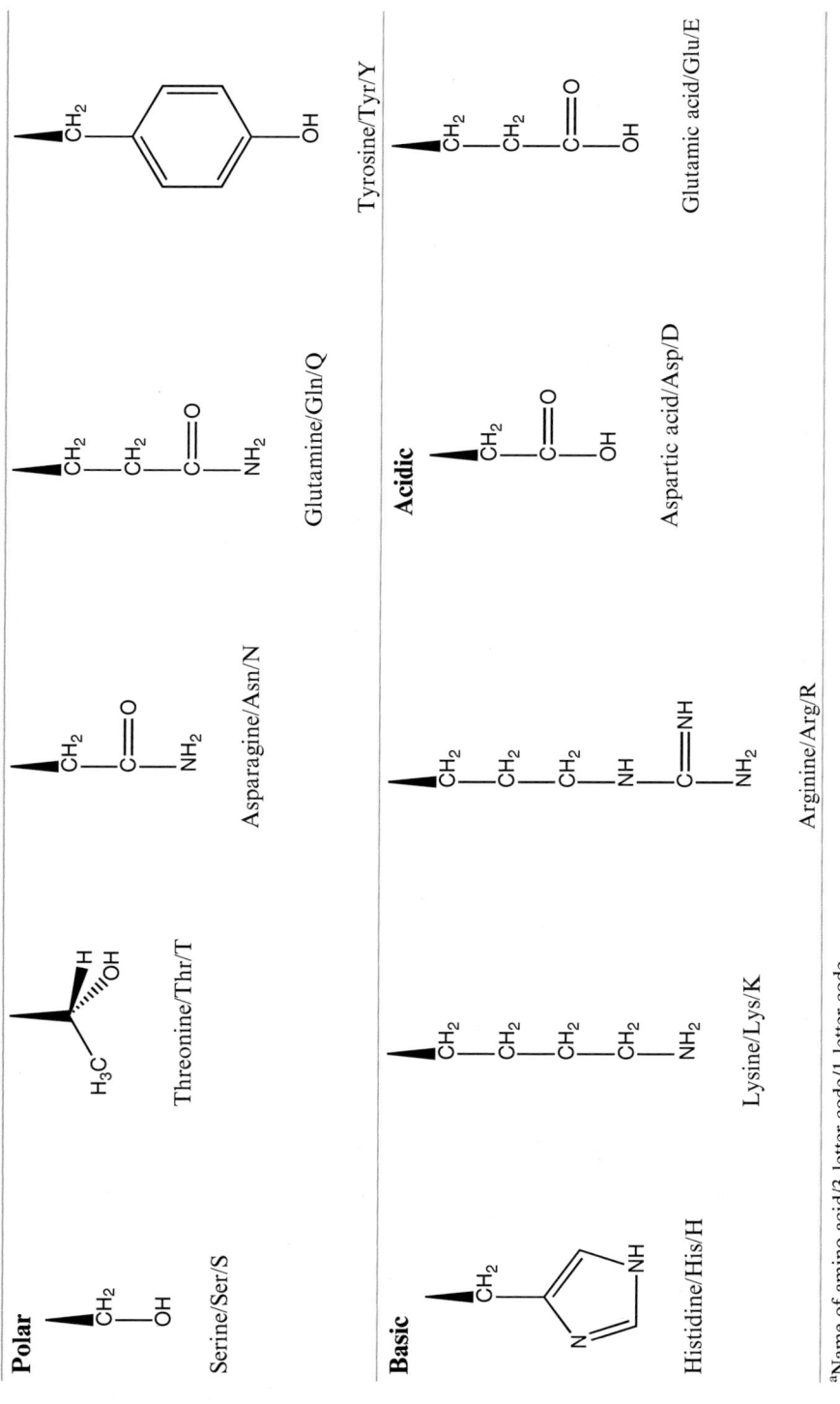

Polar

Serine/Ser/S

Threonine/Thr/T

Asparagine/Asn/N

Glutamine/Gln/Q

Tyrosine/Tyr/Y

Acidic

Aspartic acid/Asp/D

Glutamic acid/Glu/E

Basic

Histidine/His/H

Lysine/Lys/K

Arginine/Arg/R

[a]Name of amino acid/3-letter code/1-letter code

Elucidating the origin of biological homochirality has occupied a central position on studies of the origin of life since the discovery of chiral symmetry by Pasteur [23, 24]. There is, however, no consensus on how an initial small bias for handedness arose and how a slight excess of handedness led to such a dominance of biological evolution on Earth. Researchers in this area have agreed to redefine the question of how homochirality arose into a search for an advantage factor that was capable of ensuring a large quantitative excess of one of the enantiomers over the other [24]. There exist many diverse suggestions for the nature of advantage factors, but they can be globally grouped into two classes, local and global. Local advantage factors might have existed in a particular region on the Earth's surface. They could have varied from region to region and/or might have existed during a definite period of time. On the other hand, global advantage factors are caused by the parity nonconservation in weak interactions. The action of the advantage factor might, in principle, lead to an almost chirally pure state of the medium [25].

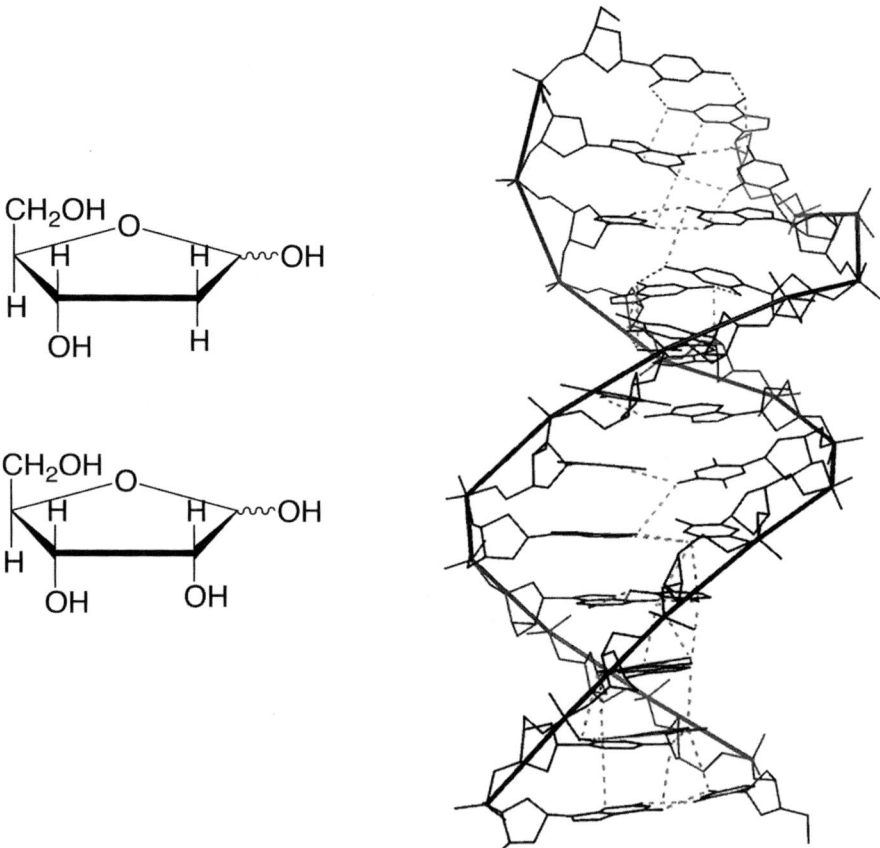

Figure 4 Structural model of a B-DNA right-handed double helix. The structural formula for 2-deoxy-D-ribofuranose (top), and D-ribofuranose (bottom) are shown on the left. They are the sugar components in DNA and RNA, respectively. The carbohydrate-phosphate "backbone" in the DNA dodecamer is highlighted in black. Hydrogen bonding between complementary bases is shown with dotted lines. Redrawn using the sequence PDB CODE = 156D reported in [49].

Although complete homochirality in the prebiotic chiral monomers may not be necessary for the initiation of prebiotic homochiral polymer chemistry, a small initial enantiomeric excess may be required. This small enantiomeric excess could be produced by some physical chiral influence. On a cosmic scale, enantioselective mechanisms depending on parity violation or circularly polarized light, are the only ones that could predetermine a particular handedness. In all other mechanisms, like a local chiral environment induced by a mineral surface, the ultimate choice would arise purely by chance [22]. That life evolved to use L-amino acids in proteins, and D-sugars in nucleic acids after it began remains a possibility, but there is considerable evidence that the formation of these biopolymers from homochiral material is necessary for them to function. Usually, in laboratory synthesis, unless special experimental conditions are used, a racemic (50:50) mixture of the D- and its L-enantiomer will be obtained when chiral molecules are formed from achiral starting materials. When a particular reagent or catalyst is added, an excess of one of the enantiomers can be used to direct the formation of more homochiral molecules. However, by which mechanism and when the first one-handed building block for life arose, still remains an open question [26].

Clay minerals have long been postulated to play a role in concentrating the products of non-biological organic processes leading to the Huxley-Darwin-Oparin-Haldane model for the origin of living systems. Because of their large surface areas and adsorption characteristics, clays can act as catalysts for the formation of RNA oligomers from activated nucleotide precursors. Furthermore, montmorillonite clay and alumina both catalyze polymerization of amino acids in peptide formation reactions [27]. Chiral selectivity has been demonstrated in experimental and theoretical studies. The layered structure of clays could also allow them to play a key role in concentrating, shielding and catalyzing the assembly of some of the most important molecules necessary for the development of living systems.

Clay minerals such as montmorillonite, kaolinite and illite (Table 1) represent a naturally abundant class of layered aluminum-silicates that are important in the context of prebiotic chemistry due to their good biocompatibility, string adsorption, ion exchange ability, and expansibility [28], ancient origin and wide distribution. Different studies have demonstrated that they can absorb a variety of biomolecules including proteins, DNA, RNA, lipids, purines, pyrimidines [29–31].

Miller-Urey type experiments have been performed in the presence of montmorillonite, resulting in an increased yield of alkylated amino acids [32], or in the preferred synthesis of some specific compounds such as glycine, alanine, or aspartic acid [33]. Greenland and coworkers [34, 35] investigated different adsorption mechanisms of amino acids on H^+-, Na^+-, and Ca^{2+}-montmorillonite. They noticed that basic amino acids, such as arginine, histidine, and lysine (Table 2) were preferentially adsorbed on Na^+- and Ca^{2+}-montmorillonite by cation exchange, whereas non-polar amino acids such as alanine, serine, leucine, or phenylalanine, and acidic amino acids like aspartic acid and glutamic acid were adsorbed on H^+-montmorillonite by proton transfer. They also noticed that the adsorption of glycine and its oligopeptides onto Ca^{2+}-montmorillonite and Ca^{2+}-illite increased with the degree of oligomerization. Hedges and Hare [36] suggested that the amino and carboxyl terminal groups of the amino acids were involved in their adsorption to kaolinite. Dashman and Stotzky [37, 38] reported that kaolinite was less efficient in the adsorption of amino acids and peptides than montmorillonite. Other clay minerals, such as serpentine, can adsorb measurable quantities of aspartic and glutamic acid [39], whereas allophane can take up appreciable amounts of alanine [40].

The adsorption of nucleic acid bases on clay minerals has also been investigated. Lawless et al. and Banin et al. [41, 42] described the adsorption of adenosine monophosphate (5'-AMP) by montmorillonite in the presence of different metal cations and variation of the medium acidity. They found a maximum adsorption on Zn^{2+}-montmorillonite at a pH value close to neutrality. Winter and Zubay [43] found that montmorillonite adsorbed more adenine than adenosine, 5'-AMP, 5'-ADP, or 5'-ATP, while hydroxyapatite preferably adsorbed 5'-AMP compared to adenine or adenosine. They also found an adsorption dependence on the acidity of the medium. Other types of clay minerals have also been examined. Hashizume and Theng [44] found that allophane exhibited a greater affinity for 5'-AMP over adenine, adenosine or ribose.

The transformation of formaldehyde, the simplest aldehyde (HCOH), into complex sugars like ribose and from there to nucleic acids (e.g., RNA) in experiments simulating the early Earth conditions was proposed by Butlerov in 1861 under the name of formose reaction [45]. Clay minerals, such as montmorillonite and brucite can catalyze the self-condensation of formaldehyde and even more important, montmorillonite can stabilize several of the formed oligomers [46, 47].

Fraser and colleagues [27] have shown that vermiculite clay gels with large interlayer spacing act as amplifiers that sequentially change the D/L ratios of solutions containing the amino acids alanine, lysine, and histidine. A proposed natural ancient scenario for their results could be a clay-rich lagoon containing simple prebiotic molecules. The continuous repetition of D/L fractionation like the one they found, over geological periods of time, could lead to clay/solution nanofilm multilayers with amino acids having a chiral excess. Alternatively, chiral excess could have originated under the physicochemical conditions of interstellar space, or could have been produced in a modified type of Miller-Urey synthesis in the presence of polarized light or dust particles with chiral centers. Such activated clay films have the capacity to concentrate and to induce significant chiral separation of amino acids. They propose that such a mechanism could be the reason for the chiral selectivity observed for the rare amino acid isovaline in the Murchinson or Orgueil meteorites [27]. The discovery of isovaline (2-amino-2-methylbutanoic acid) in the biosphere suggests an extraterrestrial origin of amino acids and has been linked to the homochirality of life on Earth [50].

4 General Conclusions

Clay minerals formed since early geological times on Earth and were widely distributed. Due to their catalytic and adsorption properties, they have played a significant role in the search for an explanation of a biology dominated by L-amino acids and by D-sugars. Clays have also offered a catalytic surface beneficial to the aggregation and preservation of the first biopolymers. Even when an extensive number of experiments have validated these functions, the specific conditions dominating the early Earth scenario are difficult to find in laboratory simulations. Perhaps a definitive answer for the specific role of clays in the prebiotic chemistry of the early Earth will never be found. Whether they offer an initial bias for handedness or just a surface that preferentially favored the domination of handedness on Earth, is a question that still requires further investigation.

Abbreviations

5'-AMP   adenosine monophosphate
5'-ADP   adenosine diphosphate
5'-ATP   adenosine triphosphate
H_2O_v water vapor

Acknowledgment Support from the CONACyT grant 105450 is acknowledged.

References

1. N. F. Comins, W. J. Kaufmann, *Discovering the Universe*, W. H. Freeman and Co., New York, 2008.
2. J. H. Seinfeld, S. N. Pandis, *Atmospheric Chemistry and Physics. From Air Pollution to Climate Change*, 2nd edn., John Wiley & Sons Inc., Chichester, UK, 2006.
3. P. Cloud, *Sci. Amer.* **1983**, *249*, 176–189.
4. T. H. Jordan, *Proc. Natl. Acad. Sci. USA* **1979**, *76*, 4192–4200.
5. T. Tanimoto, in *Global Earth Physics: A Handbook of Physical Constants*, Ed T. J. Ahrens, American Geophysical Union, Washington, D. C., 1995, pp. 214–224.
6. R. L. Rudnick, S. Gao, in *The Crust*, Vol. 3 of *Treatise on Geochemistry*, Eds H. D. Holland, K. K. Turekian, Elsevier-Pergamon, Oxford, UK, 2003, pp. 1–64.
7. C.-W. Chiu, J.-J. Lin, *Progr. Polymer. Sci.* **2012**, *37*, 406–444.
8. H. Hashizume, in *Clay Minerals in Nature: Their Characterization, Modification and Application*, Eds M. Valaskova, G. S. Martynkova, *InTech*, **2012**, pp. 191–208.
9. T. F. Bristow, R. E. LMilliken, *Clays Clay Miner.* **2011**, *59*, 339–358.
10. C. Ponnamperuma, A. Shimoyama, E. Friebele, *Orig. Life* **1982**, *12*, 9–40.
11. G. Millot, *The Geology of Clays*, Masson, Paris, 1970.
12. D. D. Eberl, *Phil. Trans. Roy. Soc.* **1984**, *A311*, 241–257.
13. S. Hillier, in *Clay Mineralogy, Encyclopaedia of Sediments and Sedimentary Rocks*, Eds G. V. Middleton, M. J. Church, M. Coniglio, L. A. Hardie F. J. Longstaffe, Kluwer Academic Publishers, Dordrecht, 2003, pp. 139–142.
14. J. D. Bernal, *Proc. Phys. Soc. B London*, **1949**, 62, 597–618.
15. R. Saladino, C. Crestini, U. Ciambecchini, F. Ciciriell, G. Costanzo, E. Di Mauro, *Chembiochem* **2004**, *5*, 1558–1566.

16. J. P. Ferris, *Orig. Life Evol. Biosph.* **2002**, *32*, 311–332.
17. J. P. Ferris, A. R. Hill, R. Lis, L. E. Orgel, *Nature* **1996**, *381*, 59–61.
18. M. Franchi, E. Gallori, *Gene* **2005**, *346*, 205–214.
19. W. Huang, J. P. Ferris, *Chem. Commun.* **2003**, *21*, 1458–1459.
20. M. M. Hanczyc, S. M. Fujikawa, J. W. Szostak, *Science* **2003**, *302*, 618–622.
21. N. E. Schore, K. P. C. Vollhardt, *Organic Chemistry Structure and Function*, 6th edn., W. H. Freeman and Co., New York, 2011, pp. 169–214.
22. L. D. Barron, *Space Sci. Rev.* **2008**, *135*, 187–201.
23. L. Pasteur, *C. R. Acad. Sci. Paris* **1848**, *26*, 535–538.
24. L. Pasteur, *Anal. Chim. Phys.* **1848**, *24*, 442–459.
25. V. A. Avetisov, V. I. Goldanskii, V. V. Kuz'min, *Physics Today* **1991**, 33–41.
26. S. P. Fletcher, *Nat. Chem.* **2009**, *1*, 692–693.
27. D. G. Fraser, H. G. Greenwell, N. T. Skipper, M. V. Smalley, M. A. Wilkinson, B. Demé, R. K. Heenan, *Phys. Chem. Chem. Phys.* **2011**, *13*, 825–830.
28. C. H. Zhou, *Appl. Clay Sci.* **2011**, *53*, 87–96.
29. P. Cai, Q. Huang, M. Li, W. Liang, *Colloids and Surfaces B: Biointerfaces* **2008**, *62*, 299–306.
30. E. Biondi, S. Branciamore, L. Fusi, S. Gago, E. Gallori, *Gene* **2007**, *389*, 10–18.
31. L. Perezgasga, A. Serrato-Díaz, A. Negrón-Mendoza, L. de Pablo Galán, F. G. Mosqueira, *Orig. Life Evol. Biosph.* **2005**, *35*, 91–110.
32. A. Shimoyama, N. Blair, C. Ponnamperuma, in *Synthesis of Amino Acids under Primitive Earth Conditions in the Presence of Clay. Origin of Life*, Ed. N. H. Noda, Center for Academic Publisher, Tokyo, 1978, pp. 95–99.
33. S. Yuasa, *Nendo Kagaku* , **1989**, *29*, 89–96.
34. D. J. Greeland, R. H. Laby, J. P. Quirk, *Trans. Faraday Soc.* **1962**, *58*, 829–841.
35. D. J. Greeland, R. H. Laby, J. P. Quirk, *Trans. Faraday Soc.* **1965**, *61*, 2024–2035.
36. J. I. Hedges, P. E. Hare, *Geochim. Cosmochim. Acta* **1987**, *51*, 255–259.
37. T. Dashman, G. Stotzky, *Soil Biol. Biochem.* **1982**, *14*, 447–456.
38. T. Dashman, G. Stotzky, *Soil Biol. Biochem.* **1984**, *16*, 51–55.
39. H. Hashizume, *Viva Origino* **2007**, *35*, 60–65.
40. H. Hashizume, B. K. G. Theng, *Clay Minerals* **1999**, *34*, 233–238.
41. J. G. Lawless, A. Banin, F. M. Church, J. Mazzurco, R. Huff, J. Kao, A. Cook, T. Lowe, J. B. Orenberg, *Orig. Life* **1985**, *15*, 77–87.
42. A. Banin, J. G. Lawless, J. Mazzurco, F. M. Church, L. Margulies, J. B. Orenberg, *Orig. Life* **1985**, *15*, 89–101.
43. D. Winter, G. Zubay, *Orig. Life Evol. Biosph.* **1995**, *25*, 61–81.
44. H. Hashizume, B. K. G. Theng, *Clays Clay Miner.* **2007**, *55*, 599–605.
45. M. A. Butlerov, *C. R. Acad. Sci. Paris* **1861**, *53*, 145–147.
46. N. W. Gabel, C. Ponnamperuma, *Nature* **1967**, *216*, 453–455.
47. R. Saladino, V. Neri, C. Crestini, *Philos. Mag.* **2010**, *90*, 2329–2337.
48. R. Miranda, *Estudio de la región carboxilo terminal de la proteína de unión a caja TATA de Taenia solium*, Bachelor thesis, Universidad Nacional Autónoma de México, México, 2013.
49. H. R. Drew, R. M. Wing, T. Takano, C. Broka, S. Tanaka, K. Itakura, R. E. Dickerson, *Proc. Natl. Acad. Sci. USA* **1981**, *78*, 2179–2183.
50. J. R. Cronin, S. Pizzarello, *Adv. Space Res.* **1999**, *23*, 293–299.

Chapter 2
Living on Acetylene.
A Primordial Energy Source

Felix ten Brink

Contents

ABSTRACT 15
1 INTRODUCTION 16
2 ACETYLENE 16
  2.1 Properties of Acetylene 17
  2.2 Sources and Bioavailability of Acetylene on Earth and Other Planets 17
3 BACTERIA LIVING ON ACETYLENE 18
  3.1 *Pelobacter acetylenicus* 19
4 ACETYLENE HYDRATASE FROM *PELOBACTER ACETYLENICUS* 20
  4.1 Biochemical and Spectroscopic Properties 20
  4.2 Molybdenum-Substituted Enzyme 21
  4.3 Crystallization 22
  4.4 Structural Overview 22
  4.5 Active Site Setup 23
  4.6 Site-Directed Mutagenesis 25
  4.7 Density Functional Theory Calculations on the Substrate
     Binding Mode and Amino Acid Protonation States 27
  4.8 Towards the Reaction Mechanism 28
5 CONCLUSIONS 32
ABBREVIATIONS AND DEFINITIONS 33
REFERENCES 34

Abstract The tungsten iron-sulfur enzyme acetylene hydratase catalyzes the conversion of acetylene to acetaldehyde by addition of one water molecule to the C≡C triple bond. For a member of the dimethylsulfoxide (DMSO) reductase family this is a rather unique reaction, since it does not involve a net electron

F. ten Brink (✉)
Laboratoire de Bioénergetique et Ingenerie des Protéines, Institut de Microbiologie de la Méditerranée, CNRS/Aix-Marseille Université, F-13402 Marseille Cedex 20, France
e-mail: ftenbrink@imm.cnrs.fr

© Springer Science+Business Media Dordrecht 2014
P.M.H. Kroneck, M.E. Sosa Torres (eds.), *The Metal-Driven Biogeochemistry of Gaseous Compounds in the Environment*, Metal Ions in Life Sciences 14, DOI 10.1007/978-94-017-9269-1_2

transfer. The acetylene hydratase from the strictly anaerobic bacterium *Pelobacter acetylenicus* is so far the only known and characterized acetylene hydratase. With a crystal structure solved at 1.26 Å resolution and several amino acids around the active site exchanged by site-directed mutagenesis, many key features have been explored to understand the function of this novel tungsten enzyme. However, the exact reaction mechanism remains unsolved. Trapped in the reduced W^{IV} state, the active site consists of an octahedrally coordinated tungsten ion with a tightly bound water molecule. An aspartate residue in close proximity, forming a short hydrogen bond to the water molecule, was shown to be essential for enzyme activity. The arrangement is completed by a small hydrophobic pocket at the end of an access funnel that is distinct from all other enzymes of the DMSO reductase family.

Keywords acetylene • hydration • iron sulfur • tungsten

Please cite as: *Met. Ions Life Sci*. 14 (2014) 15–35

1   Introduction

Acetylene (C_2H_2, IUPAC name ethyne) is only a minor trace gas in the composition of the Earth's atmosphere, but notably it can be used as carbon and energy source by several bacteria. One of these, *Pelobacter acetylenicus*, was isolated by Schink in 1985 and the acetylene-converting enzyme of *P. acetylenicus*, the acetylene hydratase (AH; EC 4.2.1.112), has been studied in great detail over the years. AH is a hydrolyase, that catalyzes the addition of one molecule of water to the C≡C triple bond of acetylene-forming acetaldehyde. Therefore, the conversion of acetylene by AH is distinct from the only other known enzymatic reaction of acetylene, the reduction of acetylene to ethylene by nitrogenase [1]. Although the addition of a molecule of water to acetylene is formally not a redox reaction, AH activity depends on the presence of a strong reducing agent like titanium(III) citrate or sodium dithionite.

2   Acetylene

In biological systems, acetylene is well known as inhibitor of microbial processes by interaction with the metal sites of several metallo enzymes, such as nitrogenase, hydrogenase, ammonia monooxygenase, methane monooxygenase, assimilatory nitrate reductase or nitrous oxide reductase [2]. Thus, acetylene has been employed for the quantification of several important biological processes for a long time. For instance, nitrogen fixation can be measured by determining the reduction of acetylene to ethylene (C_2H_4) by nitrogenase [3], and inhibition of N_2O reductase by acetylene can be used to quantify denitrification rates [2]. Acetylene itself is a highly flammable gas that forms explosive mixtures with air over a wide range of concentrations (2.4–83 % vol, material safety datasheet, Air Liquide GmbH, Germany). Set under pressure, it can polymerize spontaneously in an exothermic reaction [2].

2.1   Properties of Acetylene

The physical and chemical properties of acetylene are mainly determined by the carbon-carbon triple bond. Formed by the overlapping of one of the sp hybrid orbitals and two p orbitals of each of the sp hybridized carbon atoms the bond consists of one σ-bond and two orthogonal π-bonds. The H–C≡C bond angle is 180 °C with a bond length of 121 pm for the triple bond. The triple bond increases the electronegativity of the carbon atoms and thereby the acidity of the H–C bonds of acetylene compared to ethylene and ethane, resulting in a pK_a of 24 for acetylene compared to 44 for ethylene [2]. This acidity is manifested in the formation of heavy metal acetylides (e.g., Cu(I) acetylide or Ag(I) acetylide) by reaction of acetylene with the corresponding metal cations. In addition, alkali metal acetylides (e.g., sodium acetylide) are formed by reaction of the elemental metal with acetylene. Acetylene can also serve as a ligand to transition metals. Both σ-bonded and π-bonded systems have been described. Examples include (HC≡CH)$_2$NiBr$_2$, HC≡CH–Ni(CN)$_2$ and [π–C$_5$H$_5$–Ni]$_2$C$_2$H$_2$ [2].

In general, the chemistry of acetylene is rather rich and diverse. Reactions of acetylene include reduction and oxidation as well as electrophilic and nucleophilic additions [4]. Due to the electron configuration of the C≡C triple bond, electrophilic additions to alkynes are much slower compared to additions to alkenes. On the other hand, nucleophilic additions are much faster than those on alkenes [5]. The reason for this is that the electrons of the carbon atoms are massed in the triple bond, leaving the "backside" of the cores of the carbon atoms open for nucleophiles to attack.

2.2   Sources and Bioavailability of Acetylene on Earth and Other Planets

Today, acetylene is only a minor trace gas in the composition of the Earth's atmosphere. Depending on where the samples were taken, concentrations of acetylene between 0.02–0.08 ppbv were detected [6]. Since acetylene on Earth seems to be mainly of anthropogenic origin, with exhaust from combustion engines as main source, samples from oceanic and rural areas show less abundance of acetylene [7].

Apart from the anthropogenic acetylene on Earth, naturally occurring acetylene can be found among other prebiotic molecules in interstellar gas clouds [8]. Another place where acetylene occurs naturally in greater abundance is on Saturn's moon Titan. Titan's atmosphere is considered to be a cold model of our Earth's early atmosphere approximately 4 billion years ago [6]. Photochemical processes in Titan's upper atmosphere that create acetylene from methane are considered to be the source of this acetylene [9]. Besides an atmospheric concentration of 3.5 ppm acetylene [6], the Cassini/Huygens mission to Titan detected lakes consisting of hydrocarbons (76–79 % ethane, 7–9 % propane, 5–1 % methane, 2–3 % hydrogen cyanide, 1 % butene, 1 % butane, and 1 % acetylene) on Titan's surface [10, 11]. The similarity between the atmosphere of Titan and the assumed composition of the Earth's early atmosphere have led to speculations that photochemical reactions along with other sources like volcanic eruptions may have provided the developing life on Earth with sufficient amounts of acetylene to be a viable source of carbon and energy [6].

Bacterial growth on acetylene is facilitated by its rather high solubility in water of 47.2 mM (20 °C; 1 atm or 101 kPa) compared to other gaseous compounds like O_2, H_2 or N_2 that have solubilities of around 1 mM under these conditions [2].

3   Bacteria Living on Acetylene

The energy richness of the $C\equiv C$ triple bond and the rather high solubility of acetylene in water make it a suitable substrate for bacteria, provided an adequate source is available. The first reports on bacteria living on acetylene were published more than 80 years ago [12]. In 1979, *Nocardia rhodochrous* was grown on acetylene as sole source of carbon and energy in the presence of dioxygen [13]. One year later, an acetylene hydratase activity was found in cell-free extracts from *Rhodococcus A1* grown on acetylene by anaerobic fermentation [14]. In these cell-free extracts, acetylene was converted to acetaldehyde by addition of a water molecule. Boiling of the extracts or addition of dioxygen inhibited this activity, indicating that it originated from an oxygen-sensitive enzyme. In 1981, anaerobic oxidation of acetylene to CO_2 was found in enrichment cultures from estuarine sediments [15]. In this case, acetate was identified as major intermediate of the process. Further studies revealed that two groups of bacteria were responsible for the oxidation of acetylene to CO_2. Fermenting bacteria converted acetylene to acetaldehyde, which they dismutated to acetate and ethanol. These products were then further oxidized by sulfate-reducing bacteria [16]. According to Culbertson et al. [16], the morphology of these Gram-negative bacteria was described to be similar to *Pelobacter acetylenicus*, an acetylene-fermenting bacterium, isolated three years earlier by Schink [17]. After the isolation of the acetylene hydratase from *P. acetylenicus* [1], Rosner et al. tested new isolates and bacterial strains from culture collection in the presence of a similar enzyme by activity tests and antibody cross reaction [18]. Acetylene hydratase activity was discovered in several cell-free extracts. As for *P. acetylenicus*, it required a strong reductant and the presence of molybdate or tungstate in the growth media. However, a cross reaction with antibodies against the acetylene hydratase from *P. acetylenicus* was not found [18].

Recently, a more systematic study was conducted on the presence of acetylene hydratase activity in anoxic sediments and waters from all over the United States of America and several deep sea sites [19]. Acetylene consumption was found in only 21 % of the samples investigated, usually after an incubation time of several days to months. The lag phase was explained by selecting for and thereby enriching bacteria that thrive on acetylene. This would mean that acetylene-fermenting bacteria were rather scarce in the original samples. In a second part of the study, the authors tried to amplify genes coding for an acetylene hydratase from DNA extracted from the sediment and water samples. The use of primers synthesized from the acetylene hydratase gene of *P. acetylenicus* resulted in 63 PCR products out of 645 environmental samples (9.8 %) [19]. Since AH-like genes could not be amplified in all samples in which acetylene hydratase activity was found, it was argued that the primer may have been overly specific for the acetylene hydratase gene of *P. acetylenicus*, hence, that Miller et al. could amplify the genes from *Pelobacter*-like organisms but not from other, less closely related bacteria [19].

3.1 Pelobacter acetylenicus

Pelobacter acetylenicus was isolated 1985 by enrichment on acetylene, which can be used as sole source of carbon and energy by this bacterium [17]. Two strains are deposited at the DSMZ culture collection in Braunschweig (Germany). The type strain WoAcy1 (DSMZ 3246) was isolated from a freshwater creek sediment and strain GhAcy1 (DSMZ 3247) from a marine sediment. *Pelobacter* are Gram-negative, strictly anaerobic deltaproteobacteria that ferment low-molecular-weight organic compounds, but no sugars. Characteristic substrates are gallic acid, acetoin, polyethylene glycol, and acetylene [20]. *Pelobacter acetylenicus* is mesophilic and grows between pH 6.0 and pH 8.0. From over 40 different substrates tested, only acetylene, acetoin, ethanolamine, and choline supported growth directly, while 1,2-propanediol and glycerol could be used in the presence of acetate [17]. No growth was found on the acetylene derivatives acetylene carboxylate or acetylene dicarboxylate or the structurally related compounds ethylene and cyanide.

During growth of *P. acetylenicus*, acetylene is fermented to nearly equal amounts of ethanol and acetate and small amounts of acetaldehyde [17]. According to the metabolic scheme proposed by Schink in 1985, the energetics of the initial hydration of acetylene can be derived (equations 1 and 2) [17]:

$$C_2H_2 + H_2O \rightarrow CH_3CHO \qquad \Delta G_0' = -119.9 \text{ kJ mol}^{-1} \qquad (1)$$

The subsequent disproportionation of acetaldehyde to ethanol and acetate yields by far less energy, but is still exergonic:

$$2\,CH_3CHO + H_2O \rightarrow CH_3CH_2OH + CH_3COO^- + H^+$$

$$\Delta G_0' = -17.3\,kJ\,mol^{-1} \quad acetaldehyde \tag{2}$$

In the oxidative branch of the disproportionation acetaldehyde is transformed to acetyl phosphate via acetyl-CoA. An acetate kinase then transfers the phosphate group from acetyl phosphate to a molecule of ADP, forming ATP and acetate. According to the growth yields determined by Schink, this kinase reaction is the only energy-conserving step during fermentation of acetylene by *P. acetylenicus* [17].

4   Acetylene Hydratase from *Pelobacter acetylenicus*

Acetylene hydratase from *P. acetylenicus* is so far the only acetylene hydratase that has been intensively studied by biochemical, spectroscopic, crystallographic, and computational approaches. The first isolation and purification of acetylene hydratase was reported in 1995 [1], followed by a high-resolution crystal structure [21] and heterologous expression and site-directed mutagenesis of several amino acids around the putative active site [22].

4.1   Biochemical and Spectroscopic Properties

AH was isolated as a monomer of 73 kDa according to SDS-PAGE [1] *versus* 83.5 kDa in MALDI-TOF mass spectra [23] from the soluble fraction of broken cells. Activity of acetylene hydratase, in cell-free extracts or purified, was dependent on (i) the presence of tungstate or molybdate in the growth media and (ii) addition of a strong reductant like titanium(III) citrate ($E_0' = -480$ mV; [24]) or dithionite ($E_0' = -527$ mV; [25]) to the activity assay [1]. The pH optimum of the enzyme reduced with 2 mM titanium(III) citrate was between pH 6.0 and pH 6.5. The temperature optimum was 50 °C (2 mM Ti(III) citrate; pH 7.0), the K_m for acetylene was 14 µM [1].

According to its amino acid sequence, acetylene hydratase is a member of the DMSO reductase family. Consistent with this, the metal content of AH was determined by ICP-MS analysis to be 0.4 to 0.5 mol W per mol enzyme and 3.7 to 3.9 mol Fe per mol enzyme [1, 22, 23]. Like in all members of the DMSO reductase family, a bis-molybdopterin-guanine-dinucleotide cofactor coordinates the tungsten ion in AH (Figure 1).

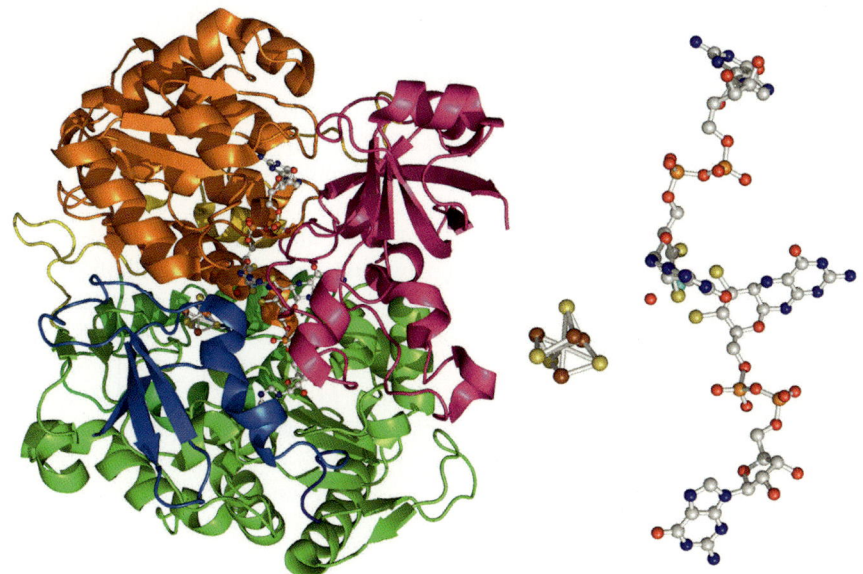

Figure 1 Acetylene hydratase from *Pelobacter acetylenicus*. Left: Overall structure of AH. Color codes in the cartoon representation: Domain I in blue, II in green, III in orange, and IV in pink. Right: Cofactors of AH W(MGD)$_2$ and [4Fe-4S] cluster. Element colors: W in cyan, O in red, S in yellow, N in blue, C in gray, P in orange, and Fe in brown, PDB code 2E7Z.

While acetylene hydratase as isolated under N$_2$/H$_2$ (94 %/6 % v/v) atmosphere was EPR-silent, EPR spectra of the enzyme reduced with dithionite showed a typical signal of a low potential ferredoxin-type [4Fe-4S] cluster with g values of $g_z = 2.048$, $g_y = 1.939$, and $g_x = 1.920$ [23]. After oxidization of the AH with one equivalent hexacyanoferrate(III) ([Fe(CN)$_6$]$^{3-}$) the EPR spectra showed the signal of a W(V) center with $g_x = 2.007$, $g_y = 2.019$, and $g_z = 2.048$. Upon further addition of hexacyanoferrate(III), the W(V) signal disappeared, as the tungsten site in the enzyme was oxidized to W(VI). Since AH isolated under N$_2$/H$_2$ needs to be reduced to be active, the dependence of AH activity on the applied redox potential was studied by Meckenstock et al. [23] in potentiometric titrations. The midpoint redox potential of the [4Fe-4S] cluster was determined to be –410 mV. The enzyme activity had a midpoint redox potential of –340 mV, meaning that AH is already active when the tungsten site is in its W(IV) state. Apparently, the redox state of the iron-sulfur center appears to be less important for enzyme activity [23].

4.2 Molybdenum-Substituted Enzyme

By growing *P. acetylenicus* in a medium containing only trace amounts of tungstate (2 nM instead of the original 800 nM in the tungstate medium) and elevated amounts of molybdate (2 μM instead of the original 6 nM) it is possible to exchange the central W ion in the active site of AH against Mo [26, 27]. The resulting molybdenum-substituted acetylene hydratase (AH(Mo)) is 10 times less active than the original acetylene hydratase in its native tungsten form (AH(W)) but still converts acetylene to acetaldehyde at a rate of 1.9 μmol min^{-1} mg^{-1} at 37 °C compared to 14.8 μmol min^{-1} mg^{-1} of the original AH(W) [27]. According to ICP-MS AH(Mo) contained 0.45–0.51 mol Mo per mol enzyme, 2.7–3.1 mol Fe per mol enzyme, and no tungsten [26–28], ruling out that the low activity of AH(Mo) derives from residual tungsten in the enzyme.

The EPR spectrum of AH(Mo), as isolated under N$_2$/H$_2$ (94 %/6 % v/v) atmosphere, showed a weak signal assigned to a Mo(V) center with $g_x = 1.978$, $g_y = 1.99$, and $g_z = 2.023$. The signal size increased upon oxidation with hexacyanoferrate(III), indicating that AH(Mo) was isolated in a partially oxidized state. Dithionite-reduced samples of AH(Mo) showed the identical signal of a ferredoxin type [4Fe-4S] cluster as AH(W) with $g_z = 2.048$, $g_y = 1.939$, and $g_x = 1.920$ [28]. When comparing the contribution of secondary structural elements to the total fold in AH(Mo) and AH(W) by circular dichroism (CD) spectroscopy, only slight differences in the amount of α-helices (14.3 % in AH(Mo) *versus* 11.3 % in AH(W)) and β-sheets (35.4 % in AH(Mo) *versus* 39.9 % in AH(W)) were found [27].

4.3   Crystallization

A high resolution X-ray structure of acetylene hydratase was solved in 2007 (PDB 2E7Z) [21]. The crystallization of AH was performed under a N$_2$/H$_2$ (94 %/6 % v/v) atmosphere at 20 °C using the sitting drop vapor diffusion method. Yellow brownish, plate-shaped crystals grew within 1–3 weeks from a 10 mg/mL solution of AH in 5 mM HEPES/NaOH pH 7.5 containing 5 mM Na$^+$ dithionite. 2 μL of the protein solution were mixed with 2.2 μL of 0.1 M Na$^+$ cacodylate, pH 6.5, containing 0.3 M Mg(acetate)$_2$, 21 % PEG 8000, and 0.04 M Na$^+$ azide. 15 % MPD was added as cryo protectant before flush freezing of the crystals in liquid N$_2$. The crystals belonged to space group C2 with $a = 120.8$ Å, $b = 72.0$ Å, $c = 106.8$ Å, and $β = 124.3°$, and contained one monomer per asymmetric unit. The native structure was solved by single wavelength anomalous dispersion (Fe absorption edge) and refined to 1.26 Å. In total the model consisted of 730 amino acid residues, 880 water molecules, two MGD cofactor molecules, and one [4Fe-4S] cluster. Additionally, two MPD molecules, one acetate molecule, and one sodium ion have been identified in the crystal structure of AH [21].

4.4   Structural Overview

As expected, from the biochemical and spectroscopic characterization of the enzyme, the structure is a monomer of 730 amino acids, containing a bis-molybdopterin-guanine-dinucleotide (bis-MGD) and a cubane [4Fe-4S] cluster. The two cofactors are buried deep inside a four domain fold, as found typically in enzymes of the DMSO reductase family (Figure 1) [21]. Domain I (residues 4-60) harbors the [4Fe-4S] cluster, ligated by the four cysteine residues Cys9, Cys12, Cys16, and Cys46. Domain II (residues 65–136 and 393–542) and III (residues 137–327) have an αβα fold with homologies to the NAD-binding fold in dehydrogenases. Each of these two domains provides hydrogen bonds, needed to bind one of the MGD cofactors. The interactions are mainly provided by variable loop regions at the C-termini of the strands of a parallel β-sheet. The coordination of both of the MGD cofactors is completed by domain IV (residues 590–730), which consists mainly of a seven stranded β-barrel fold [21].

The overall tertiary structure of acetylene hydratase and the position of the cofactors, with the two MGDs (P_{MGD} and Q_{MGD}) in an elongated conformation and the [4Fe-4S] cluster close to the Q_{MGD} is similar to all other structures of members of the DMSO reductase published so far [21]. However, the access from the surface of the protein towards the putative active site, consisting of the tungsten ion ligated by the two MGDs with the iron-sulfur cluster in close proximity is unique for an enzyme of the DMSO reductase family. In all structures of DMSO reductase family enzymes published so far, the access funnel starts at the pseudo twofold axis between domain II and III. A shift in the loop region of the residues 327–335 towards the surface of the protein and further rearrangements of the residues 336–393 seal this entrance point in AH. In other enzymes (e.g., nitrate and formate reductases), this loop region separates the [4Fe-4S] cluster from the Mo/W site. In AH, the shift of this loop opens a new access funnel towards the central tungsten ion at the intersection of domains I, II, and III, allowing the substrate to reach the central tungsten ion from a totally different direction than in other enzymes of the DMSO reductase family [21].

4.5   Active Site Setup

While the overall fold of AH is quite similar to other enzymes of the DMSO reductase family, major rearrangements are found at the active site [21]. The tungsten center in its reduced W(IV) state is coordinated by the four sulfur atoms of the dithiolene moieties of the P_{MGD} and Q_{MGD} cofactors and by one sulfur atom of a cysteine residue (Cys141), as found in the dissimilatory nitrate reductase. The sixth ligand position is taken by a tightly coordinated oxygen atom at 2.04 Å distance from the tungsten ion. Due to a rotation of the P_{MGD} cofactor, the geometry of the coordination in AH is not square pyramidal or trigonal prismatic, as typically found in enzymes of this family [29], but resembles more an octahedral or trigonal antiprismatic coordination (Figure 2) [21].

The access funnel opened by a shift in the loop region of residues 327–335 ends in a ring of six bulky hydrophobic residues (Ile14, Ile113, Ile142, Trp179, Trp293, and Trp472) forming a small hydrophobic pocket directly above the oxygen ligand and an adjacent aspartate residue (Asp13) (Figure 3). Asp13, a direct neighbor of the [4Fe-4S]-coordinating Cys12, forms a tight hydrogen bond of 2.41 Å to the oxygen ligand of the W ion. Although it was yet not possible to solve a crystal

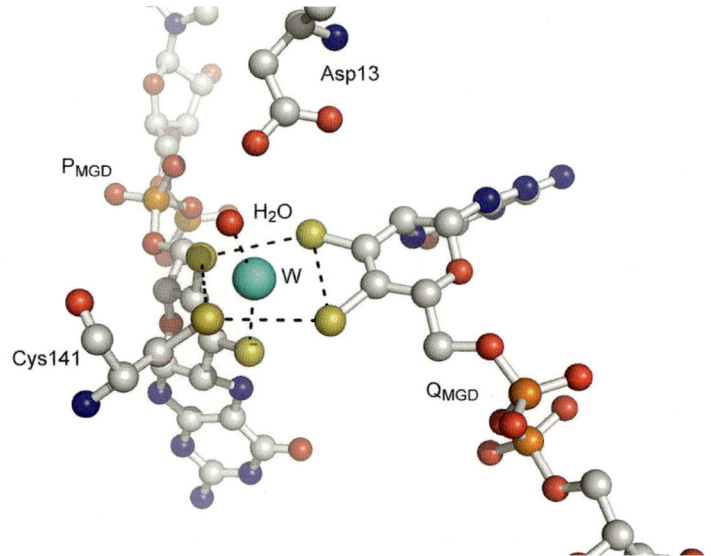

Figure 2 Active site of acetylene hydratase. The W ion is octahedrally coordinated by the two MGD cofactors (P and Q), Cys141, and a water molecule. Asp13 forms a short hydrogen bond to the water molecule. Element colors: W in cyan, O in red, S in yellow, N in blue, C in gray, P in orange.

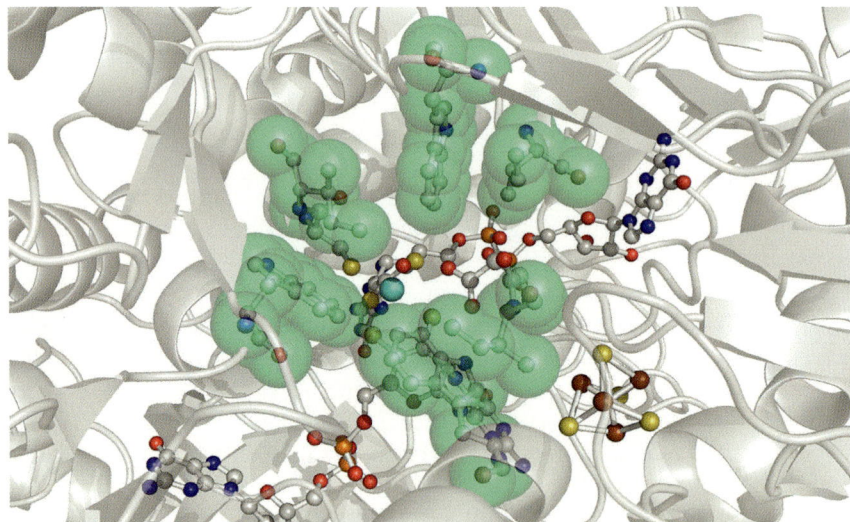

Figure 3 View down the access funnel towards the active site of AH. The six amino acid residues (Ile14, Ile113, Ile142, Trp179, Trp293, and Trp472) of the hydrophobic ring are represented as green spheres. W in cyan, O in red, S in yellow, N in blue, C in gray, P in orange, and Fe in brown.

structure with acetylene or an inhibitor bound [21, 27], an acetylene molecule docked computationally to the AH structure gave an excellent fit in the pocket of the hydrophobic ring with its carbon atoms positioned directly above the oxygen ligand and the carboxylic acid group of Asp13 [21].

Crucial for deriving a reaction mechanism for AH from the X-ray structure is the nature of the oxygen ligand of the tungsten ion. The bond length of 2.04 Å observed in the structure falls between the values expected for a hydroxo ligand (OH$^-$ 1.9–2.1 Å) and a coordinated water molecule (2.0–2.3 Å) [21]. Seiffert et al. [21] decided for a water molecule, since the close proximity of the heavy scatterer tungsten may distort the distance observed in the X-ray data by Fourier series termination and a simulation of this effect resulted in a true ligand distance of 2.25 Å.

4.6 Site-Directed Mutagenesis

The development of a protocol for the heterologous expression of AH in *Escherichia coli* allowed for further studies towards the reaction mechanism by site-directed mutagenesis [22]. Initially, the metal and cofactor content of the expressed protein was quite low compared to that of the native AH from *P. acetylenicus* (0.06 mol W and 1.22 mol Fe *versus* 0.4 mol W and 3.7 mol Fe per mol enzyme) [22]. This problem was partially solved by addition of an N-terminal chaperone-binding sequence of the soluble nitrate reductase NarG from *E. coli*. According to the literature this 30 amino acid long sequence is one of two binding sites for chaperones during insertion of the bis-MGD cofactor during protein biosynthesis [30]. The resulting fusion variant of AH had a higher content of W (0.14 mol per mol enzyme) and Fe (3.5 mol per mol enzyme) and therefore an increased activity compared to the heterologously expressed AH. Overall, when normalized to their W content, the activity of the heterologously expressed AH was nearly identical to that of the native enzyme purified from *P. acetylenicus* (Table 1) [22].

Table 1 Specific and relative activity of acetylene hydratase and variants [22].

Acetylene hydratase	W mol/mol AH	Fe mol/mol AH	Specific activity [μmol*min^{-1}*mg^{-1}]	Relative activity W[a]
P. acetylenicus AH	0.37 ± 0.04	3.69 ± 0.04	14.2 ± 0.9	38.4
E. coli AH	0.06 ± 0.02	1.22 ± 0.26	2.6 ± 0.8	43.3
E. coli AH D13A	0.09 ± 0.02	1.17 ± 0.29	0.2 ± 0.1	2.2
E. coli AH D13E	0.05 ± 0.01	1.11 ± 0.30	2.5 ± 0.3	50.0
E. coli NarG-AH	0.14 ± 0.06	3.17 ± 0.49	9.7 ± 1.9	69.3
E. coli NarG-AH K48A	0.15 ± 0.01	3.56 ± 0.31	7.2 ± 0.3	48.0
E. coli NarG-AH I142A	0.18 ± 0.02	3.20 ± 0.22	2.2 ± 0.2	12.2

[a]Relative to tungsten concentration, in nmol.

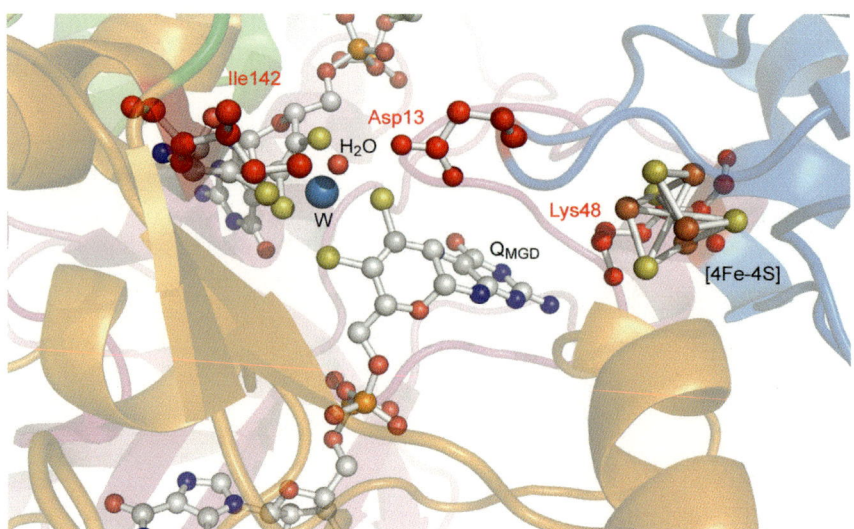

Figure 4 Amino acids exchanged by site-directed mutagenesis. Marked in red: Asp13 that forms a short hydrogen bond to the water ligand of the W ion was exchanged against alanine and glutamate, respectively. Lys48 that mediates the electron transfer between the [4Fe-4S] cluster and the Q_{MGD} cofactor was exchanged against alanine. Ile142 that is part of the hydrophobic ring and involved in positioning the substrate for the reaction was exchanged against alanine.

Three amino acids at the active site could be exchanged by site-directed mutagenesis: Asp13, Lys48, and Ile142 (Figure 4). Asp13 forms a hydrogen bond to the oxygen ligand of the W ion and was expected to be important for the reaction of AH by helping to activate the oxygen atom for the addition on the C≡C triple bond. The exchange of Asp13 against alanine resulted in a dramatic loss of activity (0.2 µmol min^{-1} mg^{-1} for the D13A variant compared to 2.6 µmol min^{-1} mg^{-1} for the expressed wild-type) while the exchange of Asp13 against glutamate had nearly no effect on the activity of AH (2.5 µmol min^{-1} mg^{-1} for the D13E variant compared to 2.6 µmol min^{-1} mg^{-1} for expressed wild-type). These results underline the important role of the carboxylic acid group at this position for the reaction of AH (Figure 5) [22].

Lys48 is located between the [4Fe-4S] cluster and the Q_{MGD} cofactor. In other enzymes of the DMSO reductase family, this residue is involved in electron transfer between the two cofactors [29]. As the reaction of AH does not involve a net electron transfer, the exchange of Lys48 against alanine did not affect the catalysis rate of the enzyme (Figure 5) [22].

Ile142 is part of the hydrophobic ring that is expected to form the substrate binding site at the end of the access funnel towards the active site [22]. The exchange of Ile142 against alanine resulted in a strong loss of activity (2.2 µmol min^{-1} mg^{-1} for the NarG-AH I142A variant compared to 9.7 µmol min^{-1} mg^{-1} for the NarG-AH fusion protein with the N-terminal chaperone-binding sequence). This finding supports the idea that the cavity within the hydrophobic ring is the substrate binding site of AH (Figure 5) [22].

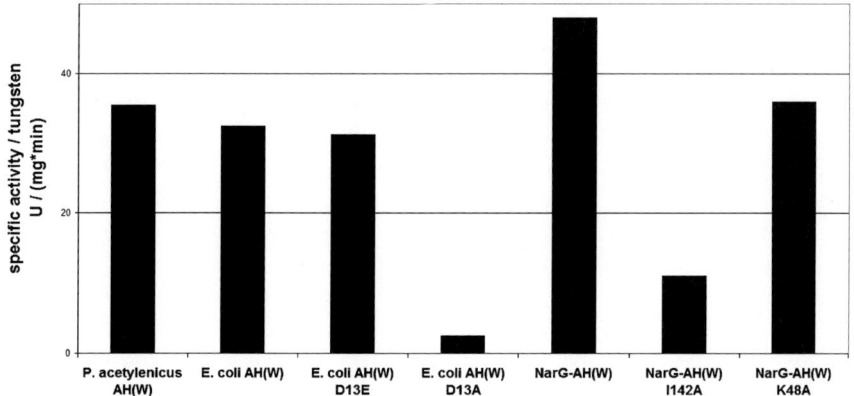

Figure 5 Activities of AH variants relative to their W content. For better comparability, the specific activity of the AH variants was divided by the W content in nmol. Data adapted from [27] (see also Table 1).

4.7 Density Functional Theory Calculations on the Substrate Binding Mode and Amino Acid Protonation States

Several groups directed a major computational effort towards the understanding of the reaction mechanism of AH. Seiffert et al. [21] used density functional theory (DFT) based atomic charge calculation to determine the titration curves of all residues in AH. Notably, out of 34 Asp and 58 Glu residues, only three showed a highly aberrant titration behavior: Asp298, Glu494, and Asp13. Asp13 that forms a hydrogen bond to the water ligand of the W center stayed protonated in the pH range 0–24, when the [4Fe-4S] cluster in close proximity was reduced. Whereas a partially deprotonated Asp13 was found at high pH values when the [4Fe-4S] cluster and the W center were fully oxidized [21].

Antony and Bayse [34], Vincent et al. [37], and Liao et al. [35] applied DFT methods to calculate the energies of acetylene adduct formation and the energetic barriers between intermediate states of several possible reaction mechanisms of model complexes that mimic the active site of AH. Antony and Bayse [34] used models of molybdenum- and tungsten-oxo dithiocarbamates (dtc) and dithiolates (dtl) that have been shown to form adducts with alkynes [31, 32] or even had AH-like activity [33]. Furthermore, a truncated model of the active site of AH consisting of a W ion coordinated by four dithiolene sulfur atoms, a cysteine sulfur and a water molecule was used to study the binding of water *versus* acetylene [34]. In these models, the free energy difference $\Delta G_{complex}$ for the formation of an acetylene metal complex favored the tungsten complexes over their molybdenum analogues, due to a more favorable interaction of the 5d W orbitals with the π molecule orbitals of acetylene. Additionally, the models showed that a thiolate ligand (Cys141 in AH) made the complex formation more exergonic than an oxo ligand in this position would do [34]. When calculating the ΔG values for the substitution of the water by acetylene, the acetylene complex was favored by ~10 kcal/mol [34].

Vincent et al. [37] used a truncated model of the AH active site, consisting of the W ion ligated by 4 dithiolene sulfur atoms, Cys141, and a water molecule and the carboxylic acid group of Asp13 hydrogen-bonded to the water molecule, to calculate the energy barriers between intermediates of several putative reaction pathways. A very similar approach was followed by Liao et al. [35]. Here, a much larger model complex of the active site of AH was used, consisting of the W ion ligated by the two pterin cofactors, Cys141, and a water molecule. Additionally, Asp13 and several other amino acids in close proximity (Cys12, Met140, Ile142, Trp179, and Arg606) were included [35]. While the reaction pathways deriving from the DFT calculation will be discussed in Section 4.8, it is noteworthy that Liao et al. [35] calculated a pK_a of 6.3 for the Asp13 residue. Thus in this model, Asp13 will most likely be deprotonated under reaction conditions in contrast to the model of Seiffert et al. [21] in which Asp13 is always protonated.

Using their established model for DFT calculations on the reaction mechanism of AH [35], Liao and Himo [36] calculated the binding and activation energies for several other compounds such as ethylene, acetonitrile, and propyne. In biochemical experiments none of these compounds was turned over by AH [22]. The DFT calculations showed that compared to acetylene, ethylene and propyne have a about ~6 kcal/mol and ~5 kcal/mol higher binding energy for the initial displacement of the water molecule at the W ion. The subsequent steps of the reaction would have a much higher barrier than in the case of acetylene, showing why these compounds are no substrates for AH (see Figure 7 in Section 4.8) [36]. Acetonitrile had a much higher binding energy than acetylene (~13 kcal/mol) but the differences of ~3 kcal/mol more for the barriers of the subsequent steps were too low to draw a firm conclusion whether acetonitrile is a substrate of AH or not [36].

4.8   Towards the Reaction Mechanism

When the crystal structure of AH was solved at a resolution of 1.26 Å, Seiffert et al. [21] proposed two alternative reaction mechanisms on the basis of their structural data. Depending on the nature of the oxygen ligand of the W ion (OH⁻ or H_2O), either a nucleophilic addition or an electrophilic Markovnikov-type addition were proposed [21]. In both cases, acetylene, located in the pocket formed by the hydrophobic ring, would not interact directly with the W ion but only with the oxygen ligand activated by the W^{IV} center and Asp13. A hydroxo ligand (OH⁻) would constitute a strong nucleophile that would yield a vinyl anion with acetylene. The basicity of the vinyl anion would be sufficient to deprotonate Asp13 to form a vinyl alcohol that would tautomerize to acetylene. A water molecule would then bind to the W ion and get deprotonated by the now basic Asp13 to restore the active site for the next reaction cycle [21].

A bound water molecule would gain a partially positive net charge by the proximity of the protonated Asp13, turning it into an electrophile that could directly attack the C≡C triple bond with a vinyl cation as intermediate (Figure 6) [21].

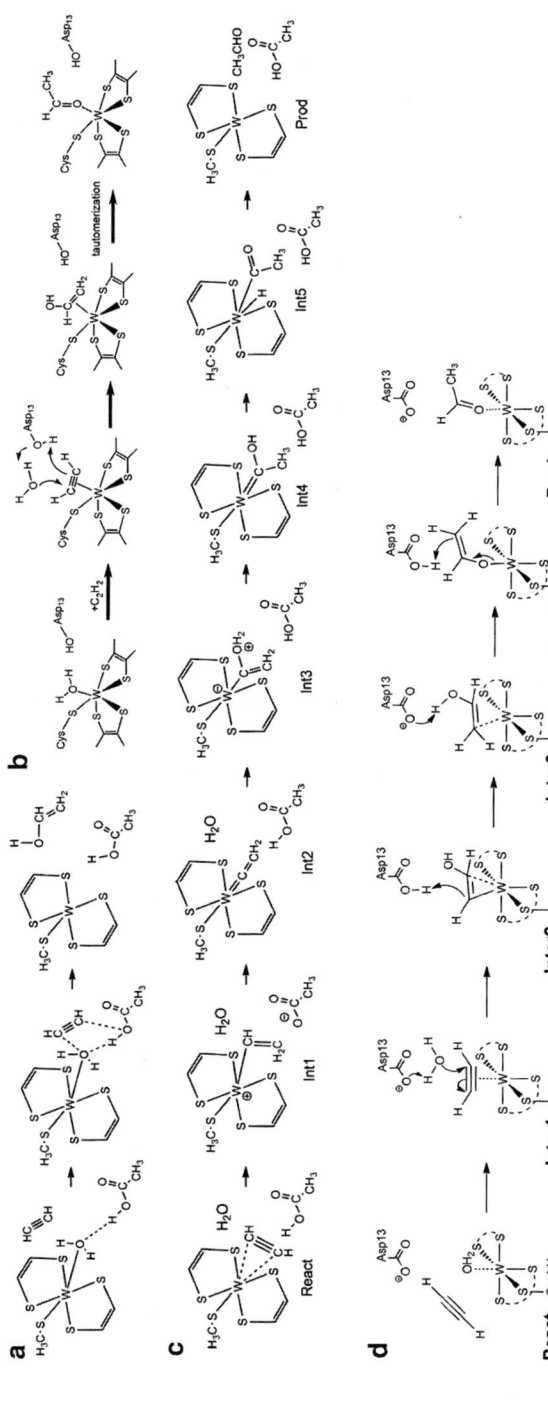

Figure 6 Pathways of several reaction mechanisms proposed for AH. (**a**) Nucleophilic attack of the activated water ligand of the W ion [21] (**b**) First shell mechanism involving a η^2-acetylene complex followed by a nucleophilic attack of a water molecule [34]. (**c**) First shell mechanism with a tungsten vinylidene complex as intermediate [37]. (**d**) Mechanism with a η^2-acetylene complex followed by several proton transfers to and from Asp13 [35]. Data adapted from [34, 35, 37].

The second shell mechanism was supported by the loss of activity when Ile142 was exchanged against alanine in the site-directed mutagenesis experiments [22]. However, in all DFT calculations performed so far, a first shell mechanism with a direct binding of acetylene to the W ion gave much lower energy barriers [34, 35, 37]. Antony and Bayse [34] calculated the substitution of the water molecule by acetylene to form a η^2-acetylene complex to be favorable by a ΔG of -10 kcal/mol. Therefore, a reaction mechanism starting with the formation of a η^2-acetylene complex followed by a nucleophilic attack by a water molecule was proposed, yielding either a η^2-complex of a vinyl alcohol or a β-hydroxovinylidene (Figure 6) [34]. Vincent et al. [37] calculated the energetic barriers for the inter-mediates of the reaction mechanism proposed by Seiffert et al. [21] and Antony and Bayse [34]. The nucleophilic attack of a water ligand on the C_α of acetylene to form a vinyl alcohol via a vinyl anion and deprotonation of Asp13 [21] was calculated to be exothermic in total (-21.4 kcal/mol) but the barrier of 43.9 kcal/mol was quite high [37]. The results for the nucleophilic attack of a water molecule on a η^2-acetylene complex were quite similar, the overall reaction was slightly exothermic (-1.9 kcal/mol) but the barrier was also quite high (41.0 kcal/mol) [37]. The nearly identical barriers for both mechanisms could be explained by the similarities between both pathways, involving a nucleophilic attack of a water molecule, a cyclic intermediate structure and a proton shuttle from Asp13 [37].

Therefore, Vincent et al. [37] proposed a new reaction mechanism with overall lower barriers. The mechanism starts with a η^2-acetylene complex that forms an end-on bound vinylidene complex (W=C=CH$_2$) by deprotonation of Asp13. The attack of a water molecule activated by a hydrogen bond to Asp13 leads to the formation of a carbene complex that will isomerize to form acetaldehyde (Figure 6) [37]. The barriers for the formation of the end-on bound vinylidene complex and the formation of the carbene complex were calculated to be 28.1 kcal/mol and 34.0 kcal/mol respectively. The isomerization to acetaldehyde via a tungsten hydride complex requires the breaking of a W-C bond with a barrier of 29 kcal/mol and the decomposition of the product (3 kcal/mol barrier) (Figure 7) [37].

Liao et al. [35] postulated another reaction mechanism based on DFT calcula-tions. Notably in this model, Asp13 is in a deprotonated state because the model includes three hydrogen bonds of Asp13 to Cys12, Trp179, and the H$_2$O ligand of the W ion, lowering the calculated pK_a of Asp13 to 6.3. In the first step of this mechanism acetylene forms a η^2-complex with the W ion by displacing the water ligand. The displaced water molecule is activated for a nucleophilic attack on the η^2-acetylene complex by a proton transfer to the ionized Asp13. The resulting vinyl anion will be protonated by Asp13, yielding a vinyl alcohol. The tautomerization needed to form acetaldehyde from the vinyl alcohol, can either occur spontaneously after release of the alcohol from the active site or will be assisted by the W ion and Asp13 [35]. The assisted tautomerization starts with a proton transfer from the OH group of the vinyl alcohol to Asp13, yielding an enolate that binds with the oxygen atom instead of a carbon atom to the W ion. The proton is then delivered back to C2 yielding the product acetaldehyde (Figure 6) [35]. The displacement of the water molecule by acetylene in the first reaction step was calculated to be exothermic

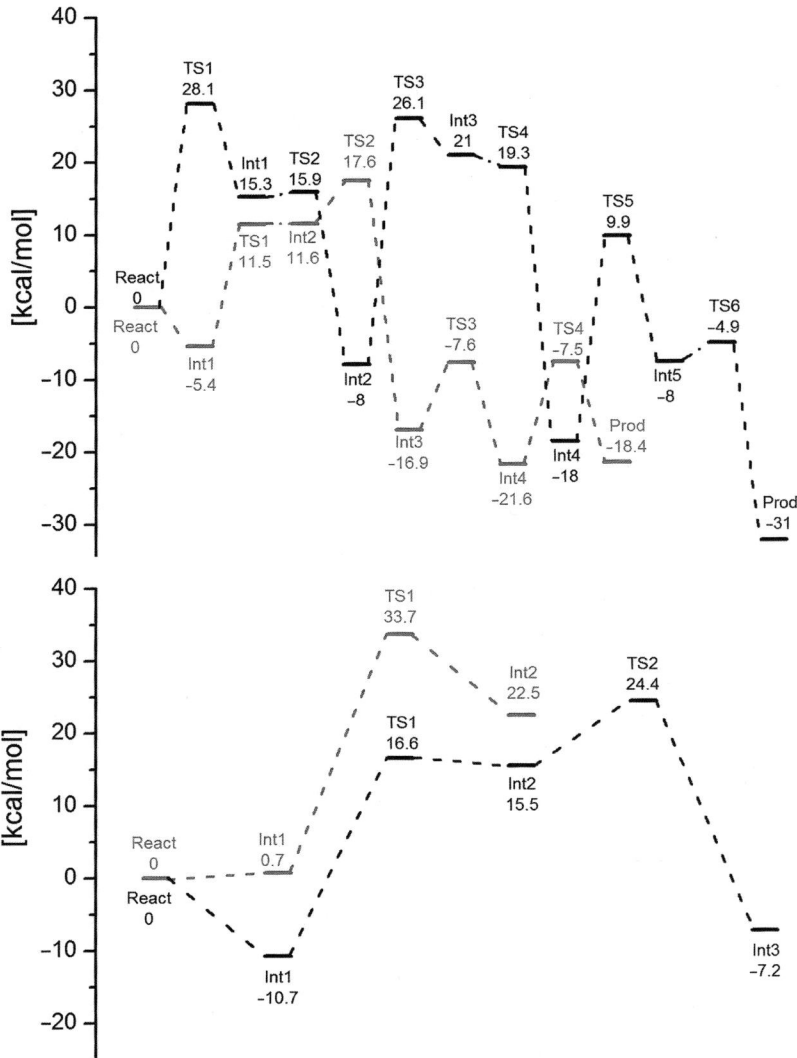

Figure 7 Potential energy profiles of transition states (TS) and intermediates (Int) in reaction mechanisms proposed for AH. **Top**: Energy profiles for the hydration of acetylene via a vinylidene complex (black line) [37] or a η^2-acetylene complex (grey line) [35]. **Bottom**: Energy profiles for the first hydration steps in the hydration of propyne (black line) and ethylene (grey line) [36]. Data adapted from [35–37].

(−5.4 kcal/mol). The barrier of the subsequent reaction steps were calculated to be 16.9 kcal for the nucleophilic attack, the protonation of the vinyl anion by Asp13 yields another 6 kcal/mol, raising the total barrier for the η^2-acetylene complex to vinyl alcohol transition to 23 kcal/mol. The following deprotonation to form an enolate has a barrier of 9.3 kcal/mol and the final reprotonation at C2 to form acetaldehyde has a barrier of 14.1 kcal/mol (Figure 7) [35]. When Liao et al. [35] used their model to calculate the barriers for the nucleophilic attack of a water ligand on acetylene [21] they found a barrier of 45.4 kcal/mol (compared to 43.9 reported by Vincent et al. [37]).

So far, a first shell mechanism is clearly favored in all DFT calculations. But since the hydrophobic ring and therefore the putative binding site for acetylene in the second shell mechanism [21] was not included in any of the models, it would be interesting to see a calculation with this feature included.

5   Conclusions

The isolation of *P. acetylenicus* [17] and the purification of a first tungsten-dependent acetylene hydratase [1] led to numerous speculations with regard to the physiological role and origin of this enzyme. Hydrolytic transformations of toxic compounds such as cyanides and nitriles were brought forward as one possibility [17]. However, neither growth of *P. acetylenicus* nor a reaction of AH with one of these compounds could be observed [1, 17, 22]. When the X-ray structure of AH was solved [21], major structural rearrangements became obvious in comparison to other members of the DMSO reductase family. The different location of the access funnel and the absence of a loop region separation of the [4Fe-4S] from the MGD cofactor in formate and nitrate reductases leads to a new active site on the opposite face of the W ion [21]. At the end of the access funnel, a ring of 6 bulky hydrophobic residues forms a binding pocket ideally suited for a small hydrophobic molecule like acetylene [21, 22].

Taking everything into account, AH appears to be an enzyme that is highly adapted to the conversion of acetylene, thus, it might be a rather old enzyme, from a past when acetylene was more abundant in the Earth's atmosphere [16]. With regard to its reaction mechanism, a clear statement is not yet possible. Unfortunately, a high resolution structure of AH with acetylene or an intermediate of the reaction bound at the active site, could not be obtained so far [21, 27]. Computational attempts to model the reaction pathway led to new insights and a deeper understanding of the atomic structure and possible substrate-active site interaction on the atomic level. For example, it was possible to demonstrate why certain potential substrates, such as ethylene, acetonitrile, and propyne do not react with AH [36], under the assumption of a reaction pathway. However, with the rapidly increasing computer power and shorter computing times, future calculations will take into account important features of AHs active site including the hydrophobic ring and the putative binding pocket for acetylene within it.

Note that an acetylene molecule can be modeled into this pocket located directly above the water ligand of the W ion at a distance of ~4 Å [21, 22]. Once these important features have been included into the calculations, a second shell mechanism might become favorable, which is clearly supported by the X-ray data and the results from site-directed mutagenesis. In the case of a first shell mechanism, the more polar reaction product, acetaldehyde will have to pass through a narrow hydrophobic gate upon leaving the active site, energetically not a favorable event. On the other hand, in a second shell mechanism, acetaldehyde would just be expelled into the substrate channel, assisting the product release by the repellant interaction of a polar molecule with a hydrophobic surrounding [27].

Abbreviations and Definitions

Å	Ångström; 1 Å $= 10^{-10}$ m
acetyl-CoA	acetyl-coenzyme A
ADP	adenosine 5'-diphosphate
AH	acetylene hydratase
AH(Mo)	acetylene hydratase, molybdenum-substituted
AH(W)	acetylene hydratase, in its native tungsten form
ATP	adenosine 5'-triphosphate
CD	circular dichroism
Da	Dalton; 1 Da $= 1$ g$\cdot$mol^{-1}
DFT	density functional theory
DMSO	dimethyl sulfoxide
DSMZ	Deutsche Sammlung von Mikroorganismen und Zellkulturen
dtc	dithiocarbamate
dtl	dithiolate
EPR	electron paramagnetic resonance
HEPES	2-[4-(2-hydroxyethyl)piperazine-1-yl]ethanesulfonic acid
ICP-MS	inductively coupled plasma-mass spectrometry
MALDI-TOF	matrix-assisted laser desorption/ionization time of flight analysis
MGD	molybdopterin-guanine-dinucleotide
MPD	2-methyl-2,4-pentanediol
NAD$^+$/NADH	β-nicotinamide adenine dinucleotide (oxidized/reduced)
NarG	nitrate reductase G subunit from *E. coli*
NarG-AH	acetylene hydratase with a fused chaperone binding sequence from NarG
PDB	Protein Data Bank
PEG	polyethylene glycol
SDS-PAGE	sodium-dodecyl sulfate polyacrylamide gel electrophoresis
U	unit; 1 U $= 1$ µmol$\cdot$min^{-1}
UV/Vis	ultraviolet/visible
v/v	volume per volume

References

1. B. M. Rosner, B. Schink, *J. Bacteriol.* **1995**, *177*, 5767–5772.
2. M. R. Hyman, D. J. Arp, *Anal. Biochem.* **1988**, *173*, 207–220.
3. W. D. Stewart, G. P. Fitzgerald, R. H. Burris, *Proc. Natl. Acad. Sci. USA* **1967**, *58*, 2071–2078.
4. P. Yurkanis-Bruice, *Organic Chemistry*, 4th ed., Pearson/Prentice Hall, Upper Saddle River, NJ, 2004, pp. 242–268.
5. F. Bohlmann, *Angw. Chem.* **1957**, *69*, 82–86.
6. R. S. Oremland, M. A. Voytek, *Astrobiology* **2008**, *8*, 45–58.
7. R. A. Whitby, E. R. Altwicker, *Atmos. Environ.* **1978**, *12*, 1289.
8. P. Thaddeus, *Philos. Trans. R. Soc. Lond. B Biol. Sci.* **2006**, *361*, 1681–1687.
9. D. Schulze-Makuch, D. H. Grinspoon, *Astrobiology* **2005**, *5*, 560–567.
10. D. Cordier, O. Mousis, J. I. Lunine, P. Lavvas, V. Vuitton, *Astrophys. J. Lett.* **2009**, *707*, L128–L131.
11. T. Tokano, *Astrobiology* **2009**, *9*, 147–164.
12. L. Birch-Hirschfeld, *Zentralbl. Bakteriol. Parasitenkd. Infektionskr. Hyg. Abt.* **1932**, *2*, 113.
13. D. Kanner, R. Bartha, *J. Bacteriol.* **1979**, *139*, 225–230.
14. J. A. M. de Bont, M. W. Peck, *Arch. Microbiol.* **1980**, *127*, 99.
15. C. W. Culbertson, A. J. Zehnder, R. S. Oremland, *Appl. Environ. Microbiol.* **1981**, *41*, 396–403.
16. C. W. Culbertson, F. E. Strohmaier, R. S. Oremland, *Orig. Life Evol. Biosph.* **1988**, *18*, 397–407.
17. B. Schink, *Arch. Microbiol.* **1985**, *142*, 295–301.
18. B. M. Rosner, F. A. Rainey, R. M. Kroppenstedt, B. Schink, *FEMS Microbiol. Lett.* **1997**, *148*, 175–180.
19. L. G. Miller, S. M. Baesman, J. Kirshtein, M. A. Voytek, R. S. Oremland, *Geomicrobiol. J.* **2013**, *30*, 501–516.
20. B. Schink, in *The Prokaryotes*, Eds M. Dworkin, S. Falkow, E. Rosenberg, K.-H. Schleifer, E. Stackebrandt, Springer, New York, Vol. 7, 2006, pp. 5–11.
21. G. B. Seiffert, G. M. Ullmann, A. Messerschmidt, B. Schink, P. M. Kroneck, O. Einsle, *Proc. Natl. Acad. Sci. USA* **2007**, *104*, 3073–3077.
22. F. tenBrink, B. Schink, P. M. Kroneck, *J. Bacteriol.* **2011**, *193*, 1229–1236.
23. R. U. Meckenstock, R. Krieger, S. Ensign, P. M. Kroneck, B. Schink, *Eur. J. Biochem.* **1999**, *264*, 176–182.
24. A. J. Zehnder, K. Wuhrmann, *Science* **1976**, *194*, 1165–1166.
25. P. A. Loach, in *Oxidation-Reduction Potentials, Absorbance Bands and Molar Absorbance of Compounds Used in Biochemical Studies*, Vol. 1, Ed G. D. Fasman, CRC Press, Cleveland, **1976**, pp. 122–130.
26. D. J. Abt, in *Tungsten-acetylene hydratase from Pelobacter acetylenicus and molybdenum-transhydroxylase from Pelobacter acidigallici: Two novel molybdopterin and iron-sulfur containing enzymes*, PhD Dissertation, University of Konstanz, Germany, **2001**.
27. F. tenBrink, in *Acetylene Hydratase from Pelobacter acetylenicus - Functional Studies on a Gas-Processing Tungsten, Iron-Sulfur Enzyme by Site Directed Mutagenesis and Crystallography*, PhD Dissertation, University of Konstanz, Germany, **2010**.
28. G. Seiffert, in *Structural and Functional Studies on two Molybdopterine and Iron-Sulfur containing Enzymes: Transhydroxylase from Pelobacter acidigallici and Aceytlene Hydratase from Pelobacter acetylenicus*, PhD Dissertation, University of Konstanz, Germany, **2007**.
29. H. Dobbek, R. Huber, *Met. Ions Biol. Syst.* **2002**, *39*, 227–263.
30. F. Sargent, *Microbiology* **2007**, *153*, 633–651.
31. E. A. Maatta, R. A. D. Wentworth, W. E. Newton, J. W. McDonald, G. D. Watt, *J. Am. Chem. Soc.* **1978**, *100*, 1320–1321.
32. J. L. Templeton, B. C. Ward, G. J. J. Chen, J. W. McDonald, W. E. Newton, *Inorg. Chem.* **1981**, *20*, 1248–1253.

33. J. Yadav, K. S. Das, S. Sarkar, *J. Am. Chem. Soc.* **1997**, *119*, 4315–4316.
34. S. Antony, C. A. Bayse, *Organometallics* **2009**, *28*, 4938–4944.
35. R. Z. Liao, J. G. Yu, F. Himo, *Proc. Natl. Acad. Sci. USA* **2010**, *107*, 22523–22527.
36. R.-Z. Liao, F. Himo, *ACS Catalysis* **2011**, *1*, 937–944.
37. M. A. Vincent, I. H. Hillier, G. Periyasamy, N. A. Burton, *Dalton Trans.* **2010**, *39*, 3816–3822.

Chapter 3
Carbon Monoxide. Toxic Gas and Fuel for Anaerobes and Aerobes: Carbon Monoxide Dehydrogenases

Jae-Hun Jeoung, Jochen Fesseler, Sebastian Goetzl, and Holger Dobbek

Contents

ABSTRACT 38
1 INTRODUCTION 38
    1.1 Chemistry of Carbon Monoxide 38
    1.2 Carbon Monoxide in the Biosphere 40
        1.2.1 Biological Cycle of Carbon Monoxide 40
        1.2.2 Use of Carbon Monoxide under Aerobic and Anaerobic Conditions 41
2 STRUCTURE AND FUNCTION OF CARBON MONOXIDE
   DEHYDROGENASES 43
    2.1 Cu,Mo-Containing Carbon Monoxide Dehydrogenases 43
        2.1.1 Structure of Cu,Mo-Carbon Monoxide Dehydrogenases 44
        2.1.2 Spectroscopic Investigations 46
        2.1.3 Enzymatic Activity 47
        2.1.4 Reaction Mechanism 48
    2.2 Monofunctional Ni,Fe-Containing Carbon Monoxide Dehydrogenases 49
        2.2.1 Function, Distribution, and Overall Structure 49
        2.2.2 Electronic States and Structure of Cluster C 50
        2.2.3 Pathways and Channels Involved in Catalysis 52
        2.2.4 Inhibited States of Cluster C 54
        2.2.5 Mechanism of Reversible Carbon Dioxide Reduction at Cluster C 55
    2.3 Bifunctional Ni,Fe-Containing Carbon Monoxide Dehydrogenases 56
        2.3.1 Classification and Distribution 56
        2.3.2 Structural Characterization of Bacterial CODH/ACS 58
        2.3.3 Substrate Binding and Reaction Mechanism 59
    2.4 Cu,Mo versus Ni,Fe: Parallels and Differences in Catalytic Strategies 62
3 CONCLUDING REMARKS AND FUTURE DIRECTIONS 64
ABBREVIATIONS 65
ACKNOWLEDGMENT 65
REFERENCES 65

J.-H. Jeoung • J. Fesseler • S. Goetzl • H. Dobbek (✉)
Institut für Biologie, Strukturbiologie/Biochemie, Humboldt-Universität zu Berlin, Unter den
Linden 6, D-10099 Berlin, Germany
e-mail: holger.dobbek@biologie.hu-berlin.de

© Springer Science+Business Media Dordrecht 2014
P.M.H. Kroneck, M.E. Sosa Torres (eds.), *The Metal-Driven Biogeochemistry
of Gaseous Compounds in the Environment*, Metal Ions in Life Sciences 14,
DOI 10.1007/978-94-017-9269-1_3

Abstract Carbon monoxide (CO) pollutes the atmosphere and is toxic for respiring organisms including man. But CO is also an energy and carbon source for phylogenetically diverse microbes living under aerobic and anaerobic conditions. Use of CO as metabolic fuel for microbes relies on enzymes like carbon monoxide dehydrogenase (CODH) and acetyl-CoA synthase (ACS), which catalyze conversions resembling processes that eventually initiated the dawn of life.

CODHs catalyze the (reversible) oxidation of CO with water to CO_2 and come in two different flavors with unprecedented active site architectures. Aerobic bacteria employ a Cu- and Mo-containing CODH in which Cu activates CO and Mo activates water and takes up the two electrons generated in the reaction. Anaerobic bacteria and archaea use a Ni- and Fe-containing CODH, where Ni activates CO and Fe provides the nucleophilic water. Ni- and Fe-containing CODHs are frequently associated with ACS, where the CODH component reduces CO_2 to CO and ACS condenses CO with a methyl group and CoA to acetyl-CoA.

Our current state of knowledge on how the three enzymes catalyze these reactions will be summarized and the different strategies of CODHs to achieve the same task within different active site architectures compared.

Keywords acetyl-CoA synthase • carboxydotrophic • iron-sulfur cluster • molybdenum • molybdopterin • nickel

Please cite as: *Met. Ions Life Sci.* 14 (2014) 37–69

1 Introduction

1.1 Chemistry of Carbon Monoxide

Carbon monoxide (CO) is a colorless, odorless, flammable gas that burns with a blue flame in air. Unlike the heavier carbon dioxide (CO_2), CO has a similar molecular weight as N_2 and O_2 and readily mixes with air. CO has a low solubility in water with approximately 0.35 L per L H_2O at 0 °C. The C-O distance of 1.06 Å (solid CO) and 1.128 Å (gaseous CO) agrees with the presence of a C-O triple bond making it isoelectronic with CN^- and N_2. The triple bonded electronic structure of CO results into a formal C-O polarization with a negative charge at C and a positive charge at O. This polarization is almost exactly cancelled out by the counter acting polarization due to the higher electronegativity of oxygen, making CO an unusual "carbonyl"-type compound with an electron-rich carbon atom.

CO is produced by incomplete combustion of carbon-containing compounds. CO can also be generated by adding concentrated sulfuric acid at 70–80 °C to formic acid (HCOOH), a reaction, which is the formal dehydration of formic acid to its anhydride CO. Conversely, CO can be converted to sodium formate by reaction with soda lye at elevated temperatures and pressures. In technical and industrial processes CO can be produced by reaction of elemental carbon with air and water vapor producing a mixture of CO and H_2 called "synthesis gas" or "water-gas" [1].

CO is an attractive partner for π-electron-rich metals due to its chemical properties. The HOMO of CO is centred at C and it therefore binds preferentially to metals with the C and not the electron lone pair at O, which is in an orbital of lower energy. Because CO is an unsaturated soft ligand, it is able to donate σ-electrons to the metal and to accept metal $d\pi$-electrons by a process termed back bonding. While electron donation along the σ-bond removes electron density from C, the accepted metal $d\pi$-electrons increase the electron density both at C and O. Thus, when CO binds to a metal its C becomes more positive and its O more negative, thereby polarizing the molecule. The stability of the formed metal-carbonyl depends on (i) the electron configuration of the metal; (ii) the other ligands coordinating the metal, and (iii) on the arrangement of the ligands. Metals of low π-basicity form carbonyls, which are sensitive to nucleophilic attack as the C is particularly positively charged and different nucleophiles, including hydroxyl ions react with metal carbonyls. In addition to forming metal carbonyls, CO inserts into metal alkyl bonds producing metal acyls. Metal carbonyl and metal acyl formation with CO are found in industrial processes and microbial physiology and enzymology [2].

Two industrial processes with analogies to the enzyme chemistry described in the following make use of the reactivity of metal carbonyls [2]. First, the water-gas shift reaction alters the CO:H_2 ratio of synthesis gas by reacting CO with water to CO_2 and H_2. When CO binds to a metal it becomes activated for the nucleophilic attack of an OH^- group. The formed metallacarboxylic acid liberates CO_2 and leaves a metal hydride species, which may be protonated to H_2. This reaction is analogous to the chemistry catalyzed by carbon monoxide dehydrogenases, except that the enzymes do not produce H_2 and keep the two electrons and two protons separated. Second, in the Monsanto- and Cativa-processes acetic acid is produced by reaction of CO with methanol. The active catalysts are precious transition metals of the 4d and 5d row of the periodic system, which catalyze a condensation reaction analogous to the reaction catalyzed at the Ni,Fe-cluster of acetyl-CoA synthases. The overall reaction is a carbonylation of methanol, which is relying on the tendency of CO to react with the metal-bound methyl group by migratory insertion. The formed metal-acetyl group is reductively eliminated to give acetic acid as product. Both processes are used to convert CO on a large scale.

CO oxidation produces CO_2, which is a linear molecule with formal C=O double bonds. CO_2 is overall non-polar, but due to the different electronegativities of C and O, it has a positively polarized C atom and negatively polarized O atoms. Reduction of CO_2 needs activation by decrease of the C–O bond order, which is also apparent by a decrease of the O–C–O bond angle from $180°$ to $133°$ in the one-electron reduced CO_2. The carbon centred lowest unoccupied molecular orbital allows reduction and nucleophilic attack at the C with transfer of electron density from the attacking nucleophile into the lowest unoccupied molecular orbital. The one-electron reduction of CO_2 is thermodynamically strongly disfavored ($E^{0'} = -1.9$ V), while the two-electron reductions to CO ($E^{0'} = -0.52$ V) and formate ($E^{0'} = -0.43$ V) [2] are more favorable. For a more comprehensive discussion of the chemistry and biochemistry of CO_2 we refer the reader to the recent review by Appel et al. [2].

1.2   Carbon Monoxide in the Biosphere

CO is a component of the past and present atmosphere. It is a ubiquitous pollutant of the present atmosphere [3], but may have facilitated early life in the primordial atmosphere by serving as carbon source [4, 5]. Ubiquitous volcanic emissions in the early Earth raised global CO concentrations to approximately 100 ppm. Nowadays, 0.05–0.35 ppm of CO are found in non-urban environments, while in dense urban regions with high traffic CO concentrations can reach 1.30 ppm [3, 6]. Concentrations of up to 5,540 ppm can still be encountered in volcanic environments [7], nurturing present day life on CO in the vicinity.

1.2.1   Biological Cycle of Carbon Monoxide

1.2.1.1   Sources of Carbon Monoxide

Natural and anthropogenic processes generate CO. They are responsible for a CO emission of 2,500–2,600 teragram (Tg) per year [3, 8]. Most CO is emitted by natural processes including atmospheric methane oxidation, natural hydrocarbon oxidation, volcanic activity, production by plants and photochemical degradation of organic matters in water, soil, and marine sediments [3, 8, 9]. Notable amounts of CO originate from the enzymatic degradation of heme [10, 11]. Anthropogenic processes such as incomplete combustion of fossil fuels and various industrial processes are responsible for the remaining annual 1,200 Tg of CO emitted to the atmosphere [3].

1.2.1.2   Removal of Carbon Monoxide

Chemical and biological processes are also responsible for CO removal. The major part of CO in the atmosphere becomes oxidized to CO_2 by rapid reaction with hydroxyl radicals in the troposphere (2,000–2,800 Tg/yr), reducing the half-life of CO in the troposphere to a few months [8, 12]. Microbes consume CO by using it as a source of energy and carbon (Section 1.2.2). Overall, soil and marine microbes reduce the global budget of CO by 20 % per year [3, 9, 13], to which soil microbes contribute with 200–600 Tg of CO removal per year [14]. CO-oxidizing bacteria have a natural enrichment in the top layer of burning charcoal piles from where several of these soil microbes have been isolated [15]. CO is also consumed by pathogenic Mycobacteria, like the tubercle bacillus *Mycobacterium tuberculosis*, which can grow on CO as sole source of carbon and energy [16]. Thus, biological processes are relevant to remove CO and keep it at low trace gas concentrations.

1.2.2 Use of Carbon Monoxide under Aerobic and Anaerobic Conditions

CO gains increasing attention in biology and medicine as a signaling molecule, acting at trace gas concentrations [10, 11]. Ironically, the toxic CO participates in various regulatory processes in vasodilatory action, oxygen-sensing, and neurotransmission [10] and may be used as therapeutic and bactericidal agent [10, 11].

CO is also a key metabolite in many microorganisms, which are using it as a source of energy and carbon [9, 17–19]. CO-utilizing microbes convert CO to CO_2 and a pair of reducing equivalents as products. The enzyme catalyzing CO oxidation is carbon monoxide dehydrogenase (CODH, CO:acceptor oxidoreductase). CODHs of aerobic and anaerobic microorganisms have different structures and contain different metals in their active sites (see Section 2) [2]. The CO-consuming microorganisms employ diverse metabolic pathways for coupling CO oxidation to various pathways of energy conservation and carbon fixation [9, 17].

1.2.2.1 Fates of Carbon Monoxide under Aerobic Conditions

Some aerobic CO-utilizing microorganisms are able to grow with CO as sole carbon source by fixing the CO_2 formed along the Calvin-Benson-Bassham cycle. We can distinguish between carboxydothrophs, which grow at elevated CO concentrations (over 10 %) and carboxydovores, which use CO at concentrations below 1000 ppm [9]. Aerobic carboxydothrophs include α-proteobacteria, firmicutes, actinobacteria, while aerobic carboxydovores include α-, β-, and γ-proteobacteria. Both groups of microorganisms fuel CO into their cellular metabolism via the Calvin-Benson-Bassham cycle [16, 20–23]. Some aerobic carboxydovores are unable to use CO_2 as sole carbon source as they lack the enzymes required for CO_2 fixation along the established pathways like the Calvin-Benson-Bassham cycle or the reverse tricarboxylic acid cycle [9]. These organisms, which include α-proteobacteria like *Stappia stellulata* and *Ruegeria* (previously *Silicibacter*) *pomeroyi* apparently use CO only as a source of energy [22, 24]. Thus, CO may be used in different ways.

Enzymatic CO oxidation supplies microorganisms with powerful reducing equivalents. The reducing equivalents generated are channeled into a respiratory chain via ubiquinones and cytochromes leading to the reduction of the terminal electron acceptor. Commonly molecular oxygen serves as terminal electron acceptor, but some microorganisms couple CO oxidation to reduction of nitrate or dinitrogen [9, 25].

1.2.2.2   Fates of Carbon Monoxide under Anaerobic Conditions

Phylogenetically diverse anaerobic bacteria and archaea, including sulfate reducers, hydrogenogens, acetogens, and methanogens employ the Wood-Ljungdahl pathway (also called reductive acetyl-CoA pathway) to generate biomass and energy from CO (Figure 1) [18, 26, 27]. This pathway allows microbes to thrive in anoxic niches in the presence of CO or CO_2 and H_2. The Wood-Ljungdahl pathway consists of two branches, namely a methyl and a carbonyl branch.

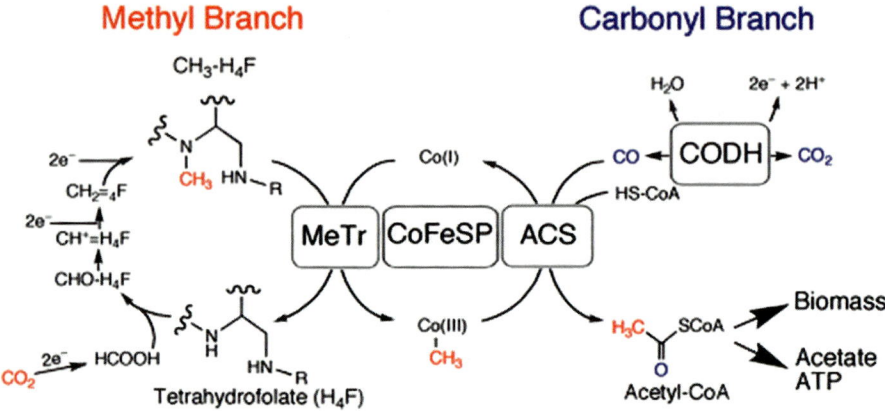

Figure 1 The Wood-Ljungdahl pathway. ACS, acetyl-CoA synthase; CODH, Ni,Fe-containing carbon monoxide dehydrogenase; CoFeSP, corrinoid iron-sulfur protein; MeTr, methyl-tetrahydrofolate:corrinoid iron-sulfur protein methyltransferase. The figure is modified from Ragsdale and Pierce [26].

The methyl branch is present in all organisms as folate-dependent one-carbon pathway. Here, CO_2 is reduced to formate, which is subsequently bound to tetrahydrofolate (H_4F), yielding 10-formyl-H_4F. After dehydration, the methenyl group is reduced stepwise, generating 5-methyl-H_4F [26]. The methyl-H_4F: corrinoid iron-sulfur protein methyltransferase (MeTr) passes the methyl group of methyl-H_4F to the corrinoid iron-sulfur protein (CoFeSP), which connects the methyl and the carbonyl branch (Figure 1).

In the carbonyl branch the CODH component of CODH/ACS catalyzes the reduction of CO_2 to CO, which is condensed to acetyl-CoA by acetyl-CoA synthase (ACS) with CoASH and the methyl group donated by CoFeSP. Acetyl-CoA can be used by the microbes as building block for cellular carbon compounds, to generate acetate (by acetogens) or as a source of energy (by acetoclastic methanogens) (Figure 1) [26].

Low potential electrons derived from anaerobic CO/CO_2 conversion ($E^{0'} = -558$ mV) [28] are transferred to various electron acceptors in CO-utilizing microorganisms [26, 27]. Obligate anaerobic sulfate reducers use the high-energy electrons to reduce sulfate to sulfide (sulfidogenesis) [29, 30], and *Sulfospirillum carboxydovorans* uses CO to reduce elemental sulfur, dimethylsulfoxide, and thiosulfate [31].

Some anaerobic carboxydotrophs like *Rhodospirillum rubrum* and *Carboxydothermus hydrogenoformans* produce molecular hydrogen (hydrogenogenesis) [32, 33]. Ferredoxins mediate the transfer of electrons gained from CO-oxidation to a membrane-bound Ni,Fe-hydrogenase, which is the site for proton reduction and proton translocation across the cytoplasmic membrane. The resulting proton motive force facilitates synthesis of ATP by ATP synthase [34, 35].

Methanogenic archaea are able to use CO in energy-conserving processes coupled to distinct metabolic pathways, which overlap each other [27]. CO oxidation by the CODH/ACS complex supplies reducing equivalents in methanogens. In the aceticlastic pathway, CO generated from acetyl-CoA cleavage by the CODH/ACS complex is used to generate electrons from CO oxidation, producing methane as final product [36–38]. In the hydrogenotrophic pathway, electrons resulting from CO oxidation are used to generate H_2, which is subsequently employed as an electron source to reduce CO_2 to methane. Finally, CO-derived reducing power serves to generate methane by reduction of methyl-coenzyme M in the methylotrophic pathway [37].

2 Structure and Function of Carbon Monoxide Dehydrogenases

2.1 Cu,Mo-Containing Carbon Monoxide Dehydrogenases

The structure and function of Cu,Mo-CODHs have been previously reviewed [2, 39, 40]. This section will focus on recent breakthroughs in the investigation of Cu,Mo-CODHs and on similarities and differences of Cu,Mo-CODHs compared to related molybdenum hydroxylases.

CODHs catalyze the (reversible) oxidation of CO to CO_2 according to equation (1):

$$CO + H_2O \rightleftharpoons CO_2 + 2H^+ + 2e^- \tag{1}$$

Cu,Mo-CODHs belong to a large family of pro- and eukaryotic molybdoflavoproteins, which share the use of a pyranopterin cofactor in their active site [41–44]. The pyranopterin cofactor is a heteroaromatic tricyclic ring system composed of a bicyclic pterin unit and a pyran ring, which is responsible for anchoring the Mo ion via an enedithiolate moiety at the pyran ring (Figure 2) [41]. In addition to the pyranopterin cofactor molybdenum hydroxylases typically employ two [2Fe2S] clusters and a flavin adenine dinucleotide (FAD) molecule as cofactors (Figure 3), but there are some variations in cofactor content [44, 45].

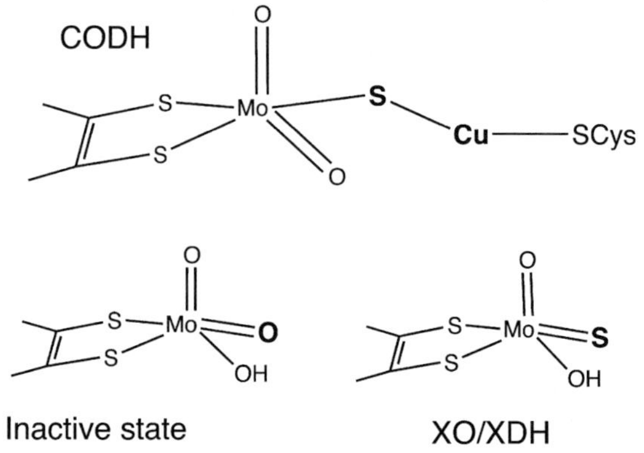

Figure 2 Active
and inactive states of
molybdenum hydroxylases.
The CODH and XO/XDH
active sites are converted
by cyanolysis of the sulfide-
ligand to the same inactive
state. CODH: Cu,
Mo-containing carbon
monoxide dehydrogenase;
XO: xanthine oxidase;
XDH: xanthine
dehydrogenase.

According to their sequence and structure Cu,Mo-CODHs belong to the molybdenum hydroxylases. Molybdenum hydroxylases typically catalyze the hydroxylation of activated C–H bonds by coupling the formal transfer of a hydride ion to a
Mo=S bond to the attack of a Mo-bound oxo/hydroxo group on the activated
carbon atom of the substrate. Variations exist in molybdenum hydroxylases isolated
from anaerobic bacteria, which can contain a terminal Mo=Se group instead of the
Mo=S group [46]. Before the Cu ion was identified, Cu,Mo-CODHs were thought
to be dependent on the presence of selenium in the active site [47–49].

Cu,Mo-CODHs share the sequence motif VAYXCSFR found in their active site
loop, where C denotes the cysteine residue binding the Cu ion, the most prominent
distinguishing feature of Cu,Mo-CODHs when compared to other molybdenum
hydroxylases [44, 49, 50]. A group of enzymes closely related to the canonical
Cu,Mo-CODHs called CoxL-II type enzymes is distinct by the lack of the Cu-binding
cysteine residue in the active site [9]. The physiological function of CoxL-II type
enzymes is not established, but the CoxL-II enzyme from *Bradyrhizobium
japonicum* USDA 110 was reported to oxidize CO, albeit 10–1000 times slower
than the Cu,Mo-CODH from *Oligotropha carboxydovorans* [51].

2.1.1   Structure of Cu,Mo-Carbon Monoxide Dehydrogenases

Cu,Mo-CODHs are typically encoded by three genes. The transcriptional order
in *Oligotropha carboxydovorans* is *coxM-coxS-coxL*, encoding for subunits
with 288, 166, and 809 amino acids, respectively. The Cu,Mo-CODH of
O. carboxydovorans is the most extensively investigated Cu,Mo-CODH and all
further discussions relate to this enzyme unless otherwise noted.

The structure of the *O. carboxydovorans* Cu,Mo-CODH has been determined in
different states and refined to a resolution of 1.09 Å [49, 50], while the structure of
the enzyme from *H. pseudoflava* has been refined to a resolution of 2.2 Å [52].
Both structures show a butterfly shaped dimer of heterotrimers (Figure 3). Each of
the three subunits harbors one type of cofactor. The large L subunit carries the
molybdopterin cytosine dinucleotide (MCD) cofactor, which is part of the active

site. The small S subunit harbors two [2Fe2S] clusters and the medium M subunit binds an FAD molecule non-covalently. The cofactors form a conduit for electron transfer from the molybdenum ion via the [2Fe2S] clusters to the FAD, from where electrons can be directly transferred to quinones in the cytoplasmic membrane [53]. Arrangement of the cofactors as well as the fold of the three subunits is like that of other molybdenum hydroxylases [42, 44].

The L subunit can be divided into two domains, both interacting with the MCD cofactor by a network of hydrogen bonds. The characteristic feature of Cu,Mo-CODHs is a unique active site loop where a cysteine residue forms a covalent bond to a Cu(I) ion (Figure 3).

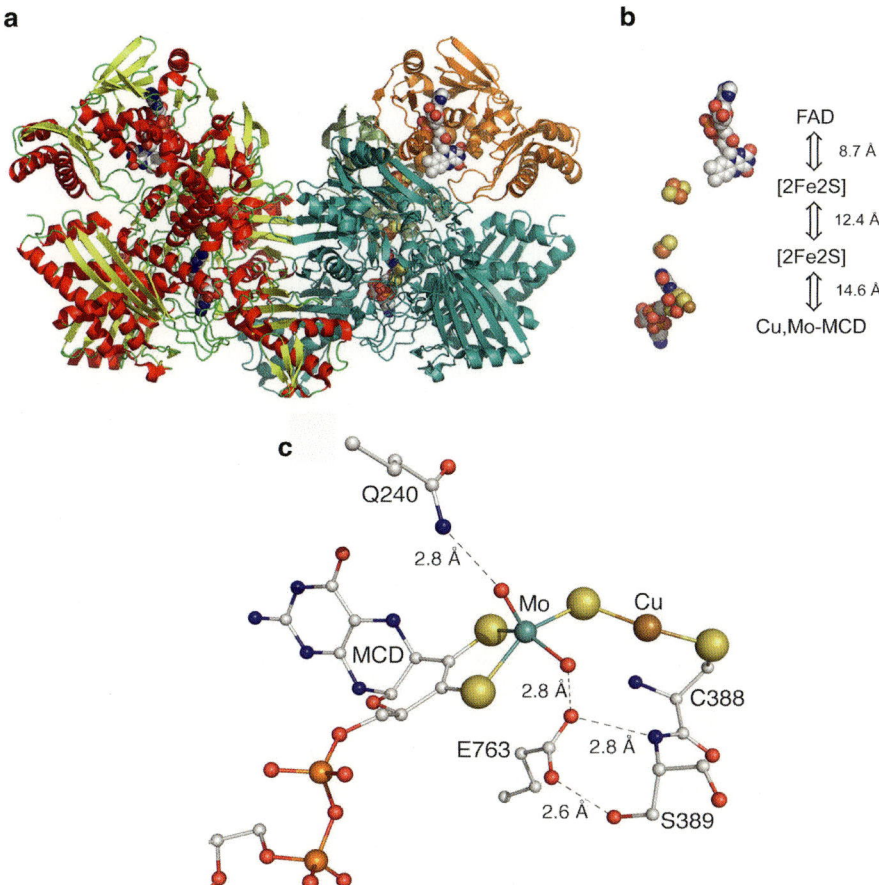

Figure 3 The structure of Cu,Mo-CODHs. (**a**) Overall structure of the dimer of trimers, $(LMS)_2$. The L subunit of the right monomer is colored in cyan, the M subunit in orange, and the S subunit in green. (**b**) Cofactors of one LMS monomer with shortest distance between the redox active sites of the cofactors. (**c**) Active site architecture including residues in the second coordination sphere of the metals.

The active site is buried within the L subunit. It is accessible from the surface through a hydrophobic channel with a diameter of 6 Å, which ends at the [CuSMo] unit of the active site. The enedithiolate unit of the pyranopterin cofactor binds Mo, whereas Cu is coordinated by the cysteine sulfur of the active site loop. Cu and Mo are bridged by a μ_2-sulfido ligand, keeping Cu and Mo in a distance of 3.74 Å in the Mo(VI) state, which increases to 3.93 Å in the Mo(IV) state [50].

The ligands of Mo have a distorted square pyramidal arrangement. The equatorial coordination sphere of the Mo ion consists of the enedithiolate moiety of the pyran ring, the bridging sulfido ligand, and one oxo/hydroxyl (or hydroxo) ligand. An additional oxo ligand is found in the apical position. A glutamate residue in *trans* to the apical oxo ligand with a Glu-Mo distance of 3.14 Å and a glutamine residue in hydrogen-bonding distance to the apical oxo ligand create the second coordination sphere of Mo (Figure 3). All elements of the first and second coordination sphere are conserved in the Mo hydroxylases.

The [2Fe2S] clusters and FAD are bound in the S and M subunit, respectively. The [2Fe2S] cluster proximal to the active site Mo is bound in a four-helix bundle domain, while the distal [2Fe2S] cluster is coordinated within a domain closely resembling the fold of a plant-type ferredoxin [49]. Both [2Fe2S] clusters are shielded from the solvent. The FAD molecule is bound between the N-terminal and middle domain of the M subunit. Access to the redox-active N5 position of the isoalloxazine ring of FAD is restricted by a tyrosine in the *O. carboxydovorans* and a tryptophan residue in the *H. pseudoflava* Cu,Mo-CODH [49, 52]. The cofactors are arranged in two independent electron transfer chains in the dimer with short cofactor-cofactor distances within each monomer allowing a rapid electron transfer [54].

2.1.2 Spectroscopic Investigations

Molybdenum cycles through three oxidation states (Mo(VI), Mo(V) and Mo(IV)) during catalysis of which only Mo(V) is paramagnetic and exhibits an EPR signal. The Mo(V) signal of Cu,Mo-CODH shows strong hyperfine coupling with Cu ($I = 3/2$) [55], which is consistent with a delocalization of the electron spin within the SOMO of Mo(V) along the entire Mo(V)-S-Cu(I) moiety [56]. This strong coupling is reproduced in a Mo-S-Cu model complex with a geometry similar to the active site of Cu,Mo-CODH [57]. The Mo(V) signal splits upon addition of [^{13}C]CO indicating that either [^{13}C]CO or product [^{13}C]CO$_2$ is a part of the signal-giving species [55]. An extended analysis of the [^{13}C]CO-^{63,65}Cu coupling using EPR and ENDOR spectroscopy provided additional evidence for the presence of a copper-carbonyl intermediate in CO oxidation [56]. That the Mo(V) signal changes in the presence of substrate has also been observed for H$_2$. When Cu,Mo-CODH reacts with H$_2$ a new EPR signal with larger g anisotropy and hyperfine coupling to the ^{63,65}Cu in the active site arises [58].

Confirmatory and complementary insights have also been obtained by X-ray absorption spectroscopy, which confirmed the presence of Cu and Mo in the active site of Cu,Mo-CODH [59]. While the crystallographic study indicates the presence of a hydroxyl group in the equatorial plane, X-ray absorption spectroscopy studies are consistent with the presence of an oxo group [59], which is also favored by computational studies on the mechanism of Cu,Mo-CODH [56, 60, 61].

2.1.3 Enzymatic Activity

Further details of the catalytic cycle were revealed by determining the activity of Cu,Mo-CODH with different substrates. At its pH optimum of 7.2 Cu,Mo-CODH oxidizes CO with a k_{cat} of 93.3 s^{-1} and a K_m for CO of 10.7 μM at 25 °C [55]. Quasi single-turnover kinetics (up to three turnovers) indicate that the likely rate limiting step of the overall reaction is the reductive half-reaction and that CO binds in a rapid initial step to the enzyme. Furthermore, Cu,Mo-CODH acts as an efficient uptake hydrogenase [62] oxidizing H$_2$ with a limiting rate constant of 5.3 s^{-1} and a K_d for H$_2$ of 525 μM [58].

Cu,Mo-CODH transfers electrons generated by CO oxidation to different physiological and artificial electron acceptors [39, 48]. In *O. carboxydovorans* cells the enzyme is attached to the inner aspect of the cytoplasmic membrane [63] and the interaction is specific for cytoplasmic membranes of CO-grown cells [64]. Electrons generated by CO-oxidation at the Cu,Mo-site are transferred efficiently from the FAD to membrane-bound quinones, agreeing with a direct electron transfer into the quinone pool under physiological conditions [53].

Cu,Mo-CODHs are inactivated by several small molecule inhibitors. Cyanide is isoelectronic and isosteric with CO and inactivates the oxidized enzyme with a half-life of about 30 mins. Cyanolysis releases concomitantly up to 1 mol of thiocyanate (SCN$^-$) and 1 mol of Cu per active site of Cu,Mo-CODH [50]. The crystal structure and spectroscopic investigations are in agreement with the formation of a Mo tri-oxo species upon cyanolysis, depleting the active site of Cu and the bridging sulfido-ligand [50, 59]. Isocyanides and CO show a similar σ-donor and π-acceptor ligand character and are isoelectronic with a non-bonding pair of electrons in the p-orbital of the terminal carbon. Isocyanides act as inhibitors, which bind to the active site of Cu,Mo-CODH by inserting into the Cu-S bond. Inhibition is concurrent with oxidation of the inhibitor, which forms a thiocarbamate derivative [50]. The inactive Mo tri-oxo species is also formed when isolated Cu,Mo-CODH reacts with CO under oxic conditions, indicating that the rapid transfer of the electrons generated by CO oxidation to the quinone pool [53] is important to render the enzyme stable under physiological conditions when turnover occurs in the presence of dioxygen.

Inactive Mo tri-oxo Cu,Mo-CODH can be reconstituted by addition of sulfide and copper under anoxic, reducing conditions, resulting in about 50 % functional enzyme species [65]. A reconstitution protocol similar to that of Resch, Dobbek, and Meyer [65] was used to substitute copper for silver, generating a Ag,Mo-CODH [66]. Most surprisingly, activity of the enzyme was regained and resulted in approx. 30 % functional enzyme after reconstitution. The silver-substituted enzyme has a limiting rate constant of 8.1 s^{-1} in CO oxidation [66], but unlike the Cu-containing enzyme it is unable to oxidize H$_2$ [58].

2.1.4   Reaction Mechanism

The structural, spectroscopic, and computational investigations agree with the following mechanism for Cu,Mo-CODHs (Figure 4). Catalytic CO oxidation starts with the Mo(VI)/Cu(I) state, where CO binds to Cu(I) generating a Cu(I)-CO species [56]. Binding to the electron-rich Cu(I) populates the π^* orbital of CO, thereby activating it for the nucleophilic attack of the equatorial Mo=O oxygen on the carbon of Cu-bound CO. Based on the crystal structure of an isocyanide-inhibited state of Cu,Mo-CODH the possibility of a thiocarbonate intermediate was discussed [50]. However, the formation of a C-S bond with CO in the catalytic cycle is most likely thermodynamically unfavorable, as indicated by computational studies [60, 61]. Thus, CO oxidation likely leaves the Mo-S-Cu moiety intact and involves a Mo(VI)-S-Cu(I)-C-O- metallacycle in the next step [67], which breaks down to form CO_2 and Mo(IV). Electron transfer from the active site to external acceptors, closes the catalytic cycle. Electrons generated by CO or H_2 oxidation are in a first step taken up by Mo reducing Mo(VI) to Mo(IV). Electrons are then transferred via the two different [2Fe2S] clusters to the FAD, where they are donated to external electron acceptors, regenerating the enzyme for another turnover.

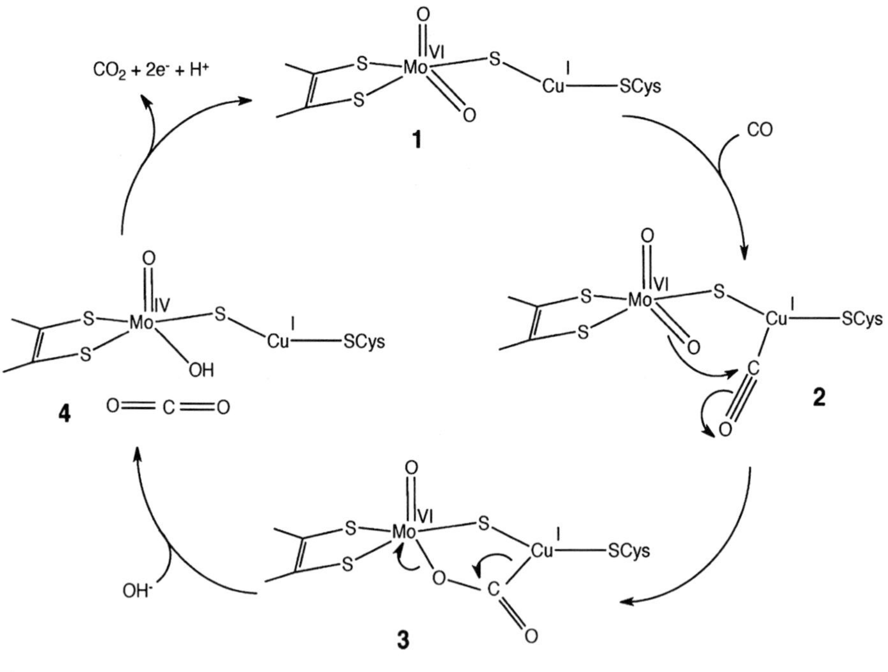

Figure 4 Mechanism of CO oxidation by Cu,Mo-CODHs. (**1**) The Mo(VI)-Cu(I) state is ready for CO activation. (**2**) Cu is binding CO, activating it for a nucleophilic attack by the Mo-bound oxo ligand. (**3**) The intermediary metallacycle rearranges and water binding may support CO_2 liberation. (**4**) The Mo(IV)-Cu(I) active site releases two electrons and a proton to regenerate the Mo(VI)-Cu(I) state 1.

Cu,Mo-CODHs are structurally very similar to other Mo-hydroxylases from which they differ mainly by the additional Cu ion. However, the first and second coordination sphere of Mo, including the stereochemistry of the ligands are practically the same in Cu,Mo-CODHs [50], xanthine oxidoreductase [68], and quinolone 2-oxidoreductase [69]. Nevertheless, typical Mo-hydroxylases hydroxylate C-H groups, whereas Cu,Mo-CODHs oxidize CO, raising the question on the parallels in their catalytic requirements. A recent comparison of the orbital contributions found remarkably similar electronic structures for central intermediates in CO and C-H bond activation [67]. Cu,Mo-CODH stabilizes the CO_2 bound intermediate by C-Cu σ → Mo-S π* and Mo-S π → C-Cu σ* charge transfer, whereas C-H bond activation in xanthine oxidase involves C-H σ → Mo-S π* and Mo-S π → C-H σ* charge transfers [67]. According to the orbital contributions, Cu,Mo-CODHs employ the Cu(I) ion as a substitute for the hydrogen bound to C in C-H groups, or in other words, Cu(I)-CO resembles a C-H unit activated for a nucleophilic attack.

H_2 is chemically unlike CO, but H_2 oxidation by Cu,Mo-CODH also depends on the presence of Cu(I) [50]. Accordingly, catalytic H_2 oxidation by Cu,Mo-CODH was postulated to occur via formation of a copper hydride species in the active site [50], which is consistent with recent spectroscopic results on a Mo(V) species generated by H_2 incubation of Cu,Mo-CODH [58]. Thus, by extending the Mo=S moiety to a Mo-S-Cu(I) unit, the CO as well as H_2-oxidizing activities were implemented into a Mo hydroxylase, a small change with a large effect.

2.2   Monofunctional Ni,Fe-Containing Carbon Monoxide Dehydrogenases

2.2.1   Function, Distribution, and Overall Structure

Since the discovery and initial isolation of Ni,Fe-CODHs more than three decades ago [70, 71], several reviews summarized the current state of research [2, 72–78]. Here, we focus on past and present advancements in the field of monofunctional Ni,Fe-CODHs.

Ni,Fe-CODHs are either monofunctional or bifunctional in a complex with acetyl-CoA synthase and catalyze the reversible oxidation of CO (equation 1). Monofunctional Ni,Fe-CODHs are used by anaerobic bacteria to employ CO as energy source and the enzymes isolated from *Carboxydothermus hydrogenoformans* (*Ch*) [34] and *Rhodospirillum rubrum* (*Rr*) [79] have been investigated in some detail. The importance of monofunctional Ni,Fe-CODHs is reflected in the genome of *C. hydrogenoformans*, where five different gene clusters contain structural genes encoding Ni,Fe-CODHs [80]. Two monofunctional Ni,Fe-CODHs (CODH I_{Ch} and CODH II_{Ch}) catalyze the oxidation of CO [34], whereas a third bifunctional Ni,Fe-CODH (CODH III_{Ch}) is found in a stable complex with ACS and supports autotrophic carbon assimilation (Section 2.3) [81].

Crystal structures of monofunctional Ni,Fe-CODHs isolated from *C. hydrogenoformans* (CODH II$_{Ch}$) [82] and *R. rubrum* (CODH$_{Rr}$) [83] have been determined. The overall structure is a mushroom-shaped homodimer with five metal clusters, of which three are cubane-type [4Fe4S] clusters (two cluster B and one cluster D) and two are the active site clusters C (Figure 5).

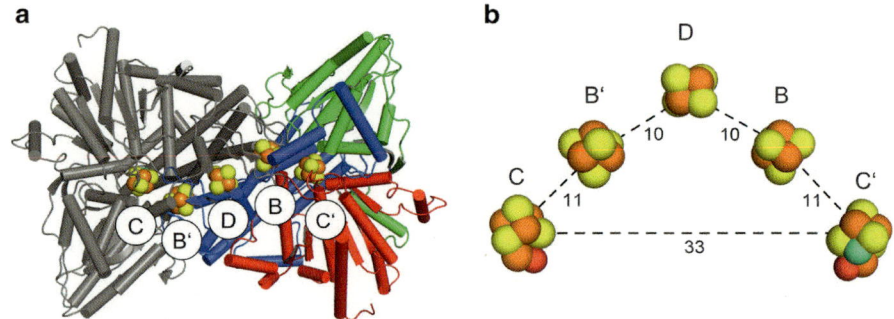

Figure 5 Homodimeric structure of monofunctional CODH II$_{Ch}$. (**a**) Cartoon-representation of dimeric Ni,Fe-CODH. The two subunits of CODH are shown with different colors, where one subunit is highlighted in blue, green, and red for the N-terminal, middle and C-terminal domain, respectively, and the other in grey. The metal clusters encountered are depicted as spheres (Fe is colored in orange, S in yellow, Ni in cyan, and O in red). (**b**) Cluster arrangement in CODH. Cluster D is connecting the two subunits covalently and is in electron transfer distance to clusters B and B'. Cluster C/C' is situated on the end of the electron transfer chain, in close distance to cluster B of the opposing subunit. The distances between Fe atoms of individual clusters are given in Ångstrom.

Cluster D is coordinated by two cysteines of each monomer at the dimer interface covalently linking both monomers. Cluster B is positioned within typical biological electron transfer distances [54, 84] of 10 Å from cluster D and 11 Å from the active site cluster C.

2.2.2 Electronic States and Structure of Cluster C

Ni,Fe-CODHs catalyze the oxidation of CO as well as the reduction of CO_2 efficiently. CO oxidation at cluster C of CODH II$_{Ch}$ occurs with a k_{cat} of 31,000 s^{-1} and a K_M for CO of 18 μM [34]. The specificity constant for CO oxidation (k_{cat}/K_M: 1.7×10^9 M^{-1} s^{-1}) is approaching the diffusion limit and the reaction is fully reversible.

Spectroscopic studies of the catalytic cycle revealed four distinct electronic states of cluster C (C_{ox}, C_{red1}, C_{red2}, and C_{int}) [74] (Figure 6). The diamagnetic and catalytically inactive, oxidized C_{ox}-state can be reductively activated at potentials below –200 mV [85] yielding the one-electron reduced C_{red1} state [74]. C_{red1} is paramagnetic ($S = 1/2$ spin state) with EPR signals at $g_{av} = 1.82$, $g_{av} = 1.87$ and $g_{av} = 1.86$, as found for the cluster C of CODH/ACS from *M. thermoacetica* (CODH$_{Mt}$), CODH$_{Rr}$, and CODH I$_{Ch}$, respectively [86–89]. Two electrons derived from CO oxidation are transferred to cluster C, yielding

Figure 6 Redox states of cluster C. Electronic states are given together with their spin-states and approximate mid-point potentials. The n in spin state of C_{int} denotes 0 or an integer [93].

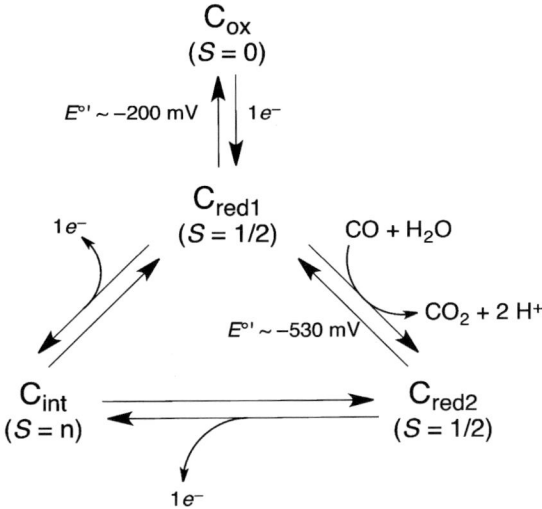

C_{red2}. The operational midpoint potential of C_{red2} ($E^{\circ\prime} = -530$ mV) [87, 90] coincides with the value obtained for the CO_2/CO couple ($E^{\circ\prime} = -558$ mV) [28]. C_{red2} shows a minor shift in the EPR spectrum compared to C_{red1} with $g_{av} = 1.86$ (g_1 1.97, g_2 1.87, and g_3 1.75) [87, 88]. Where the two additional electrons reside in the C_{red2} state is still under debate and involvement of Ni^0, a hydride-bound Ni^{2+} and a Ni-Fe bond (dative metal-metal bond) have been proposed [74, 91, 92]. The half-cycle of the reaction is followed by intramolecular transfer of one electron to cluster B, forming a transient intermediate state devoid of any paramagnetic signature (C_{int}) [93]. Subsequent electron release restores cluster C in the CO-reactive state (C_{red1}).

The structure of cluster C has been deduced from the crystal structure of CODH II_{Ch}, poised to a redox potential of -320 mV with dithiothreitol [94]. At this potential, a state equivalent to C_{red1} was expected and the structure revealed a [Ni4Fe4S-OH_x] composition. Cluster C consists of a distorted [Ni3Fe4S] heterocubane connected to an Fe(II) in exo position. The Fe(II) ion (Fe_1) is also known as ferrous component II (FC_{II}), which is coordinated by a water or hydroxyl ligand (OH_x) with a distance of 2.7 Å to Ni (Figure 7).

Structural insights into CO/CO_2 activation were gained from a CO_2-bound structure. Crystals of CODH II_{Ch} were incubated in a solution containing Ti(III)-citrate adjusted to a reduction potential of -600 mV in the presence of 45 mM $NaHCO_3$, equivalent to a concentration of solvated CO_2 of 0.45 mM at pH 8.0. The presence of CO_2 and a potent electron donor creates turnover conditions under which the crystals were allowed to react for several minutes before they were shock-frozen [94]. The structure of the -600 mV + CO_2 state unraveled a [$NiFe_4S_4(CO_2)$] cluster, where a carboxylate bridges Ni and Fe_1 acting as a μ_2-η^2-ligand. The carbon atom of CO_2 completes the square-planar coordination geometry of Ni and one of the oxygens of CO_2 replaces the hydroxo/water ligand at Fe_1. CO_2 is in an activated state as evident from a O-C-O-bending angle of

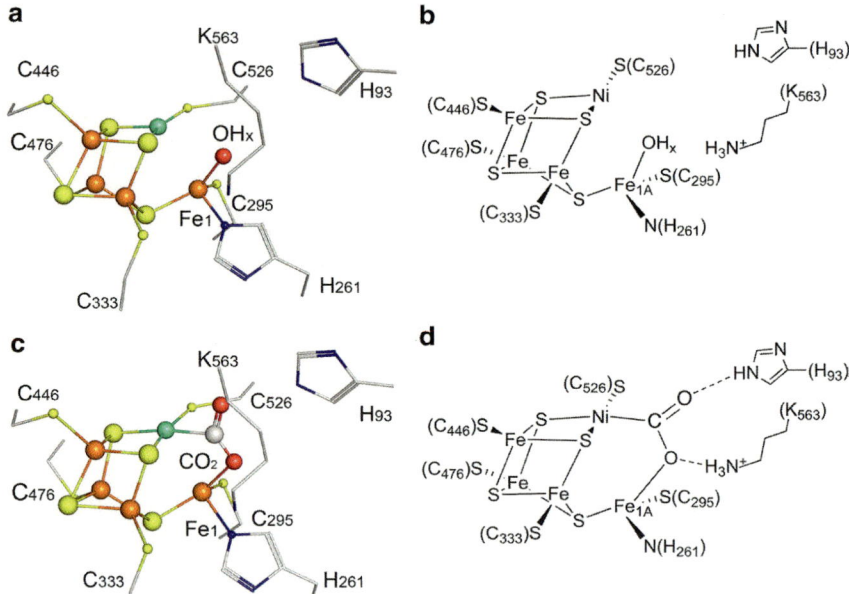

Figure 7 Structure of cluster C from CODH II$_{Ch}$. (**a**) Ball-and-stick model of active site cluster C with conserved ligands in the second coordination sphere in the −320 mV state. (**b**) Schematic drawing of the [Ni4Fe4S-OH$_x$] cluster C. (**c**) [Ni4Fe4S(CO$_2$)] cluster observed in the −600 mV + CO$_2$ state. (**d**) Schematic view of cluster C with bound CO$_2$. The secondary positions of Fe$_1$ (Fe$_{1B}$) and Cys$_{295}$ were omitted in the schematic drawings for clarity.

approximately 133°. The Ni,Fe-bound CO$_2$ is additionally stabilized by hydrogen bonds to a lysine and a histidine residue [94].

2.2.3 Pathways and Channels Involved in Catalysis

The high turnover number of Ni,Fe-CODHs demand a rapid and guided channeling system for substrates and products. CO and CO$_2$, water as well as protons and electrons must be able to rapidly reach and egress from the active site.

Four different paths reach out from cluster C: a gas channel (CO/CO$_2$), a proton relay, a water network, and an electron transfer chain (Figure 8). Monofunctional Ni,Fe-CODHs employ two sets of gas channels. One is conserved and coincides with the tunnel that is connecting cluster C of Ni,Fe-CODH and the active site cluster A of ACS in bifunctional Ni,Fe-CODHs (Section 2.3). The other channel is unique to monofunctional Ni,Fe-CODHs and is directed to the solvent, allowing rapid progress and egress of CO/CO$_2$ from the active site towards the solvent [95]. Recently, molecular dynamics and density functional theory calculations pointed to an additional, dynamically formed gas channel, through which CO$_2$ may diffuse from the solvent to cluster C of CODH/ACS [96]. Simulations imply, that upon CO$_2$ reduction the extended hydrogen network prohibits CO leakage through the gas channel.

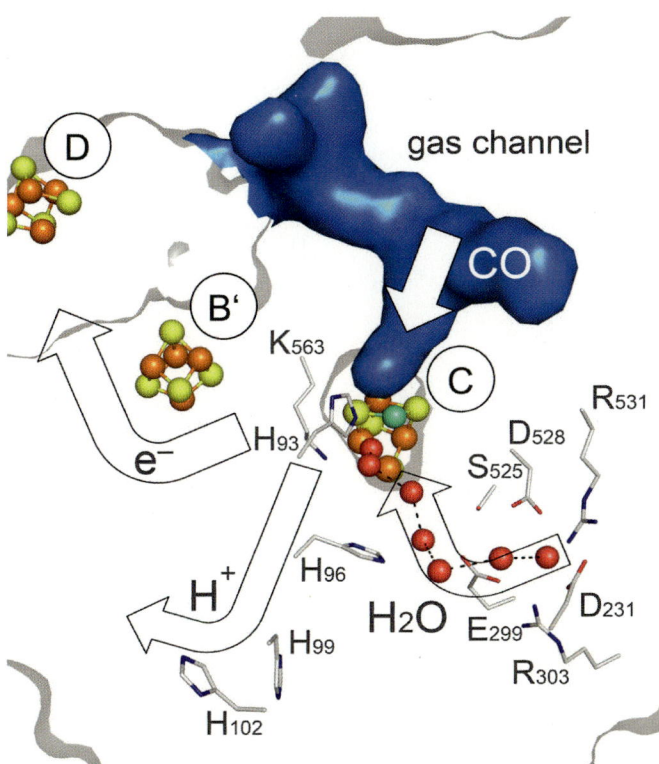

Figure 8 Channels involved in substrate/product transfer in monofunctional CODH II$_{Ch}$. Hydrophobic channels around cluster C have been calculated with the program Hollow [159] and are shown as blue surface. Metal clusters are depicted as spheres and colored in cyan for Ni, orange for Fe, and yellow for S. Water molecules are represented by red spheres. Charged and hydrophilic residues form a water channel network. The electron transfer network from clusters C↔B'↔D is indicated by an arrow. The proton relay shuttle is compromised of histidine residues H$_{96}$, H$_{99}$, and H$_{102}$, where the last residue has direct contact to the protein surface. The surface is contoured in grey.

Transfer of water is guided by charged and hydrophilic residues that form a network from the protein surface to cluster C. Protons leave the active site through His$_{93}$ and three consecutive histidine residues (96, 99, and 102), of which the last residue of the chain is in contact with the solvent. Electrons may be transferred along the previously mentioned FeS-cluster cascade (C↔ B'↔D) before reaching external electron acceptors. Whether cluster D is part of the electron transfer pathway has not been established yet. If it participates in electron transfer, electrons generated at one active site could be transferred to the other active site of the dimer, thereby facilitating an intramolecular reductive activation.

2.2.4 Inhibited States of Cluster C

Light was shed upon a potential binding mode for CO by structural studies of cyanide-inhibited cluster C [97, 98]. Surprisingly, cyanide-inhibited structures for CODH/ACS$_{Mt}$ and CODH II$_{Ch}$ revealed different binding modes (Figure 9).

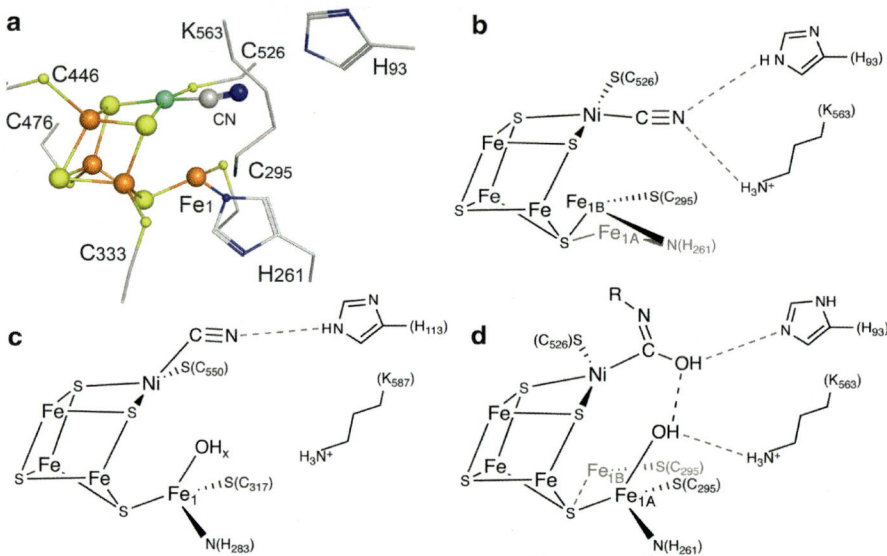

Figure 9 Inhibited states of CODH. (**a**) Ball-and-stick model of cyanide-inhibited CODH II$_{Ch}$ [98]. (**b**) Schematic drawing of cyanide-inhibited CODH II$_{Ch}$. (**c**) Cyanide-inhibited cluster C of ACS/CODH$_{Mt}$ [97]. (**d**) n-butyl-isocyanate bound to cluster C of CODH II$_{Ch}$ [95]. Major and minor conformations for Fe$_1$ are depicted in black and grey colors, respectively.

In the structure of CODH II$_{Ch}$-CN, cyanide binds to the open apical coordination site of Ni in an approximately linear fashion (Ni-C-N angle of 170°) (Figure 9a and 9b) [98]. The OH$_x$ ligand on Fe$_1$ is lost and Fe$_1$ shifts from the A position (Fe$_{1A}$) to an alternative position B (Fe$_{1B}$). In untreated CODH II$_{Ch}$ crystals the Fe$_1$ atom occupies the alternative positions with a ratio of 60/30 % for Fe$_{1A}$/Fe$_{1B}$, respectively [94], while after CN treatment the cluster has inverted occupancies (10/70 %) [98]. Cyanide binding is stabilized by hydrogen bonds with Lys$_{563}$ and His$_{93}$.

A different coordination was found for the structure of CN-inhibited CODH/ACS$_{Mt}$: cyanide binds to Ni completing a distorted tetrahedral coordination with a distinctly bent conformation (Ni-C-N angle of 114°, Figure 9c) [97]. As this bent angle has not been expected from inorganic model complexes [99], the authors suggested that a conserved isoleucine residue in close proximity to the Ni atom prevents a linear apical binding mode [78, 97]. A patch of electron density in the structure of the CODH component of the acetyl-CoA decarbonylase/synthase (ACDS) complex from archaea *Methanosarcina barkeri* above the Ni ion was modeled as a CO molecule [100]. The modeled CO molecule has a similarly bent coordination as the CN ligand in the CODH/ACS$_{Mt}$-CN complex. However, CO had not been added to the crystals of the CODH component and the authors suggested it to be a stable adduct left from enzyme isolation, where it would need to be stably bound for several days. However, a stable Ni-CO complex is difficult to reconcile with the rapid equilibrium binding of CO to the active site of Ni,Fe-CODHs and their high turnover numbers [101].

Additional information on the reactivity of cluster C became available when the structure of CODH II$_{Ch}$ incubated with n-butyl isocyanide (nBIC) was solved. nBIC inhibits Cu,Mo-CODH [50] and acts as a rapid binding competitive inhibitor and slow-turnover substrate of Ni,Fe-CODH [102]. nBIC reacted with the OH$_x$ ligand of Fe$_1$, yielding the product n-butyl isocyanate bound to cluster C [95]. Several H bonds with the OH$_x$ ligand of Fe$_1$ as well as with the side chains of Lys$_{563}$ and His$_{93}$ stabilize the bound reaction product. In summary, studies of inhibited enzyme states as well as incubation with slow turnover substrates revealed valuable information on possible substrate binding modes and the stabilization of reaction intermediates.

Structural insights were recently complemented by electrochemical studies of inhibited states of CODH II$_{Ch}$ [89]. An overview of key findings from electrochemistry can be found in Chapter 4 of this volume.

2.2.5 Mechanism of Reversible Carbon Dioxide Reduction at Cluster C

A reaction mechanism for reversible CO oxidation/CO$_2$ reduction has been deduced by combining insights from high-resolution crystal structures of CODH II$_{Ch}$ [94] with insights into the electronic structure from spectroscopic studies [86–88, 90] (Figure 10).

In the active state for CO oxidation (C$_{red1}$), cluster C has a hydroxyl ligand bound to ferrous Fe$_1$ (**1**). In this state, CO may either bind in the apical coordination site and relocate to the equatorial site at the Ni ion, or it binds directly to the equatorial position. Hydrogen bonds with His$_{93}$ stabilize the bound CO and increase the polarization of the Ni-bound carbonyl, preparing the carbon atom of CO for a nucleophilic attack by the OH$_x$ group (**2**). The resulting carboxylate bridges Ni and Fe$_1$ and is stabilized by electrostatic and H-bonding interactions with Lys$_{563}$ and protonated His$_{93}$ (**3**). Addition of a water molecule may assist in provoking the release of CO$_2$ and two protons. The newly added water molecule is bound to Fe$_1$ and cluster C is in the two electron reduced C$_{red2}$ state (**4**). Transfer of two electrons via cluster B restores cluster C and closes the catalytic cycle. Advances during the last years defined the major steps of the mechanism [94, 95, 98]. To further refine the mechanism a combined spatial and electronic structure description of all states will be necessary.

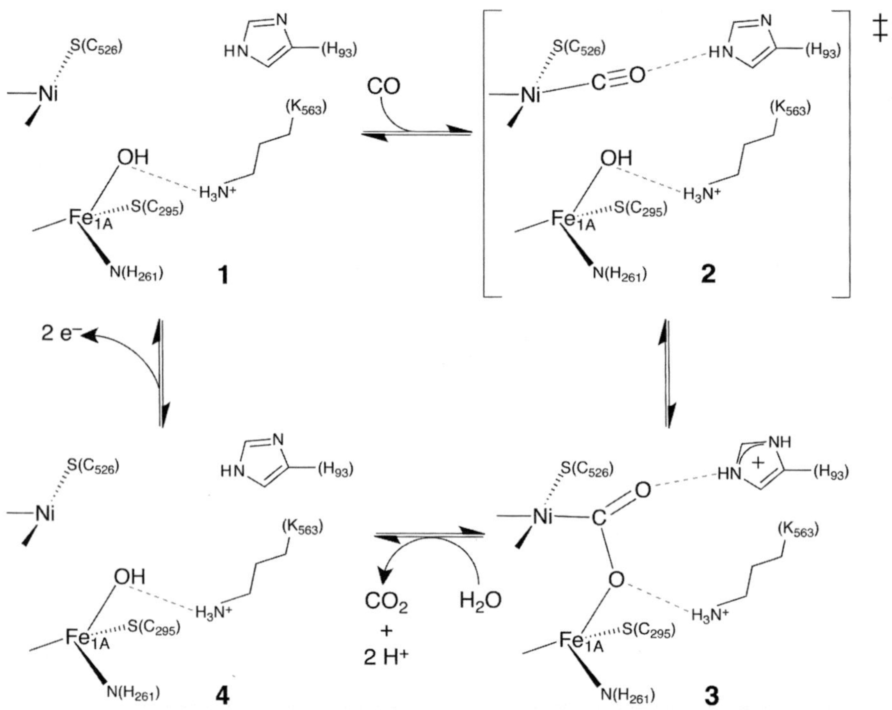

Figure 10 Mechanism of CO oxidation at cluster C. (**1**) A hydroxyl group is bound to Fe_1 in the active state (C_{red1}). (**2**) CO enters the active site from above and binds to Ni completing a square-planar coordination. A hydrogen bond with histidine H_{93} stabilizes this transient state, which orients the nucleophilic hydroxyl group in close proximity to react with CO. (**3**) The oxidation product CO_2 is bound to cluster C in a bridging conformation. (**4**) Uptake of a water molecule supports subsequent release of CO_2. Cluster C contains two additional electrons compared to the C_{red1} state (C_{red2}). Successive release of two electrons restores cluster C in the catalytically competent C_{red1} state.

2.3 Bifunctional Ni,Fe-Containing Carbon Monoxide Dehydrogenases

2.3.1 Classification and Distribution

The structure and function of bifunctional carbon monoxide dehydrogenases have been reviewed thoroughly during the last decade [17, 19, 26, 76, 77, 103–108]. Here, the structural and mechanistic aspects of the bifunctional enzyme complex are described, focusing on the ACS component. CODH/ACS catalyzes the final step of the Wood-Ljungdahl pathway by coupling both enzyme activities. The CODH component is responsible for the reversible reduction of CO_2, while ACS catalyzes the reversible condensation of CO, CoASH and a methyl group to acetyl-CoA [26] (equation 2).

$$CH_3\text{-}Co(III)CoFeSP + CO + CoASH \rightleftharpoons CH_3C(O)SCoA + Co(I)CoFeSP + H^+$$

$$(2)$$

Based on subunit composition, Lindahl et al. [74, 109] proposed four classes of Ni,Fe-containing CODHs, of which classes I, II, and III are bifunctional (Figure 11). Class I and II enzyme complexes, which are also termed acetyl-CoA decarbonylase/synthase (ACDS), are found in methanogenic archaea. Class I enzymes are used by obligate autotrophic methanogens like *Methanobacterium thermoautotrophicum* to generate acetyl-CoA from CO_2, CoASH, and H_2 [110]. Class II enzymes are used by facultative chemo-autotrophic methanogens like *Methanosarcina frisia* and *M. thermophila*, which are able to catabolize acetate generating CO_2, CoASH, and a methyl group bound to tetrahydrosarcinapterin, a tetrahydrofolate analogue used by archaea [38]. Class III CODH/ACS is found in acetogens [26]. As illustrated in Figure 11, the monofunctional CODH homodimers (class IV) consist of one subunit (β), which is homologous to the CODH components of class I/II (subunit α) and class III (subunit β). The class III enzymes consist of two autonomous proteins, an $(\alpha\beta)_2$ dimer of dimers (ACS/CODH), and a γδ heterodimer (CoFeSP). The α subunit (ACS component) is homologous to the β subunit of class I/II, but lacks a 30 kDa region at the N-terminus. CoFeSP is homologous to the γ and δ subunits of the archaeal enzyme complex (class I/II),

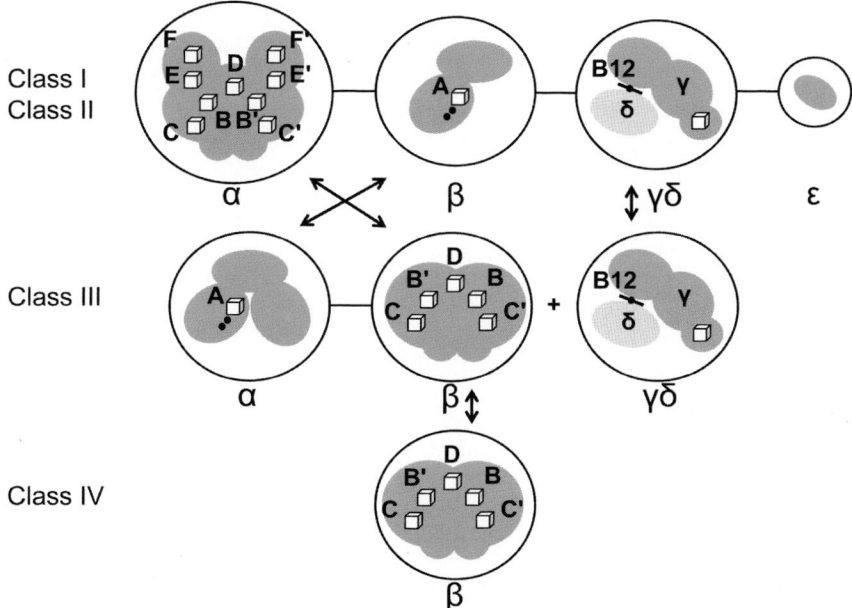

Figure 11 Subunit composition of Ni,Fe-containing CODHs. Connected circles indicate multiprotein complexes consisting of the corresponding subunits. Homology is indicated by arrows. The figure was adapted from Lindahl [74]. Details are given in the text.

which is composed of five different subunits (α, β, γ, δ, and ε). The CODH component of class I/II has two extra clusters (clusters E and F) per monomer, which presumably facilitate the electron transfer between CODH and a ferredoxin [100]. The function of the ε subunit is still unclear, and it may be involved in electron transfer from CODH to FAD [100].

2.3.2 Structural Characterization of Bacterial CODH/ACS

The overall structure and active site of the Ni,Fe-CODH component of bifunctional CODH/ACS (Figure 12) [111, 112] are similar to the ones described for monofunctional Ni,Fe-CODH (Section 2.2) (Figure 5). The ACS component consists of three structural domains, which are connected by flexible linker regions [81, 111, 112]. The N-terminal domain contains a Rossmann fold and interacts with Ni,Fe-CODH. The active site of ACS, termed cluster A, is a cubane type [4Fe4S]-cluster bridged to a binuclear Ni,Ni site located in the C-terminal domain of ACS (Figure 12).

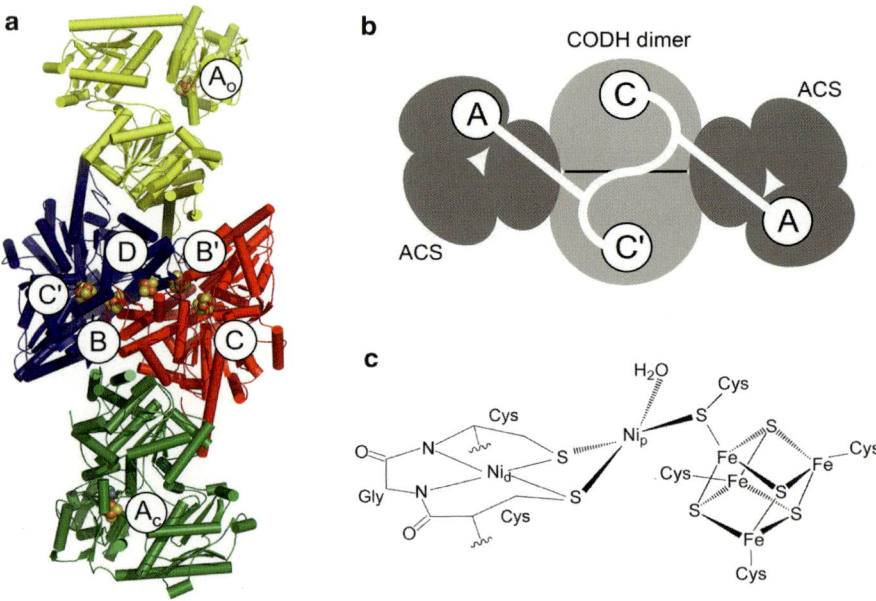

Figure 12 The structure of bacterial CODH/ACS. (**a**) Cartoon representation of the overall structure of the $\alpha_2\beta_2$ CODH/ACS complex from *Moorella thermoacetica* (PDB 1OAO) [111]. Metal clusters are presented as balls and sticks and are labeled with A to D. CODH subunits are colored in blue and red and ACS subunits are colored in yellow for the Ni-Ni-containing cluster A (A_o) and green for the Zn,Ni-containing cluster A (A_c). (**b**) Schematic representation of the gas channel connecting the CODH/ACS active sites. (**c**) Schematic representation of the Ni-Ni containing cluster A, based on PDB 1RU3 [81]. Details are given in the text.

The Ni ion proximal to the [4Fe4S]-cluster (Ni_p) is labile [113] and readily substituted by Cu and Zn [114]. Ni_p is thiolate-bridged to the distal Ni (Ni_d) that is coordinated by two cysteines and a glycine residue in a square planar thiolato- and carboxamido-type N_2S_2 coordination environment.

Due to the long distance between clusters A and C (~70 Å) [111, 112] a direct electron transfer between both clusters is unlikely. A potential gas channel has been characterized that connects cluster C and A by crossing the N-terminal domain of ACS (Figure 12b) [111, 112, 115–121]. This channel facilitates the diffusion of CO between both clusters and prevents the loss of CO [119, 120]. However, ACS can also directly take up CO from the solvent [116]. A direct uptake is also consistent with the presence of a monomeric ACS in *C. hydrogenoformans*, which is produced when exogenous CO is available as substrate [81].

An "open" and a "closed" conformation of ACS have been reported, that directly affect the state of the gas channel [81, 111, 112]. Compared to the closed state, in which cluster A is buried within the ACS domain interface, the middle and C-terminal domains are rotated by ~50° in the open state [111]. On the one hand this rotation exposes cluster A to the solvent and makes Ni_p accessible to methylated CoFeSP. On the other hand the gas channel is blocked by a single helix of the N-terminal ACS domain, restricting CO diffusion. The gas channel is open in the closed conformation of ACS and allows condensation of CO, CoA, and the ACS-bound methyl group to occur.

2.3.3   Substrate Binding and Reaction Mechanism

Different mechanisms for the condensation reaction have been proposed, varying in the oxidation states of the Ni ions and the sequence of substrate binding.

We have currently no direct evidence where the substrates bind. However, some indirect evidence is pointing to Ni_p as the presumable position of CO activation. The gas channel for CO diffusions opens directly above the proximal Ni site [111, 112] and there is evidence that Ni_p must be present for methyl group transfer from CoFeSP to ACS [122, 123]. The Arg-rich interdomain cavity formed by the three ACS subunits was proposed as CoA binding site [124, 125].

A mechanism in which both Ni ions serve as binding sites for the substrates and the [4Fe4S] cluster acts as an electron reservoir appears plausible based on the composition of cluster A [112, 126]. However, mechanisms favoring one Ni are also supported by computational studies [127] and model complexes [128–130]. Although a role of the [4Fe4S] cluster for electron transfer appears plausible it is unlikely, as electron transfer to and from the [4Fe4S] cluster was shown to be about 200-fold slower than methyl group transfer [131].

In the absence of low potential reductants, oxidized ACS is inactive and EPR-silent [87]. The [4Fe4S] cluster of oxidized ACS is in the [4Fe4S]$^{2+}$ oxidation state [90, 132, 133]. Both Ni ions have a square-planar coordination and are likely present as Ni^{2+} [111–113, 133–136]. Reductive activation by one [101, 137] or two electrons [107, 122] has been proposed to occur before or with the methylation, but how electrons are transferred to cluster A is still unclear.

Reduced, CO-treated ACS exhibits a characteristic EPR spectrum with g-values of 2.074 and 2.028, referred to as NiFeC signal [138, 139]. The corresponding NiFeC species was suggested to be $[4Fe4S]^{2+}$-$(Ni_p^+$-CO$)$-(Ni_d^+) [137, 140] and its relevance for the mechanism of ACS is controversially discussed. The formation and decay of the NiFeC EPR signal are equal or faster than the overall rate of acetyl-CoA formation, thus the NiFeC species is catalytically competent [137, 141]. Furthermore, NiFeC is the predominant metal-carbonyl species formed when ACS reacts with CO [141]. A Ni_p^+ species that resembles the CO binding state of ACS was generated by photolysis of the Ni_p^+-CO state, exhibiting a low barrier for recombination with CO [142]. On the other hand, Gencic and coworkers [143] argued that the NiFeC species is due to a CO-inhibited state, which is in equilibrium with the free form of cluster A.

The nature of the NiFeC species is important because it defines the possible mechanism. Although the condensation reaction catalyzed by ACS does not produce or consume electrons, the binding of a methyl cation temporarily oxidizes the cluster A by two electrons. Two major mechanisms have been proposed to account for the electronic states during catalysis (Figure 13).

The **paramagnetic reaction mechanism** includes the NiFeC species as a CO-bound intermediate (Figure 13a). CO and the methyl group bind in random order to the reduced Ni_p^+ [101]. Bender et al. [142] proposed that during catalysis a diamagnetic resting state (Ni_p^{2+}) predominates over the thermodynamically unfavorable Ni_p^+ species, which is formed when the one-electron reduction is coupled to the binding of CO, pulling the equilibrium towards Ni_p^+-CO formation. Transmethylation generates a reactive methyl-Ni_p^{3+} species [37] that readily accepts an electron to generate the stable methyl-Ni_p^{2+} state, which is in agreement with the EPR-silent CH_3-ACS species [122]. Carbonyl insertion into the Ni-C bond generates an acetyl-Ni^{2+} intermediate, which is nucleophilically attacked by CoA. Thiolysis of the acetyl-Ni^{2+} intermediate liberates two electrons. One electron is used to regenerate the Ni_p^{1+} state, while the other electron might go into an internal electron transfer pathway to reduce the unstable methylated Ni^{3+}-species [142].

In the **diamagnetic reaction mechanism** (Figure 13b) Ni_p is supposed to cycle between Ni^0 and Ni^{2+}, while the [4Fe4S] cluster remains in the oxidized state, rendering only diamagnetic species catalytically relevant [111, 122]. The formation of the two electron reduced Ni_p^0 state is in agreement with recent density functional theory calculations, suggesting a proton coupled electron transfer mechanism prior to methylation [144]. The methyl group and CO can bind in random order to Ni_p^0, generating a Ni_p^0-CO or Ni_p^{2+}-CH_3 species. In contrast to the paramagnetic mechanism, no further reduction of the reactive Ni^{3+} state is needed. The two electrons liberated at thiolysis may be used to regenerate the Ni_p^0 state. One problem of this

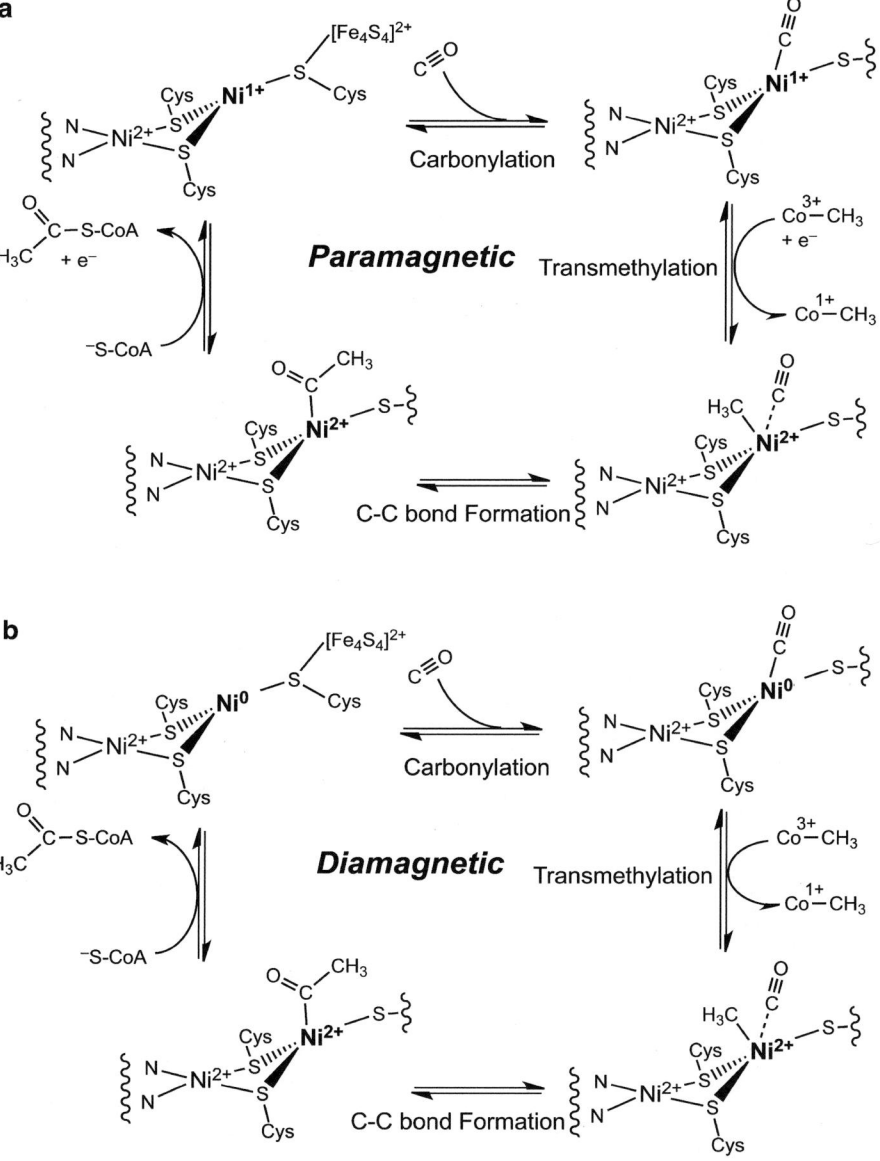

Figure 13 Proposed reaction mechanisms of ACS. The paramagnetic mechanism (**a**) includes the paramagnetic NiFeC species as intermediate, while the diamagnetic mechanism (**b**) integrates only diamagnetic intermediates, with Ni cycling between Ni^0 and Ni^{2+}. Although the sequence of substrate binding is random, both mechanisms are shown as ordered reactions for clarity. See the text for details.

mechanistic proposal may be that Ni_p^0 would need to co-exist next to an oxidized [4Fe4S] cluster, a combination whose stability has been doubted [145].

Recently, the diamagnetic mechanism was extended by including the conformational states of ACS [143]. In the extended mechanism the protein conformation controls coordination of ligands at cluster A through a conserved phenylalanine residue that moves into and out of the coordination environment of Ni_p. A phenylalanine to alanine exchange produced an ACS variant with a substantially decreased reactivity with methylated CoFeSP, while its reactivity with CO was increased. These results suggest that phenyalanine sterically determines the favored substrate order (methyl group first) and prevents CO inhibition.

Thus, despite the existing computational, spectroscopic, kinetic, and crystallographic studies the mechanism of acetyl-CoA formation at cluster A remains unresolved and new approaches to the question are required.

2.4   Cu,Mo versus Ni,Fe: Parallels and Differences in Catalytic Strategies

Cu,Mo-CODHs and Ni,Fe-CODHs catalyze the oxidation of CO, but they achieve this with two very different active site architectures (Figure 14): one CODH contains Cu and Mo, the other Ni and Fe; one active site has two metals and one μ_2-sulfido-ligand, the other has five metal ions and four μ_3-sulfido-ligands; in Cu,Mo-CODH one of the metals is coordinated by an organic cofactor and one cysteine, in Ni,Fe-CODHs we find no organic cofactor but five cysteines and one histidine as coordinating ligands. Thus, in contrast to enzymes like serine proteases and Zn-dependent carboanhydrases, CODHs show no signs of convergent evolution to create non-homologous, isofunctional enzymes.

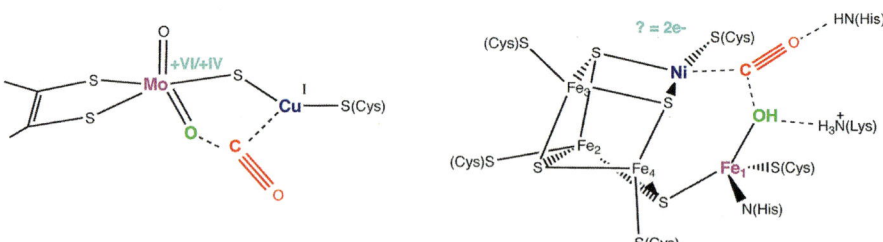

Figure 14 Catalytic strategies of Cu,Mo- and Ni,Fe-CODH in comparison. Cu (Cu,Mo-CODH) and Ni (Ni,Fe-CODH) bind and activate CO. The activated nucleophilic water is bound to Mo (Cu,Mo-CODH) and Fe_1 (Ni,Fe-CODH). Electrons are taken up by Mo (Cu,Mo-CODH), and it is unknown where the electrons are stored in cluster C.

To reveal the parallels among these seemingly different solutions to the same problem, we will focus on the catalytic requirements for performing a CO oxidation (Figures 4 and 10):

(i) CO oxidation begins with binding and activating CO. By using Cu(I) and a Ni(II) ion in a sulfur-rich coordination, both enzymes employ for this task a metal with sufficient π-electron donor character and have an open coordination, site for CO binding. Both nickel and copper are assumed to retain their redox states during the catalytic cycle, restricting their role to bind and polarize CO.

(ii) The next task is to bind and activate water, which is achieved by Mo(VI) (Cu,Mo-CODH) and Fe(II) (Ni,Fe-CODH). Water is "activated" by deprotonation to give either a hydroxyl ligand (Ni,Fe-CODH) or an oxo ligand (Cu,Mo-CODH). Deprotonation of the water ligand is necessary to increase its nucleophilicity. The activated water must be sufficiently close and properly oriented to react with the C atom of the bound CO, making the relative arrangement of the open coordination sites for CO and water crucial.

(iii) Stabilization of bound CO_2 is likely achieved in both types of CODHs by coordination to both metals. Ni,Fe-CODHs catalyze CO_2 reduction with turnover numbers of up to 50 s^{-1}, with observed rate constants depending on the electron donor and the presence of electron mediators [146]. However, as only Ni,Fe-CODHs are able to reduce CO_2, activation of CO_2 may not be important or achievable for Cu,Mo-CODHs. Residues in the second coordination sphere stabilize CO_2, as in the structure of CO_2-bound Ni,Fe-CODH, where the oxygen atoms of CO_2 are in hydrogen-bond distance to two protein side chains [94].

(iv) In contrast to the water-gas shift reaction, where the two electrons and two protons generated by CO oxidation are directly released as H_2 [147], CODHs keep the protons and electrons separated. In Cu,Mo-CODHs Mo(VI) acts as the electron acceptor taking both electrons originating from CO oxidation.

It is not clear where the electrons are stored in the C_{red2} state of Ni,Fe-CODHs and different electron acceptors within cluster C appear possible. The proposed oxidation state of the C_{red1} state is $\{[Ni^{2+}Fe^{2+}]:[Fe_3S_4]^-\}$, which can be reconciled with the S = 1/2 spin state and Mössbauer data [74]. Two-electron uptake could theoretically involve the formal reduction of Ni^{2+} to Ni^0, the formation of a Ni-hydride or a Ni-Fe bridging hydride ion or, as recently suggested [91], formation of a Ni-Fe bond. Both, the observation of an S = 1/2 EPR signal for the C_{red2} state, as well as the minor changes observed upon reduction of C_{red1} to C_{red2} by Mössbauer spectroscopy, argue against an uptake of the electrons by the $[Fe_3S_4]$ moiety of cluster C. Certainly, further spectroscopic and computational investigations will be required to fully understand the differences in electronic structure between C_{red1} and C_{red2}.

Both types of CODHs are capable of CO oxidation, but the rates with which they achieve this task differ by approximately two orders of magnitude. While Cu, Mo-CODH has a limiting turnover number of close to 100 s^{-1} [50, 55], the most active Ni,Fe-CODHs achieve turnover numbers of 30,000–40,000 s^{-1} [34]. It is not yet clear what limits turnover rates in Cu,Mo-CODHs, however two observations may give a hint. First, Hille and coworkers showed that the reductive half-reaction is likely limiting the overall rate, therefore the chemistry at the Cu-S-Mo site and not the electron flow from the FAD to external electron acceptors is probably the limiting catalytic factor [55]. Second, Mo-hydroxylases have typically turnover numbers in the range of $1-100 \, s^{-1}$ [148] and there may be an intrinsic barrier in one of the key steps of the Mo chemistry that limits turnover in this enzyme family including Cu,Mo-CODHs.

3   Concluding Remarks and Future Directions

Our understanding how the oxidation of CO is catalyzed by enzymes greatly increased during the last 10–15 years. The composition and architecture of the responsible metal clusters have been firmly established and mechanistic proposals have been refined by additional spectroscopic and computational studies. Of course, several questions about the enzyme mechanisms remain unanswered and we expect to see further progress in the next years.

One especially promising area of research not covered in this text is the assembly of the metal sites and, although some progress has been made in the last years, we still know considerably less about how the active sites of CODH and ACS are assembled, when compared to other metalloenzymes, such as nitrogenases [149–152], hydrogenases [153–155], and ureases [156].

A deeper understanding of common principles and differences between CODHs and their related enzymes will probably advance the field, once we better understand their evolution. Judged by the employed metals, use by aerobic *versus* anaerobic microbes and the phylogenetic distribution of homologous proteins, one would guess that ancestral Ni,Fe-CODHs existed already very early in the evolution, likely in pre-LUCA times, whereas Cu,Mo-CODHs are evolutionary young inventions. Further phylogenetic analysis may yield clues how the catalytic strategies of the enzymes developed over time.

The complete genome sequences of prokaryotes show a vast amount of enzymes, whose function we do not know and cannot even guess at the moment. Given the rich chemistry of CO, it would be surprising if organisms living with CO would not use more than the two reactions described in this chapter to harness the energy stored in CO. One especially attractive way is C-C coupling of CO in the presence of H_2 similar to the Fischer-Tropsch reaction, which has recently been shown as a side-reaction of V- and Mo-containing nitrogenases [157, 158]. We are therefore optimistic that our knowledge on the use of CO by microbes will further expand in the years to come.

Abbreviations

ACS	acetyl-CoA synthase
ACDS	acetyl-CoA decarbonylase/synthase
CoA	coenzyme A
CODH	carbon monoxide dehydrogenase
CoFeSP	corrinoid iron-sulfur protein
Ch	*Carboxydothermus hydrogenoformans*
Cu,Mo-CODH	Cu- and Mo-containing carbon monoxide dehydrogenase
ENDOR	electron nuclear double resonance
EPR	electron paramagnetic resonance
FAD	flavin adenine dinucleotide
H_4F	tetrahydrofolate
HOMO	highest occupied molecular orbital
LUCA	last universal common ancestor
MCD	molybdopterin cytosine dinucleotide
MeTr	methyltransferase
Mt	*Moorella thermoacetica*
Ni,Fe-CODH	Ni- and Fe-containing carbon monoxide dehydrogenase
nBIC	n-butyl isocyanide
PDB	Protein Data Bank
Rr	*Rhodospirillum rubrum*
SOMO	singly occupied molecular orbital
Tg	teragram
XO	xanthine oxidase

Acknowledgment Research in our laboratory has been supported by the German funding agency DFG through individual project grants (DO-785/1, DO-785/5, DO-785/6) and the Cluster of Excellence "Unifying Concepts in Catalysis - UniCat" (EXC 314).

References

1. R. Burch, *Phys. Chem. Chem. Phys.* **2006**, *8*, 5483–5500.
2. A. M. Appel, J. E. Bercaw, A. B. Bocarsly, H. Dobbek, D. L. DuBois, M. Dupuis, J. G. Ferry, E. Fujita, R. Hille, P. J. A. Kenis, C. A. Kerfeld, R. H. Morris, C. H. F. Peden, A. R. Portis, S. W. Ragsdale, T. B. Rauchfuss, J. N. H. Reek, L. C. Seefeldt, R. K. Thauer, G. L. Waldrop, *Chem. Rev.* **2013**, *113*, 6621–6658.
3. M. A. K. Khalil, J. P. Pinto, M. J. Shearer, *Chemosphere: Global Change Sci.* **1999**, *1*, xi–xiii.
4. C. Huber, G. Wachtershauser, *Science* **1997**, *276*, 245–247.
5. W. Martin, M. J. Russell, *Phil. Trans. Roy. Soc. B* **2007**, *362*, 1887–1925.
6. P. J. Crutzen, L. T. Gidel, *J. Geophys. Res.* **1983**, *88*, 6641–6661.
7. T. G. Sokolova, A. M. Henstra, J. Sipma, S. N. Parshina, A. J. M. Stams, A. V. Lebedinsky, *FEMS. Microbiol. Ecol.* **2009**, *68*, 131–141.

8. M. A. K. Khalil, R. A. Rasmussen, *Chemosphere* **1990**, *20*, 227–242.

9. G. M. King, C. F. Weber, *Nat. Rev. Microbiol.* **2007**, *5*, 107–118.

10. S. W. Ryter, L. E. Otterbein, *Bioessays* **2004**, *26*, 270–280.

11. B. Y. Chin, L. E. Otterbein, *Curr. Opin. Pharmacol.* **2009**, *9*, 490–500.

12. Y. Lu, M. A. K. Khalil, *Chemosphere* **1993**, *26*, 641–655.

13. R. Conrad, *Microbiol. Rev.* **1996**, *60*, 609–640.

14. G. M. King, *Chemosphere: Global Change Sci.* **1999**, *1*, 53–63.

15. O. Meyer, H. G. Schlegel, *Annu. Rev. Microbiol.* **1983**, *37*, 277–310.

16. S. W. Park, E. H. Hwang, H. Park, J. A. Kim, J. Heo, K. H. Lee, T. Song, E. Kim, Y. T. Ro, S. W. Kim, Y. M. Kim, *J. Bacteriol.* **2003**, *185*, 142–147.

17. S. W. Ragsdale, *Ann. N. Y. Acad. Sci.* **2008**, *1125*, 129–136.

18. G. Fuchs, *Ann. Rev. Microbiol.* **2011**, *65*, 631–658.

19. S. W. Ragsdale, *Crit. Rev. Biochem. Mol. Biol.* **2004**, *39*, 165–195.

20. B. Kruger, O. Meyer, *Arch. Microbiol.* **1984**, *139*, 402–408.

21. C. M. Lyons, P. Justin, J. Colby, E. Williams, *J. Gen. Microbiol.* **1984**, *130*, 1097–1105.

22. G. M. King, *Appl. Environ. Microbiol.* **2003**, *69*, 7266–7272.

23. S. E. Hoeft, J. S. Blum, J. F. Stolz, F. R. Tabita, B. Witte, G. M. King, J. M. Santini, R. S. Oremland, *Int. J. Syst. Evol. Microbiol.* **2007**, *57*, 504–512.

24. M. A. Moran, A. Buchan, J. M. Gonzalez, J. F. Heidelberg, W. B. Whitman, R. P. Kiene, J. R. Henriksen, G. M. King, R. Belas, C. Fuqua, L. Brinkac, M. Lewis, S. Johri, B. Weaver, G. Pai, J. A. Eisen, E. Rahe, W. M. Sheldon, W. Y. Ye, T. R. Miller, J. Carlton, D. A. Rasko, I. T. Paulsen, Q. H. Ren, S. C. Daugherty, R. T. Deboy, R. J. Dodson, A. S. Durkin, R. Madupu, W. C. Nelson, S. A. Sullivan, M. J. Rosovitz, D. H. Haft, J. Selengut, N. Ward, *Nature* **2004**, *432*, 910–913.

25. K. Frunzke, O. Meyer, *Arch. Microbiol.* **1990**, *154*, 168–174.

26. S. W. Ragsdale, E. Pierce, *Biochim. Biophys. Acta* **2008**, *1784*, 1873–1898.

27. E. Oelgeschlaeger, M. Rother, *Arch. Microbiol.* **2008**, *190*, 257–269.

28. D. A. Grahame, E. Demoll, *Biochemistry* **1995**, *34*, 4617–4624.

29. R. Rabus, T. A. Hansen, F. Widdel, *The Prokaryotes - A Handbook on the Biology of Bacteria*, Eds S. F. M. Dworkin, H. Rosenberg, K.-H. Schleifer, E. Stackebrandt, Springer, New York, 2006.

30. J. Sipma, A. M. Henstra, S. N. Parshina, P. N. L. Lens, G. Lettinga, A. J. M. Stams, *Crit. Rev. Biotechnol.* **2006**, *26*, 41–65.

31. A. Jensen, K. Finster, *Anton. Leeuw. Int. J. G.* **2005**, *87*, 339–353.

32. R. L. Kerby, P. W. Ludden, G. P. Roberts, *J. Bacteriol.* **1995**, *177*, 2241–2244.

33. V. A. Svetlichny, T. G. Sokolova, M. Gerhardt, M. Ringpfeil, N. A. Kostrikina, G. A. Zavarzin, *Syst. Appl. Microbiol.* **1991**, *14*, 254–260.

34. V. Svetlitchnyi, C. Peschel, G. Acker, O. Meyer, *J. Bacteriol.* **2001**, *183*, 5134–5144.

35. J. D. Fox, Y. P. He, D. Shelver, G. P. Roberts, P. W. Ludden, *J. Bacteriol.* **1996**, *178*, 6200–6208.

36. C. Welte, U. Deppenmeier, *Biochim. Biophys. Acta* **2013**, *1873*, 1130–1147.

37. R. K. Thauer, *Microbiology* **1998**, *144*, 2377–2406.

38. D. A. Grahame, *J. Biol. Chem.* **1991**, *266*, 22227–22233.

39. H. Dobbek, L. Gremer, O. Meyer, R. Huber, in *Handbook of Metalloproteins*, Eds A. Messerschmidt, R. Huber, T. Poulos, K. Wieghardt, John Wiley & Sons, Ltd, Chichester, 2001, Vol. 2, pp. 1136–1147.

40. R. Hille, *Dalton Trans.* **2013**, *42*, 3029–3042.

41. M. J. Romao, M. Archer, I. Moura, J. J. Moura, J. LeGall, R. Engh, M. Schneider, P. Hof, R. Huber, *Science* **1995**, *270*, 1170–1176.

42. M. J. Romao, *Dalton Trans.* **2009**, 4053–4068.

43. H. Dobbek, R. Huber, in *Metal Ions in Biological Systems*, Vol. 39, Eds A. Sigel, H. Sigel, Marcel Dekker, Inc, New York, 2002, pp. 165–196.

44. H. Dobbek, *Coord. Chem. Rev.* **2011**, *255*, 1104–1116.

45. A. Magalon, J. G. Fedor, A. Walburger, J. H. Weiner, *Coord. Chem. Rev.* **2011**, *255*, 1159–1178.
46. N. Wagener, A. J. Pierik, A. Ibdah, R. Hille, H. Dobbek, *Proc. Natl. Acad. Sci. USA* **2009**, *106*, 11055–11060.
47. O. Meyer, K. V. Rajagopalan, *J. Bacteriol.* **1984**, *157*, 643–648.
48. O. Meyer, L. Gremer, R. Ferner, M. Ferner, H. Dobbek, M. Gnida, W. Meyer-Klaucke, R. Huber, *Biol. Chem.* **2000**, *381*, 865–876.
49. H. Dobbek, L. Gremer, O. Meyer, R. Huber, *Proc. Natl. Acad. Sci. USA* **1999**, *96*, 8884–8889.
50. H. Dobbek, L. Gremer, R. Kiefersauer, R. Huber, O. Meyer, *Proc. Natl. Acad. Sci. USA* **2002**, *99*, 15971–15976.
51. M. J. Lorite, J. Tachil, J. Sanjuan, O. Meyer, E. J. Bedmar, *Appl. Environ. Microbiol.* **2000**, *66*, 1871–1876.
52. P. Hänzelmann, H. Dobbek, L. Gremer, R. Huber, O. Meyer, *J. Mol. Biol.* **2000**, *301*, 1221–1235.
53. J. Wilcoxen, B. Zhang, R. Hille, *Biochemistry* **2011**, *50*, 1910–1916.
54. C. C. Page, C. C. Moser, X. Chen, P. L. Dutton, *Nature* **1999**, *402*, 47–52.
55. B. Zhang, C. F. Hemann, R. Hille, *J. Biol. Chem.* **2010**, *285*, 12571–12578.
56. M. Shanmugam, J. Wilcoxen, D. Habel-Rodriguez, G. E. Cutsail, M. L. Kirk, B. M. Hoffman, R. Hille, *J. Am. Chem. Soc.* **2013**, *135*, 17775–17782.
57. C. Gourlay, D. J. Nielsen, J. M. White, S. Z. Knottenbelt, M. L. Kirk, C. G. Young, *J. Am. Chem. Soc.* **2006**, *128*, 2164–2165.
58. J. Wilcoxen, R. Hille, *J. Biol. Chem.* **2013**, *288*, 36052–36060.
59. M. Gnida, R. Ferner, L. Gremer, O. Meyer, W. Meyer-Klaucke, *Biochemistry* **2003**, *42*, 222–230.
60. M. Hofmann, J. K. Kassube, T. Graf, *J. Biol. Inorg. Chem.* **2005**, *10*, 490–495.
61. P. E. M. Siegbahn, A. F. Shestakov, *J. Comput. Chem.* **2005**, *26*, 888–898.
62. B. Santiago, O. Meyer, *FEMS Microbiol. Lett.* **1996**, *142*, 309–310.
63. M. Rohde, F. Mayer, S. Jacobitz, O. Meyer, *FEMS Microbiol. Lett.* **1985**, *28*, 141–144.
64. F. Spreitler, C. Brock, A. Pelzmann, O. Meyer, J. Kohler, *Chembiochem.* **2010**, *11*, 2419–2423.
65. M. Resch, H. Dobbek, O. Meyer, *J. Biol. Inorg. Chem.* **2005**, *10*, 518–528.
66. J. Wilcoxen, S. Snider, R. Hille, *J. Am. Chem. Soc.* **2011**, *133*, 12934–12936.
67. B. W. Stein, M. L. Kirk, *Chem. Comm.* **2014**, *50*, 1104–1106.
68. K. Okamoto, K. Matsumoto, R. Hille, B. T. Eger, E. F. Pai, T. Nishino, *Proc. Natl. Acad. Sci. USA* **2004**, *101*, 7931–7936.
69. I. Bonin, B. M. Martins, V. Purvanov, S. Fetzner, R. Huber, H. Dobbek, *Structure* **2004**, *12*, 1425–1435.
70. G. B. Diekert, E. G. Graf, R. K. Thauer, *Arch. Microbiol.* **1979**, *122*, 117–120.
71. H. L. Drake, S. I. Hu, H. G. Wood, *J. Biol. Chem.* **1980**, *255*, 7174–7180.
72. J. G. Ferry, *Ann. Rev. Microbiol.* **1995**, *49*, 305–333.
73. S. W. Ragsdale, *Subcell. Biochem.* **2000**, *35*, 487–518.
74. P. A. Lindahl, *Biochemistry* **2002**, *41*, 2097–2105.
75. C. L. Drennan, J. W. Peters, *Curr. Opin. Struct. Biol.* **2003**, *13*, 220–226.
76. A. Volbeda, J. C. Fontecilla-Camps, *J. Biol. Inorg. Chem.* **2004**, *9*, 525–532.
77. C. L. Drennan, T. I. Doukov, S. W. Ragsdale, *J. Biol. Inorg. Chem.* **2004**, *9*, 511–515.
78. Y. Kung, C. L. Drennan, *Curr. Opin. Chem. Biol.* **2011**, *15*, 276–283.
79. D. Bonam, P. W. Ludden, *J. Biol. Chem.* **1987**, *262*, 2980–2987.
80. M. Wu, Q. Ren, A. S. Durkin, S. C. Daugherty, L. M. Brinkac, R. J. Dodson, R. Madupu, S. A. Sullivan, J. F. Kolonay, D. H. Haft, W. C. Nelson, L. J. Tallon, K. M. Jones, L. E. Ulrich, J. M. Gonzalez, I. B. Zhulin, F. T. Robb, J. A. Eisen, *PLoS Genet.* **2005**, *1*, e65.
81. V. Svetlitchnyi, H. Dobbek, W. Meyer-Klaucke, T. Meins, B. Thiele, P. Romer, R. Huber, O. Meyer, *Proc. Natl. Acad. Sci. USA* **2004**, *101*, 446–451.
82. H. Dobbek, V. Svetlitchnyi, L. Gremer, R. Huber, O. Meyer, *Science* **2001**, *293*, 1281–1285.

83. C. L. Drennan, J. Heo, M. D. Sintchak, E. Schreiter, P. W. Ludden, *Proc. Natl. Acad. Sci. USA* **2001**, *98*, 11973–11978.
84. H. B. Gray, J. R. Winkler, *Proc. Natl. Acad. Sci. USA* **2005**, *102*, 3534–3539.
85. J. Feng, P. A. Lindahl, *Biochemistry* **2004**, *43*, 1552–1559.
86. Z. G. Hu, N. J. Spangler, M. E. Anderson, J. Q. Xia, P. W. Ludden, P. A. Lindahl, E. Munch, *J. Am. Chem. Soc.* **1996**, *118*, 830–845.
87. P. A. Lindahl, E. Munck, S. W. Ragsdale, *J. Biol. Chem.* **1990**, *265*, 3873–3879.
88. N. J. Spangler, P. A. Lindahl, V. Bandarian, P. W. Ludden, *J. Biol. Chem.* **1996**, *271*, 7973–7977.
89. V. C. Wang, M. Can, E. Pierce, S. W. Ragsdale, F. A. Armstrong, *J. Am. Chem. Soc.* **2013**, *135*, 2198–2206.
90. P. A. Lindahl, S. W. Ragsdale, E. Munck, *J. Biol. Chem.* **1990**, *265*, 3880–3888.
91. P. A. Lindahl, *J. Inorg. Biochem.* **2012**, *106*, 172–178.
92. P. Amara, J. M. Mouesca, A. Volbeda, J. C. Fontecilla-Camps, *Inorg. Chem.* **2011**, *50*, 1868–1878.
93. D. M. Fraser, P. A. Lindahl, *Biochemistry* **1999**, *38*, 15706–15711.
94. J. H. Jeoung, H. Dobbek, *Science* **2007**, *318*, 1461–1464.
95. J. H. Jeoung, H. Dobbek, *J. Biol. Inorg. Chem.* **2012**, *17*, 167–173.
96. P. H. Wang, M. Bruschi, L. De Gioia, J. Blumberger, *J. Am. Chem. Soc.* **2013**, *135*, 9493–9502.
97. Y. Kung, T. I. Doukov, J. Seravalli, S. W. Ragsdale, C. L. Drennan, *Biochemistry* **2009**, *48*, 7432–7440.
98. J. H. Jeoung, H. Dobbek, *J. Am. Chem. Soc.* **2009**, *131*, 9922–9923.
99. P. M. Sheridan, L. M. Ziurys, *J. Chem. Phys.* **2003**, *118*, 6370–6379.
100. W. Gong, B. Hao, Z. Wei, D. J. Ferguson, Jr., T. Tallant, J. A. Krzycki, M. K. Chan, *Proc. Natl. Acad. Sci. USA* **2008**, *105*, 9558–9563.
101. J. Seravalli, S. W. Ragsdale, *J. Biol. Chem.* **2008**, *283*, 8384–8394.
102. M. Kumar, S. W. Ragsdale, *J. Am. Chem. Soc.* **1995**, *117*, 11604–11605.
103. T. C. Brunold, *J. Biol. Inorg. Chem.* **2004**, *9*, 533–541.
104. D. A. Grahame, *Trends Biochem. Sci.* **2003**, *28*, 221–224.
105. G. Bender, E. Pierce, J. A. Hill, J. E. Darty, S. W. Ragsdale, *Metallomics* **2011**, *3*, 797–815.
106. E. L. Hegg, *Acc. Chem. Res.* **2004**, *37*, 775–783.
107. P. A. Lindahl, *J. Biol. Inorg. Chem.* **2004**, *9*, 516–524.
108. S. W. Ragsdale, *J. Inorg. Biochem.* **2007**, *101*, 1657–1666.
109. P. A. Lindahl, B. Chang, *Orig. Life Evol. Biosph.* **2001**, *31*, 403–434.
110. S. Lange, G. Fuchs, *Eur. J. Biochem.* **1987**, *163*, 147–154.
111. C. Darnault, A. Volbeda, E. J. Kim, P. Legrand, X. Vernede, P. A. Lindahl, J. C. Fontecilla-Camps, *Nat. Struct. Biol.* **2003**, *10*, 271–279.
112. T. I. Doukov, T. M. Iverson, J. Seravalli, S. W. Ragsdale, C. L. Drennan, *Science* **2002**, *298*, 567–572.
113. W. K. Russell, C. M. V. Stalhandske, J. Q. Xia, R. A. Scott, P. A. Lindahl, *J. Am. Chem. Soc.* **1998**, *120*, 7502–7510.
114. J. Seravalli, Y. M. Xiao, W. W. Gu, S. P. Cramer, W. E. Antholine, V. Krymov, G. J. Gerfen, S. W. Ragsdale, *Biochemistry* **2004**, *43*, 3944–3955.
115. T. I. Doukov, L. C. Blasiak, J. Seravalli, S. W. Ragsdale, C. L. Drennan, *Biochemistry* **2008**, *47*, 3474–3483.
116. X. S. Tan, P. A. Lindahl, *J. Biol. Inorg. Chem.* **2008**, *13*, 771–778.
117. X. S. Tan, H. K. Loke, S. Fitch, P. A. Lindahl, *J. Am. Chem. Soc.* **2005**, *127*, 5833–5839.
118. X. S. Tan, A. Volbeda, J. C. Fontecilla-Camps, P. A. Lindahl, *J. Biol. Inorg. Chem.* **2006**, *11*, 371–378.
119. E. L. Maynard, P. A. Lindahl, *J. Am. Chem. Soc.* **1999**, *121*, 9221–9222.
120. J. Seravalli, S. W. Ragsdale, *Biochemistry* **2000**, *39*, 1274–1277.
121. X. Tan, C. Sewell, Q. Yang, P. A. Lindahl, *J. Am. Chem. Soc.* **2003**, *125*, 318–319.

122. D. P. Barondeau, P. A. Lindahl, *J. Am. Chem. Soc.* **1997**, *119*, 3959–3970.
123. W. S. Shin, M. E. Anderson, P. A. Lindahl, *J. Am. Chem. Soc.* **1993**, *115*, 5522–5526.
124. T. Shanmugasundaram, G. K. Kumar, H. G. Wood, *Biochemistry* **1988**, *27*, 6499–6503.
125. S. W. Ragsdale, H. G. Wood, *J. Biol. Chem.* **1985**, *260*, 3970–3977.
126. J. Seravalli, W. W. Gu, A. Tam, E. Strauss, T. P. Begley, S. P. Cramer, S. W. Ragsdale, *Proc. Natl. Acad. Sci. USA* **2003**, *100*, 3689–3694.
127. C. E. Webster, M. Y. Darensbourg, P. A. Lindahl, M. B. Hall, *J. Am. Chem. Soc.* **2004**, *126*, 3410–3411.
128. P. T. Matsunaga, G. L. Hillhouse, *Angew. Chem. Int. Edit.* **1994**, *33*, 1748–1749.
129. G. C. Tucci, R. H. Holm, *J. Am. Chem. Soc.* **1995**, *117*, 6489–6496.
130. D. Sellmann, D. Haussinger, F. Knoch, M. Moll, *J. Am. Chem. Soc.* **1996**, *118*, 5368–5374.
131. X. S. Tan, C. Sewell, Q. W. Yang, P. A. Lindahl, *J. Am. Chem. Soc.* **2003**, *125*, 318–319.
132. J. Q. Xia, Z. G. Hu, C. V. Popescu, P. A. Lindahl, E. Munck, *J. Am. Chem. Soc.* **1997**, *119*, 8301–8312.
133. M. R. Bramlett, A. Stubna, X. S. Tan, I. V. Surovtsev, E. Munck, P. A. Lindahl, *Biochemistry* **2006**, *45*, 8674–8685.
134. T. C. Harrop, P. K. Mascharak, *Coord. Chem. Rev.* **2005**, *249*, 3007–3024.
135. W. W. Gu, S. Gencic, S. P. Cramer, D. A. Grahame, *J. Am. Chem. Soc.* **2003**, *125*, 15343–15351.
136. T. Funk, W. W. Gu, S. Friedrich, H. X. Wang, S. Gencic, D. A. Grahame, S. P. Cramer, *J. Am. Chem. Soc.* **2004**, *126*, 88–95.
137. J. Seravalli, M. Kumar, S. W. Ragsdale, *Biochemistry* **2002**, *41*, 1807–1819.
138. S. W. Ragsdale, H. G. Wood, W. E. Antholine, *Proc. Natl. Acad. Sci. USA* **1985**, *82*, 6811–6814.
139. S. W. Ragsdale, L. G. Ljungdahl, D. V. Dervartanian, *Biochem. Biophys. Res. Commun.* **1982**, *108*, 658–663.
140. R. P. Schenker, T. C. Brunold, *J. Am. Chem. Soc.* **2003**, *125*, 13962–13963.
141. S. J. George, J. Seravalli, S. W. Ragsdale, *J. Am. Chem. Soc.* **2005**, *127*, 13500–13501.
142. G. Bender, T. A. Stich, L. F. Yan, R. D. Britt, S. P. Cramer, S. W. Ragsdale, *Biochemistry* **2010**, *49*, 7516–7523.
143. S. Gencic, K. Kelly, S. Ghebreamlak, E. C. Duin, D. A. Grahame, *Biochemistry* **2013**, *52*, 1705–1716.
144. A. Chmielowska, P. Lodowski, M. Jaworska, *J.Phys. Chem. A* **2013**, *117*, 12484–12496.
145. P. Amara, A. Volbeda, J. C. Fontecilla-Camps, M. J. Field, *J. Am. Chem. Soc.* **2005**, *127*, 2776–2784.
146. S. A. Ensign, *Biochemistry* **1995**, *34*, 5372–5381.
147. D. S. Newsome, *Catalysis Reviews - Science and Engineering* **1980**, *21*, 275–318.
148. R. Hille, *Chem. Rev.* **1996**, *96*, 2757–2816.
149. M. W. Ribbe, Y. Hu, K. O. Hodgson, B. Hedman, *Chem. Rev.* **2013**.
150. Y. Hu, M. W. Ribbe, *J. Biol. Chem.* **2013**, *288*, 13173–13177.
151. Y. Hu, M. W. Ribbe, *Biochim. Biophys. Acta* **2013**, *1827*, 1112–1122.
152. Y. Hu, M. W. Ribbe, *Methods Mol. Biol.* **2011**, *766*, 3–7.
153. J. W. Peters, J. B. Broderick, *Annu. Rev. Biochem.* **2012**, *81*, 429–450.
154. Y. Nicolet, J. C. Fontecilla-Camps, *J. Biol. Chem.* **2012**, *287*, 13532–13540.
155. A. Böck, P. W. King, M. Blokesch, M. C. Posewitz, *Adv. Microb. Physiol.* **2006**, *51*, 1–71.
156. M. A. Farrugia, L. Macomber, R. P. Hausinger, *J. Biol. Chem.* **2013**, *288*, 13178–13185.
157. C. C. Lee, Y. L. Hu, M. W. Ribbe, *Science* **2010**, *329*, 642–642.
158. Y. L. Hu, C. C. Lee, M. W. Ribbe, *Science* **2011**, *333*, 753–755.
159. B. K. Ho, F. Gruswitz, *BMC Struct. Biol.* **2008**, *8*, 49.

Chapter 4
Investigations of the Efficient Electrocatalytic Interconversions of Carbon Dioxide and Carbon Monoxide by Nickel-Containing Carbon Monoxide Dehydrogenases

Vincent C.-C. Wang, Stephen W. Ragsdale, and Fraser A. Armstrong

Contents

ABSTRACT 72
1 DIRECT CARBON DIOXIDE/CARBON MONOXIDE INTERCONVERSIONS
  IN BIOLOGY 72
2 NICKEL-CONTAINING CARBON MONOXIDE DEHYDROGENASES 75
3 PROTEIN FILM ELECTROCHEMISTRY 80
4 CARBON MONOXIDE DEHYDROGENASES AS ELECTROCATALYSTS 82
  4.1 The Electrocatalytic Voltammograms of Class IV Enzymes 82
  4.2 The Electrocatalytic Voltammograms of Class III Enzymes 84
5 POTENTIAL-DEPENDENT REACTIONS WITH INHIBITORS 85
  5.1 How Class IV Carbon Monoxide Dehydrogenases Respond to Cyanide 86
  5.2 How Class IV Carbon Monoxide Dehydrogenases Respond to Cyanate 87
  5.3 How Class IV Carbon Monoxide Dehydrogenases Respond to Sulfide
      and Thiocyanate 89
6 DEMONSTRATIONS OF TECHNOLOGICAL SIGNIFICANCE 91
7 CONCLUSIONS 95
ABBREVIATIONS 95
ACKNOWLEDGMENTS 96
REFERENCES 96

V.C.-C. Wang • F.A. Armstrong (✉)
Department of Chemistry, Inorganic Chemistry Laboratory, University of Oxford,
South Parks Road, Oxford, OX1 3QR, UK
e-mail: fraser.armstrong@chem.ox.ac.uk

S.W. Ragsdale
Department of Biological Chemistry, University of Michigan Medical School,
Ann Arbor, MI 48109-0606, USA
e-mail: sragsdal@umich.edu

© Springer Science+Business Media Dordrecht 2014   71
P.M.H. Kroneck, M.E. Sosa Torres (eds.), *The Metal-Driven Biogeochemistry
of Gaseous Compounds in the Environment*, Metal Ions in Life Sciences 14,
DOI 10.1007/978-94-017-9269-1_4

Abstract Carbon monoxide dehydrogenases (CODH) play an important role in utilizing carbon monoxide (CO) or carbon dioxide (CO_2) in the metabolism of some microorganisms. Two distinctly different types of CODH are distinguished by the elements constituting the active site. A Mo-Cu containing CODH is found in some aerobic organisms, whereas a Ni-Fe containing CODH (henceforth simply Ni-CODH) is found in some anaerobes. Two members of the simplest class (IV) of Ni-CODH behave as efficient, reversible electrocatalysts of CO_2/CO interconversion when adsorbed on a graphite electrode. Their intense electroactivity sets an important benchmark for the standard of performance at which synthetic molecular and material electrocatalysts comprised of suitably attired abundant first-row transition elements must be able to operate. Investigations of CODHs by protein film electrochemistry (PFE) reveal how the enzymes respond to the variable electrode potential that can drive CO_2/CO interconversion in each direction, and identify the potential thresholds at which different small molecules, both substrates and inhibitors, enter or leave the catalytic cycle. Experiments carried out on a much larger (Class III) enzyme CODH/ACS, in which CODH is complexed tightly with acetyl-CoA synthase, show that some of these characteristics are retained, albeit with much slower rates of interfacial electron transfer, attributable to the difficulty in making good electronic contact at the electrode. The PFE results complement and clarify investigations made using spectroscopic investigations.

Keywords carbon monoxide dehydrogenase • CO_2 reduction • electrocatalyst • metalloenzyme • protein film electrochemistry

Please cite as: *Met. Ions Life Sci.* 14 (2014) 71–97

1 Direct Carbon Dioxide/Carbon Monoxide Interconversions in Biology

Finding new ways of using renewable energy to reduce carbon dioxide (CO_2) to fuels and thus supplement biological photosynthetic CO_2 fixation, is a major scientific challenge with huge implications for future civilizations. Green plants and many microorganisms generally fix CO_2 by photosynthesis – a process that is familiar to everyone: however, some microorganisms use a totally different system for fixing CO_2 and exploit pathways in which carbon monoxide (CO) is an important intermediate.

Moreover, some anaerobic sulfate-reducing bacteria, such as *Desulfotomaculum carboxydivorans*, hydrogenogens such as *Carboxydothermus hydrogenoformans* (*Ch*), some phototrophs, such as *Rhodospirillum rubrum* (*Rr*), and CO-utilizing microorganisms and aerobic carboxidotrophs, such as *Oligotropha carboxidovorans,* are able to use CO as their sole carbon and energy source. These alternative pathways have been documented in various review articles [1–6]. The metalloenzymes involved are called carbon monoxide dehydrogenases (CODH) and they provide particularly useful insight into the chemical principles underlying catalysis of carbon dioxide activation. The primary reaction being catalyzed is the half-cell reaction that is written as equation (1).

$$CO_2 + 2\,H^+ + 2\,e^- \rightleftharpoons CO + H_2O \tag{1}$$

Two chemically distinct types of CODH are distinguished by the structure of their active site. Some aerobes use enzymes containing a μ-sulfido bridged Cu-S-Mo(pyranopterin) center [5, 7]: on the other hand, some anaerobic micro-organisms, such as *Moorella thermoacetica* (*Mt*) and *Carboxydothermus hydrogenoformans* use enzymes in which the active site is a distorted [Ni3Fe-4S] cubane-like cluster with a pendant (dangling) Fe [8]. The entire [Ni4Fe-4S] cluster is referred to as the 'C-cluster' and the corresponding enzymes are known as Ni-CODHs. The Ni-CODH family is extended to include the class of enzymes known as CODH/ACS, in which a Ni-CODH is strongly associated with an acetyl-CoA synthase. In acetogenic bacteria, CODH-ACS catalyzes C-C bond formation using CO (formed from CO_2) and a methyl group (donated by a corrinoid iron sulfur protein) to form acetyl-CoA (equation 2). In methanogenic archaea, Ni-CODH coupled to acetyl-CoA decarbonylase/synthase (ACDS) cleaves acetyl-CoA to CO_2 and methane [9].

$$CO + [CH_3\text{-Co(III)FeSP}]^{2+} + CoASH \rightleftharpoons CH_3COSCoA + [Co(I)FeSP]^+ + H^+ \tag{2}$$

This article focuses on the anaerobic Ni-CODHs, which have been divided into four classes in terms of the metabolic role and subunit components of the enzymes, as depicted, cartoonwise, in Figure 1 [10]. The classification can be confusing, as Roman numerals are used to denote both class and isozymes within a class, but hopefully this will not present a problem.

Class I enzymes consist of five subunits (αβγδε) and are found in autotrophic methanogens, where acetyl-CoA is synthesized from CO_2 and H_2. Class II enzymes are found in acetoclastic methanogens and catalyze the decarbonylation of acetyl-CoA to CO_2, coenzyme A (CoA), and a methyl group which is transferred to tetrahydrosarcinapterin. Class III enzymes are found in homoacetogens and are sub-classified as two independent enzymes, a $\alpha_2\beta_2$ tetramer (CODH/ACS) and a γδ heterodimer (CoFeSP): the catalytic center for synthesizing acetyl-CoA in tetra-meric CODH/ACS is the same as found in Class I enzymes. Finally, Class IV enzymes consist only of the CODH entity, a monofunctional α_2 dimer that catalyzes the conversion of CO to CO_2 in the anaerobic CO-dependent energy metabolism of bacteria and archaea. The structure of a Class IV CODH is shown in Figure 2, along with the structure of a Class III CODH/ACS.

Important examples of all five Ni-CODHs are provided by the thermophilic bacterium *Carboxydothermus hydrogenoformans* which can grow on CO as the sole carbon and energy source [11]. The first of these, referred to as CODH I_{Ch} (the Roman letter does not refer to the classification mentioned above) is involved in energy conversion and delivers electrons derived from CO oxidation to a hydrogenase that evolves H_2 [12]. In *Rhodospirillum rubrum*, an enzyme very closely related to CODH I_{Ch} forms part of a well-characterized CODH-hydrogenase

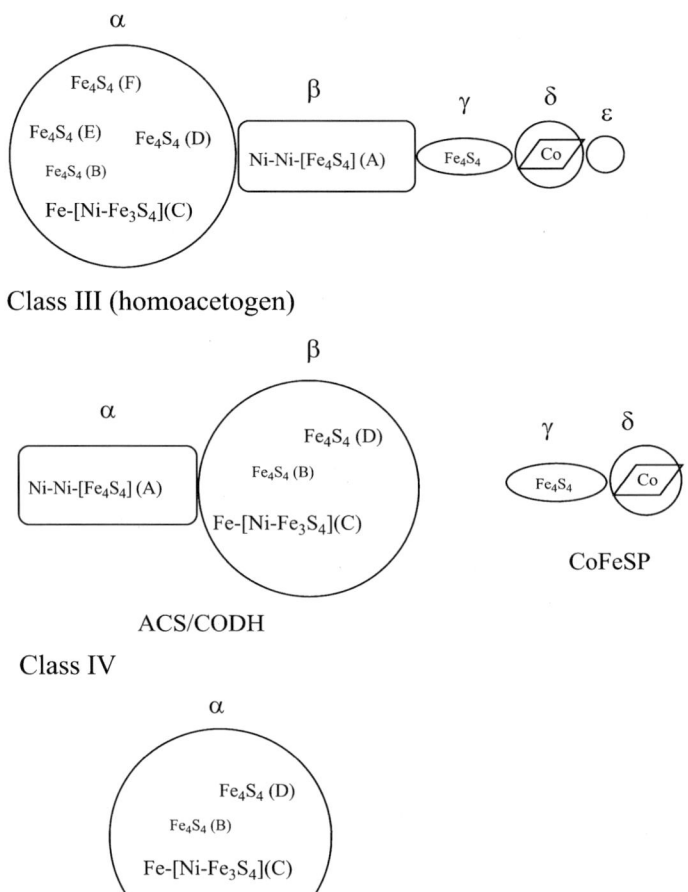

Figure 1 Four different classes of anaerobic NiFe-containing CODH are classified by the subunit composition and biological roles. The upper case letters denote the type of metal cluster. The A-cluster is the active site for acetyl-CoA synthesis. The C-cluster is the active site in which CO/CO_2 conversion occurs. Several $[Fe_4S_4]$ clusters are required for electron transfers in enzymes (B-cluster and D-cluster) and Class I and Class II enzymes have two additional [4Fe-4S] clusters (E and F clusters) in the enzyme complex. Co refers to the cobalt-corrinoid serving as a methyl carrier.

complex [13]. In contrast, the role of CODH II_{Ch} remains unclear despite this enzyme being the most structurally well characterized by X-ray crystallography [8, 14, 15]. The sequence identity and similarity between CODH I_{Ch} and CODH II_{Ch} are 58.3 % and 73.9 %, respectively. The third member, CODH $III_{Ch,}$ is complexed tightly with ACS as CODH/ACS, as mentioned above, and catalyzes acetyl-CoA synthesis [16].

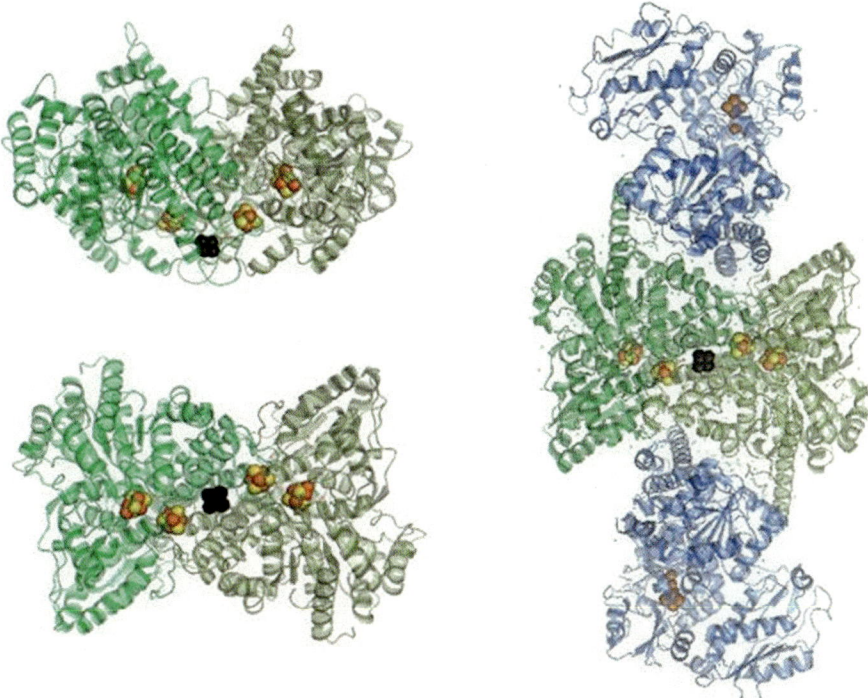

Figure 2   3D-structures of carbon monoxide dehydrogenases that have been studied by protein film electrochemistry. **Upper left**: CODH II$_{Ch}$ (PDB code 3B53) showing how a relay of FeS clusters leads from the exposed D-cluster (black) to two C-clusters, each one housed in one of the subunits. **Lower left**: Same structure viewed facing up from the D-cluster. **Right**: CODH/ACS (PDB code 2Z8Y) shown with the CODH viewed in same way as for lower-left view of CODH II$_{Ch}$.

Of the remaining examples, CODH IV$_{Ch}$ is suggested to be associated with a multisubunit enzyme complex for oxidative stress response based on genomic analysis, while the biological role of CODH V$_{Ch}$ remains unclear.

The ease by which these enzymes interconvert CO_2 and CO has attracted intense interest from both chemists and biochemists, who have applied a variety of spectroscopic and structural methods in efforts to establish a firm mechanistic understanding. The aim of this chapter is to describe how the application of protein film electrochemistry (PFE) has added to this understanding [17–19]. But first we will summarize some of the structural and spectroscopic information that has been available now for several years.

2   Nickel-Containing Carbon Monoxide Dehydrogenases

Several crystal structures of NiFe-containing CODH (Class IV) or CODH/ACS (Class III) from different organisms have been solved [4, 8, 15, 20–22]. All Class IV enzymes have a dimeric structure (Figure 2) in which each monomer contains a unique active site (called the C-cluster) which is a [Ni4Fe-4S] cluster (or [Ni3Fe-4S]) cubane cluster linked to an extra-cuboidal pendant or 'dangling' Fe. A [4Fe-4S] cluster (called the B-cluster) is located about 10 Å from each C-cluster, although these are coordinated by the other subunit. Finally, a single [4Fe-4S] cluster (called the D-cluster) lying about 10 Å from each B-cluster and close to the protein surface, is coordinated by both subunits. The distances suggest immediately that the B-cluster and D-cluster convey electrons between the C-cluster and an external physiological redox partner, which is probably a ferredoxin.

Several redox states of the C-cluster involved in the mechanism of CO/CO_2 interconversion by CODH have been identified [23]. These are known as C_{ox}, C_{red1}, C_{int}, and C_{red2} in order of decreasing oxidation level. Various structures of the active site of CODH II_{Ch} are shown in Figure 3 including two obtained from enzyme crystallized at different reduction potentials (–320 mV, and –600 mV with CO_2 present) [8]. The –320 mV structure (Figure 3a) shows that an O-donor

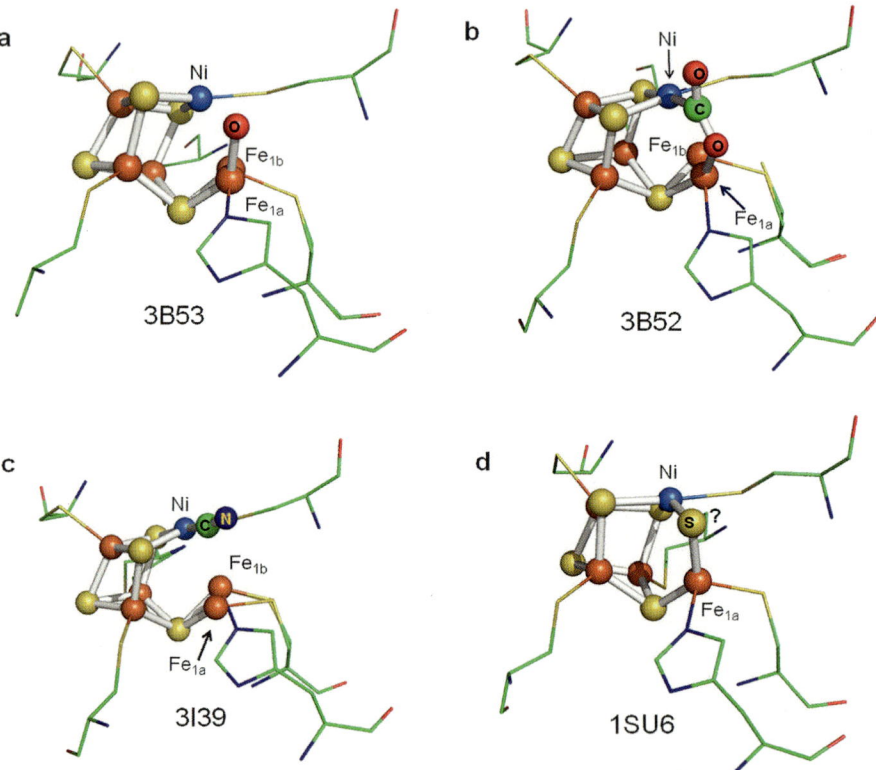

Figure 3 The active site of CODH II_{Ch} captured in crystal structures obtained under different conditions. (**a**) –320 mV, (**b**) –600 mV with CO_2, (**c**) –320 mV with cyanide, and (**d**) CO-reduced CODH. Two positions are found for the pendant iron atom in the crystal structure, which are labelled Fe_{1a} and Fe_{1b} respectively. The PDB codes are shown in each case. Color codes for atoms: nickel, light blue; iron, orange; oxygen, red; carbon, green; nitrogen, blue; sulfur, yellow.

(it is assumed this is hydroxide) binds to the pendant Fe atom and the Ni atom is coordinated by three sulfido ligands from the [3Fe-4S] core with a distorted T-shaped coordination geometry. The structure of CO_2-bound CODH II_{Ch} obtained by incubating CODH II_{Ch} with $NaHCO_3$ at -600 mV (Figure 3b) reveals further that the C-atom from CO_2 binds to the Ni-atom at a distance of 1.96 Å, completing a distorted square-planar geometry, and one of the O-atoms of the bound CO_2 binds to the pendant Fe at a distance of 2.05 Å.

These results suggest a mechanism in which CO oxidation involves nucleophilic attack on CO-bound Ni by a OH^- ligand that is coordinated to the pendant Fe, thereby yielding the Ni-(CO(OH)) intermediate that is detected. In the reverse reaction, the pendant Fe atom abstracts an O-atom from the C-coordinated CO_2 via a proton-coupled two-electron process, and leaves CO bound to Ni with OH^- on the pendant Fe. This picture of the mechanism is included in Scheme 1 [4]. Spectroscopic data add further structural definition to this description, while PFE data discussed below show how the reactions with substrates and inhibitors depend on potential.

From potentiometric redox titrations, it is known that the inactive and EPR-silent state, C_{ox}, is reduced by one electron to give the active state C_{red1}, which exhibits a characteristic EPR signal, $g_{av} \sim 1.82$ ($g = 2.01$, 1.80, 1.65) [24]. Based on the structural evidence described above, CO binds to the Ni atom in the C_{red1} state to start the catalytic cycle for CO oxidation, the eventual result of which is that CO_2 is released leaving the two-electron reduced state C_{red2}. The C-cluster is re-reduced back to C_{red1} by two one-electron transfers, via an EPR-silent state known as C_{int}. It is very significant that C_{red2} displays an EPR signal with $g_{av} \sim 1.86$ ($g = 1.97$, 1.86, 1.75) [24].

Based on earlier Mössbauer spectroscopy data, Lindahl suggested that C_{ox} should be assigned as $[Ni^{2+}Fe_p^{3+}]:[3Fe-4S]^{1-}$ (Fe_p = pendant Fe site) and C_{red1} should be assigned as $[Ni^{2+}Fe_p^{2+}]:[3Fe-4S]^{1-}$ [24]. The fact that C_{red1} and C_{red2} differ by two electrons but both show an EPR spectrum with $g_{av} < 2$ suggests that the underlying electronic structure of the [3Fe-4S] core fragment is unchanged between the two states. At least one electron must therefore be transferred at the Ni subsite, since it is unclear how a two-electron change could be accommodated at the pendant Fe that is also coordinated by sulfide, histidine-N, and cysteine-S. The question then arises: what formal redox changes actually occur at Ni? If the pendant Fe remains as Fe^{2+}, both electrons must be transferred at the Ni subsite. Two alternative descriptions of the Ni atom in the C_{red2} state have been suggested: Ni(0) or the isoelectronic protonated site formulated as a nickel hydrido species Ni(II)-H [8, 25]. Further light on this question stems from Ni K- or L-edge X-ray absorption spectroscopy (XAS) studies on CODH$_{Rr}$ and CODH II_{Ch} in differing redox states [26–28]. Results from as-isolated CODH II_{Ch}, CO-treated CODH II_{Ch}, dithionite-reduced CODH II_{Ch}, indigo-carmine-oxidized CODH$_{Rr}$, dithionite-reduced CODH II_{Rr} and CO-treated CODH$_{Rr}$, all suggest that the Ni subsite in the C-cluster in these samples is present as Ni(II) regardless of redox state.

The reduction potential for the two-electron interconversion between C_{red1} and C_{red2} is approximately -520 mV according to EPR potentiometric titrations [29]. The reduced B-cluster exhibits the typical EPR signal of a $[4Fe-4S]^{1+}$ cluster

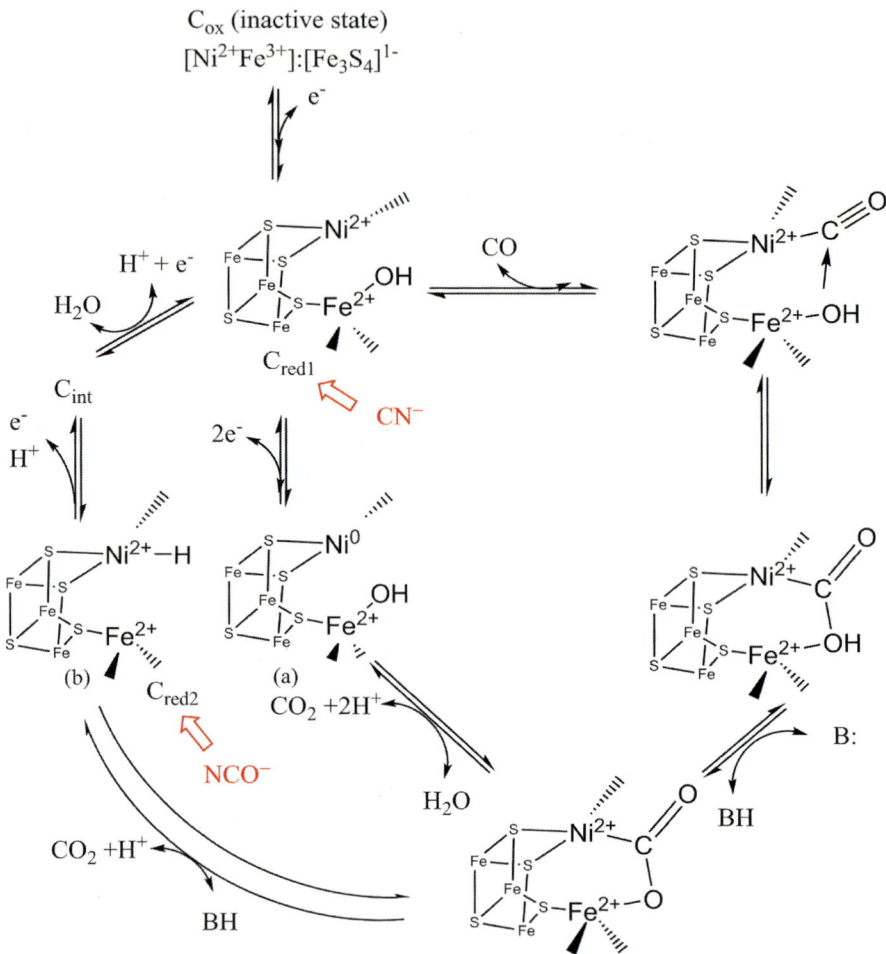

Scheme 1 Two possible mechanisms of Ni-Fe CODH proposed by (a) Dobbek [8] and Lindahl [23], in which the Ni subsite in C_{red2} is formally Ni(0) or (b) as proposed by Fontecilla-Camps et al. [25], in which the Ni subsite in C_{red2} is bonded to a H atom, making what is formally a Ni(II)-H species. The base 'B' refers to the amino acid that accepts/donates the proton, possibly His or Lys. The red colors represent substrate-mimic inhibitors in which cyanide (CN^-) binds to the C_{red1} state and cyanate (NCO^-) binds to the C_{red2} state.

(g_{av} ~1.94) and the reduction potential for the $[4Fe-4S]^{2+/1+}$ couple is –440 mV in the enzyme from *Moorella thermoacetica* and –418 mV in the enzyme from *Rhodospirillum rubrum* [29, 30]. Significantly, the D-cluster remained undetected until the first crystal structure of CODH was solved. One possibility is that this cluster had escaped detection because the reduced $[4Fe-4S]^{1+}$ state has S > ½, but studies of $CODH_{Rr}$ by magnetic circular dichroism (MCD) and resonance Raman spectroscopy suggest that the D-cluster remains in the oxidized $[4Fe-4S]^{2+}$ state even at –520 mV [31]. This information will be important when we consider the results from PFE experiments later.

Cyanide (CN$^-$), isoelectronic with CO, is a well-studied inhibitor of CODH [32–34]. Several CODH crystal structures from different species with CN$^-$ or CO bound to the Ni atom in the C-cluster have been determined: that of CN$^-$-bound CODH II$_{Ch}$ is included in Figure 3 [14]. A consensus is not achieved: compared to the almost linear NC-Ni arrangement in CN$^-$-bound CODH II$_{Ch}$, the CN$^-$-bound CODH/ACS$_{Mt}$ and CO(formyl)-bound ACDS/CODH$_{Mb}$ reveal, respectively, a *bent* NC-Ni structure with bond angle ~114° and a *bent* OC-Ni structure with bond angle ~107° [20, 35]. The OH$^-$ ligand remains on the pendant Fe in both structures. Compared with the structure of CO$_2$-bound CODH II$_{Ch}$, the N-atom in CN$^-$-bound CODH/ACS$_{Mt}$ and the O-atom in CO-bound ACDS/CODH$_{Mb}$ each overlay with the corresponding O-atom from CO$_2$ in CO$_2$-bound CODH II$_{Ch}$ whereas the C-atom in both CN$^-$-bound ACS/CODH$_{Mt}$ and CO-bound ACDS/CODH$_{Mb}$ is displaced from its position in CO$_2$-bound CODH II$_{Ch}$ [8]. The shift in C-atom position could facilitate the nucleophilic attack by OH$^-$. Interestingly, the crystal structure of CODH II$_{Ch}$ incubated with *n*-butyl isocyanide at –320 mV reveals that the C-atom from the plausible product "*n*-butyl-isocyanate" binds to the Ni with a distorted tetrahedral coordination geometry [36]. A second *n*-butyl-isocyanide is found in the putative gas channel in CODH II$_{Ch}$.

Introduction of CN$^-$ to a sample of CODH/ACS$_{Mt}$ poised in the C$_{red1}$ state results in a new EPR signal with $g_{av} = 1.72$ ($g = 1.87, 1.78, 1.55$) [23, 33, 34, 37]. In contrast, no change in the EPR spectrum is observed when CN$^-$ is introduced to the C$_{red2}$ state [37]. Further studies of ^{13}CN$^-$-bound CODH/ACS$_{Mt}$ by electron nuclear double resonance (ENDOR) spectroscopy revealed a doublet peak in the C$_{red1}$ state – early evidence that CN$^-$ (^{13}C nuclear spin I = ½) binds directly to the C-cluster [38, 39]. Experiments to evaluate if cyanate (NCO$^-$) is an inhibitor indicated that it binds also to C$_{red1}$ even though (since it is an analogue of CO$_2$ not CO) it should display behavior opposite to that of CN$^-$: this result was puzzling, but as we explain below, it was clarified by PFE experiments.

Some crystal structures of CODH II$_{Ch}$ revealed a μ-sulfido ligand bridging the Ni atom and the pendant Fe atom (Figure 3d), and this observation led to suggestions that an additional (5th) sulfide is necessary for catalysis [15, 40]. In place of inorganic sulfide (S^{2-}), the thiolate sulfur from Cys531 was found to bridge the Ni atom and the pendant Fe atom in the crystal structure of CODH$_{Rr}$ [21]. However, crystal structures from recombinant CODH II expressed in *E. coli* and of CODH from *Moorella thermoacetica* and *Methanosarcina barkeri* (*Mb*) revealed no presence of a 5th μ-sulfido ligand [20, 35]. The CO oxidation activities of CODH measured by solution assays in the presence of sodium sulfide showed varying results between different species and redox conditions [33, 41]. The role of sulfide is considered later when we discuss the PFE results.

Scheme 1 highlights a challenge for investigating enzymes such as CODHs that catalyze redox reactions via rapid passage through a series of intermediates, the presence (lifetime) of each of which depends on how fast or favorable are the electron transfers that determine their existence. Direct coordination of a reactant or inhibitor, or release of a product, should be selective for a particular oxidation state: the question is, how to control and fine-tune the relative amounts of each state during catalysis and simultaneously observe the effect on rate when reactant or inhibitor are introduced. Not all inhibitors bind to active states: some exert their influence by facilitating redox processes that lead to inactive states. Here, an immediate problem is posed by the fact that C_{ox} is EPR-silent, thus denying us a substantial amount of spectroscopic information; the same is true for C_{int}, which is particularly elusive.

3 Protein Film Electrochemistry

Protein film electrochemistry refers to a suite of dynamic electrochemical techniques that address enzyme molecules directly attached to an electrode surface, allowing catalytic activity to be recorded (as current) as a continuous function of electrode potential and/or time [42–45]. The importance of steady-state activity relating directly to steady-state current at any given potential is a unique aspect of PFE. The concept is depicted (idealistically) in Figure 4. The electrode provides a wide and continuous range of potential that is very difficult to achieve using conventional methods; for example, catalytic CO_2 reduction by CODH requires

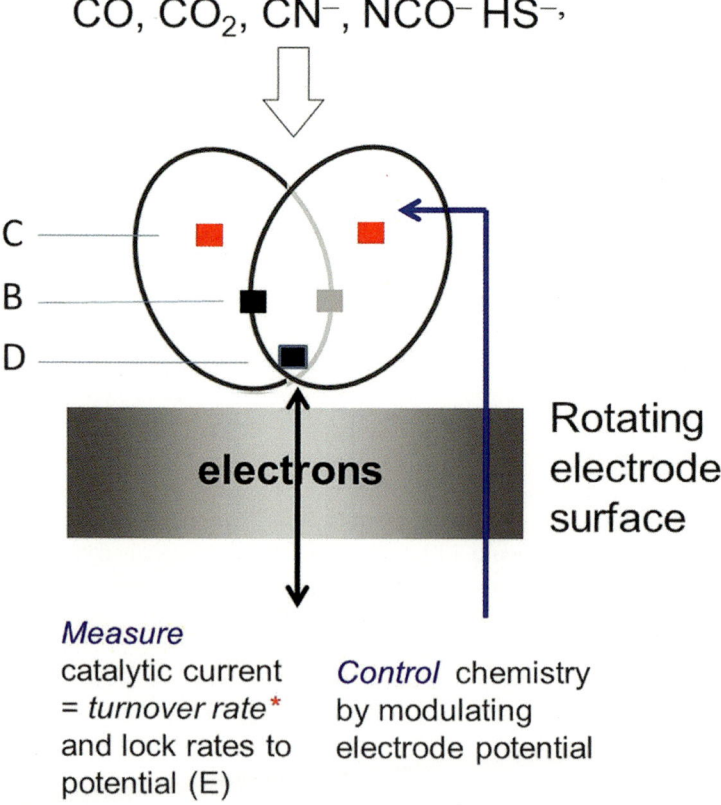

Figure 4 Cartoon depicting the concept of protein film electrochemistry, as applied to CODH. Squares represent the FeS clusters D, B, and C.

potentials more negative than –0.5 V at pH = 7.0, and it is difficult to obtain such a negative potential by chemical mediators or titrants, let alone provide a continuous scale. A pyrolytic graphite 'edge' (PGE) electrode is particularly suitable: it is easily polished by abrasion (an aqueous alumina slurry) to expose a pristine surface and it has a very wide potential region, extending from approximately +1 V to –1 V *versus* the standard hydrogen electrode (SHE) at neutral pH.

Many enzymes are now established to be excellent electrocatalysts – defined here as catalysts for half-cell reactions that occur at an electrode [46, 47]. Enzymes are typically bound at the electrode by simple adsorption (exploiting a rough electrode surface that contacts the enzyme at different points and maximizes hydrophobic and electrostatic interactions) or by chemical attachment that may involve engineering the enzyme to place specific residues in suitable positions for linkage formation. It is essential that the enzyme contacts the electrode surface in such a way that electron relay centers within the enzyme, such as the FeS clusters in CODHs, lie within a sufficiently short distance to allow fast electron tunnelling. For class IV CODHs, the obvious entry and exit site is the exposed D-cluster, as emphasized in Figure 2.

Dynamic electrochemical techniques include cyclic voltammetry (the basic search tool, in which current is monitored as the potential is scanned) and chronoamperometry (initiating a reaction at a fixed electrode potential, either by stepping to that potential or injecting a reactant, and monitoring the current as a function of time). Because a change in current corresponds directly to a change in activity, chronoamperometry is an excellent technique for measuring how fast inhibitors are bound or released, all under strictly controlled potential conditions.

In terms of operational advantage, PFE normally needs a relatively tiny amount of enzyme sample, a few pmole/cm^2 – which is orders of magnitude smaller than required for spectroscopy. Because the enzyme is immobilized on the electrode (area typically 0.03 cm^2) the same sample can be successively exposed to solutions having different pH values and inhibitor/substrate concentrations. For studying air-sensitive enzymes such as Ni-CODH, the sealed electrochemical cell is housed in a glove box. The electrode is normally rotated at variable high speed, which in addition to assisting the transport of reactants to the enzyme, ensures that dispersion of product can be controlled – the latter factor allowing product inhibition to be investigated easily. If the substrate is a gas, this is introduced into the headspace of the electrochemical cell and it is easy to replace substrates in the solution by purging with inert gas.

Cyclic voltammetry provides a direct, continuous 'spectrum' of the potential dependence. Importantly, when the enzyme catalysis approaches electrochemical reversibility, the trace cuts across the potential axis at the equilibrium value and just a tiny potential bias either side of this value switches the reaction between reduction and oxidation. Such reversibility is found with many enzymes. The ratio of oxidation and reduction currents measured at appropriate potential values yields the so-called 'catalytic bias' – the tendency of the enzyme to operate preferentially in a particular direction [47]. Overlay of scans in each direction shows that electrocatalysis is at steady state, whereas hysteresis is a sign that the enzyme alters its activity on a timescale that is slow relative to the rate that the potential is scanned.

The fact that the catalytic activity is observed as current and measured across a range of potentials means that each enzyme produces a spectrum-like fingerprint of its characteristics, but PFE rarely allows determination of absolute steady-state catalytic rates because the electroactive coverage is generally unknown. In order to determine the electroactive coverage of an enzyme it is necessary to detect so-called 'non-turnover' signals due to stoichiometric electron exchanges with relay centers such as FeS clusters. These signals are peak-like in each scan direction and the electroactive coverage is estimated from the areas under the peaks. Turnover frequency k_{cat} is then calculated from the current (i) using the relationship

$$i = k_{cat} \, nFA\Gamma \tag{3}$$

where n is the number of electrons transferred, F is the Faraday constant, A is the electrode area, and Γ is the electroactive coverage (equation 3). As a guide, it is considered very difficult to observe a non-turnover signal if Γ is below 10^{-12} moles cm^{-2}. Consequently, with $A = 0.03$ cm^2, a current of 10 μA corresponds to a lower limit for k_{cat} in the region of 2000 s^{-1}.

The difficulty in determining absolute steady-state activities contrasts with the ease by which transient reaction rates are measured at a constant potential. The rate of a change in activity is measured from the time course of the current following a perturbation such as potential step or injection of reagent. The amplitude of the change is proportional to catalytic activity and electroactive coverage, but the rate should be independent of these variables. The chronoamperometry experiment examines, directly, the rate of change of rate – a measurement that is equivalent to recording the second derivative of a normal kinetic plot of concentration *versus* time.

4   Carbon Monoxide Dehydrogenases as Electrocatalysts

Based upon the structure of CODH II$_{Ch}$, it is clear from Figure 2 that the D-cluster should be a feasible entry/exit point for electrons, and the enzyme should be electroactive when attached to a rough electrode surface at which productive orientations are statistically likely.

4.1   The Electrocatalytic Voltammograms of Class IV Enzymes

As shown in Figure 5a, a cyclic voltammogram of CODH I$_{Ch}$ adsorbed on a PGE electrode displays reversible electrocatalysis: under a gas mixture of composition 50 % CO/50 % CO$_2$, the trace cuts through the potential axis at the expected equilibrium potential and a detectable net current in each direction is easily achieved with a minimal overpotential [17, 18]. Under the same conditions, CODH II$_{Ch}$ shows very little CO_2 reduction, but a much larger reduction current is observed if the atmosphere is changed to 100 % CO_2 (as shown in Figure 6 in Sect. 5.1), showing that the marked bias toward CO oxidation displayed by

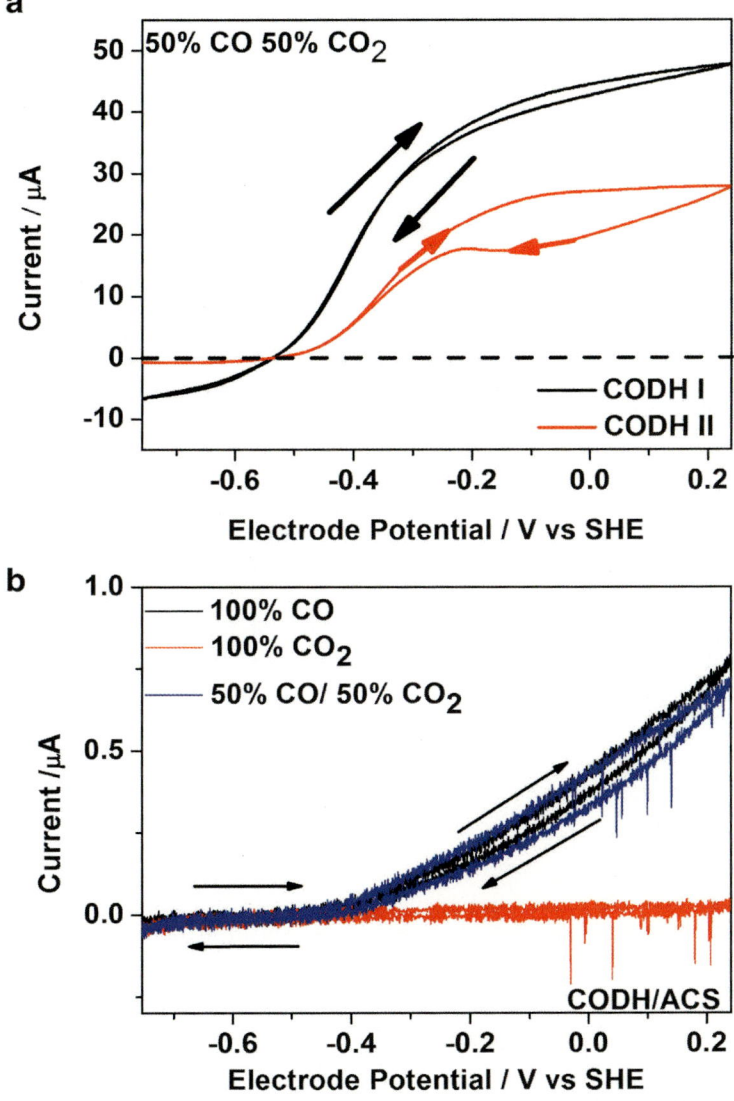

Figure 5 Electrocatalysis by carbon monoxide dehydrogenases adsorbed on a PGE rotating disk electrode in the presence of CO_2 and CO. (**a**) CODH I$_{Ch}$ and CODH II$_{Ch}$ under 50 % CO and 50 % CO_2. Reprinted with permission from [19]; copyright 2013 Wiley-VCH Verlag GmbH & Co. KGaA. (**b**) CODH/ACS under different gas ratios. Experimental conditions: 25 °C, 0.2 M MES buffer (pH 7.0), electrode rotation rate 3500 rpm, and scan rate 2 mV s^{-1} (a) and scan rate 1 mV s^{-1} (b).

CODH II_{Ch} is due, at least in part, to CO being a much stronger product inhibitor of CO_2 reduction for that isozyme [19]. Features of the complex waveshape are explained in terms of a model that takes into consideration the potential at which electrons enter and leave the enzyme (via the D-cluster) and potential-dependent interconversions between different catalytic states [47]. The potential of the center that serves as the electron entry/exit site is an important determinant of catalytic bias, and maximum reversibility is obtained when this value matches that of the reaction being catalyzed. For CODH I_{Ch} and CODH II_{Ch}, where the obvious electron entry/exit site is the D-cluster, it is noteworthy that the reduction potential of the D-cluster has not been determined. However, a rationale for the observation of reversible electrocatalysis in terms of the model shows that this value must lie close to –0.5 V.

At high potentials, both CODH I_{Ch} and CODH II_{Ch} become inactivated as the active site transforms into the C_{ox} state: this is seen particularly clearly for CODH II_{Ch} which shows a re-activation 'peak' at approximately –0.2 V [19]. Generally, inactivation is slow and its rate is independent of potential (consistent with the rate being determined by a chemical process rather than electron transfer) whereas re-activation is fast and controlled by electron transfer, hence the sharp transition.

Before proceeding further, it is worthwhile drawing comparisons with conventional methods of measuring activity. The standard procedure for determining the activity of CO oxidation by Ni-CODH is to measure the color change that occurs when colorless oxidized methyl-viologen (MV^{2+}) is reduced to the blue reduced MV^+ radical by electron transfer from CO-reduced CODH [48–51]. Measuring CO_2 reduction is much more problematic because it is necessary to use a fast electron donor that is more reducing than CO: it is obviously difficult to generate and stabilize such a donor in aqueous solution. Both V_{max} and K_M are sensitive to the driving force (electrode potential) and this fact is easy to see from cyclic voltammograms measured for a range of substrate concentrations [52].

4.2 The Electrocatalytic Voltammograms of Class III Enzymes

Figure 5b shows the same kind of experiment carried out with a film of CODH/ACS_{Mt} on a PGE electrode. This enzyme is much larger than CODH I_{Ch} or II_{Ch}, the two CODH subunits being flanked by two ACS subunits (Figure 2). In contrast to the smaller enzymes, the cyclic voltammogram of CODH/ACS_{Mt} shows no CO_2 reduction, even when no CO is present (100 % CO_2); instead the scan reveals a CO oxidation current that increases slowly as the potential is raised, indicative of sluggish electron transfer. The CO oxidation current at 0 V is about two orders of magnitude smaller than that typically observed for CODH I_{Ch} and CODH II_{Ch} under similar conditions. The sluggish electron transfer may reflect the much larger size of CODH/ACS_{Mt} as the flanking ACS subunits could prevent the D-cluster from achieving such a close approach to the electrode surface. What is not clear though is why CO_2 reduction is not observed, when it is fully expected in view of the physiological function of the enzyme. One possibility is that electrons cannot use the D-cluster which may become obstructed or held too far from the electrode, but instead enter the catalytic cycle at a higher potential than with CODH I$_{Ch}$ or CODH II$_{Ch}$. The reducing power is thus dissipated and electron transfer to the C-cluster becomes an unfavorable step.

To identify whether ACS influences the performance of CODH, an experiment was performed in which several chemical reagents, such as sodium dodecyl sulfate (which separates CODH and ACS partially), 1,10-phenanthroline, (which inhibits the active site in ACS), and acetyl-CoA (the product of the reaction carried out by CODH/ACS$_{Mt}$) were added [53–55]. However, an electrocatalytic current due to CO$_2$ reduction was still not observed, so this observation remains a puzzle. The turnover frequency for CO$_2$ reduction by CODH/ACS$_{Mt}$ is normally determined by measuring the CO product binding to hemoglobin and a rate constant of approximately 1.3 s^{-1} was reported [56]. However, several studies have led to the picture that during synthesis of acetyl-CoA, CO arising from reduction of CO$_2$ diffuses through a channel in the enzyme complex to reach the A-cluster without escaping [57, 58].

5 Potential-Dependent Reactions with Inhibitors

The catalytic current provides an important observable (commonly called a 'handle') for investigating the kinetics and potential dependence of the reactions with inhibitors, at a level of kinetic detail that is difficult to achieve by conventional methods. Some of the opportunities for inhibitors to bind are shown in Scheme 1 and elaborated upon (in an electrochemical sense) in Scheme 2. Protein film electrochemistry is able to reveal and clarify the different ways that these small molecules interrupt catalysis. We will discuss next how cyanide, isoelectronic with CO, mainly inhibits CO oxidation, whereas cyanate, isoelectronic with CO$_2$, targets CO$_2$ reduction. Sulfide is also an inhibitor but is unusual because it acts only under oxidizing conditions. Thiocyanate (SCN$^-$) is also considered.

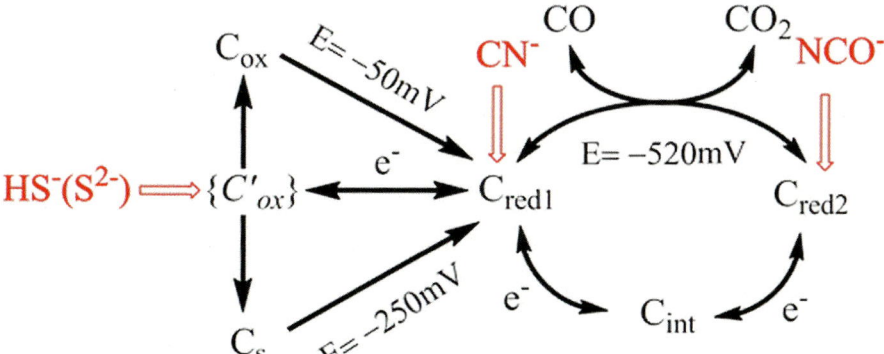

Scheme 2 Summary of the interceptions of the catalytic cycle of CODH by small molecule inhibitors. The potential –520 mV is the standard potential for the CO$_2$/CO half cell reaction at pH 7.0. The potentials –50 mV and –250 mV are the values observed for re-activation of CODH I with and without sulfide. Reprinted with permission from [18]; copyright 2013 American Chemical Society.

5.1 How Class IV Carbon Monoxide Dehydrogenases Respond to Cyanide

Qualitative details of the enzyme's inhibition by cyanide are revealed in the cyclic voltammograms shown in Figure 6, which were recorded for CODH I_{Ch} and CODH II_{Ch} [18, 19]. A complication here is that cyanide is lost quite rapidly from the electrochemical cell as HCN, but key characteristics are easily seen and extracted. The important observations are summarized as follows. When CN^- is

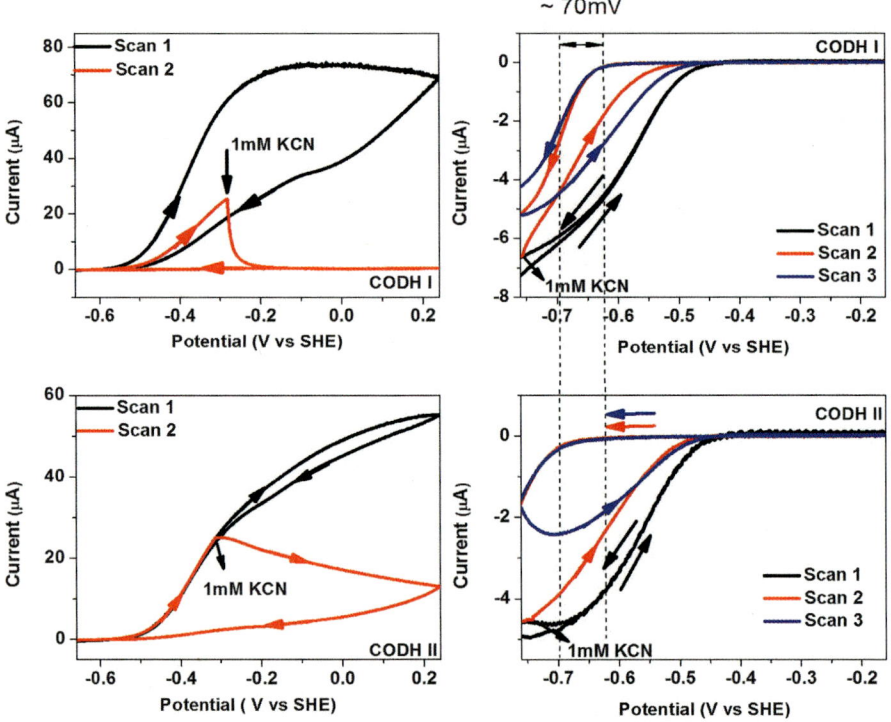

Figure 6 A potential-domain dynamic picture of the effect of cyanide on catalytic properties of CODH. **Left panels**: Inhibition of CO oxidation activity of CODH I_{Ch} and CODH II_{Ch} by cyanide under 100 % CO. An aliquot of KCN stock solution (giving a final concentration of 1 mM in the electrochemical cell) was injected during the second cycle. Conditions: 25 °C, 0.2 M MES buffer (pH = 7.0), electrode rotation rate, 3500 rpm, scan rate, 1 mV sec^{-1}. **Right panels**: Inhibition of CO_2 reduction activity of CODH I_{Ch} and CODH II_{Ch} by CN^- under 100 % CO_2, showing that CN^- does not bind under strongly reducing conditions. An aliquot of KCN stock solution (giving a final concentration of 1 mM in the electrochemical cell) was injected during the second cycle. Note that a more negative potential (by approximately 70 mV) is required to reductively reactivate CODH II_{Ch}. Conditions: 25 °C, 0.2 M MES buffer (pH = 7.0), electrode rotation rate, 3500 rpm, scan rate, 1 mV sec^{-1}. Reproduced by permission from [19]; copyright 2013 Wiley-VCH Verlag GmbH & Co. KGaA.

injected at –0.3 V, during CO oxidation, the current drops: the reaction is more rapid for CODH I_{Ch}. Injecting CN^- instead at the limit of very negative potential, at which CO_2 reduction is observed, then scanning in a positive direction, CN^- binds once the electrode potential becomes more positive than –750 mV, resulting in complete inhibition of CO_2 reduction (within a narrow potential window). Continuing the scan to more positive potential, it is clear that CO oxidation is completely blocked. The activity remains at zero during the remaining cycle, until the electrode potential passes down through –0.6 V, at which point the CO_2 reduction activity starts to increase strongly. The rapid release of cyanide, when the electrode potential is taken below –0.6 V, is an observation that is unique to PFE, because conventional kinetic assays rarely explore such highly reducing conditions and lack the ability to record the activity simultaneously as the potential is scanned.

The data are interpreted in terms of the differing dominance of the two major redox states of the C-cluster, C_{red1}, and C_{red2}, as the potential is varied. At potentials above –0.6 V, C_{red1} dominates and CN^- is tightly bound, blocking catalysis. At electrode potentials below –0.6 V, C_{red2} becomes the dominant catalytic species, and recovery of activity that occurs during the scan in this very negative potential region reveals that CN^- is released. During the experiment, which is carried out at a very slow scan rate, the cyanide is lost as HCN which is removed in the gas flow; hence the effect diminishes with time. The preferential binding to C_{red1} perfectly complements the EPR data which showed that only the spectrum of this state is affected by CN^- [37]. The kinetics of binding and release of CN^- are measured quantitatively by chronoamperometric experiments, in which the time dependence of the increase or decrease in current is recorded at a constant potential. Some exemplary experiments are shown in Figure 7.

The top panels of Figure 7 show the rates at which CN^- reacts to inhibit CO oxidation and CO_2 reduction by CODH II_{Ch}. The lower panels show a more complicated series of injections and potential steps, to compare the rates of reductive reactivation of CODH I_{Ch} and CODH II_{Ch} upon stepping from +0.14 V to –0.76 V. Reductive release of CN^- from CODH II_{Ch} is much faster and occurs within the step time.

5.2 How Class IV Carbon Monoxide Dehydrogenases Respond to Cyanate

Inhibition by cyanate (NCO^-), an analogue of CO_2, shows the opposite trend to that observed for CN^-: it binds tightly at potentials below –0.5 V but is released at higher potentials (Figure 8) [18, 19]. The effect is such that CODH functions as a unidirectional catalyst in the presence of cyanate, and CO oxidation is virtually unaffected apart from the requirement for a small overpotential before the activity climbs. The sharp potential dependence shows that NCO^- reacts selectively with

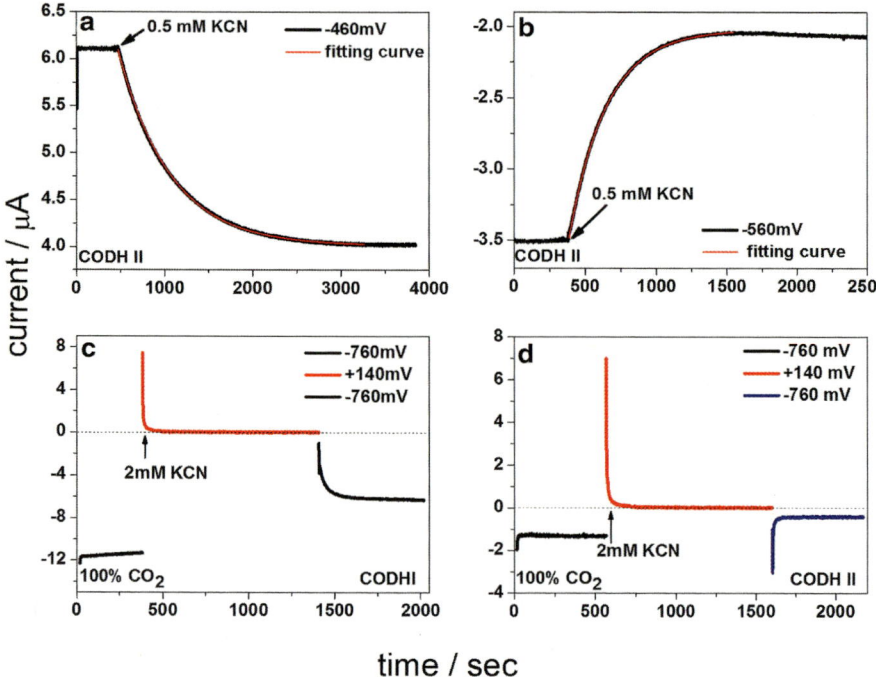

Figure 7 Chronoamperometric measurements of the inactivation (**a** and **b**) and re-activation (**c** and **d**) rate of cyanide-inhibited CODH I$_{Ch}$ (c) and CODH II$_{Ch}$ (a, b, d). The inactivation rate of CODH II$_{Ch}$ was measured at –460 mV (CO oxidation, (a)) and at –560 mV (CO$_2$ reduction, (b)). A final concentration of 0.5 mM cyanide in the electrochemical cell was used to measure the half-life time for inactivation. Cyanide release from CODH II$_{Ch}$ (d) at –760 mV is much faster than the instrumental response. Conditions: 25 °C, 0.2 M MES buffer (pH = 7.0), and rotation rate 3500 rpm. Reproduced by permission from [19]; copyright 2013 Wiley-VCH Verlag GmbH & Co. KGaA.

C$_{red2}$, the small overpotential reflecting the additional energy required to dislodge the inhibitor by oxidation.

The binding and release of CN$^-$ and NCO$^-$ are orders of magnitude slower than the catalytic turnover frequencies of their substrate counterparts CO and CO$_2$, thus showing that these analogues, while having very similar molecular shapes and size, fall far short of being really good analogues [19]. The differences must ultimately relate to acid-base (proton-transfer) properties, overall electrostatic effects and subtle effects of frontier orbitals. Comparisons between CODH I$_{Ch}$ and CODH II$_{Ch}$ show that rates of both inhibition by CN$^-$ and re-activation are higher for CODH II$_{Ch}$ even though it binds CN$^-$ more tightly in a thermodynamic sense, mirroring the stronger binding of CO that makes it such a potent product inhibitor of this isozyme [19].

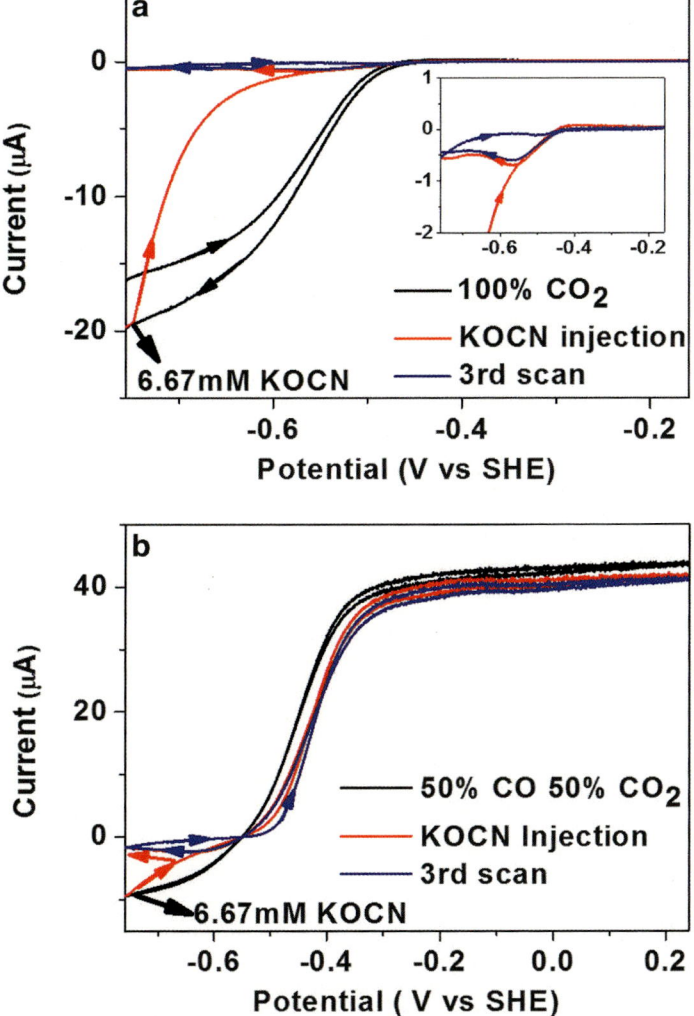

Figure 8 Inhibition of CODH I_{Ch} by cyanate. Potassium cyanate was injected (final concentration in the solution is 6.67 mM) into the electrochemical cell under gas atmospheres 100 % CO_2 (Figure 8a) and 50 % CO, 50 % CO_2 (Figure 8b). Conditions: 25 °C, 0.2 M MES buffer (pH = 7.0), rotation rate 3500 rpm and scan rate 1 mV s⁻¹. Reprinted by permission from [18]; copyright 2013 American Chemical Society.

5.3 How Class IV Carbon Monoxide Dehydrogenases Respond to Sulfide and Thiocyanate

Sulfide inhibits $CODH_{Ch}$ rapidly when the potential is more positive than –50 mV but no reaction is observed at more negative potentials, at which CO oxidation occurs unaffected by its presence [18]. Sulfide is not a competitive inhibitor like

CN^-, but instead reacts to stabilize an inactive state at a higher oxidation level (Figure 9). This state must be analogous to but not identical with C_{ox} because the reactivation potential of the μ-sulfido product is much more negative. It is therefore unlikely that sulfide can be an activator of CODH catalysis, as was suggested by earlier work [33, 40]. An interesting observation is made when thiocyanate, NCS^-, is introduced. In marked contrast to cyanate, NCO^-, NCS^- does not inhibit CO_2 reduction; instead there is partial inhibition of CO oxidation and an oxidized inactive state is formed that does not activate at the same potential as C_{ox} (in CODH I), but at a more negative potential, closer to that observed when sulfide is present. One possibility is that NCS^- also reacts to stabilize an inactive oxidized state by leaving a bridging sulfur, either as sulfide or an intact S-bound NCS^-.

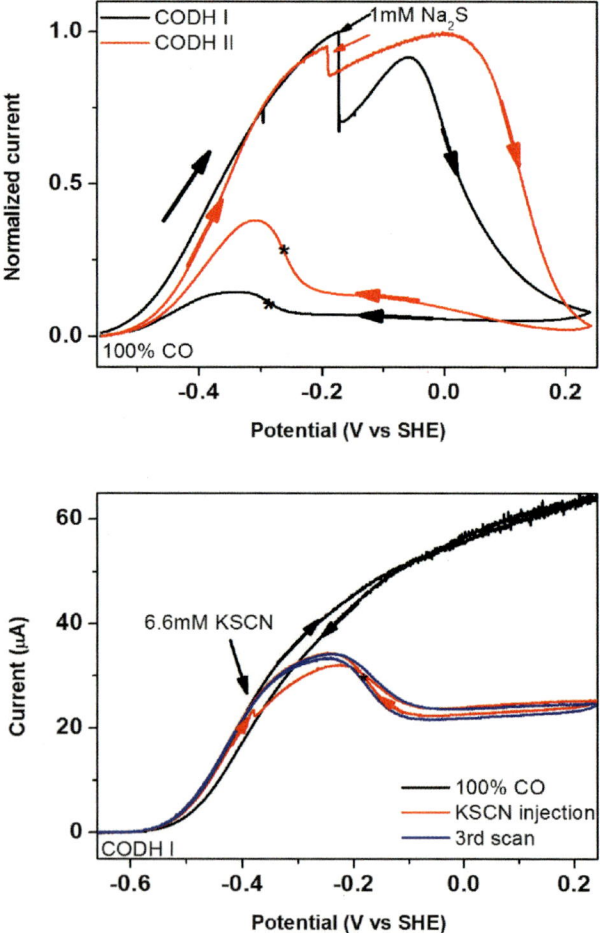

Figure 9 Reactions of CODH with sulfide and thiocyanate. **Upper Panel**: Cyclic voltammograms showing the reaction of CODH I$_{Ch}$ and CODH II$_{Ch}$ with sulfide: An aliquot of Na$_2$S stock solution (giving 1 mM final concentration) was injected into the electrochemical cell. The panel is reconstructed using data from [18] and [19]. **Lower Panel**: Cyclic voltammograms showing the inhibition of CODH I$_{Ch}$ by thiocyanate: An aliquot of KSCN stock solution (giving 6.6 mM final concentration) was injected into the electrochemical cell. Experimental conditions: 25 °C, 0.2 M MES buffer (pH 7.0), rotation rate 3500 rpm, scan rate: 1 mV s^{-1}.

The main differences between CODH I_{Ch} and CODH II_{Ch} are the stronger inhibition of CODH II_{Ch} by CO product and cyanide [19]. These discoveries may shed light on the possible role of CODH II_{Ch} in biological systems. Interestingly, sulfide is a similar inhibitor of [NiFe]-hydrogenases, binding to and stabilizing the inactive Ni(III) form rather than acting competitively in the active potential region. One aspect here is that sulfide is a good bridging ligand, which CN^- is not, so it can form the Ni-S-Fe linkage that is observed in some crystal structures.

Figure 10 summarizes the 'redox spectrum' for activities and inhibition of CODH I_{Ch}.

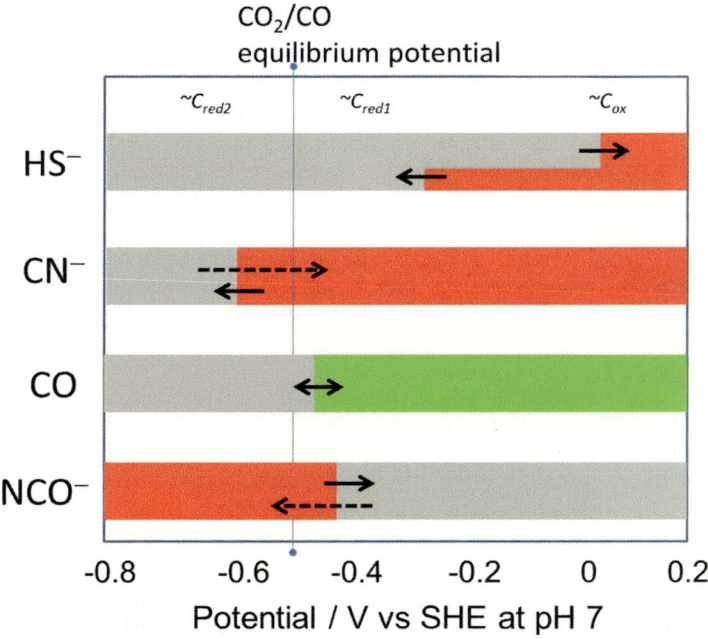

Figure 10 Potential dependence of binding of inhibitors to CODH I_{Ch}. In red potential regions, the agent is an inhibitor. In green potential zones, the agent is a substrate that is transformed. In gray zones, there is little or no interaction. The filled arrows indicate faster reactions compared to those indicated by dashed arrows. Reprinted with permission from [18]; copyright 2013 American Chemical Society.

6 Demonstrations of Technological Significance

The high electrocatalytic activities of Class IV CODHs match those observed with many hydrogenases – a fact that inspired an aesthetically interesting experiment to mimic the physiological process in which CO oxidation is coupled to H_2 evolution [59]. This process is the biological counterpart of the industrially important 'water gas shift' reaction that requires high temperatures. Notably, CO as a component of syngas, can be used to synthesize liquid hydrocarbons by reaction with H_2 in the Fischer-Tropsch process [60]. In the enzyme-coupled version, a hydrogenase (Hyd-2) from *Escherichia coli* and CODH I_{Ch} are both adsorbed onto conducting graphite platelets, formed by grinding a piece of pyrolytic graphite. In microorganisms, CO-reduced CODH I_{Ch} produces electrons that are transferred one at a time via a ferredoxin to a membrane-bound hydrogenase for H_2 evolution [48]. In the system of two enzymes on the graphite platelets in aqueous suspension, CO-reduced CODH releases electrons to Hyd-2 for H_2 evolution via conduction across the graphite platelets. The turnover frequency is 2.5 s^{-1} at 30 °C, which is comparable to catalysts operated at high temperatures in industry. Results are shown in Figure 11.

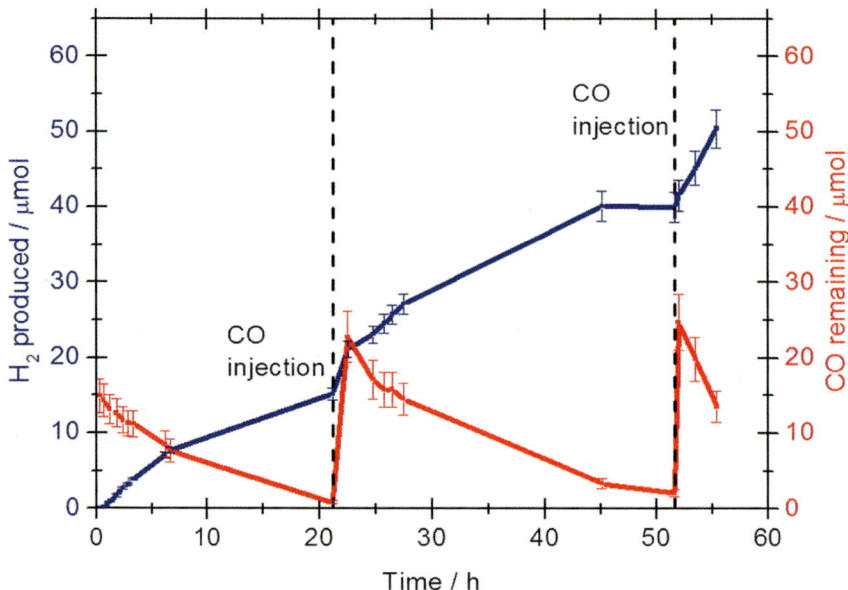

Figure 11 Water gas shift reaction catalyzed by Hyd-2 hydrogenase from *Escherichia coli* and CODH I_{Ch} co-adsorbed on conducting graphite platelets. Plot shows production of H_2 (blue line) and depletion of CO (red line) over the course of 55 h, with fresh CO injections of 600 μL at the points indicated. Reprinted with permission from [59]; copyright 2009 American Chemical Society.

The development of efficient photo-catalysts for CO_2 reduction plays an important role in converting CO_2 to carbon-based fuels in a clean and sustainable way. Experiments have been carried out in which CODH I_{Ch} is attached to semiconductor nanoparticles, such as TiO_2 and CdS, to study the visible light-driven reduction of CO_2 to CO as illustrated in Figure 12 [61, 62]. For TiO_2 (anatase) the bandgap is 3.1 eV which lies in the UV region, so a Ru-photosensitizer was co-attached to the nanoparticle along with CODH I_{Ch}. Upon illumination, the turnover frequency for CO production is 0.14 s^{-1} at pH 6 and 30°C. In the CdS system, a photosensitizer is not necessary because the band gap lies in the visible region. For the hybrid CODH-CdS nanorod system, the turnover frequency for CO production is 1.23 s^{-1} [62].

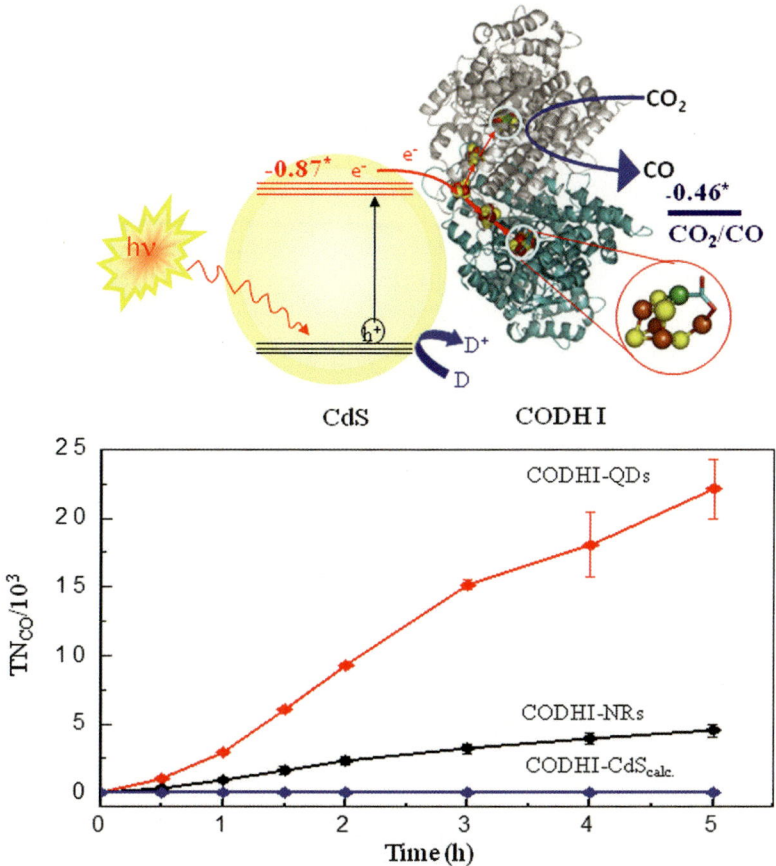

Figure 12 UpperPanel: Cartoon showing a hybrid system for light-driven CO_2 reduction using CODH I_{Ch} on CdS nanoparticles. Numbers refer to potentials in V *versus* SHE. D represents an electron donor. **LowerPanel**: Comparisons of the performance of CODH I_{Ch} adsorbed on various CdS nanoparticles: NR = nanorods, QD = quantum dots, CdS_{calc} = calcined sample. Adapted with permission from [62]; copyright 2012 Royal Society of Chemistry.

We have already discussed how CODH I_{Ch} and CODH II_{Ch} attached to graphite electrodes show high activities of CO oxidation and CO_2 reduction with a minimal overpotential [17]. The catalytic bias is largely determined by the fact that the reduction potential of the D-cluster lies in the region of the potential for the CO_2/CO redox couple [47]. When TiO_2 or CdS are used as electrode materials, the electrocatalytic activity is biased instead strongly in favor of CO_2 reduction, an observation that is explained by the potential dependence of electron availability in the material. Semiconductors have a low carrier density and, in simplistic physics terms, the availability of electrons for transfer into a catalyst is governed by the position of the flatband potential E_{FB}. Both TiO_2 and CdS are *n*-type semiconductors and a potential more negative than E_{FB} is required for electrons to become concentrated close to the surface and available for transfer. At potentials well above E_{FB} no current flows. Results comparing CODH I_{Ch}

Figure 13 Cyclic voltammograms of electrocatalysis by CODH I$_{Ch}$ adsorbed at PGE (**upper panel**), CdS thin film (**middle panel**), and TiO$_2$ thin film (**lower panel**) electrodes, scanned in 0.2 M MES, pH 6.0, scan rate 10 mV s^{-1}. Voltammograms recorded under 100 % CO$_2$ are depicted in blue, and those recorded under 50 % CO, 50 % CO$_2$ are shown in red. Adapted with permission from [63]; copyright 2013 American Chemical Society.

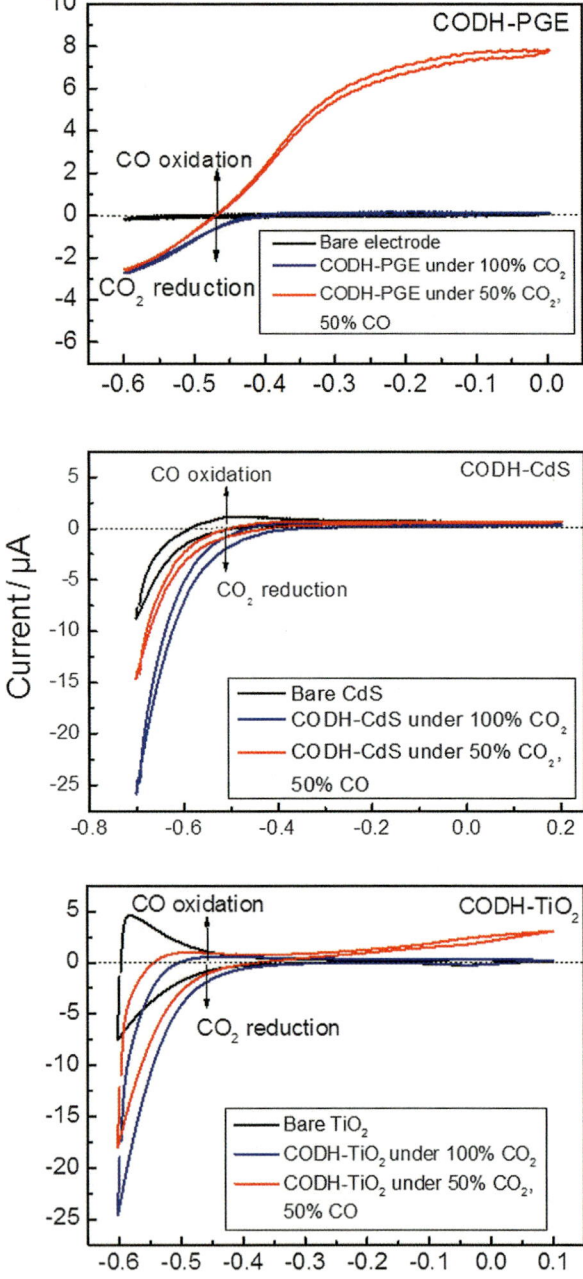

attached to PGE, CdS thin film and TiO_2 thin film electrodes are shown in Figure 13. In contrast to the result observed for PGE, only CO_2 reduction is observed at CdS and TiO_2.

The experiments demonstrate how to rectify reversible electrocatalysts for the fuel-forming direction since the most efficient catalysts for solar fuel production should operate close to reversible potentials [63].

7   Conclusions

The linear CO_2 molecule is difficult to reduce, but enzymes provide a superb example of how this chemistry has been perfected by evolution. The active site of CODH reveals simple tricks that could be mimicked by chemists, although it is going to be challenging to synthesize the precise, outer-shell environment that nature so easily rolls out during transcription. Protein film electrochemistry has proved to be a valuable tool for separating the different reactions as a smooth function of potential, as it has been demonstrated for hydrogenases. The PFE results have yielded important information – in the 'potential domain' – to complement and clarify the picture built up from structure and spectroscopy.

Abbreviations

acetyl-CoA	acetyl-coenzyme A
ACDS	acetyl-CoA decarbonylase synthase
Ch	*Carboxydothermus hydrogenoformans*
CoA	coenzyme A
CODH	carbon monoxide dehydrogenase
CODH/ACS	carbon monoxide dehydrogenase/acetyl-CoA synthase
CoFeSP	corrinoid-iron-sulfur protein
E_{FB}	flatband potential
EPR	electron paramagnetic resonance
Mb	*Methanosarcina barkeri*
MES	2-(N-morpholino)ethanesulfonic acid
Mt	*Moorella thermoacetica*
PDB	Protein Data Base
PFE	protein film electrochemistry
PGE	pyrolytic graphite edge
Rr	*Rhodospirillum rubrum*
SHE	standard hydrogen electrode
XAS	X-ray absorption spectroscopy

Acknowledgments The authors thank the UK Research Councils (BBSRC (grants H003878-1 and BB/I022309) and EPSRC (Supergen 5, EH/H019480/1)), and NIH (GM39451) for supporting their research. Vincent Wang thanks the Ministry of Education, Taiwan (R.O.C) for financial support through a scholarship for study abroad.

References

1. S. W. Ragsdale, E. Pierce, *Biochim. Biophys. Acta* **2008**, *1784*, 1873–1898.
2. J. G. Ferry, *Annu. Rev. Microbiol.* **2010**, *64*, 453–473.
3. E. Oelgeschlaeger, M. Rother, *Arch. Microbiol.* **2008**, *190*, 257–269.
4. M. Can, F. A. Armstrong, S. W. Ragsdale, *Chem. Rev.* **2014**, *114*, 4119–4174.
5. O. Meyer, H. G. Schlegel, *Annu. Rev. Microbiol.* **1983**, *37*, 277–310.
6. G. Bender, E. Pierce, J. A. Hill, J. E. Darty, S. W. Ragsdale, *Metallomics* **2011**, *3*, 797–815.
7. H. Dobbek, L. Gremer, R. Kiefersauer, R. Huber, O. Meyer, *Proc. Natl. Acad. Sci. USA* **2002**, *99*, 15971–15976.
8. J. H. Jeoung, H. Dobbek, *Science* **2007**, *318*, 1461–1464.
9. D. A. Grahame, *J. Biol. Chem.* **1991**, *266*, 22227–22233.
10. P. A. Lindahl, B. Chang, *Orig. Life Evol. Biosph.* **2001**, *31*, 403–434.
11. S. W. Ragsdale, *Crit. Rev. Biochem. Mol. Biol.* **2004**, *39*, 165–195.
12. B. Soboh, D. Linder, R. Hedderich, *Eur. J. Biochem.* **2002**, *269*, 5712–5721.
13. J. D. Fox, Y. P. He, D. Shelver, G. P. Roberts, P. W. Ludden, *J. Bacteriol.* **1996**, *178*, 6200–6208.
14. J. H. Jeoung, H. Dobbek, *J. Am. Chem. Soc.* **2009**, *131*, 9922–9923.
15. H. Dobbek, V. Svetlitchnyi, J. Liss, O. Meyer, *J. Am. Chem. Soc.* **2004**, *126*, 5382–5387.
16. V. Svetlitchnyi, H. Dobbek, W. Meyer-Klaucke, T. Meins, B. Thiele, P. Romer, R. Huber, O. Meyer, *Proc. Natl. Acad. Sci. USA.* **2004**, *101*, 446–451.
17. A. Parkin, J. Seravalli, K. A. Vincent, S. W. Ragsdale, F. A. Armstrong, *J. Am. Chem. Soc.* **2007**, *129*, 10328–10329.
18. V. C. C. Wang, M. Can, E. Pierce, S. W. Ragsdale, F. A. Armstrong, *J. Am. Chem. Soc.* **2013**, *135*, 2198–2206.
19. V. C. C. Wang, S. W. Ragsdale, F. A. Armstrong, *ChemBioChem* **2013**, *14*, 1845–1851.
20. Y. Kung, T. I. Doukov, J. Seravalli, S. W. Ragsdale, C. L. Drennan, *Biochemistry* **2009**, *48*, 7432–7440.
21. C. L. Drennan, J. Y. Heo, M. D. Sintchak, E. Schreiter, P. W. Ludden, *Proc. Natl. Acad. Sci. USA* **2001**, *98*, 11973–11978.
22. C. Darnault, A. Volbeda, E. J. Kim, P. Legrand, X. Vernede, P. A. Lindahl, J. C. Fontecilla-Camps, *Nat. Struct. Biol.* **2003**, *10*, 271–279.
23. P. A. Lindahl, *Angew. Chem. Int. Ed.* **2008**, *47*, 4054–4056.
24. P. A. Lindahl, *Biochemistry* **2002**, *41*, 2097–2105.
25. P. Amara, J. M. Mouesca, A. Volbeda, J. C. Fontecilla-Camps, *Inorg. Chem.* **2011**, *50*, 1868–1878.
26. W. W. Gu, J. Seravalli, S. W. Ragsdale, S. P. Cramer, *Biochemistry* **2004**, *43*, 9029–9035.
27. G. O. Tan, S. A. Ensign, S. Ciurli, M. J. Scott, B. Hedman, R. H. Holm, P. W. Ludden, Z. R. Korsun, P. J. Stephens, K. O. Hodgson, *Proc. Natl. Acad. Sci. USA* **1992**, *89*, 4427–4431.
28. C. Y. Ralston, H. X. Wang, S. W. Ragsdale, M. Kumar, N. J. Spangler, P. W. Ludden, W. Gu, R. M. Jones, D. S. Patil, S. P. Cramer, *J. Am. Chem. Soc.* **2000**, *122*, 10553–10560.
29. P. A. Lindahl, E. Munck, S. W. Ragsdale, *J. Biol. Chem.* **1990**, *265*, 3873–3879.
30. N. J. Spangler, P. A. Lindahl, V. Bandarian, P. W. Ludden, *J. Biol. Chem.* **1996**, *271*, 7973–7977.
31. J. L. Craft, P. W. Ludden, T. C. Brunold, *Biochemistry* **2002**, *41*, 1681–1688.
32. J. Seravalli, M. Kumar, W. P. Lu, S. W. Ragsdale, *Biochemistry* **1995**, *34*, 7879–7888.
33. S. W. Ha, M. Korbas, M. Klepsch, W. Meyer-Klaucke, O. Meyer, V. Svetlitchnyi, *J. Biol. Chem.* **2007**, *282*, 10639–10646.

34. S. A. Ensign, M. R. Hyman, P. W. Ludden, *Biochemistry* **1989**, *28*, 4973–4979.
35. W. Gong, B. Hao, Z. Wei, D. J. Ferguson, T. Tallant, J. A. Krzycki, M. K. Chan, *Proc. Natl. Acad. Sci. USA* **2008**, *105*, 9558–9563.
36. J.-H. Jeoung, H. Dobbek, *J. Biol. Inorg. Chem.* **2012**, *17*, 167–173.
37. M. E. Anderson, P. A. Lindahl, *Biochemistry* **1994**, *33*, 8702–8711.
38. M. E. Anderson, V. J. Derose, B. M. Hoffman, P. A. Lindahl, *J. Am. Chem. Soc.* **1993**, *115*, 12204–12205.
39. V. J. DeRose, J. Telser, M. E. Anderson, P. A. Lindahl, B. M. Hoffman, *J. Am. Chem. Soc.* **1998**, *120*, 8767–8776.
40. H. Dobbek, V. Svetlitchnyi, L. Gremer, R. Huber, O. Meyer, *Science* **2001**, *293*, 1281–1285.
41. J. Feng, P. A. Lindahl, *J. Am. Chem. Soc.* **2004**, *126*, 9094–9100.
42. K. A. Vincent, A. Parkin, F. A. Armstrong, *Chem. Rev.* **2007**, *107*, 4366–4413.
43. C. Léger, P. Bertrand, *Chem. Rev.* **2008**, *108*, 2379–2438.
44. K. A. Vincent, F. A. Armstrong, *Inorg. Chem.* **2005**, *44*, 798–809.
45. C. Léger, S. J. Elliott, K. R. Hoke, L. J. C. Jeuken, A. K. Jones, F. A. Armstrong, *Biochemistry* **2003**, *42*, 8653–8662.
46. F. A. Armstrong, J. Hirst, *Proc. Natl. Acad. Sci. USA* **2011**, *108*, 14049–14054.
47. S. V. Hexter, T. Esterle, F. A. Armstrong, *Phys. Chem. Chem. Phys.* **2014**, *16*, 11822–11833.
48. V. Svetlitchnyi, C. Peschel, G. Acker, O. Meyer, *J. Bacteriol.* **2001**, *183*, 5134–5144.
49. S. W. Ragsdale, J. E. Clark, L. G. Ljungdahl, L. L. Lundie, H. L. Drake, *J. Biol. Chem.* **1983**, *258*, 2364–2369.
50. W. S. Shin, P. A. Lindahl, *Biochim. Biophys. Acta* **1993**, *1161*, 317–322.
51. J. Y. Heo, C. R. Staples, C. M. Halbleib, P. W. Ludden, *Biochemistry* **2000**, *39*, 7956–7963.
52. M. J. Lukey, A. Parkin, M. M. Roessler, B. J. Murphy, J. Harmer, T. Palmer, F. Sargent, F. A. Armstrong, *J. Biol. Chem.* **2010**, *285*, 20421–20421.
53. J. Q. Xia, J. F. Sinclair, T. O. Baldwin, P. A. Lindahl, *Biochemistry* **1996**, *35*, 1965–1971.
54. W. Shin, P. A. Lindahl, *J. Am. Chem. Soc.* **1992**, *114*, 9718–9719.
55. S. W. Ragsdale, H. G. Wood, *J. Biol. Chem.* **1985**, *260*, 3970–3977.
56. M. Kumar, W. P. Lu, S. W. Ragsdale, *Biochemistry* **1994**, *33*, 9769–9777.
57. E. L. Maynard, P. A. Lindahl, *J. Am. Chem. Soc.* **1999**, *121*, 9221–9222.
58. J. Seravalli, S. W. Ragsdale, *Biochemistry* **2000**, *39*, 1274–1277.
59. O. Lazarus, T. W. Woolerton, A. Parkin, M. J. Lukey, E. Reisner, J. Seravalli, E. Pierce, S. W. Ragsdale, F. Sargent, F. A. Armstrong, *J. Am. Chem. Soc.* **2009**, *131*, 14154–14155.
60. O. O. James, A. M. Mesubi, T. C. Ako, S. Maity, *Fuel Process. Technol.* **2010**, *91*, 136–144.
61. T. W. Woolerton, S. Sheard, E. Pierce, S. W. Ragsdale, F. A. Armstrong, *Energy Environ. Sci.* **2011**, *4*, 2393–2399.
62. Y. S. Chaudhary, T. W. Woolerton, C. S. Allen, J. H. Warner, E. Pierce, S. W. Ragsdale, F. A. Armstrong, *Chem. Commun.* **2012**, *48*, 58–60.
63. A. Bachmeier, V. C. C. Wang, T. W. Woolerton, S. Bell, J. C. Fontecilla-Camps, M. Can, S. W. Ragsdale, Y. S. Chaudhary, F. A. Armstrong, *J. Am. Chem. Soc.* **2013**, *135*, 15026–15032.

Chapter 5
Understanding and Harnessing Hydrogenases, Biological Dihydrogen Catalysts

Alison Parkin

Contents

ABSTRACT 100
1 INTRODUCTION 100
2 DIHYDROGEN CYCLES AND HYDROGENASES 100
   2.1 The Global Dihydrogen Cycle 101
   2.2 Dihydrogen Cycling in Microbial Communities 102
   2.3 Solar Dihydrogen Economy 105
      2.3.1 Hydrogenase Photoelectrolysis Devices 107
      2.3.2 Photosynthetic Dihydrogen 109
3 [NiFe] HYDROGENASES 111
   3.1 Structure and Function 111
      3.1.1 Mechanisms for Dioxygen Tolerance 111
      3.1.2 [NiFeSe] Hydrogenases 114
   3.2 Biosynthesis 114
4 NICKEL-FREE HYDROGENASES: [FeFe] AND [Fe] ENZYMES 115
   4.1 [FeFe] Hydrogenase Structure and Function 115
   4.2 [FeFe] Hydrogenase Biosynthesis 117
   4.3 [Fe] Hydrogenases 117
5 INSIGHTS INTO HYDROGENASE MECHANISM
   FROM SMALL MOLECULE MIMICS 119
6 GENERAL CONCLUSIONS 121
ABBREVIATIONS AND DEFINITIONS 121
ACKNOWLEDGMENT 122
REFERENCES 122

A. Parkin (✉)
Department of Chemistry, University of York, Heslington, York, YO10 5DD, UK
e-mail: alison.parkin@york.ac.uk

© Springer Science+Business Media Dordrecht 2014
P.M.H. Kroneck, M.E. Sosa Torres (eds.), *The Metal-Driven Biogeochemistry of Gaseous Compounds in the Environment*, Metal Ions in Life Sciences 14, DOI 10.1007/978-94-017-9269-1_5

Abstract It has been estimated that 99 % of all organisms utilize dihydrogen (H_2). Most of these species are microbes and their ability to use H_2 as a metabolite arises from the expression of H_2 metalloenzymes known as hydrogenases. These molecules have been the focus of intense biological, biochemical, and chemical research because hydrogenases are biotechnologically relevant enzymes.

Keywords dihydrogen • [FeFe] • hydrogenase • hydrogen technology • [NiFe]

Please cite as: *Met. Ions Life Sci.* 14 (2014) 99–124

1   Introduction

For humankind, dihydrogen (H_2) is often discussed in terms of a 'fuel of the future'. In the context of biogeochemistry this description could not be more inappropriate: microbial H_2 metabolism is an ancient process catalyzed by enzymes known as hydrogenases. In modern scientific research these H_2-activating complexes are of interest not only because of their biological importance but also because of their technological relevance to the construction of a carbon-free energy economy.

Dihydrogen is an ideal replacement for fossil fuels: reaction with molecular oxygen (O_2) yields water as the only product, the energy per unit mass of fuel is high (specific enthalpy of combustion 120 MJ $(kg\ fuel)^{-1}$ for H_2 compared to 50 MJ $(kg\ fuel)^{-1}$ for natural gas [1]), and H_2-powered motor vehicle technology already exists. However, finding a sustainable and scalable method for H_2 production and a replacement for the precious metal platinum in H_2 fuel applications is challenging.

Hydrogenases are therefore studied in the context of understanding how to construct, tune, and utilize highly efficient and active H_2 catalysts in H_2 energy devices, with the added inspiration that these biological molecules are built from Earth abundant elements. A major issue is the O_2 reactivity of the enzymes, which is being explored using a variety of complementary biochemical techniques. This chapter will therefore introduce the H_2 enzymes and H_2 cycling in both a biological and technological context, to fully overview our current understanding of how both microbes and humans harness the chemical function of hydrogenases.

2   Dihydrogen Cycles and Hydrogenases

Hydrogenases are sub-classified into three different types based on the active site metal content, yielding the terms iron-iron hydrogenase, nickel-iron hydrogenase, and iron hydrogenase. All hydrogenases catalyze reversible H_2 uptake *in vitro*, but while the [FeFe] and [NiFe] hydrogenases are true redox catalysts, driving H_2 oxidation and proton (H^+) reduction (equation 1), the [Fe] hydrogenases catalyze the reversible heterolytic cleavage of H_2 shown by reaction (2).

$$H_2 \rightleftharpoons 2\,H^+ + 2\,e^- \tag{1}$$

$$H_2 \rightleftharpoons H^+ + H^- \tag{2}$$

The biochemical and chemical details of this activity are described in more detail in Sections 3, 4, and 5, while this section sets into context the biological and technological significance of enzymatic H_2 catalysis.

2.1   The Global Dihydrogen Cycle

It has been proposed that biological H_2 uptake is of such fundamental importance that it could have been a key reaction in the origin of life on Earth [2]. The inorganic elements required to construct the reaction centers of the H_2-oxidizing hydrogenases: sulfur, Ni, and Fe, would have been readily available in the oceanic hydrothermal vents where life may first have arisen [2]. The evolutionary relationship between the essential respiratory enzyme Complex I and [NiFe] hydrogenase also suggests that biological H_2 catalysis is an ancient process [3]. There is the possibility that H_2 availability in the early Earth atmosphere was as high as 30 % [4, 5]. This has large implications on the possible scale of biological carbon fixation in Earth's pre-photosynthetic era, because anaerobic methanogens, a major component of life on the planet during the Archaean period (2.5 billion years ago), produce methane by coupling hydrogenase-catalyzed H_2 splitting with carbon dioxide (CO_2) reduction (equation 3) [6, 7]. Indeed the same biochemistry may even be responsible for life elsewhere in the solar system [8].

$$CO_2 + 4\,H_2 \rightarrow CH_4 + 2\,H_2O \tag{3}$$

In the modern era, H_2 is not a major constituent of the atmosphere of Earth, with an average abundance of only approximately 0.5 parts-per-million in dry air (0.5 μmole H_2 per mole of air) [9]. This level varies with seasonality and geographical location, and comparing different studies is complex because the amount of atmospheric H_2 is so low that accurate measurement using gas chromatography is sensitive to even the metallic material of the air calibration cylinder [10].

Regardless of precise H_2 levels, across different atmospheric H_2 studies there is a consensus that both biological and abiological processes contribute significantly to the global H_2 cycle (Table 1). The largest source of environmental H_2 is the atmospheric photochemical process of hydrocarbon dissociation, and microbial H_2 production only ranks as the fourth largest contributor [9]. However, biological processes are the dominant sink for atmospheric H_2 illustrating that overall the most important physiological role of H_2 is as a biological fuel. Within microbial environments cellular processes which result in H_2 production are nearly always linked with either inter or intracellular H_2 uptake (Section 2.2).

Table 1 Major contributions to the global H_2 cycle[a].

	Ranges resulting from summarizing different studies [$Tg\ H_2\ yr^{-1}$][b]
Sources	
Fossil fuel combustion	11–20
Biomass and biofuel burning	8–20
Photochemical production	30–77
Biological production	3–12
Sinks	
Microbial soil deposition	55–88
Reaction with OH	15–19

[a]Table adapted from Vollmer et al. [9].
[b]The unit 'Tg' refers to teragram, 1×10^{12} g.

A future H_2 fuel economy may have a significant effect on atmospheric H_2 levels since any leaks from storage containers or inefficiencies in fuel conversion processes will give rise to an increase in the amount of H_2 in the air. This is often discussed in terms of the possible problems that could be caused by dihydrogen's properties as a greenhouse gas [5, 9, 10]. However, the function of H_2 as a virulence factor in disease-causing bacteria such as *Salmonella* is only just being explored [11, 12], and it is worth considering that the role of H_2 as a microbial fuel may also provide important guidance for future air quality legislation.

2.2 Dihydrogen Cycling in Microbial Communities

The term 'hydrogenase' was coined in 1931 by Stephenson and Stickland to describe enzymatic activity identified in bacteria isolated from river mud [13, 14]. The H_2 uptake property of different soil microbes remains a focus of research to this day [9, 15, 16], but H_2 activity is also an essential metabolic reaction in environmental niches ranging from the human gut [11] to photosynthetic algal or cyanobacterial mats [17], oceanic hydrothermal vents [18], and sulfidic geothermal springs [19]. We now know that hydrogenases are produced by microbes from the three domains of Life: Archaea ([Fe] and [NiFe] hydrogenases), Bacteria ([NiFe] and [FeFe] hydrogenases), and Eukarya ([FeFe] hydrogenases) [7, 20].

Given that the evolution of H_2 biocatalysis pre-dates photosynthesis, it is unsurprising to note that the majority of H_2-dependent metabolic processes occur under anaerobic conditions. For example in carbon monoxide (CO)-rich, O_2-free environments the bacterial species *Carboxydothermus hydrogenoformans* functions as a 'hydrogenogenic' organism, meaning it couples CO oxidation (equation 4), catalyzed by a carbon monoxide dehydrogenase enzyme (see also Chapter 2) with hydrogenase-catalyzed H_2 production, reaction (5) [21]. The net reaction (equation 6) is therefore the same as the commercial water-gas-shift process, and by 'wiring' the two redox enzymes together the bacteria harnesses the small exothermic energy output ($\Delta G^0_{298} = -28\,\text{kJ}\,(\text{mol CO})^{-1}$) to adenosine triphosphate (ATP) synthesis [21].

$$CO + H_2O \rightarrow CO_2 + 2\,H^+ + 2\,e^- \tag{4}$$

$$2\,H^+ + 2\,e^- \rightarrow H_2 \tag{5}$$

$$CO + H_2O \rightarrow CO_2 + H_2 \tag{6}$$

Alternatively, the acetogenic bacterium *Acetobacterium woodii* can grow using a H_2 and CO_2 mixture as the sole growth substrate because of the action of a bifunctional enzyme that contains both a hydrogenase and a formate dehydrogenase active site. Overall this H_2-dependent CO_2 reductase complex catalyzes the conversion of H_2 and CO_2 to formic acid, reaction (7). From an energy technology viewpoint this can be seen as a microbial method for converting H_2 into an easily storable liquid fuel via CO_2 sequestration [22].

$$H_2 + CO_2 \rightarrow HCOOH \tag{7}$$

Although most hydrogenases are not expressed or functional in oxygenic conditions, it is not the case that oxygenic hydrogenotrophic (H_2-converting) growth is impossible. For example, *Ralstonia eutropha* (*R. eutropha*) is a well-studied 'Knallgas' bacterium which relies on H_2 uptake (equation 8) in air when growing under autotrophic conditions, whereby H_2 is used as a source of energy via the activity of a membrane-bound [NiFe] hydrogenase (Figure 1) [23].

In terms of respiratory energy output hydrogenase O_2 functionality is very beneficial; H_2 is one of the strongest reducing agents (equation 5) with a standard reduction potential ($E^0_{298}(H^+/H_2)$) of 0 V and O_2 is one of the strongest oxidizing agents (equation 9) ($E^0_{298}(O_2/H_2O) = +1.23$ V). Coupling these processes across the cytoplasmic membrane therefore gives the cell access to a high membrane-potential and 'wiring' H_2 oxidation to O_2 reduction in such a way means that the bacterial respiration is analogous to a H_2/O_2 fuel cell.

$$H_2 \rightarrow 2\,e^- + 2\,H^+ \tag{8}$$

$$\tfrac{1}{2}\,O_2 + 2\,H^+ + 2\,e^- \rightarrow H_2O \tag{9}$$

$$H_2 + \tfrac{1}{2}\,O_2 \rightarrow H_2O \tag{10}$$

Dihydrogen is often such a valuable microbial resource that in mixed populations like the gut microbiota H_2-producing and H_2-uptake organisms will symbiotically complement one another, with the gaseous, fast-diffusing nature of H_2 promoting such *intra*cellular exchanges. The balance between the uptake and production parts of the gut H_2 cycle is important because fermentative bacteria generate H_2 as a method for disposing of reducing equivalents. Any H_2 buildup forces these microbes to

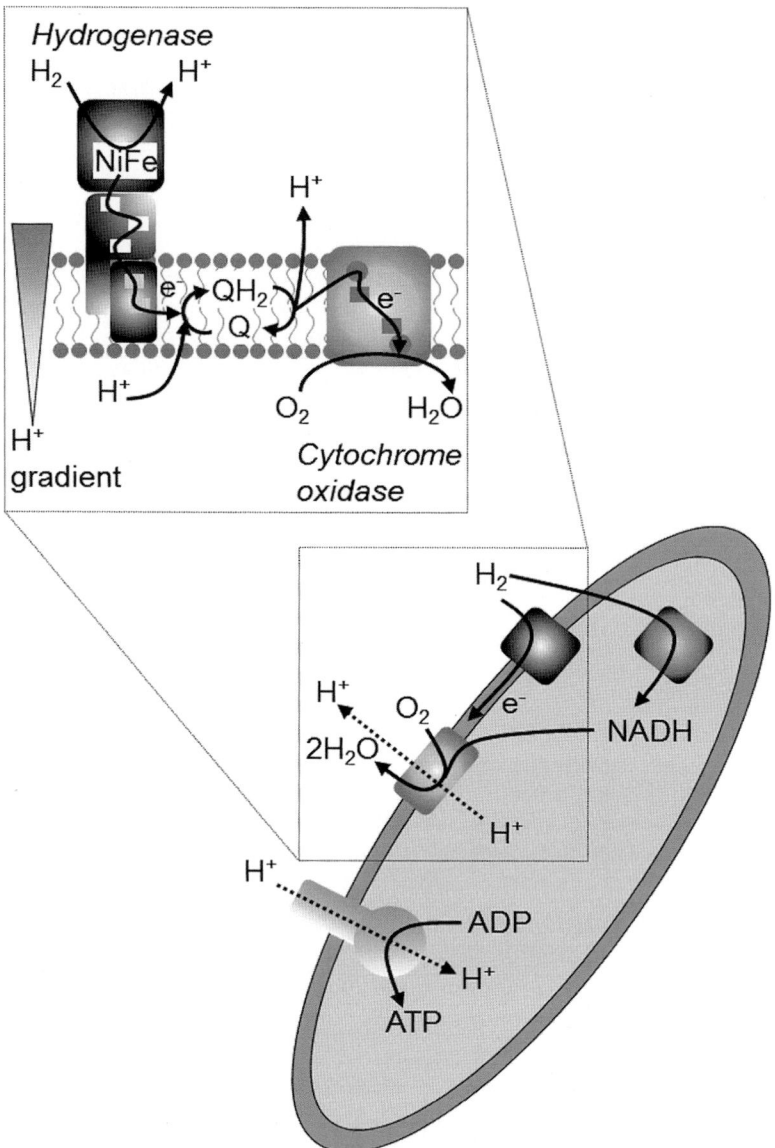

Figure 1 How oxygenic dihydrogen metabolism at a membrane-bound hydrogenase or a cytoplasmic bidirectional hydrogenase is coupled to dioxygen reduction at cytochrome oxidase contributing to the generation of a proton gradient which drives ATP synthesis in *Ralstonia eutropha*. The expanded section gives more detail on how the quinone pool mediates membrane electron transfer, where Q indicates quinone and QH_2 indicates dihydroquinone. Figure is inspired by [81] and [82].

accumulate reduced compounds such as ethanol, unbalancing the whole microbiota in a way that can result in human illness [24–26]. It is thought that their relative populations of methanogens, acetogens, and sulfate-reducing bacteria may even be explained by their hydrogenase activity [27].

*Inter*cellular H_2 cycling can also be important, with many H_2-producing microbes generating H_2-uptake hydrogenases to permit recycling of this potent fuel. The H_2 is not necessarily generated by a hydrogenase since both nitrogenases (N_2-fixing enzymes, see Chapter 7 in [84]) and hydrogenases produce H_2 when supplied with a source of electrons and protons. Cyanobacteria are photosynthetic organisms which produce hydrogenases and are capable of H_2 production when exposed to light after a period under anaerobic dark conditions. However, it is often a nitrogenase which catalyzes a large proportion of cyanobacterial H_2 production, and in these N_2-fixing strains a [NiFe] hydrogenase acts as a complementary H_2-uptake enzyme so that no net H_2 production is detected [17, 28]. Conversely, in *Escherichia coli* (*E. coli*), three [NiFe] hydrogenases have been characterized, and one is a H_2-producing enzyme while the other two recycle the gas (Figure 2) [29].

2.3 Solar Dihydrogen Economy

Although human cells do not have the appropriate deoxyribonucleic acid to express hydrogenases, thanks to our grasp of modern molecular biology techniques we have the ability to utilize these enzymes via microbial production and purification. A major driver of this work is to understand hydrogenases in the context of how to develop catalysts for a solar H_2 economy, an energy cycle which involves using the Sun's energy to split H_2O into H_2 and O_2 (Figure 3). Paradoxically humans often need least energy when the Sun shines the brightest, and producing H_2 is an effective method for storing solar energy in the form of a 'clean' fuel.

Harnessing solar power is desirable because Earth receives energy from the Sun at the rate of approximately 120,000 TW (1 TW = 1×10^{12} W = 1×10^{12} J s^{-1}), and when the human population reaches 10 billion it is projected that the rate of our energy use will be close to 20 TW [30, 31]. It is estimated that coupling current Si solar electricity technology with modern commercial water electrolyzers would give an overall solar energy H_2O splitting efficiency of 10–11 % [30]. This technology is therefore sustainable in terms of energetic and environmental considerations, but unfortunately it is not scalable or affordable because of the reliance of commercial electrolyzers on rare earth materials [32]. Proof-of-concept devices have been assembled to show that hydrogenases, which are constructed from common earth elements, can be utilized to replace platinum as the H_2-producing catalyst in solar H_2 devices. The two main approaches have been to either fabricate molecular photo-electrocatalysis devices using purified enzyme [33], or to harness living photosynthetic H_2-producing microorganisms. Sections 2.3.1 and 2.3.2 focus on how an understanding of hydrogenase biology and biochemistry can lead to advances in these methods of solar H_2 production. The challenges of H_2 storage are

E. coli H$_2$-uptake enzymes:

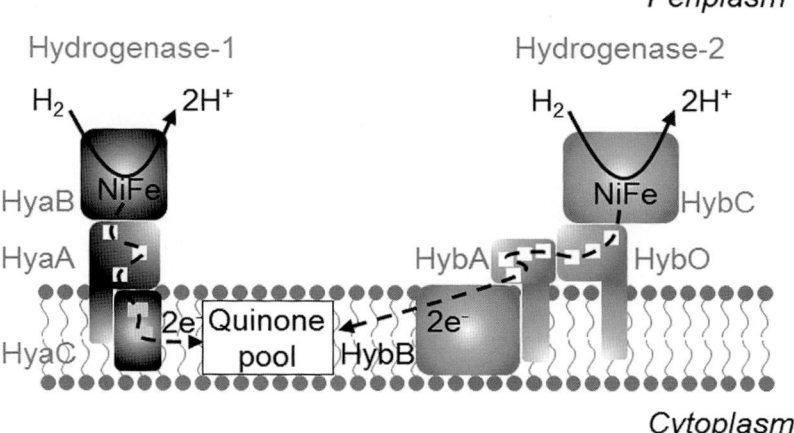

H$_2$-production formate hydrogenlyase complex:

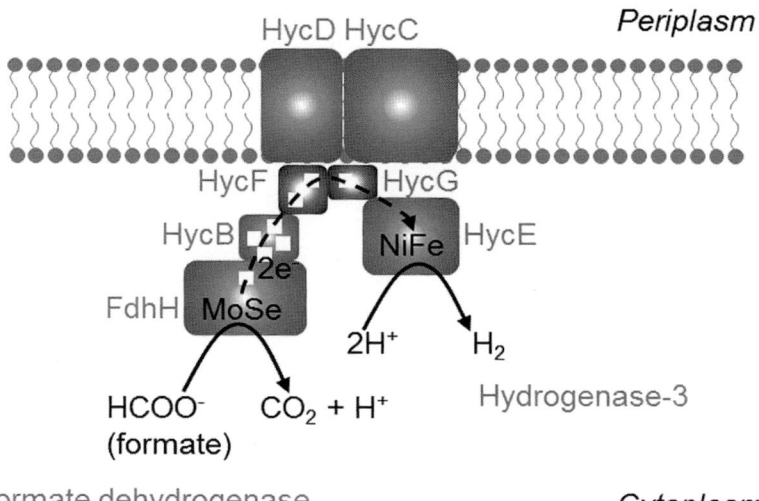

Figure 2 The enzymatic origin of microbial intracellular H$_2$ cycling within *Escherichia coli*, a bacterium capable of producing two membrane-bound dihydrogen uptake hydrogenases which extend into the periplasm (hydrogenase-1 and hydrogenase-2) and one dihydrogen-producing enzyme which extends into the cytoplasm. While hydrogenase-1 and -2 are 'wired' to the quinone pool, hydrogenase-3 is part of a larger formate hydrogenlyase complex. Adapted from [62].

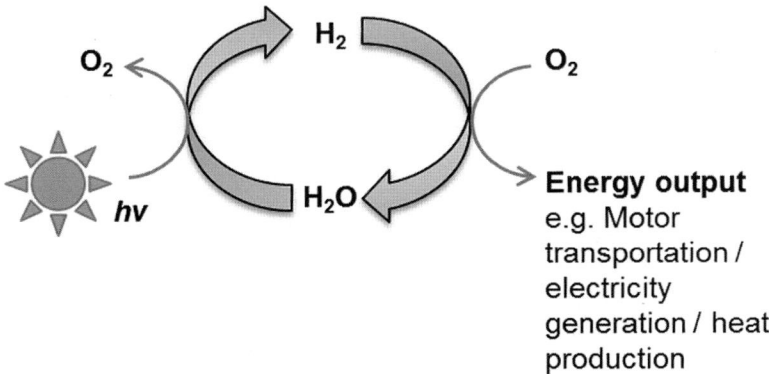

Figure 3 The fundamental notion of a possible future solar-H_2 economy. For more detailed images see [83].

not discussed, but it is acknowledged that this is problematic. Ultimately, linking H_2 production to CO_2 sequestration, or using H_2 as a carbon-free additive to natural gas may be necessary.

2.3.1 Hydrogenase Photoelectrolysis Devices

In solar fuel systems there are three key processes: (i) photoexcitation, (ii) reduction initiated by the photo-induced excited electron (e.g., H_2 production, equation 11), and (iii) oxidation initiated by the photo-induced hole (e.g., O_2 production, equation 12), see Figure 4.

$$2\,H^+ + 2\,e^- \rightarrow H_2 \tag{11}$$
$$2\,H_2O \rightarrow O_2 + 4\,H^+ + 4\,e^- \tag{12}$$

In molecular photoelectrolysis devices fuel-producing catalysts such as hydrogenases perform a dual role, both enhancing the rate of the reduction reactions and providing an interface to enhance electron-hole separation, thus preventing recombination. A key consideration in choosing an effective catalyst is the overpotential, the difference between the potential which must be applied to the catalyst for it to function at a significant rate and the equilibrium potential for the redox couple.

Because [NiFe] and [FeFe] hydrogenases catalyze the reduction of protons with a minimal overpotential, excited electrons only need to be at a reduction potential slightly more negative than $E^\circ(H^+/H_2)$ to drive H_2 production at a bimetallic hydrogenase. A variety of different visible-light photoexcitation centers have been directly coupled to [NiFe] or [FeFe] hydrogenases to drive H_2 production from water: organic dyes [34], biological photosystem I [35, 36], and ruthenium-dyes [37]. Alternatively, using external electrical circuitry, a carbon felt electrode with [FeFe] hydrogenase adsorbed onto the surface was wired to a porphyrin-sensitized nanoparticulate

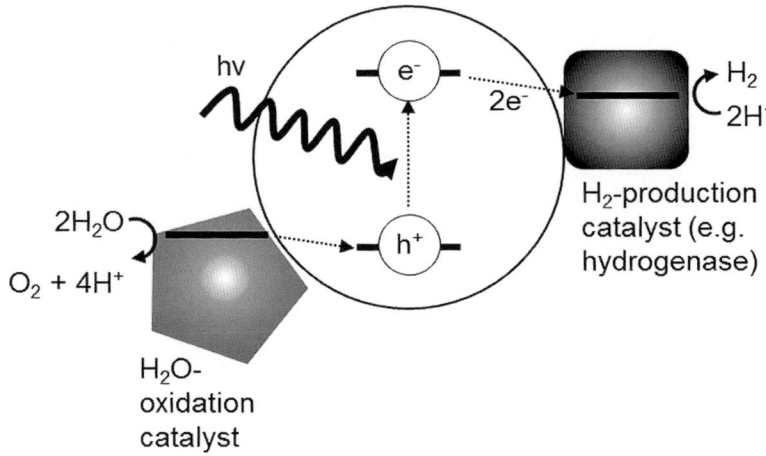

Figure 4 Schematic of molecular solar H_2-production using a hydrogenase.

titanium dioxide (TiO_2) electrode. The two electrodes were then placed into a container of electrolyte solution, separated by a proton-exchange membrane. When this photoelectrochemical cell was illuminated a photocurrent was induced as electrons flowed from the porphyrin-TiO_2 anode to the hydrogenase cathode [38].

The advantage of using hydrogenase is that rapid H_2 production rates are reported; for example, a H_2-catalyst:organic-dye system was nearly 700 times faster when a hydrogenase was used instead of a cobalt H_2 catalyst [34]. The problem with all of the photo H_2 hydrogenase experiments conducted to date is that the oxidation reaction, which completes the circuit by passing electrons to the photo-induced holes, has not been water oxidation; true water splitting has therefore not been achieved. Instead sacrificial electron donors have been used, such as reduced nicotinamide adenine dinucleotide (NADH) in the porphyrin-TiO_2 cell. There are several reasons why achieving photo-driven hydrogenase H_2 production *and* H_2O oxidation is challenging [33]. Hydrogenase-specific problems include the fact that the majority of hydrogenases are inactivated by O_2; a large amount of protein encases the active site so the maximum density of catalytic centers on a surface is low for a hydrogenase compared to a nano-particulate catalyst; there is no well-established methodology for 'wiring' hydrogenases onto surfaces [33]. To tackle these enzyme-related problems, solar H_2 devices incorporating hydrogenase-inspired synthetic analogues have also been constructed [33], and hydrogenase-mimics are described in more detail in Section 5.

Regardless of the nature of the H_2 catalyst, to achieve solar H_2O-splitting the photoexcitation process must generate both a photo-excited electron of electrochemical potential more negative than $E(2H^+/H_2)$, and a photo-induced hole of electrochemical potential more positive than $E(O_2/H_2O)$. Finding light-capture materials which both absorb light in a part of the spectrum that is not extensively blocked by the ozone layer and generate electrons and holes of the right energy to match water splitting is an ongoing challenge. To ensure an efficient rate of solar energy capture the rate of water oxidation must be fast. Finding efficient H_2O-oxidation catalysts which are built from earth-abundant elements and stable over a long time is difficult because water is such an inherently stable molecule and the product, O_2, is extremely reactive.

2.3.2 Photosynthetic Dihydrogen

Solar fuel production is the same as artificial photosynthesis and instead of building synthetic molecular H_2-producing devices an alternative approach is to re-tune microbial photosynthesis to achieve efficient solar H_2 production [39]. The advantage of using a living system is that it is capable of self-repair and self-replication and it can be assembled from inexpensive growth media, possibly even sewage. Hydrogen-producing, photosynthetic cyanobacterial and algal strains exist which can thrive under growth conditions that are inhospitable for other agricultural species, e.g., very high salt. Microbial solar H_2 culture sites would therefore not have to compete for areas of land or water supplies which could be used for farming [40, 41]. True water-splitting is achieved in oxygenic photosynthesis via the action of photosystem II which catalyzes H_2O oxidation to O_2 at a manganese center (Figure 5 and Chapter 2 in [84]). Because H_2 production utilizes electrons directly produced by the photosystems, the rate of fuel production is not limited by the rate of the CO_2-fixing Calvin cycle, which controls C-biofuel production (Figure 5). The challenge is to achieve high efficiency conversion of solar energy into H_2 fuel production rather than rapid rates of microbial growth.

As described in Section 2.2, cyanobacteria are [NiFe] hydrogenase-producing microbes which are capable of photosynthetic H_2 production. In non-N_2-fixing cyanobacterial strains like *Synechocystis* the bidirectional hydrogenase will produce H_2 under low O_2 using photosynthetically-generated nicotinamide adenine dinucleotide phosphate (NADPH) and also reduced ferredoxin (Figure 5) [28, 42] but at high O_2 levels the Calvin cycle dominates. One method for boosting cyanobacterial hydrogenase solar H_2 production is therefore to perturb the metabolic flux so that photo-generated NADPH is diverted away from CO_2-fixation and to the native hydrogenase instead [28]. This can be achieved by genetically manipulating the bacteria to over-express the hydrogenase and 20-fold increases in hydrogenase production have been achieved using this approach in *Synechocystis* [43]. Alternatively, because cyanobacterial H_2 production is inhibited by O_2, attempts are being made to genetically introduce new [NiFe] hydrogenases that function in air, such as the H_2 enzyme from the aerobe *R. eutropha* [28].

Compared to cyanobacteria the cellular organization of algal photosynthesis is slightly more complex because of the compartmentalization of the thylakoid membranes into chloroplasts. Rather than [NiFe] enzymes, algae produce [FeFe] hydrogenases and it is only the reduced ferredoxin which can act as the electron donating redox partner (Figure 5). In contrast to [NiFe] hydrogenases, all native [FeFe] hydrogenases are irreversibly inactivated by O_2 but they have a greater catalytic bias for H_2 production over H_2 oxidation. The most commonly studied algae for H_2 production is *Chlamydomonas reinhardtii*, and increases in photosynthetic H_2 production levels have been achieved by taking different approaches to overcoming O_2 inhibition. Firstly, if the hydrogenase machinery is left unchanged then

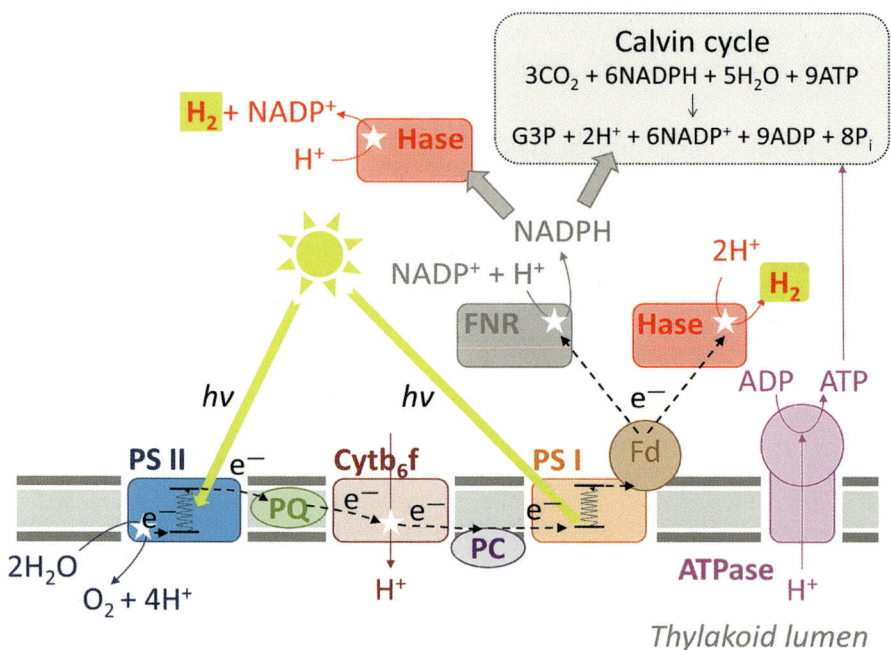

Figure 5 The possible routes for achieving photosynthetic hydrogenase (Hase) catalyzed dihydrogen production. Whereas some cyanobacterial species produce [NiFe] hydrogenases which can be reduced by both NADPH and reduced ferredoxin (Fd), the [FeFe] hydrogenases produced by some algal species only receive electrons from ferredoxin. In all cases the rate of biohydrogen production will not be limited by the Calvin cycle, unlike photosynthetic production of carbon fuels. Within the thylakoid membrane the light-driven oxygen-producing reaction which induces hydrogenase inhibition and down regulation is catalyzed by photosystem II (PS II), which is wired to the second solar excitation center, photosystem I (PS I), via the mobile redox carriers plastoquinone (PQ) and plastocyanin (PC) and the proton pumping protein complex cytochrome b_6f (Cytb$_6$f). Other abbreviations: 'ADP' is adenosine 5'-diphosphate; 'FNR' is ferredoxin-NADP$^+$ reductase; 'G3P' is glyceraldehyde 3-phosphate; 'ATPase' is adenylpyrophosphatase. For further details see [41], [40] and [39].

to boost solar H_2 activity internally induced anaerobicity must occur. When cell cultures are deprived of sulfur the rate of photosynthesis is reduced to the level of cellular respiration and the algae consume the O_2 produced by photosystem II at the rate it is made [41, 44]. Low O_2 conditions are therefore established leading to expression of the *hydA1* hydrogenase gene and sustained H_2 evolution at the expense of CO_2 assimilation. Alternatively, rather than using sulfur-restricted growth conditions, the same metabolic outcome has been induced by creating mutant cells with diminished sulfate transport activities [41]. A second approach is to re-engineer algal [FeFe] hydrogenases to make them O_2-insensitive, and this requires developments in screening methodologies as well as genetic experimentation.

Engineering cyanobacteria or algae to act as catalytic solar H_2 units requires a 'systems biology' approach, e.g., the biology of the whole cell must be considered. We need an in depth understanding of the genetic and biochemical control of hydrogenase assembly, a chemical understanding of the reactivity and redox partners of different hydrogenases, and a big picture model of metabolic flux throughout the whole microbe. The following sections will review our current understanding of hydrogenase biosynthesis and biochemistry.

3 [NiFe] Hydrogenases

3.1 Structure and Function

Based on phylogenetic analysis and biological functionality there are six distinct groups of [NiFe] hydrogenases (Groups 1, 2a, 2b, 3, 4, and 5) [15, 20]. Although structural studies on isolated enzymes have focused on the Group 1 periplasmic membrane-bound uptake enzymes, all [NiFe] hydrogenases appear to contain the same well conserved bimetallic active site architecture shown in Figure 6. From X-ray crystallography we know that the Ni is anchored to the protein via four cysteine (Cys) ligands, two are terminally coordinated and two bridge the bimetallic center [45]. Electron paramagnetic resonance (EPR) studies have shown that the Ni oxidation state changes during the catalytic cycle, with accompanying changes in the nature of the ligand bound in the fifth Ni coordination site. In a hydroxide-bound Ni^{3+} state the enzymes are oxidatively inactivated, but upon reductive activation H_2 can bind to the Ni^{2+} state to yield a Ni^{3+}-H^- hydrido complex. Conversely, the active site Fe remains in the +2 EPR-silent oxidation state although the Fourier transform infrared (FTIR) signature of the one carbon monoxide (CO) and two cyanide (CN^-) ligands can be used to follow the formation of reaction intermediates.

The [NiFe] site is buried within a large polypeptide chain and the minimal functional unit of a [NiFe] hydrogenase consists of this active site subunit bound to a second, smaller protein that contains FeS center(s) that 'wire' the H_2 catalytic center to external redox partners (Figure 6). In light of the possible application of using hydrogenases in solar fuel devices much of the recent work on the [NiFe] enzymes has focused on learning how these enzymes can function in air. The only [NiFe] hydrogenase sub-class which does not contain examples of aerobically functional enzymes is Group 4, the membrane-associated, H_2-evolving multi subunit [NiFe] hydrogenases. In Groups 1, 3, and 5 the O_2 tolerance (the ability to sustain H_2 catalysis in the presence of O_2) is thought to originate due to the structure of the electron transfer relay.

3.1.1 Mechanisms for Dioxygen Tolerance

Oxygen-sensitive Group 1 [NiFe] hydrogenases contain a chain of three iron sulfur (FeS) clusters, a central [3Fe4S] site flanked by two [4Fe4S] centers. Upon O_2 exposure all H_2 catalytic activity is lost and a "Ni-A" Ni^{3+} state can be formed which is very slow to reactivate due to the presence of reactive oxygen species. In

a   [NiFe] hydrogenase large and small subunit composition

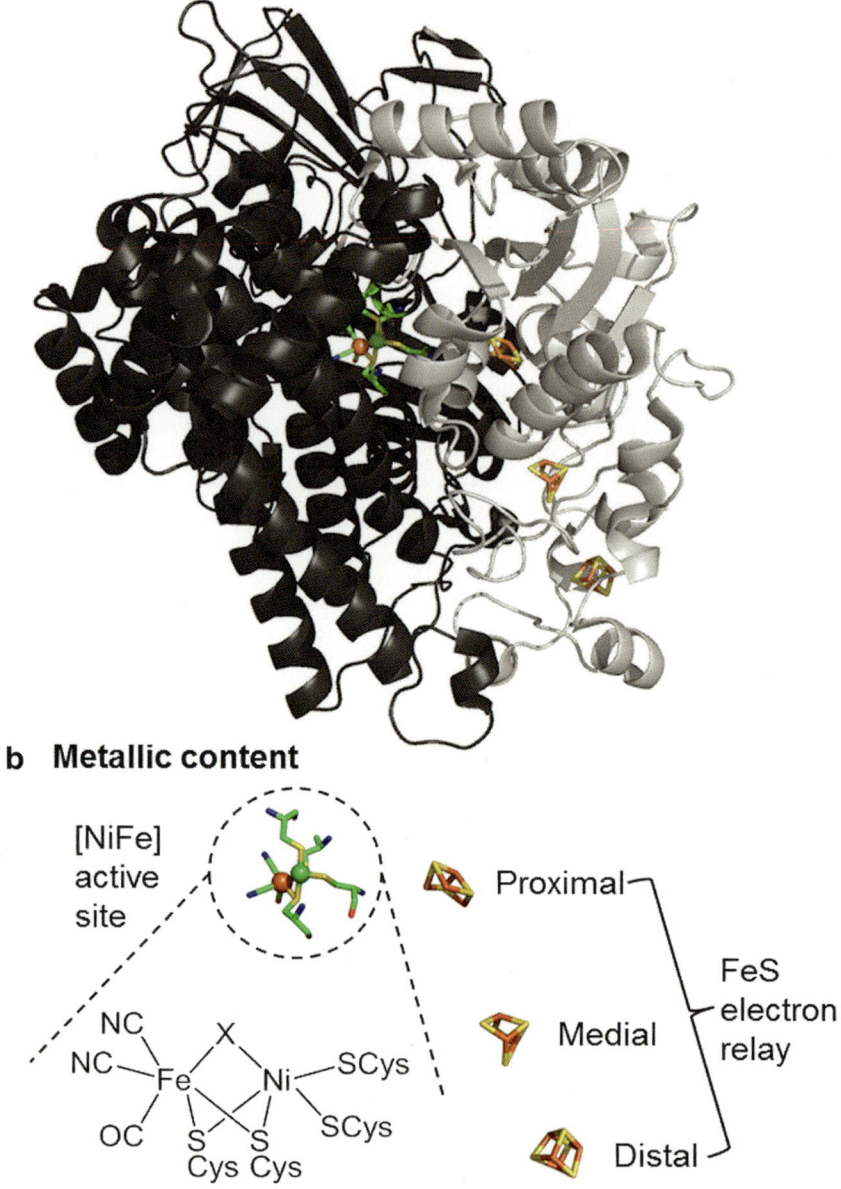

b   Metallic content

Figure 6 The structure of a [NiFe] hydrogenase. (**a**) Depicts the minimal functional unit for activity, consisting of the large (dark grey ribbon) protein subunit encasing the bimetallic active site and the small (light grey ribbon) protein subunit encasing the FeS relay. Image was generated using the *Escherichia coli* hydrogenase-1 structure, PDB code 3UQY. (**b**) A more detailed view of the metallic centers. Color code: yellow depicts sulfur; blue depicts nitrogen; bright red depicts oxygen; green ball depicts Ni, green sticks depict hydrocarbon; orange-red ball and sticks depict iron.

contrast, to accompany the [3Fe4S] and [4Fe4S] clusters which sit medial and distal from the [NiFe] active site, O_2-tolerant Group 1 hydrogenases contain a novel [4Fe3S] site which is closest ('proximal') to the H_2 reaction center [46, 47]. It is thought that the ability of the [4Fe3S] to access three different redox states ensures that electrons can be rapidly provided to 'neutralize' inhibitory O_2 via a four-electron four-proton reduction to yield two H_2O (equation 9). The O_2-tolerant Group 1 [NiFe] hydrogenases therefore only form the rapidly reactivating "Ni-B" OH^--bound inhibited Ni^{3+} state following exposure to O_2 in the presence of H_2.

Protein film electrochemistry has been a particularly valuable tool with which to dissect the O_2 tolerance mechanism of these enzymes. Recent experiments using this technique have also shown that a Group 1 hydrogenase which functions in high O_2 will become an efficient H_2 producer below pH 4 [48]. This is a transformative result because O_2-tolerant Group 1 [NiFe] hydrogenases show very little H_2 production activity at neutral pH, therefore to date all biotechnological applications have used the enzymes as H_2-oxidizing catalysts. The ideas developed in generating optimized enzyme-carbon electrodes for membrane-free H_2/O_2 fuel cells, including cross linking to 3-D carbon nanotube surfaces [49, 50], can now be applied in developing new Group 1 O_2-tolerant solar H_2 enzyme devices for operation at pH ≤ 4.

The 'Group 5' sub-class of [NiFe] hydrogenases were first categorized in 2010 and in human terms these are therefore the newest [NiFe] enzymes. They were originally identified in *Streptomyces* species as enzymes with the unique ability to oxidize atmospheric H_2, e.g., they are functional in 20 % O_2 with an apparent Michaelis constant for H_2 of less than 100 ppm by volume [15, 51, 52]. The aerobic functionality of these enzymes is so impressive that the term 'O_2-insensitive' rather than 'O_2-tolerant' has been used to describe these hydrogenases [53]. From the amino acid sequence, these enzymes are again expected to contain three FeS clusters, but only [4Fe4S] centers are predicted [53]. The distal cluster is thought to be coordinated by three Cys and one histidine, which is also standard for a Group 1 [NiFe] hydrogenase. More usually for a [4Fe4S] site, the medial cluster is predicted to be ligated by four Cys. The proximal cluster shows no evidence of additional Cys to stabilize a [4Fe3S] site, and instead a [4Fe4S] center coordinated by three Cys and one aspartate is proposed. There are currently no crystal structures of these enzymes, and no EPR, FTIR or electrochemical data, and these enzymes will doubtless be a hot topic of hydrogenase research in the coming years.

In Group 3 [NiFe] hydrogenases, the physiologically bidirectional enzymes, there are examples of hydrogenases which are also functional in O_2 [54]. In parallel with the Group 1 [NiFe] hydrogenases, the O_2-tolerance mechanism again originates from the ability of the enzyme to act as an oxidase, reducing O_2 to H_2O or H_2O_2 [54]. Elucidating the mechanism of this reactivity is complicated because the hydrogenase forms a super-complex with a second enzyme to generate a bifunctional unit capable of catalyzing H_2 activation or evolution under physiological conditions, with concerted binding and reduction or oxidation of additional soluble substrates like coenzyme F420 or NAD(P)H [55]. In the Group 3 hydrogenases from *Synechocystis* sp. PCC6803 and *R. eutropha* there are 8 FeS clusters including 2Fe sites, therefore deconvoluting the reactivity of these centers is not trivial [54, 56].

In contrast to FeS-mediated O_2 tolerance, the Group 2b [NiFe] hydrogenase H_2 sensor enzymes are able to control the transcription of hydrogenase genes in aerobic H_2-oxidizing bacteria because they contain a narrow gas channel which blocks the access of O_2 to the active site [46]. This was proved by decreasing the O_2 tolerance of *R. eutropha* enzyme variants containing wider gas channels. Attempts to engineer O_2 tolerance in an O_2-sensitive Group 1 enzyme via mimicry of this blocking mechanism were partially successful [57] and the same approach has been taken to try and make [FeFe] hydrogenases less sensitive to O_2 inhibition [58]. In addition to gas channels, it is also possible that water and proton channels may assist in controlling O_2 tolerance mechanisms in hydrogenases which reduce inhibitory O_2 to H_2O [46]. This idea arises from amino acid and structural comparisons which show that conserved residues appear to form these channels.

3.1.2   [NiFeSe] Hydrogenases

When grown on a selenium-containing medium, some sulfate-reducing or methanogenic microorganisms are able to construct hydrogenases which encode a selenocysteine (SeCys) rather than a standard S-Cys as a ligand to the Ni in the active site of Group 1 or Group 3 [NiFe] hydrogenases [59]. These enzymes are designated [NiFeSe] hydrogenases. In the absence of Se, the [NiFeSe]-producing microbes will generate a 'normal' Cys-containing [NiFe] hydrogenase homologue, but when Se is available the genetic regulation instead favors expression of the seleno H_2 enzyme, suggesting a biochemical advantage to expressing the [NiFeSe] enzyme. Corresponding to this result, in dye assays the activity of [NiFeSe] hydrogenases are higher than for [NiFe] hydrogenases.

Electrochemical studies of Group 1 [NiFeSe] hydrogenases have also shown a greater catalytic bias towards H_2 production over H_2 oxidation, which is unusual for a Group 1 [NiFe] enzyme [60]. The H_2-producing activity is also sustained in the presence of O_2, although the H_2 uptake activity is not [60]. The [NiFeSe] H_2 enzymes have therefore been applied in solar H_2-producing devices but not H_2/O_2 fuel cells [34, 37]. Relative to their [NiFe] counterparts, the Group 1 [NiFeSe] hydrogenases contain a different electron transport relay as well as a different active site. However, the enzymatic reactivity is unchanged when [NiFe] hydrogenase variants are constructed which contain the same FeS cluster content as a [NiFeSe] enzyme (three [4Fe4S] sites) [61]. This suggests that the Se plays a predominant role in tuning the novel activity of [NiFeSe] hydrogenases.

3.2   *Biosynthesis*

The maturation of [NiFe] hydrogenases has been most extensively studied for the membrane-bound hydrogenase (MBH) enzyme from *R. eutropha* and hydrogenase-1 from *E. coli*, both of which are Group 1 O_2-tolerant enzymes. The fact that the *E. coli* enzyme is only expressed under anaerobic conditions, whereas the *R. eutropha* enzyme is constructed in air, provides a useful insight into the additional complexities of assembling a hydrogenase in the presence of O_2 [46, 47, 62]. The consensus is that assembly of the [NiFe] active site is achieved via six key accessory proteins which are expressed from *hyp* assembly genes [63]. First the CN^- ligand is synthesized from carbamoyl phosphate via a HypE and HypF ATP-dependent process. Two CN^- are then integrated into a Fe-CO HypC-HypD complex, which may synthesize the CO from CO_2; alternatively, exogenous ^{13}CO can also be incorporated into the hydrogenase assembly [63, 64]. The $Fe(CN^-)_2(CO)$ complex is then delivered to the large protein subunit where it is anchored by the Cys ligands which bridge the bimetallic active site in the final hydrogenase product. Lastly, the Ni (transported into the cell via a 'Nik' transporter) is delivered to the large subunit via the action of HypA and HypB. The Ni is locked into place via Cys ligation.

Relative to O_2-sensitive Group 1 [NiFe] hydrogenases, assembly of *E. coli* hydrogenase-1 also depends on HyaE and HyaF, assembly proteins predicted to play a role in proximal cluster [3Fe4S] synthesis [47]. In contrast, a set of 6 further *hox* genes are required for aerobic assembly of *R. eutropha* MBH. Proteins HoxL and HoxV are thought to participate in active site cluster assembly, but HoxO, HoxQ, and HoxR play a role in the assembly and incorporation of FeS clusters [46, 65].

4 Nickel-Free Hydrogenases: [FeFe] and [Fe] Enzymes

4.1 [FeFe] Hydrogenase Structure and Function

The minimal functional unit for an algal [FeFe] hydrogenase is just the 'H-cluster' active site embedded in a single protein chain. However, the majority of bacterial [FeFe] hydrogenases more strongly resemble the [NiFe] hydrogenases, with a requirement for additional FeS clusters to conduct electrons between the active site to the surface of the protein (Figure 7) [45]. Putative gas and water channels have also been identified.

The name 'diiron' hydrogenase is misleading because the catalytic H-cluster actually comprises six Fe atoms, four of which form a standard [4Fe4S] cluster that is directly connected to the [FeFe] site via a S-Cys linkage. The separate Fe atoms of the [FeFe] center are labelled 'proximal' (Fe_P) or 'distal' (Fe_D) to denote their position relative to the [4Fe4S] cluster. A disulfur bridge-head dithiomethylamine ligand links the Fe_P and Fe_D atoms, and coordination of the Fe_P site is completed by one terminal CO and one terminal CN^- and a bimetallic bridging CO which moves towards Fe_D upon enzymatic reduction. Including the [4Fe4S] linkage the Fe_P is therefore coordinated by a total of 6 ligands. In addition to the bridging ligands, the Fe_D site is also thought to be able to hold up to three further ligands, one terminal CN^- and one terminal CO are integral to the enzyme structure, and the final coordination site can hold H_2O/OH^- or inhibitory CO, and is also proposed as the initial binding position for incoming H_2 [45].

a   [FeFe] hydrogenase

b   H-cluster

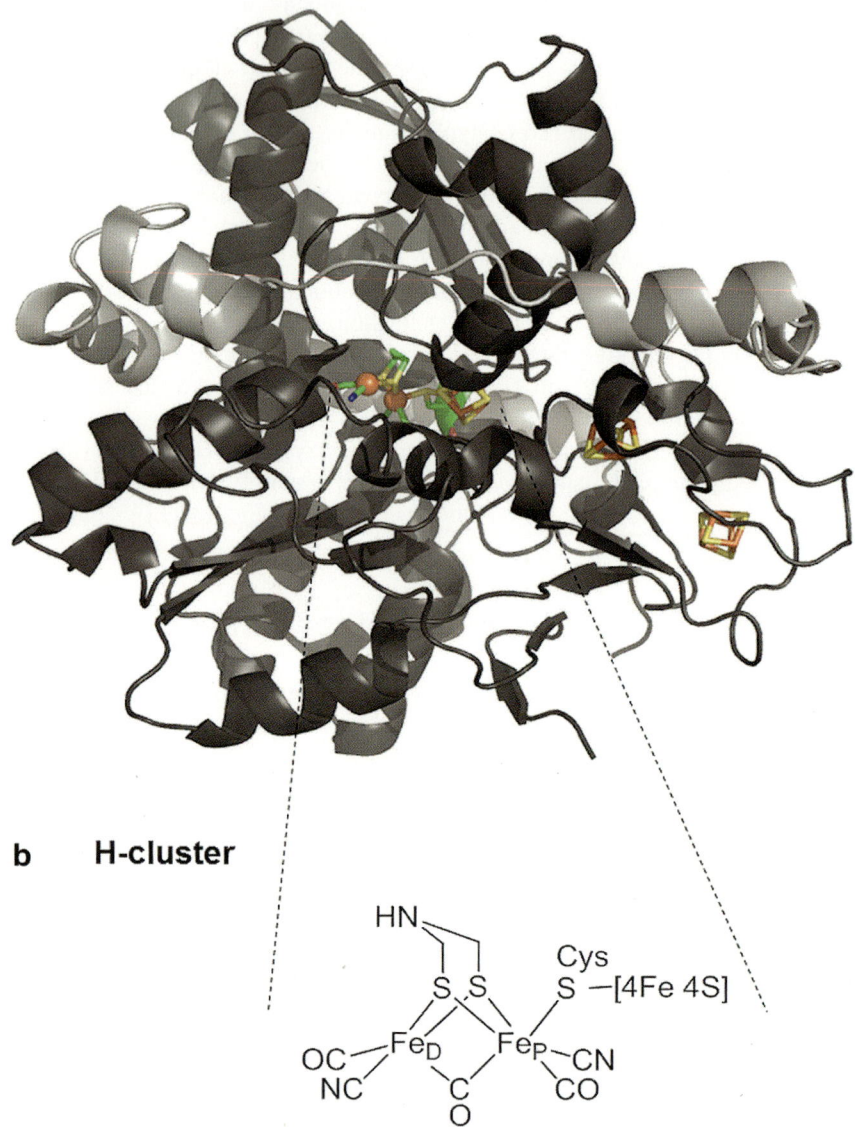

Figure 7 (**a**) The structure of the [FeFe] hydrogenase from *Desulfovibrio desulfuricans*, PDB code 1HFE. This enzyme is analogous to the [NiFe] hydrogenase shown in Figure 6 as it contains three iron sulfur clusters. Smaller [FeFe] hydrogenases are known which only require the 6Fe 'H-cluster' for activity. This enzyme is also a heterodimer, although other [FeFe] hydrogenases have a monomeric minimal functional unit. The large (dark grey ribbon) protein subunit co-ordinates all the metal centers. Color code: yellow depicts sulfur; blue depicts nitrogen; bright red depicts oxygen; green depicts hydrocarbon; orange-red ball and sticks depict iron. (**b**) The chemical structure of the H-cluster active site.

The highly conserved nature of the H-cluster is reflected in the fact that different [FeFe] enzymes have similar FTIR and EPR spectroscopic properties, and the 'H_{ox}' mixed valence $[4Fe4S]^{2+}(Fe_P^{1+}Fe_D^{2+})$ state, which will bind H_2, can be used to fingerprint all [FeFe] hydrogenases analyzed to date. Both the [4Fe4S] and [FeFe] components of the H-cluster are redox-active, and a particular challenge has been to understand the states formed under highly reducing conditions. Electrochemical studies have complemented the structural and spectroscopic characterizations and comparisons of inhibition by CO and formaldehyde as a function of potential particularly highlights how the enzyme reactivity can change dramatically at very low potentials [66]. A 'super-reduced'$[4Fe4S]^{1+}(Fe_P^{1+}Fe_D^{1+})$ is thought to play a role in the catalytic cycle [45].

The [FeFe] hydrogenases are incredibly active, for example in a H_2 production assay the *Clostridium pasteurianum* enzyme had a measured turnover rate of 3400 μmol min^{-1} mg^{-1} (in 1 min 1 g of enzyme would produce 83 L of atmospheric pressure H_2) [45]. This high activity correlates with irreversible inhibition by O_2, which damages the [4Fe4S] cluster of the active site rather than the [2Fe] center [67]. Although O_2 binding can be blocked by CO inhibition, there is currently no proposed mechanism for how to re-engineer an [FeFe] hydrogenase to 'neutralize' inhibitory O_2 and therefore remain active in air.

4.2 [FeFe] Hydrogenase Biosynthesis

The mechanism of [FeFe] hydrogenase biosynthesis is well understood and it is possible to engineer *E. coli* to produce a fully mature [FeFe] hydrogenase by transferring into the cell just four genes [45]. First, *hydA* encodes the apo-protein and this will be assembled to contain the [4Fe4S] portion of the H-cluster due to the action of the standard *E. coli* FeS "housekeeping" proteins. Second, the genes *hydE*, *hydF*, and *hydG* are required because these encode the specific maturases which synthesize and attach the [2Fe] part of the H-cluster to the [4Fe4S] cubane. The CO and CN$^-$ ligands are synthesized from tyrosine via the action of the radical S-adenosyl methionine enzyme HydG [45, 68].

4.3 [Fe] Hydrogenases

The [Fe] hydrogenases were only discovered in 1990 and they have also been labelled as 'H_2-forming methylenetetrahydromethanopterin dehydrogenases' and 'iron-sulfur cluster-free' hydrogenases [7]. These titles reveal key details about both the function and structure of these enzymes. First, the cytochrome-free hydrogenotrophic methanogens which express this enzyme use it to catalyze the reversible transfer of hydride from H_2 to methylenetetrahydromethanopterin (Figure 8). This reaction is one step in the overall metabolic reduction of CO_2 to methane. Secondly, the only metal center in each protein subunit is the single iron

a [Fe] hydrogenase

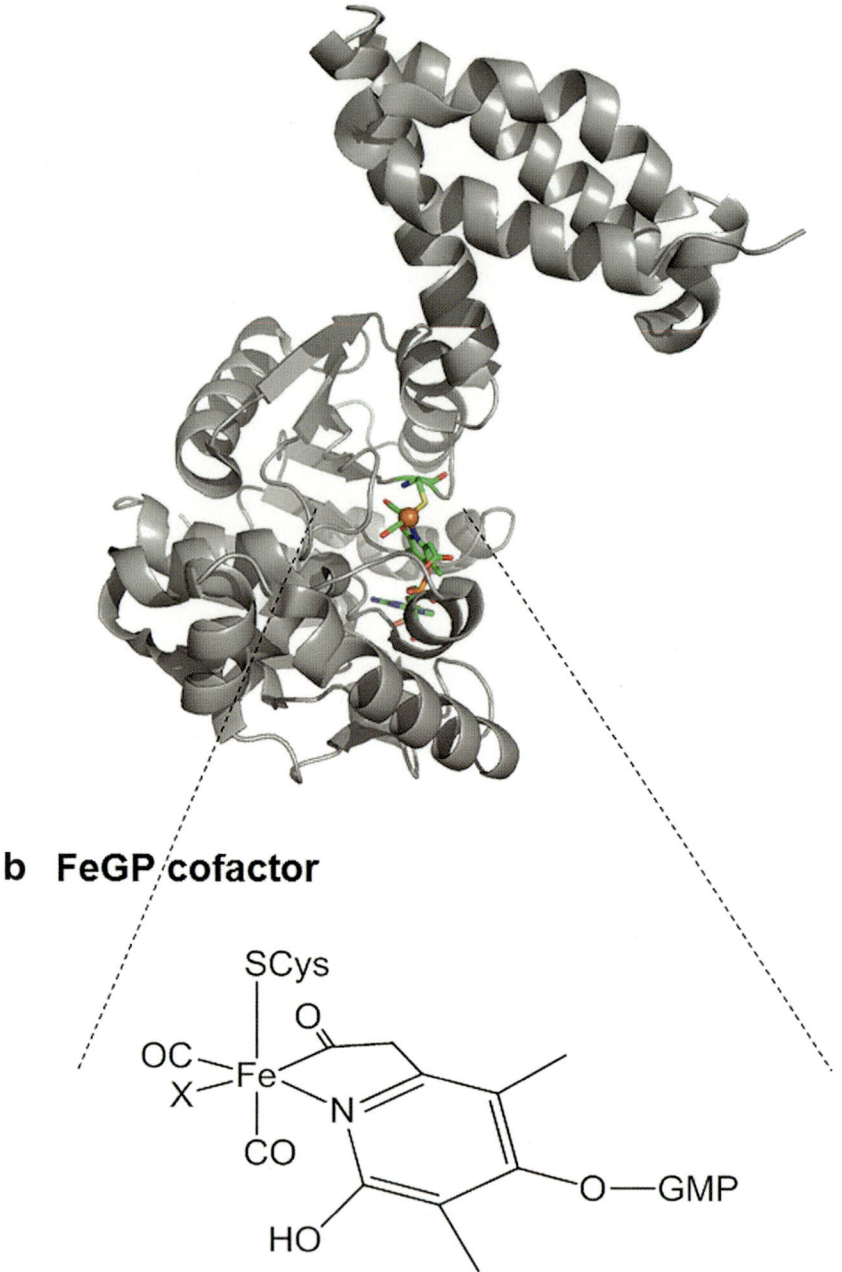

b FeGP cofactor

Figure 8 (**a**) The structure of the *Methanocaldococcus jannaschii* [Fe] hydrogenase, PDB code 3DAG. The grey ribbon shows the monomeric protein structure encasing the active site. Color code: yellow depicts sulfur; blue depicts nitrogen; bright red depicts oxygen; green depicts hydrocarbon; orange-red ball and sticks depict iron. (**b**) The chemical structure of the dihydrogen-activating Fe guanylylpyridinol catalytic site. X depicts the recently identified putative dihydrogen binding site [70].

atom at the H_2-activating Fe guanylylpyridinol (FeGP) catalytic site. There are thus two moles of Fe per mole of 76 kDa homodimer. Several [Fe] hydrogenase crystal structures have now been resolved, revealing details about both the geometry of the active site and inhibitor binding [69, 70]. Based on these studies it is predicted that upon H_2 binding the Fe is octahedrally coordinated by a Cys sulfur, two *cis*-CO ligands, the pyridinol nitrogen and acyl carbon of the FeGP cofactor, with H_2 *trans* to the acyl group.

The high O_2 sensitivity of this class of hydrogenase complicates experimental work, and EPR cannot be used to characterize this class of enzymes because the Fe remains in a low spin diamagnetic Fe^{2+} oxidation state. However a recent study comprising IR, NMR, and mass spectrometry analysis of the isolated cofactor probed the origin of several parts of the active site and showed that the CO ligands are derived from CO_2 [71]. Complementary deletion strain studies have also shown that in addition to the *hmd* gene which encodes the [Fe] hydrogenases seven *hcg* genes are essential for synthesis of a functional enzyme [72]. However, relative to the biosynthesis of other hydrogenases, the mechanism for construction of the FeGP cofactor is poorly understood and the biochemical activity of each accessory protein is yet to be determined.

5   Insights into Hydrogenase Mechanism from Small Molecule Mimics

Although within the context of Biology the presence of CO as an active site ligand is a unique requirement of hydrogenases, within a chemical context CO is an extremely common component of organometallic compounds. We rationalize that CO is required to coordinate to Fe in the hydrogenase active site because its π-acceptor bonding properties aid the stabilization of Fe in low oxidation states, thus making Fe behave more like Pt [73]. Based on such a mechanistic understanding of the reactivity properties of hydrogenase active sites it has been possible to synthesize mimic compounds which both resemble hydrogenases and are active H_2 catalysts [45, 74].

An advantage of studying synthetic small compound enzyme analogues rather than complete enzymes is the greater number of analytical methods which can be readily applied to understanding the structure and mechanism. For example, proton nuclear magnetic resonance and neutron-scattering analysis techniques both directly image hydride moieties but these experimental techniques cannot be readily used to study large enzymes. *In silico* hydrogenase molecules can also be created and computational mechanistic studies have been very valuable in interpreting enzymatic spectral data and isolating reaction intermediates which would be too short-lived to trap *in vitro* [45, 75].

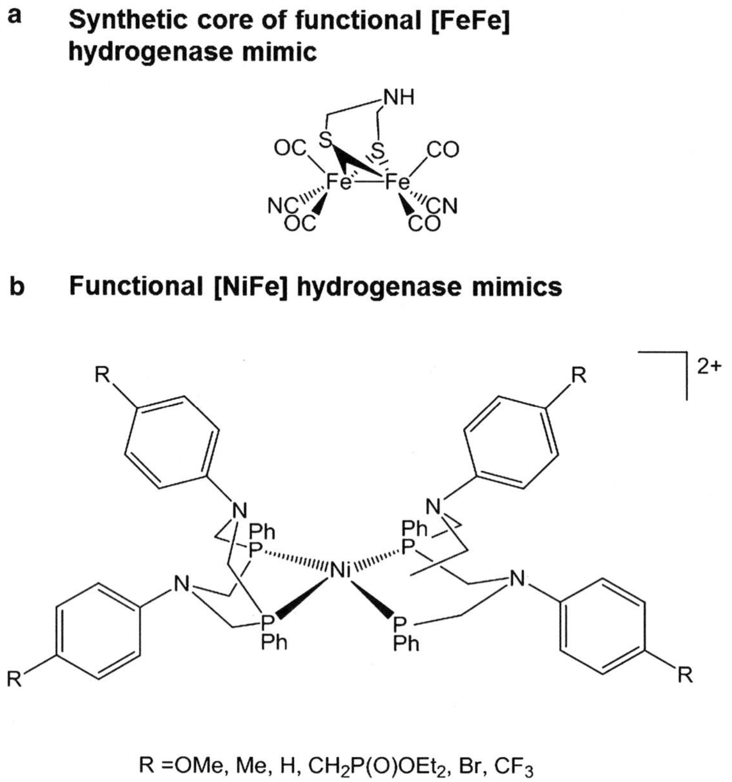

a   **Synthetic core of functional [FeFe] hydrogenase mimic**

b   **Functional [NiFe] hydrogenase mimics**

R =OMe, Me, H, $CH_2P(O)OEt_2$, Br, CF_3

Figure 9 Structures of catalytically active synthetic small molecule hydrogenase mimics. (**a**) derives from reference [78]; (**b**) derives from reference [80].

Historically [FeFe] hydrogenases have been the focus of synthetic analogue studies [74], but highly active Ni enzyme analogues have been created recently [76, 77]. Studies on both systems have highlighted the mechanistic importance of building a proton transfer site close to the H_2-activating metal center in a H_2 catalyst such as a hydrogenase. In the [FeFe] hydrogenases it has long been postulated that the active site bridging dithiolate must be a dithiomethylamine rather than containing a central bridging carbon or oxygen because it was proposed that the nitrogen would play an essential role in holding a proton during catalysis. This theory was finally proven in a recent synthetic analogue-based study [78]. Three different [FeFe] enzyme active site mimics, possessing either a nitrogen- or carbon- or oxygen-capped dithiolate ligand, were transported into a [4Fe4S] cluster containing hydrogenase apo-protein and only the nitrogen-containing molecule yielded a catalytically active H-cluster-containing enzyme (Figure 9). Similarly, in Ni model compounds the engineering of a second coordination sphere that can function as a proton relay has dramatically accelerated the rate of H_2 evolution (Figure 9) [79, 80].

6 General Conclusions

Biological H_2 catalysis is a widespread and important microbial reaction and the action of hydrogenases can be harnessed in biotechnological devices. In the past 10 years, applied hydrogenase research has moved away from enzymatic fuel cells, which utilize hydrogenases as H_2 uptake catalysts, to focusing on solar H_2 production. Of the three active site classes of hydrogenases the [NiFe] enzymes are the most widely studied but the [FeFe] enzymes have been best characterised in terms of small molecule mimicry. The [Fe] hydrogenases are the most recently discovered hydrogenase class and the least understood. It has been important to use a complementary biological and biochemical experimental toolkit to unravel the structure, function and reactivity of hydrogenases, as well as details of how these enzymes are assembled *in vivo*.

Although the [NiFe], [FeFe], and [Fe] hydrogenases are phylogenetically unrelated enzymes, they are examples of convergent evolution because all three enzyme classes contain H_2-activating sites based around a minimal chemical unit of Fe coordinated by CO and sulfur ligands. The presence of CO (and CN^-) as a structural element of an enzymatic active site is a unique feature of hydrogenases, and biosynthesis of the catalytic center requires metal transport, ligand synthesis, pre-assembly and transport. In addition to this complex process, both [NiFe] and [FeFe] hydrogenases also require FeS cluster assembly. Mapping the mechanism of hydrogenase biosynthesis is important in enabling the design of heterologous expression systems which may ultimately make sustained photo-biological H_2O splitting a feasible method for large scale H_2 production.

Abbreviations and Definitions

ATP	adenosine 5′-triphosphate
Cys	cysteine
E_{298}^0	standard reduction potential at 298 K
E. coli	*Escherichia coli*
EPR	electron paramagnetic resonance
Fe_D	distal iron
Fe_p	proximal iron
FeGP	Fe guanylylpyridinol
FeS	iron sulfur
FTIR	Fourier transform infrared
IR	infrared
MBH	membrane bound hydrogenase
MJ	megaJoule
NADH	nicotinamide adenine dinucleotide (reduced)
NADPH	nicotinamide adenine dinucleotide phosphate (reduced)

NMR   nuclear magnetic resonance
R. eutropha *Ralstonia eutropha*
SeCys   selenocysteine

Acknowledgment A. Parkin would like to acknowledge the support provided by the University of York.

References

1. P. Atkins, T. Overton, J. Rourke, M. Weller, F. Armstrong, *Shriver and Atkins' Inorganic Chemistry,* 5th edn., Oxford University Press, 2010.
2. W. Nitschke, S. E. McGlynn, E. J. Milner-White, M. J. Russell, *Biochim. Biophys. Acta Bioenerg.* **2013**, *1827*, 871–881.
3. A. Volbeda, J. Fontecilla-Camps, in *A Structural Perspective on Respiratory Complex I*, Ed. L. Sazanov, Springer Netherlands, 2012, pp. 109–121.
4. F. Tian, O. B. Toon, A. A. Pavlov, H. De Sterck, *Science* **2005**, *308*, 1014–1017.
5. R. Wordsworth, R. Pierrehumbert, *Science* **2013**, *339*, 64–67.
6. P. Kharecha, J. Kasting, J. Siefert, *Geobiology* **2005**, *3*, 53–76.
7. R. K. Thauer, A.-K. Kaster, M. Goenrich, M. Schick, T. Hiromoto, S. Shima, *Annu. Rev. Biochem.* **2010**, *79*, 507–536.
8. F. H. Chapelle, K. O'Neill, P. M. Bradley, B. A. Methe, S. A. Ciufo, L. L. Knobel, D. R. Lovley, *Nature* **2002**, *415*, 312–315.
9. M. K. Vollmer, S. Walter, J. Mohn, M. Steinbacher, S. W. Bond, T. Röckmann, S. Reimann, *Atmos. Chem. Phys.* **2012**, *12*, 6275–6289.
10. A. Jordan, B. Steinberg, *Atmos. Meas. Tech.* **2011**, *4*, 509–521.
11. R. Lamichhane-Khadka, S. L. Benoit, S. E. Maier, R. J. Maier, *Open Biol.* **2013**, *3*.
12. L. Maier, R. Vyas, Carmen D. Cordova, H. Lindsay, Thomas Sebastian B. Schmidt, S. Brugiroux, B. Periaswamy, R. Bauer, A. Sturm, F. Schreiber, C. von Mering, Mark D. Robinson, B. Stecher, W.-D. Hardt, *Cell Host Microbe* **2013**, *14*, 641–651.
13. M. Stephenson, L. H. Stickland, *Biochem. J.* **1931**, *25*, 205–200.
14. M. Stephenson, L. H. Stickland, *Biochem. J.* **1931**, *25*, 215–220.
15. P. Constant, S. P. Chowdhury, L. Hesse, J. Pratscher, R. Conrad, *Appl. Environ. Microbiol.* **2011**, *77*, 6027–6035.
16. K. Aschenbach, R. Conrad, K. Reháková, J. Doležal, K. Janatková, R. Angel, *Front. Microbiol.* **2013**, *4*, doi: 10.3389/fmicb.2013.00359.
17. H. Bothe, O. Schmitz, M. G. Yates, W. E. Newton, *Microbiol. Mol. Biol. Rev.* **2010**, *74*, 529–551.
18. J. M. Petersen, F. U. Zielinski, T. Pape, R. Seifert, C. Moraru, R. Amann, S. Hourdez, P. R. Girguis, S. D. Wankel, V. Barbe, E. Pelletier, D. Fink, C. Borowski, W. Bach, N. Dubilier, *Nature* **2011**, *476*, 176–180.
19. R. E. Macur, Z. J. Jay, W. P. Taylor, M. A. Kozubal, B. D. Kocar, W. P. Inskeep, *Geobiology* **2013**, *11*, 86–99.
20. P. M. Vignais, B. Billoud, *Chem. Rev.* **2007**, *107*, 4206–4272.
21. E. Oelgeschläger, M. Rother, *Arch. Microbiol.* **2008**, *190*, 257–269.
22. K. Schuchmann, V. Müller, *Science* **2013**, *342*, 1382–1385.
23. O. Lenz, M. Ludwig, T. Schubert, I. Bürstel, S. Ganskow, T. Goris, A. Schwarze, B. Friedrich, *ChemPhysChem* **2010**, *11*, 1107–1119.
24. F. E. Rey, M. D. Gonzalez, J. Cheng, M. Wu, P. P. Ahern, J. I. Gordon, *PNAS* **2013**, *110*, 13582–13587.
25. F. Carbonero, A. C. Benefiel, H. R. Gaskins, *Nat. Rev. Gastroenterol. Hepatol.* **2012**, *9*, 504–518.

26. J. K. DiBaise, H. Zhang, M. D. Crowell, R. Krajmalnik-Brown, G. A. Decker, B. E. Rittmann, *Mayo Clin. Proc.* **2008**, *83*, 460–469.
27. A. Strocchi, J. Furne, C. Ellis, M. D. Levitt, *Gut* **1994**, *35*, 1098–1101.
28. S. A. Angermayr, K. J. Hellingwerf, P. Lindblad, M. J. Teixeira de Mattos, *Curr. Opin. Biotechnol.* **2009**, *20*, 257–263.
29. C. Pinske, M. Jaroschinsky, F. Sargent, G. Sawers, *BMC Microbiol.* **2012**, *12*, 134.
30. R. E. Blankenship, D. M. Tiede, J. Barber, G. W. Brudvig, G. Fleming, M. Ghirardi, M. R. Gunner, W. Junge, D. M. Kramer, A. Melis, T. A. Moore, C. C. Moser, D. G. Nocera, A. J. Nozik, D. R. Ort, W. W. Parson, R. C. Prince, R. T. Sayre, *Science* **2011**, *332*, 805–809.
31. T. A. Faunce, W. Lubitz, A. W. Rutherford, D. MacFarlane, G. F. Moore, P. Yang, D. G. Nocera, T. A. Moore, D. H. Gregory, S. Fukuzumi, K. B. Yoon, F. A. Armstrong, M. R. Wasielewski, S. Styring, *Energ. Environ. Sci.* **2013**, *6*, 695–698.
32. F. E. Osterloh, B. A. Parkinson, *MRS Bulletin* **2011**, *36*, 17–22.
33. P. D. Tran, V. Artero, M. Fontecave, *Energ. Environ. Sci.* **2010**, *3*, 727–747.
34. T. Sakai, D. Mersch, E. Reisner, *Angew. Chem. Int. Ed.* **2013**, *52*, 12313–12316.
35. C. E. Lubner, P. Knörzer, P. J. N. Silva, K. A. Vincent, T. Happe, D. A. Bryant, J. H. Golbeck, *Biochemistry* **2010**, *49*, 10264–10266.
36. H. Krassen, A. Schwarze, B. R. Friedrich, K. Ataka, O. Lenz, J. Heberle, *ACS Nano* **2009**, *3*, 4055–4061.
37. E. Reisner, D. J. Powell, C. Cavazza, J. C. Fontecilla-Camps, F. A. Armstrong, *J. Am. Chem. Soc.* **2009**, *131*, 18457–18466.
38. M. Hambourger, M. Gervaldo, D. Svedruzic, P. W. King, D. Gust, M. Ghirardi, A. L. Moore, T. A. Moore, *J. Am. Chem. Soc.* **2008**, *130*, 2015–2022.
39. M. L. Ghirardi, M. C. Posewitz, P.-C. Maness, A. Dubini, J. Yu, M. Seibert, *Annu. Rev. Plant Biol.* **2007**, *58*, 71–91.
40. D. C. Ducat, J. C. Way, P. A. Silver, *Trends Biotechnol.* **2011**, *29*, 95–103.
41. R. Razeghifard, *Photosynth. Res.* **2013**, *117*, 207–219.
42. K. Gutekunst, X. Chen, K. Schreiber, U. Kaspar, S. Makam, J. Appel, *J. Biol. Chem.* **2014**, *289*, 1930–1937.
43. M. Ortega-Ramos, T. Jittawuttipoka, P. Saenkham, A. Czarnecka-Kwasiborski, H. Bottin, C. Cassier-Chauvat, F. Chauvat, *PLoS One* **2014**, *9*, e89372.
44. A. Hemschemeier, A. Melis, T. Happe, *Photosynth. Res.* **2009**, *102*, 523–540.
45. W. Lubitz, H. Ogata, O. Rüdiger, E. Reijerse, *Chem. Rev.* **2014**, *114*, 4081–4148.
46. J. Fritsch, O. Lenz, B. Friedrich, *Nat. Rev. Microbiol.* **2013**, *11*, 106–114.
47. A. Parkin, F. Sargent, *Curr. Opin. Chem. Biol.* **2012**, *16*, 26–34.
48. B. J. Murphy, F. Sargent, F. A. Armstrong, *Energ. Environ. Sci.* **2014**, *7*, 1426–1433.
49. S. Krishnan, F. A. Armstrong, *Chem. Sci.* **2012**, *3*, 1015–1023.
50. L. Xu, F. A. Armstrong, *Energ. Environ. Sci.* **2013**, *6*, 2166–2171.
51. P. Constant, S. P. Chowdhury, J. Pratscher, R. Conrad, *Environ. Microbiol.* **2010**, *12*, 821–829.
52. C. Greening, M. Berney, K. Hards, G. M. Cook, R. Conrad, *PNAS* **2014**, *111*, 4257–4261.
53. C. Schäfer, B. Friedrich, O. Lenz, *Appl. Environ. Microbiol.* **2013**, *79*, 5137–5145.
54. L. Lauterbach, O. Lenz, *J. Am. Chem. Soc.* **2013**, *135*, 17897–17905.
55. M. Horch, L. Lauterbach, O. Lenz, P. Hildebrandt, I. Zebger, *FEBS Lett.* **2012**, *586*, 545–556.
56. C. L. McIntosh, F. Germer, R. Schulz, J. Appel, A. K. Jones, *J. Am. Chem. Soc.* **2011**, *133*, 11308–11319.
57. P.-P. Liebgott, A. L. de Lacey, B. Burlat, L. Cournac, P. Richaud, M. Brugna, V. M. Fernandez, B. Guigliarelli, M. Rousset, C. Léger, S. Dementin, *J. Am. Chem. Soc.* **2010**, *133*, 986–997.
58. T. Lautier, P. Ezanno, C. Baffert, V. Fourmond, L. Cournac, J. C. Fontecilla-Camps, P. Soucaille, P. Bertrand, I. Meynial-Salles, C. Leger, *Faraday Discuss.* **2011**, *148*, 385–407.
59. C. S. A. Baltazar, M. C. Marques, C. M. Soares, A. M. DeLacey, I. A. C. Pereira, P. M. Matias, *Eur. J. Inorg. Chem.* **2011**, *2011*, 948–962.
60. A. Parkin, G. Goldet, C. Cavazza, J. C. Fontecilla-Camps, F. A. Armstrong, *J. Am. Chem. Soc.* **2008**, *130*, 13410–13416.

61. R. M. Evans, A. Parkin, M. M. Roessler, B. J. Murphy, H. Adamson, M. J. Lukey, F. Sargent, A. Volbeda, J. C. Fontecilla-Camps, F. A. Armstrong, *J. Am. Chem. Soc.* **2013**, *135*, 2694–2707.
62. L. Forzi, R. G. Sawers, *Biometals* **2007**, *20*, 565–578.
63. S. T. Stripp, B. Soboh, U. Lindenstrauss, M. Braussemann, M. Herzberg, D. H. Nies, R. G. Sawers, J. Heberle, *Biochemistry* **2013**, *52*, 3289–3296.
64. I. Bürstel, P. Hummel, E. Siebert, N. Wisitruangsakul, I. Zebger, B. Friedrich, O. Lenz, *J. Biol. Chem.* **2011**, *286*, 44937–44944.
65. J. Fritsch, E. Siebert, J. Priebe, I. Zebger, F. Lendzian, C. Teutloff, B. Friedrich, O. Lenz, *J. Biol. Chem.* **2014**, *289*, 7982–7993.
66. C. E. Foster, T. Krämer, A. F. Wait, A. Parkin, D. P. Jennings, T. Happe, J. E. McGrady, F. A. Armstrong, *J. Am. Chem. Soc.* **2012**, *134*, 7553–7557.
67. S. T. Stripp, G. Goldet, C. Brandmayr, O. Sanganas, K. A. Vincent, M. Haumann, F. A. Armstrong, T. Happe, *PNAS* **2009**, *106*, 17331–17336.
68. J. M. Kuchenreuther, W. K. Myers, D. L. M. Suess, T. A. Stich, V. Pelmenschikov, S. A. Shiigi, S. P. Cramer, J. R. Swartz, R. D. Britt, S. J. George, *Science* **2014**, *343*, 424–427.
69. S. Shima, O. Pilak, S. Vogt, M. Schick, M. S. Stagni, W. Meyer-Klaucke, E. Warkentin, R. K. Thauer, U. Ermler, *Science* **2008**, *321*, 572–575.
70. H. Tamura, M. Salomone-Stagni, T. Fujishiro, E. Warkentin, W. Meyer-Klaucke, U. Ermler, S. Shima, *Angew. Chem. Int. Ed.* **2013**, *52*, 9656–9659.
71. M. Schick, X. Xie, K. Ataka, J. Kahnt, U. Linne, S. Shima, *J. Am. Chem. Soc.* **2012**, *134*, 3271–3280.
72. T. J. Lie, K. C. Costa, D. Pak, V. Sakesan, J. A. Leigh, *FEMS Microbiol. Lett.* **2013**, *343*, 156–160.
73. G. J. Kubas, *Chem. Rev.* **2007**, *107*, 4152–4205.
74. C. Tard, C. J. Pickett, *Chem. Rev.* **2009**, *109*, 2245–2274.
75. P. E. M. Siegbahn, J. W. Tye, M. B. Hall, *Chem. Rev.* **2007**, *107*, 4414–4435.
76. M. L. Helm, M. P. Stewart, R. M. Bullock, M. R. DuBois, D. L. DuBois, *Science* **2011**, *333*, 863–866.
77. S. Ogo, K. Ichikawa, T. Kishima, T. Matsumoto, H. Nakai, K. Kusaka, T. Ohhara, *Science* **2013**, *339*, 682–684.
78. G. Berggren, A. Adamska, C. Lambertz, T. R. Simmons, J. Esselborn, S. Atta, S. Gambarelli, J. M. Mouesca, E. Reijerse, W. Lubitz, T. Happe, V. Artero, M. Fontecave, *Nature* **2013**, *499*, 66–69.
79. Y. A. Small, D. L. DuBois, E. Fujita, J. T. Muckerman, *Energ. Environ. Sci.* **2011**, *4*, 3008–3020.
80. R. M. Bullock, A. M. Appel, M. L. Helm, *Chem. Comm.* **2014**, *50*, 3125–3143.
81. A. Pohlmann, W. F. Fricke, F. Reinecke, B. Kusian, H. Liesegang, R. Cramm, T. Eitinger, C. Ewering, M. Potter, E. Schwartz, A. Strittmatter, I. Vosz, G. Gottschalk, A. Steinbuchel, B. Friedrich, B. Bowien, *Nat. Biotech.* **2006**, *24*, 1257–1262.
82. A. D. Poulpiquet, A. Ciaccafava, S. Benomar, M.-T. Giudici-Orticoni, E. Lojou, *Carbon Nanotube-Enzyme Biohybrids in a Green Hydrogen Economy*, Ed. S. Suzuki, InTech, http://www.intechopen.com/books/syntheses-and-applications-of-carbon-nanotubes-and-their-composites/carbon-nanotube-enzyme-biohybrids-in-a-green-hydrogen-economy, 2013.
83. D. Black, "Solar Fuels and Artificial Photosynthesis", Royal Society of Chemistry, www.rsc.org/solar-fuels, 2012.
84. *Sustaining Life on Planet Earth: Metalloenzymes Mastering Dioxygen and Other Chewy Gases*, Eds P. M. H. Kroneck, M. E. Sosa Torres; Vol. 15 of *Metal Ions in Life Sciences*, Eds A. Sigel, H. Sigel, R. K. O. Sigel; Springer International Publishing AG, Cham, Switzerland, **2015**, in press.

Chapter 6
Biochemistry of Methyl-Coenzyme M Reductase: The Nickel Metalloenzyme that Catalyzes the Final Step in Synthesis and the First Step in Anaerobic Oxidation of the Greenhouse Gas Methane

Stephen W. Ragsdale

Contents

ABSTRACT 126
1 INTRODUCTION 126
   1.1 Nickel Enzymes Involved in Metabolism of Environment-
      and Energy-Relevant Gases 126
   1.2 Methyl-Coenzyme M Reductase and Its Involvement
      in Generation and Utilization of Methane 127
   1.3 Ramifications of Methanogenesis in Energy
      and the Environment 128
   1.4 Discoveries Underpinning Recent Studies
      of Methyl-Coenzyme M Reductase 128
2 STRUCTURE AND PROPERTIES OF METHYL-COENZYME M
   REDUCTASE AND ITS BOUND COENZYME F_{430} 129
   2.1 Structure, Properties, and Reactivity of Coenzyme F_{430} 129
   2.2 Structure, Properties, and Reactivity
      of Methyl-Coenzyme M Reductase 130
3 REDOX AND COORDINATION PROPERTIES OF THE NICKEL CENTER
   IN METHYL-COENZYME M REDUCTASE 133
   3.1 Coordination and Oxidation States of the Free F_{430}
      Cofactor and Its Pentamethyl Ester Derivative 133
   3.2 Coordination and Oxidation States of the Nickel Center 134
4 THE CATALYTIC MECHANISM OF METHYL-COENZYME M REDUCTASE . . 137
5 SUMMARY AND PROSPECTS FOR FUTURE SCIENCE AND TECHNOLOGY .. 140
ABBREVIATIONS 142
ACKNOWLEDGMENTS 142
REFERENCES 142

S.W. Ragsdale (✉)
Department of Biological Chemistry, University of Michigan Medical School,
Ann Arbor, MI 48109-0606, USA
e-mail: sragsdal@umich.edu

© Springer Science+Business Media Dordrecht 2014
P.M.H. Kroneck, M.E. Sosa Torres (eds.), *The Metal-Driven Biogeochemistry
of Gaseous Compounds in the Environment*, Metal Ions in Life Sciences 14,
DOI 10.1007/978-94-017-9269-1_6

Abstract Methane, the major component of natural gas, has been in use in human civilization since ancient times as a source of fuel and light. Methanogens are responsible for synthesis of most of the methane found on Earth. The enzyme responsible for catalyzing the chemical step of methanogenesis is methyl-coenzyme M reductase (MCR), a nickel enzyme that contains a tetrapyrrole cofactor called coenzyme F_{430}, which can traverse the Ni(I), (II), and (III) oxidation states. MCR and methanogens are also involved in anaerobic methane oxidation. This review describes structural, kinetic, and computational studies aimed at elucidating the mechanism of MCR. Such studies are expected to impact the many ramifications of methane in our society and environment, including energy production and greenhouse gas warming.

Keywords F_{430} • methane oxidation • methanogenesis • nickel • tetrapyrrole

Please cite as: *Met. Ions Life Sci*. 14 (2014) 125–145

1 Introduction

1.1 Nickel Enzymes Involved in Metabolism of Environment- and Energy-Relevant Gases

Methyl-coenzyme M reductase (MCR) is one of the eight known Ni enzymes that catalyze the utilization and/or production of gases (methane, CO, CO_2, H_2, ammonia, and O_2) that play important roles in the global biological carbon, nitrogen, and oxygen cycles [1]. While MCR is involved in generating and metabolizing methane, CO dehydrogenase (CODH) catalyzes the two-electron interconversion of CO and CO_2, and acetyl-CoA synthase (ACS) promotes the synthesis of acetyl-CoA from CO and a methyl group and coenzyme A (CoA). Ni acireductone dioxygenase facilitates the production of CO/formate, hydrogenase the generation/utilization of hydrogen gas, urease the production of ammonia, and superoxide dismutase (SOD) the dismutation of two molecules of superoxide into hydrogen peroxide and O_2.

Among these enzymes, the Ni sites exhibit extreme plasticity in terms of metal coordination and oxidation-reduction potentials. For example, the Ni center in CODH is part of an FeS cluster, the proximal Ni in ACS appears to shift between tetrahedral and square planar while the distal Ni resembles the planar SOD active site which ligates Ni with the sulfur atoms of two Cys residues and two peptide backbone nitrogens, and the Ni in MCR is ligated by the planar nitrogens of a tetrapyrrole ring that switches among octahedral, square pyramidal and planar geometries. Furthermore, the Ni centers among the various enzymes must be able to catalyze redox processes with potentials that span from +890 mV to −160 mV [2], while in MCR, CODH, and ACS, it must be able to reach potentials as low as −600 mV [3]; thus, Ni centers in proteins perform redox chemistry over a potential range of ~1.5 V!

Because natural environments generally contain only trace amounts of soluble nickel and because the active sites are complex and often buried within the proteins, there is an additional requirement for Ni transporters [4], molecular and metallochaperones [5], sensors and regulators of the levels of enzymes involved in Ni homeostasis [6].

1.2 Methyl-Coenzyme M Reductase and Its Involvement in Generation and Utilization of Methane

Given that MCR is the only enzyme known to produce methane, that the vast abundance (90–95%) of methane is biogenic [7], and that methanogens are responsible for all biological methane production on earth [8], this enzyme is responsible for synthesizing globally one billion tons of methane per year [9]. The enzymatic synthesis of methane, reaction (1), by MCR is the defining reaction of methanogens, microbes that are the founding members of the third domain of life (the Archaea) [10–12] and have an evolutionary history of at least 3 billion years [13, 14].

$$CH_3\text{-SCoM} + CoBSH \rightarrow CH_4 + CoB\text{-SS-CoM} \qquad \Delta G^{0'} = -30 \, kJ/mol \, [13]$$

$$(1)$$

Methanogens are found only in anaerobic environments, widely distributed among aquatic sediments (ponds, marshes, swamps, rice soils, lakes, and oceans), the intestinal tract of animals (including the intestines of humans and the rumen of herbivores), sewage digesters, landfills, heart wood of living trees, decomposing algal mats, oil wells, and mild-ocean ridges. Methanogens are also found in extreme environments such as hot springs and submarine hydrothermal vents as well as in the "solid" rock of the earth's crust, kilometers below the surface [14].

Methanogens obtain their energy by the conversion of simple one- and two-carbon compounds into methane. This process is a key component of the global carbon cycle in which fermentative bacteria degrade natural polymers like lignin and cellulose to simple compounds like H_2, CO_2, formate, and acetate, which are converted by methanogens to CH_4. By its utilization of H_2, whose build up in anaerobic environments inhibits the biodegradation of organic compounds, methanogenesis stimulates the global carbon cycle.

Methane serves as a source of energy and cell carbon for methanotrophic bacteria, which oxidize this energy-rich gas through either an aerobic pathway initiated by methane monooxygenase [15] or an anaerobic process (anaerobic oxidation of methane, AOM) [16]. This latter process, which occurs on a scale of ~ 0.3 billion tons per year, is performed by methanogen-related archaeal groups (ANMR-1, -2 or -3) that coexist in a consortium with sulfate- or nitrate-reducing bacteria in methane-rich marine sediments and symbiotically couple methane oxidation to sulfate [16–19] or nitrate reduction [20]. AOM is often termed reverse methanogenesis because it involves a pathway that involves an MCR-like Ni protein [17, 21] and the other enzymatic steps involved in methanogenesis. The first reaction in AOM is the direct reverse of reaction (1) and, given that the reverse reaction is highly endothermic ($\Delta G^{0'} = +30\,\text{kJ/mol}$), coupling to sulfate or nitrate reduction is required to make the process thermodynamically feasible.

1.3   Ramifications of Methanogenesis in Energy and the Environment

Methane, the major component of natural gas, has been in use in human civilization since ancient times (6000 to 2000 bce) where it appears to have provided fuel for the "eternal fires" of the ancient Persians. Methane from animal waste was used for heating in Assyria. By 200 bce, the Chinese were drilling wells with bamboo poles to depths of 150 meters and, in the 12th century, Marco Polo mentioned in writings of his travels the use of covered sewage tanks to generate power in China.

Today, methane accounts for 22% of U.S. energy consumption [22, 23], with slightly more than half of homes using natural gas as their heating fuel. It is the simplest organic compound and has the highest energy content of any carbon-based fuel. Widely mined and used as a fuel for heating and cooking, methane also is used by the chemical industry to produce synthesis gas, to generate electricity, and to serve as a vehicle fuel in the form of compressed or liquid natural gas. Methane is considered a clean fuel because it emits less sulfur, carbon, and nitrogen than coal or oil, and leaves little ash.

Methane utilization also has environmental ramifications because it is a potent greenhouse gas whose levels have doubled over the past two centuries [24]. This rapid increase in concentration has created a mismatch between the sources and sinks of methane, causing an increasing amount of this gas to escape into the atmosphere. This is a source of concern because methane is 21 times more effective at trapping heat in the atmosphere than the major greenhouse gas, carbon dioxide [25].

It is hoped that obtaining a better understanding of the chemistry and biology related to methane metabolism will lead to biotechnological advances for decreased reliance on fossil fuels and remediation of greenhouse gas-caused climate change.

1.4   Discoveries Underpinning Recent Studies of Methyl-Coenzyme M Reductase

In what is generally accepted as the earliest scientific studies of methanogenesis, the Italian physicist Alessandro Volta and Father Carlo Campi performed many experiments around the turn of the 19th century on the "combustible air" from marshy soil [26]. Volta made a combustion chamber, a sort-of calorimeter, to measure the relative proportions of "marshy air" and regular air that produce the most energy (often measured as the loudness of the bangs upon ignition) and to determine which catalysts (e.g., iron filings in sulfuric acid) best promote the explosions.

At the turn of the 20th century it was shown that the "combustible air" was methane and that methane is cyclically produced and utilized by microbes as energy and carbon sources [27, 28]. Methods were developed to isolate and culture these anaerobic microbes in pure culture [27–31] and to work with cell-free extracts and purify enzymes from methanogens [32]. A recent *Methods in Enzymology* volume describes current methods used to study the microbiology, biochemistry, ecology, and molecular genetics of methanogens [33]. By following those methods, growing anaerobic microbes and working with anaerobic enzymes really is not that difficult.

During the last three decades, the individual steps in the pathway of methane formation have been elucidated and shown to involve many novel enzymes and cofactors [34]. The first purified enzyme clearly defined to be involved in methanogenesis was MCR, the topic of this review, which catalyzes the chemical step of methane synthesis via reaction (1). As described below, the structure and mechanism of MCR has been examined from various angles, though a consensus about the individual steps in its reaction mechanism has not been reached.

2   Structure and Properties of Methyl-Coenzyme M Reductase and Its Bound Coenzyme F_{430}

2.1   *Structure, Properties, and Reactivity of Coenzyme F_{430}*

The activity of MCR requires a bright yellow nickel cofactor called F_{430}, which has two major absorption bands with maxima at 430 nm ($\varepsilon_{430} = 23,100$ M^{-1} cm^{-1}) and 274 nm ($\varepsilon_{274} = 20,000$ M^{-1} cm^{-1}) [35–37] (Figure 1). Actually it is only yellow in its oxidized Ni(II) or Ni(III) state; in the active Ni(I) state the cofactor is green. All methanogens appear to contain F_{430} [38], which apparently is only the cofactor for this enzyme. Because the structure, biosynthesis, and redox properties of F_{430} have been extensively reviewed elsewhere [34, 39–41], I will only briefly describe its properties.

In 1980, this cofactor was isolated and shown to contain Ni and to be a tetrapyrrole [35, 37]. NMR and X-ray crystallographic methods revealed this cofactor to be a hydrocorphin, containing only five double bonds (of these only four are conjugated), thus, earning it the distinction of being the most reduced tetrapyrrole in nature [42–44]. Recognition that F_{430} is a component of MCR came when Wolfe and coworkers released and isolated [63]Ni-F_{430} from the purified enzyme [45]. The redox potential of the F_{430}-Ni(II)/F_{430}-Ni(I) pair is between -600 and -700 mV (*versus* the normal hydrogen electrode, NHE) [46, 47].

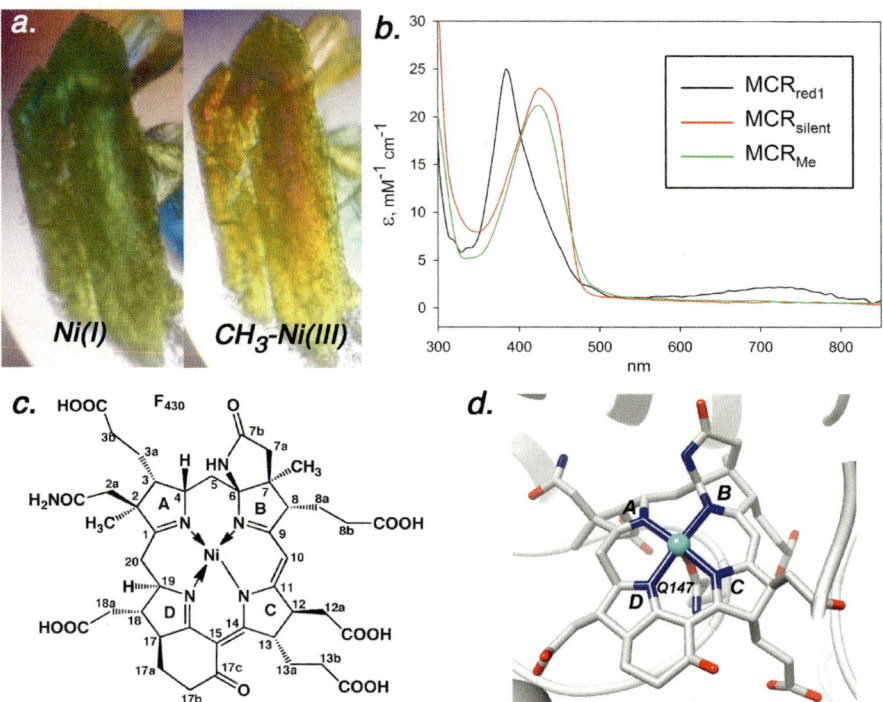

Figure 1 Properties of coenzyme F_{430}. (**a**) Photograph of methyl-coenzyme M reductase (MCR) crystals in the MCR_{red1} and MCR_{Me} states. (**b**) UV-visible spectra of bound F_{430} in its three accessible oxidation states. (**c**) The structure of free F_{430}. (**d**) Structure of bound F_{430} (PDB ID 1hbu [52]).

2.2 Structure, Properties, and Reactivity of Methyl-Coenzyme M Reductase

The first structure of the enzyme was of MCR isoenzyme I from *Methanother-mobacter marburgensis* [48]. Highly refined crystal structures of various MCR-substrate and MCR-product complexes have since been obtained. All of these studies reveal the quarternary structure of MCR as a 270 kDa dimeric association of three different subunits in an $(\alpha\beta\gamma)_2$ arrangement with F_{430} sitting at the base of a narrow channel (~50 Å) that accommodates the substrates/product.

Structures have been determined for the MCRs from *M. marburgensis* [48], *Methanosarcina barkeri,* and *Methanopyrus kandleri* [49] and from an anaerobic methane oxidizer (ANME-1) [50]. All but one (that of the methyl-Ni(III) enzyme, PDB code 3pot [51]) contain nickel in its inactive Ni(II) state and but one contain coenzyme B (CoBSH, mercapto-heptanoylthreonine phosphate) and coenzyme M (CoMSH, mercaptoethanesulfonate) in the active site (PDB codes 1hbn, 1hbo, 1hbu, 1e6y, 1e6v) [18, 49, 52]. Another crystal structure has the bound heterodisulfide product, CoBS-SCoM (MCR$_{silent}$, PDB code 1hbm) [48, 52]. Structures have also been determined of the enzyme in the presence of CoBSH analogs, which differ in the number of carbon atoms in the heptanoyl group of the native substrate [53].

MCR (Figure 2) forms a functional dimer, $(\alpha\beta\gamma)_2$, which is composed of a series of α helices shaped like an American football with dimension of 120 Å by 85 Å by 80 Å [52] . The α, α', β, and β' subunits associate tightly into a core to which the γ and γ' subunits attach on either side.

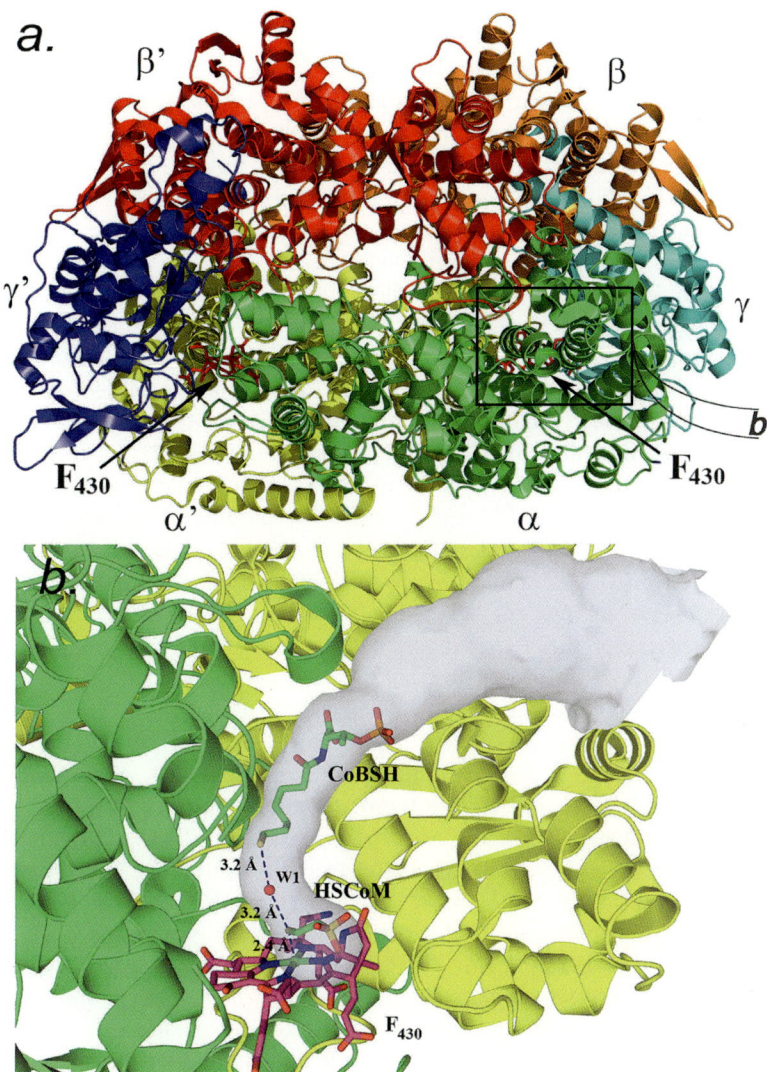

Figure 2   Structure of methyl-coenzyme M reductase. (**a**) The functional dimer, $(\alpha\beta\gamma)_2$. (**b**) The substrate channel. Reprinted by permission from [65]; copyright 2012 World Scientific Publishing Co.

The nickel ions in the two F_{430}-containing active sites are separated by 50 Å. Each active site is accessible from the surface only through a 50 Å long channel, which is 25 Å in diameter at the protein surface and narrows to 8 Å over the last 16 Å of the tunnel that leads to a pocket at the base of the channel where F_{430} binds (Figure 2b). The side chain oxygen of $Gln^{\alpha}147$ binds to the nickel atom of F_{430} on its bottom face. The protein fold, the conformation, and the binding modes of F_{430} are well conserved in all structures that have been reported, including those of the Ni(II) [54] and methyl-Ni(III) forms of the enzyme [51].

In MCR_{ox1}-silent state, $CoMS^-$ binds almost parallel to the tetrapyrrole plane, with its negatively charged sulfonate group interacting with an Arg residue. CoBSH binds perpendicular to CoMSH, with its threonine phosphate moiety interacting with residues near the lip of the tunnel and the seven-carbon mercapto-heptanoyl chain threaded through the hydrophobic tunnel, ending with its thiolate sulfur of CoBSH 8.6 Å from the Ni. The heterodisulfide product of the MCR reaction, CoBS-SCoM, binds with the CoBSH moiety in the same position as $CoBS^-$ and one of its sulfonate oxygens of $CoMS^-$ axially coordinated to nickel.

Because binding of CoBSH would block binding of methyl-CoM, an ordered mechanism was proposed in which methyl-SCoM binds first, then CoBSH, with a conformational change induced by substrate binding to bring the two substrates in appropriate position to react [52]. The ordered mechanism was demonstrated by steady-state and presteady-state kinetic studies [55]. A CoMSH-induced conformational change was indicated by the MCR_{red1}-silent structure [52], yet elegant NMR, EPR, and ^{19}F-ENDOR studies from the groups of Thauer and Jaun using CoMSH and the $S-CF_3$ analog of CoBSH demonstrated that binding of the second substrate to either the Ni(I) or Ni(II) states of MCR induces a 2 Å movement of CoBSH closer to the Ni [56]. A structural comparison of MCR in the presence of various mercaptoalkanoyl analogs of CoBSH supports this concept and suggests that rearrangement of the Tyr residues near the Ni center might trigger deeper penetration of CoBSH into MCR [53].

High-resolution crystal structures revealed that MCR contains five posttranslationally modified amino acids, including 1-N-methylhistidine (α257), 5-(S)-methylarginine (α271), 2-(S)-methylglutamine (α400), S-methylcysteine (α452), and thioglycine (α445) from the α subunits within the substrate channel [57]. The methyl groups were shown to be introduced by S-adenosylmethionine (SAM)-dependent reactions [58]. Mass spectrometric studies revealed that, while thioglycine (from $Gly^{\alpha}445$) has been found in all methanogens in which it has been examined and proposed to play a redox role during catalysis [52], various organisms (including MCR I and MCR II from *M. marburgensis*, MCR I from *Methanocaldococcus jannaschii*, and *Methanoculleus thermophilus*, MCR from *Methanococcus voltae*, *Methanopyrus kandleri,* and *Methanosarcina barkeri*, and two MCRs of the ANME-1 methanotrophic archaea cluster isolated from Black Sea mats) contain different modified methylated residues [59].

3 Redox and Coordination Properties of the Nickel Center in Methyl-Coenzyme M Reductase

The F_{430} cofactor in MCR can assume the Ni(I), Ni(II), and Ni(III) oxidation states and square planar four-, square pyramidal five-, and octahedral six-coordinate states. Both the free and bound cofactors have been studied. These states have been identified and characterized by UV-visible, X-ray absorption, resonance Raman, MCD, and EPR and related methods, including electron spin echo envelope modulation (ESEEM), electron nuclear double resonance (ENDOR) and HYSCORE.

3.1 Coordination and Oxidation States of the Free F_{430} Cofactor and Its Pentamethyl Ester Derivative

One protocol for obtaining the free F_{430} cofactor is to denature MCR with perchloric acid or trichloroacetic acid [40]. F_{430} is then released with Ni in the 2+ oxidation state. The isolated cofactor is a pentaacid, which is often derivatized to the pentamethyl ester, F_{430M}, for analysis because of its higher stability, easier purification, and solubility in non-coordinating organic solvents [60].

Ni(II)-F_{430M} has an absorption maximum at 420 nm. The Ni(II)-F_{430M} state is EPR-silent with either spin states $S = 0$ or $S = 1$ depending on the presence and types of axial ligands. Using X-ray absorption (EXAFS) studies, Hamilton and coworkers showed that both enzyme-bound and isolated F_{430} in aqueous solution contain 6-coordinate pseudo-octahedral Ni(II) at low temperature (<250K) that undergoes conversion at higher temperatures to a 4-coordinate, square-planar form [61]. There was no evidence for a 5-coordinate form of free F_{430} [40, 61].

Ni(II)-F_{430M} can be efficiently reduced with sodium amalgam in tetrahydrofuran to generate Ni(I)-F_{430M} [47], which is analogous to the catalytically active form of F_{430} in MCR. F_{430} itself can be reduced by Ti(III) citrate in aqueous solution at pH 10.0, indicating that the Ni(II)/(I) couple has a redox potential between -600 and -700 mV [3]. Similar to F_{430}, Ni(I)-F_{430M} has absorption peaks at 383 nm and 759 nm [47]. Like Cu(II), Ni(I) is a $3d^9$ metal ion; thus Ni(I)-F_{430M} is paramagnetic with an $S = \frac{1}{2}$ spin state and an EPR spectrum (g values of 2.065, 2.074 and 2.250) that exhibits hyperfine splitting from coupling to the tetrapyrrole nitrogen atoms (^{14}N nuclear spin, I, equals 1). Of significance to the MCR mechanism, Ni(I)-F_{430M} can react with methyl iodide, methyl tosylate and methyl sulfonium salts, in which the methyl group is activated, to generate methane formation via a methyl-Ni intermediate [40, 62–64]. Similar reactions have been performed with the enzyme, as described below.

The Ni(III) state of F_{430M} can also be generated electrochemically to give an EPR spectrum with g values of 2.020 and 2.211, characteristic of a tetragonally-distorted $S = \frac{1}{2}$ system with a nickel d_{z^2} ground state [40].

3.2   Coordination and Oxidation States of the Nickel Center

The bound cofactor in MCR also interconverts among various states, which are designated by their oxidation or ligation states; for example, MCR_{red1m} indicates that MCR has been reduced to the Ni(I) state and incubated with CH_3-SCoM. The "1" distinguishes this from another characterized Ni(I) state, called MCR_{red2}. Recent reviews include tables that provide the g values from the EPR spectra and the coordination environments of the known MCR states [65, 66].

The active Ni(I) form of the enzyme, termed MCR_{red1}, has g values of 2.068, 2.082, 2.274 [67] and a visible spectrum with absorption maxima at 383 nm and 759 nm [47]. The Ni(I) has a $3d^9$ configuration with the unpaired electron residing mostly in the $d_{x^2-y^2}$ orbital with a hyperfine coupling value of 25–30 MHz from the pyrrole nitrogens [66]. MCR_{red1} can be generated within the cell by replacing the 80%/20% (H_2/CO_2) gas phase with 100% H_2 [68, 69] or 100% CO [70] prior to harvesting the cells or by treating the MCR_{ox1} state with sodium sulfide (Na_2S) [71]; alternatively, it can be generated in $vitro$ by adding CH_3-SCoM to the MCR_{red2} state [72].

There is no crystal structure of this form of the protein, due to its extreme lability; however, it has been characterized by X-ray absorption spectroscopy, which provides precise nickel-ligand bond lengths and geometries. As shown in Figure 3, which compares the Ni(I), Ni(II), and methyl-Ni(III) states, in the MCR_{red1} state, the Ni(I) is pentacoordinate, being ligated by the four nitrogen atoms from the tetrapyrrole and the oxygen atom from the side chain of $Gln^{\alpha'}147$) [73, 74]. Several MCR_{red1} sub-types have been described: MCR_{red1c}, MCR_{red1m}, which are generated by incubating the Ni(I) form of the enzyme with CoMSH or methyl-SCoM, respectively, and MCR_{red1a}, generated in the absence of a CoM derivative. The EPR spectra of these states differ slightly in that the MCR_{red1m} and MCR_{red1c} forms exhibit sharper peaks and, thus, better resolved splitting of the S-shaped resonance at $g = 2.07$ [73].

Another Ni(I) state, MCR_{red2}, has been identified and, based on high-frequency EPR studies and density functional theory (DFT) computations, the MCR_{red2a} and MCR_{red2r} states were resolved [56, 75, 76]. In the MCR_{red2r} state, the EPR spectrum is markedly altered (g_1 increases from the ~2.06 value seen in other MCR_{red} states to 2.175) and the nitrogen hyperfine splitting value (~14 MHz) of one of the pyrrole nitrogens (from the A ring) is reduced relative to that of the other three nitrogens (~24 MHz). These unusual features indicate a significant distortion of the tetrapyrrole macrocycle that displaces the nitrogen in the A ring out of the plane containing the other three nitrogens in the tetrapyrrole ring and removal of the glutamine oxygen ligand. By ^{33}S isotope labeling and pulsed EPR studies, it was shown that the thiolate sulfur of $CoMS^-$ ligates the Ni in the MCR_{red2r} state [56, 77]. An exchangeable proton with a strikingly large (42–43 MHz) hyperfine coupling (^1H nuclear spin $I = 1/2$) was identified in MCR_{red2a}, indicating that this form of the protein contains a 1.6 to 1.7 Å Ni-H bond (perhaps a nickel hydride). In the MCR_{red2r} state, the proton coupling is reduced to 29 MHz, consistent with the ionizable proton from CoMSH interacting with Ni and a tetrapyrrole nitrogen [76].

Figure 3 Comparison of the Ni(II), Ni(I), and methyl-Ni(III) states of methyl-coenzyme M reductase, based on X-ray absorption spectroscopy. Ni-ligand bond lengths are shown for the axial and equatorial (tetrapyrrole) ligands. See the text for details. Reprinted by permission from [74]; copyright 2009 American Chemical Society.

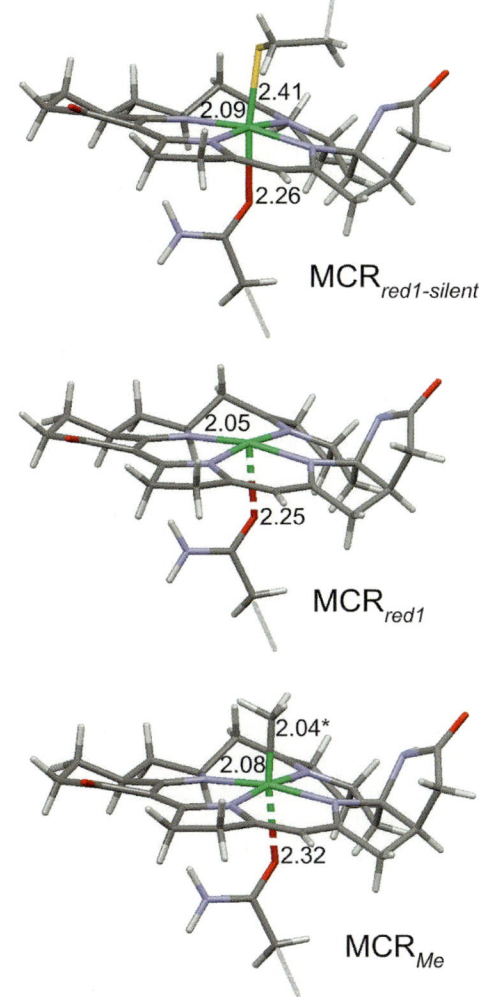

The inactive Ni(II) state (called MCR$_{silent}$) is the form that has been best characterized crystallographically [52]. This state, carrying a $3d^8$ metal ion, is EPR-silent. When cells are harvested directly from the growth medium (i.e., without activation by H_2 or CO), this is the predominant state of the enzyme. This also is the form that accumulates as the Ni(I) states undergo oxidation; in fact, because the Ni(II)/(I) redox couple has such a low potential, the Ni(I) forms decay to Ni(II) even inside anaerobic chambers maintained at <1 ppm O_2. In the MCR$_{silent}$ state, the Ni center is six-coordinate, with four nitrogen ligands from the F$_{430}$ cofactor, one oxygen ligand from a glutamine residue from opposite alpha subunit (Gln$^{α'}$147) and another axial ligand, e.g., the thiolate sulfur of CoMS$^-$ or one of the sulfonate oxygens of CoBS-SCoM [48].

When the gas phase during growth is switched to 80% N_2/20% CO_2 before cell harvesting, another form called MCR_{ox1} is generated [72]. MCR_{ox1} also can be generated by adding Na_2S to the growing culture prior to harvest [71] or by adding polysulfide to MCR_{red2} *in vivo* and *in vitro* [71]. MCR_{ox1} is much more resistant to O_2 than MCR_{red1}, in agreement with the results of EPR, ENDOR, and HYSCORE experiments that this is an oxidized form of the enzyme containing a Ni(III)-thiolate (from $CoMS^-$) in resonance with a high-spin Ni(II)-thiyl radical [78, 79]. Addition of sodium sulfite to growing cultures or to MCR_{red2} *in vitro* results in formation of a light-sensitive state, MCR_{ox2}, while addition of O_2 to MCR_{red2} generates the MCR_{ox3} state [67].

While it has been unambiguously shown that Ni(I) is the active state that initiates catalysis, it is not clear if the product of reaction between MCR and the two substrates generates Ni(II) and a methyl radical or an organometallic methyl-Ni(III) intermediate. To characterize the alkyl-Ni(III) species, MCR_{red1} has been reacted with brominated substrate analogs, like bromopropanesulfonate to generate Ni(III)-propanesulfonate or methyl iodide to give methyl-Ni(III) [80]. Whereas it had been thought that the alkyl-Ni(III) species would be highly unstable, it was found to be relatively stable and could be characterized by various spectroscopic and structural methods, as described next.

Although an organometallic methyl-Ni(III) intermediate has been proposed to be a catalytic intermediate in methane synthesis [63, 64, 81], such an intermediate has never been trapped during the reaction of MCR with native substrates. On the other hand, reaction of MCR_{red1} with methyl halides (even in the absence of CoBSH) quantitatively generates the methyl-Ni(III) state, often called MCR_{Me}. [82, 83]. This reaction likely occurs by a nucleophilic attack of Ni(I) on methyl iodide to generate the iodide anion and the Ni(III)-CH_3 species. Formation of the methyl-Ni(III) species was confirmed by EPR spectroscopy and the covalent linkage between the methyl group and the nickel center was confirmed by high resolution ENDOR and HYSCORE experiments using different isotopes of methyl iodide [82, 83]. X-ray absorption spectroscopic and X-ray crystallographic studies of the alkyl-Ni(III) state of MCR reveal a six-coordinate Ni center with an upper axial Ni-C bond at 2.04 Å, four Ni-N bonds at 2.08 Å, and a lower axial Ni-O interaction at 2.32 Å, unambiguously establishing the organometallic nature of the methyl-Ni(III) species [74, 84]. The MCR_{Me} can then react with CoMSH (and other thiolates) to generate MCR_{red1} and CH_3-SCoM (or other alkyl thioethers). Similarly, reaction of the MCR_{Me} species with CoMSH and CoBSH produces methane at a k_{cat} of 1.1 s^{-1}, which is similar to the steady-state k_{cat} for methane formation from natural substrates (4.5 s^{-1} at 25°C) – consistent with the catalytic intermediacy of the methyl-Ni(III) species [82].

The Ni(I) in MCR_{red1} can react with a variety of compounds with activated alkyl groups in the absence of CoBSH. For example, reaction with the potent inhibitor, bromopropanesulfonate (BPS), generates MCR_{PS} [84], which, like MCR_{Me}, exhibits UV-visible features that emulate those of the Ni(II) protein and an EPR spectrum with g values at 2.223 and 2.115. On the basis of EPR, ENDOR, and HYSCORE spectroscopic studies, MCR_{PS} was assigned as an organometallic

Ni(III)-propyl sulfonate in resonance with a Ni(II)-propyl sulfonate radical [85, 86]. MCR_{PS} can be converted back to the MCR_{red1} state by reaction with the low potential reductant, Ti(III) citrate, or with various thiolates, including analogs of CoMSH and CoBSH with methylene group(s) longer than the native substrate (CoB_8SH or CoB_9SH), but not with CoBSH itself [84, 87].

4 The Catalytic Mechanism of Methyl-Coenzyme M Reductase

The mechanism of the MCR-catalyzed conversion of methyl-SCoM to methane, reaction (1), requires CoBSH as the two-electron [88] and possibly also as the proton donor [9]. Two leading mechanisms have been proposed (Figure 4); however, because none of the intermediates shown in this figure have been isolated, one cannot yet conclusively state which mechanism is correct. Various experimental measurements set ground-rules for the correct mechanism. The first mechanistic constraint is that initiation of the MCR reaction requires the F_{430} cofactor to be in the Ni(I) oxidation state [71, 72]. Rule 2 is that net inversion of stereoconfiguration occurs during the reduction of methyl-CoM [89]. The third mechanistic constraint is that catalysis involves a ternary complex mechanism and has an absolute requirement for CoBSH for even a single catalytic turnover [55]. Fourth, the carbon kinetic isotope effect from $^{12}CH_3$-SCoM *versus* $^{13}CH_3$-SCoM is 1.04, indicating that the rate-limiting step is breakage of the C-S bond of CH_3-SCoM [90]. The secondary kinetic isotope effect is 1.19 in the methyl group of CD_3-SCoM, indicating that the methyl group goes from tetrahedral to trigonal planar upon reaching the transition state of the rate-limiting step [91]. The two guiding mechanisms described below follow those four rules. Now, what is needed is to trap the intermediates in the mechanism.

The two major proposals for the MCR catalytic mechanism differ in the nature of the first intermediate. Mechanism 1 involves a methyl-Ni(III) intermediate [48], while mechanism 2 proposes a methyl radical and a CoMS-Ni(II) complex [92]. Additionally, it is hypothesized that substrate binding initiates a conformational change that triggers the catalytic cycle. This concept is supported by ^{19}F-ENDOR studies [56] and transient kinetic studies [93], and by the observation that inactivation of MCR_{red1} by bromoethanesulfonate (BES) requires CoBSH [94].

In mechanism 1, MCR_{red1} Ni(I) attacks the methyl group of methyl-SCoM in an S_{N^2} nucleophilic substitution reaction to form a methyl-Ni(III) intermediate and $CoMS^-$ (Figure 4). Concurrently, a proton is transferred from CoBSH to the $CoMS^-$ leaving group. Methyl-Ni(III) accepts one electron from CoMSH to form a radical cation ($CoMS^•H^+$) and methyl-Ni(II). Proton transfer from $CoMS^•H^+$ to the methyl-Ni(II) species is proposed to follow, thus generating methane and Ni(II). The $CoBS^-$ thiolate then attacks the $CoMS^•H^+$ thiyl radical cation to give a disulfide radical anion ($CoBS^•SCoM^-$) with a two-center three-electron bond, which subsequently reduces Ni(II) back to Ni(I), leaving the neutral heterodisulfide CoBS-SCoM as the final product.

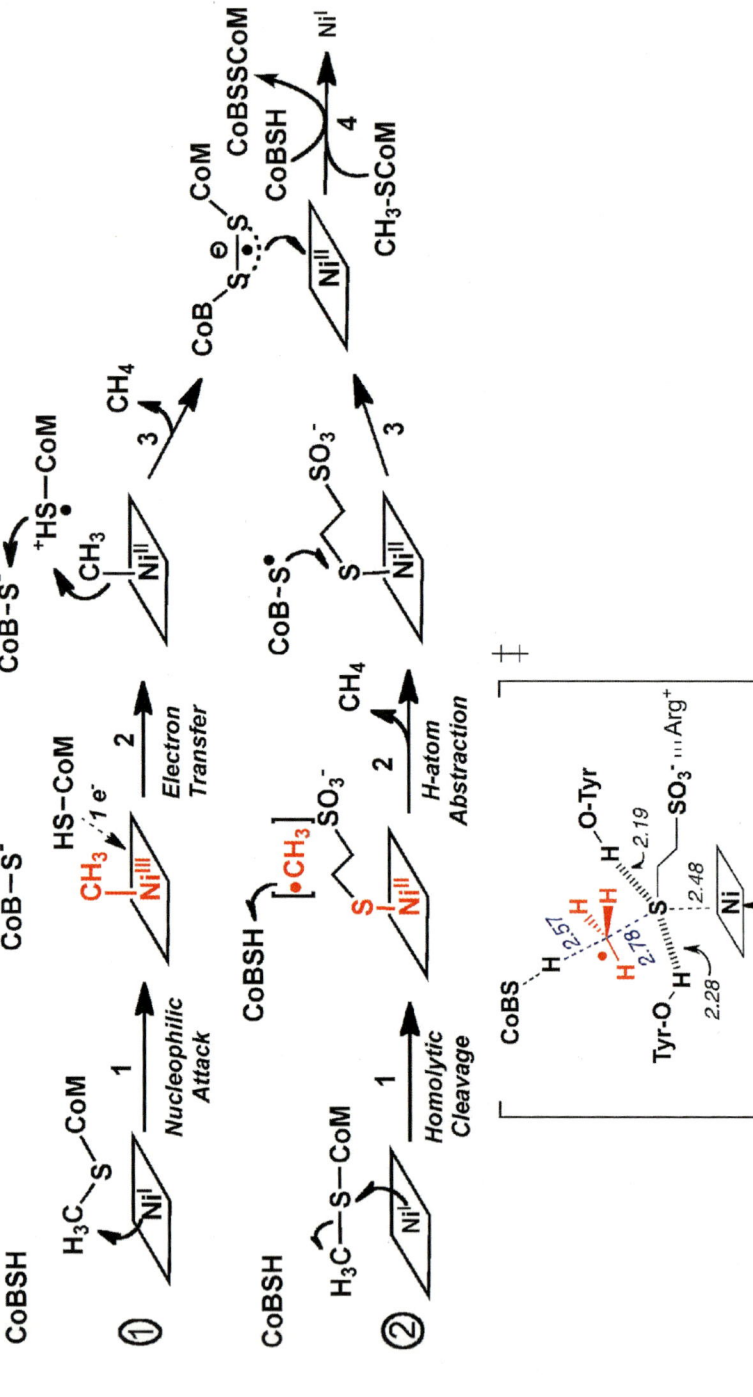

Figure 4 Comparison of the two leading proposals for the methyl-coenzyme M reductase catalytic mechanism. Also shown below the mechanisms is the proposed first transition state for C–S bond cleavage, based on [102]. See text for details.

Mechanism 1 is reminiscent of other well characterized enzymatic systems, such as methionine synthase and methyl-SCoM methyltransferase, where high-energy methyl-N and methyl-S bonds undergo a nucleophilic substitution by Co(I) yielding stable methyl-cob(III)alamin intermediates that act as methyl donors in subsequent reactions [95, 96]. Experimental evidence in support of mechanism 1 includes the inversion of stereoconfiguration during the reduction of ethyl-CoM [89], the generation of an alkyl-Ni(III) bond during the reaction with the inhibitor 3-bromopropane sulfonate [84, 97, 98] and related brominated analogs of methyl-SCoM [51, 74, 82], and the formation of methyl-Ni(II)F_{430} species in the reaction of free cofactor F_{430} with methyl iodide and methyl bromide [40]. In addition, the reaction of MCR with ethyl-CoM exhibits a significantly lower catalytic efficiency relative to that with methyl-SCoM, thus implying steric hindrance during the S_{N^2} substitution.

Based on hybrid DFT calculations, mechanism 1 is suggested to not be thermodynamically feasible because of the high energy barrier (45 kcal/mol) for breaking the C-S bond of methyl-SCoM relative to that to form a relatively weak methyl-Ni bond [92]. Thus, an alternative mechanism, mechanism 2, was proposed in which a nucleophilic attack of Ni(I) on methyl-SCoM forms Ni(II)-SCoM and a methyl radical (instead of methyl-Ni(II)), which is immediately quenched by a hydrogen atom transfer from CoBSH [92, 99, 100]. Then, as in mechanism 1, the resulting thiyl radical (CoBS$^\bullet$) reacts with Ni(II)-SCoM to give a disulfide anion radical (CoBS$^\bullet$SCoM$^-$) that transfers an electron to Ni(II) to regenerate Ni(I), thus closing the reaction cycle. DFT calculations predict the energy barrier for the first and second step of methanogenesis catalyzed by MCR to be 10 kcal/mol and 20 kcal/mol, respectively, thus making the C-S bond cleavage rate-limiting in methane formation [100]. Recent carbon kinetic isotope effect studies (i.e., ^{12}CH$_3$-SCoM/^{13}CH$_3$-SCoM = 1.04) also indicate that C-S bond cleavage is rate-limiting [91].

Mechanism 2 includes as a key intermediate a form of the enzyme (at least in terms of coordination geometry at Ni) that has been well-studied by crystallography and spectroscopy, i.e., a hexacoordinate Ni(II)-SCoM species [77, 101]. One limitation in mechanism 2 as it was originally proposed is that rotation of a free methyl radical ($^\bullet$CH$_3$) would scramble the stereochemistry at carbon. This is inconsistent with the experimental result that the reaction with ethyl-SCoM leads to net inversion of configuration at carbon. In order to explain a net inversion of configuration for the released methane molecule, the hydrogen atom transfer from CoBSH has to be faster than rotation of the methyl radical in the active site. This would likely be possible only if hydrogen atom abstraction is concerted with formation of a transient methyl radical as recently suggested [102]. A proposed transition state for the methyl radical intermediate in the concerted methyl radical mechanism is shown in Figure 4, with bond distances proposed in the computational studies. Recent kinetic isotope effect studies are consistent with such a transition state. A secondary kinetic isotope effect of 1.19 per D in the methyl group of CD$_3$-SCoM was interpreted to indicate a change in geometry of the methyl group from tetrahedral (sp^3) to trigonal planar (sp^2) upon reaching the transition state of the rate-limiting step (cleavage of the C-S bond of methyl-SCoM) [91].

In the context of reverse methanogenesis, the anaerobic oxidation of methane catalyzed by MCR [103], according to mechanism 1, methane oxidation would involve an electrophilic attack of Ni(II)F_{430} on methane, resulting in release of a proton that is captured by CoMS$^{\bullet}$ radical. Considering the pK_a of methane to be around 50, this scenario is unlikely because Ni(II) of the F_{430} cofactor is not considered to be electrophilic enough to perform such an attack. Similarly, mechanism 2 requires, in the initial step, a hydrogen atom abstraction from methane by a CoBS$^{\bullet}$ thiyl radical. The dissociation energy of the C-H bond in methane (439 kJ/mol) is 74 kJ/mol higher than that of the S-H bond, thus making such a process thermodynamically unfavorable.

Regardless of the mechanism followed by MCR, the structural and spectroscopic evidence exists to support coupling of the two active sites of MCR. The crystal structure revealed that the two active sites are interconnected through the α and α' subunits, thus, any conformational change in one of the active sites would likely be transmitted to the other site. Evidence for the conformational change induced by the substrate binding is based on the crystal structure analysis of Ni(II)-MCR [52], EPR studies [56], and transient kinetic data [104]. Additionally, it was found that at most 50% of the enzyme can be converted from the MCR$_{red1}$ active state into the MCR$_{red2}$ state upon addition of CoMSH and CoBSH [105], suggesting that only half of the sites can be used in the catalysis at any given time. Thus, coupling of the endergonic and the exergonic steps of the catalytic cycle could be envisioned as a strategy employed by MCR to lower the activation energy of the rate-limiting step.

5   Summary and Prospects for Future Science and Technology

Methanogens are masters of CO_2 reduction. They conserve energy by coupling the reduction of CO_2 to CH_4, the primary constituent of natural gas, which accounts for 22 percent of the U.S. energy consumption. Uncovering the mechanistic and molecular details of how methane is formed is critical since it is an important fuel and the second most prevalent greenhouse gas.

Methane is considered a clean fuel because it emits less sulfur, carbon, and nitrogen than coal or oil, and leaves little ash. It is the simplest organic compound and has the highest energy content of any carbon-based fuel. Widely mined and used as a fuel for heating and cooking, methane also is used by the chemical industry to produce synthesis gas, to generate electricity, and to serve as a vehicle fuel in the form of compressed or liquid natural gas. Methanogenesis is also a key component of the global carbon cycle, serving as a hydrogen sink and energy source for methanotrophic organisms.

A major focus has been aimed at MCR because it catalyzes the rate-limiting step in methane formation and oxidation. It has been a long-term goal to identify and characterize each catalytic intermediate in the MCR reaction mechanism. Besides uncovering a novel bioinorganic mechanism, this information has important social, economic and environmental ramifications. This knowledge could lead to the development of biomimetic models that could serve as catalysts for highly efficient fuel production. By understanding MCR, we will significantly increase control over production of one of the most used and widely important fuels.

It is only recently that researchers have developed the ability to characterize the intermediates in the MCR catalytic cycle. Most spectroscopic studies require a highly homogeneous enzyme that is upwards of 80% in its active state; furthermore, intermediates must accumulate to similarly high percentages. Over the decades of study of MCR, the fraction of active enzyme has risen from less than 1% to nearly 100%. The use of substrate analogs has allowed transient intermediates that barely accumulate to be trapped in relatively high yield.

Given that methods have been developed to obtain high amounts of active enzyme, all the weapons in the arsenal of the spectroscopist and mechanistic enzymologist can be launched at solving the MCR mechanism. Determination of the rate constants for each of the steps in the mechanism should be achievable as well as characterization of each of the intermediates.

Computational studies are having an important impact on studies of the enzyme and these studies can be benchmarked with appropriate experimental data. Genetic systems have been developed that should be poised to make variants of active-site residues that are proposed to be mechanistically important.

There is much optimism for future studies of MCR. Yet, challenges remain. For example, although methods have been developed to generate, stabilize and crystallize the active state of MCR, to my knowledge, no X-ray beam line in the world is currently able to retain the Ni(I) state during data collection. This is a significant problem that is likely to also be affecting our interpretations of the structures of other O_2-sensitive metalloenzymes. The problem is just more easily identified in MCR because the oxidation of MCR is easier to follow by simply observing the change in color of the crystals, from green to a bright yellow color as crystals are mounted and data collection begins.

Studying anaerobic methane oxidation by MCR is as important as understanding methane formation. This is because methane is widely available and could serve as an important liquid fuel [106, 107]. In the chemical industry, methane is used to produce synthesis gas (syngas, a mixture of CO and H_2) and as a fuel for electricity generation. Methane is also used as a vehicle fuel in the form of compressed (CNG) or liquid (LNG) natural gas, especially in Asia and South America. The current state-of-the-art process for the conversion of natural gas to liquid fuels utilizes Fischer-Tropsch chemistry, which is limited by high capital costs and low conversion efficiencies. A biotechnological process based on enzymatic methane oxidation could provide a platform for avoiding the high costs of gas storage and distribution and combine high energy density with broad compatibility across all modes of transportations. To achieve such a goal, it is important to use the various available genetic, biochemical, chemical and microbiological tools to understand the process of methane oxidation and to overcome the current limitations.

Abbreviations

ACS	acetyl-CoA synthase
AOM	anaerobic oxidation of methane
bce	before the common era = before Christ
BES	bromoethanesulfonate
BPS	bromopropanesulfonate
CNG	compressed natural gas
CoA	coenzyme A
CoBSH	coenzyme B mercaptoheptanoylthreonine phosphate
CODH	carbon monoxide dehydrogenase
CoMSH	coenzyme M mercaptoethanesulfonate
DFT	density functional theory
ENDOR	electron nuclear double resonance
EPR	electron paramagnetic resonance
ESEEM	electron spin echo modulation
EXAFS	extended X-ray absorption fine structure
HYSCORE	hyperfine sublevel correlation spectroscopy
LNG	liquid natural gas
MCD	magnetic circular dichroism
MCR	methyl-coenzyme M reductase
NHE	normal hydrogen electrode
NMR	nuclear magnetic resonance
SAM	S-adenosylmethionine
SOD	superoxide dismutase

Acknowledgments I thank those students, postdoctoral fellows and collaborators who have been working on the biochemistry of methane formation, with special thanks to Dariusz Sliwa for helping to generate Figure 1 for this paper. I gratefully acknowledge support (DE-FG02-08ER15931) from the Chemical Sciences, Geosciences and Biosciences Division, Office of Basic Energy Sciences, Office of Science, U.S. Department of Energy and from ARPA-E (DE-AR0000426).

References

1. S. W. Ragsdale, *J. Inorg. Biochem.* **2007**, *101*, 1657–1666.
2. A. F. Miller, *Acc. Chem. Res.* **2008**, *41*, 501–510.
3. C. Holliger, A. J. Pierik, E. J. Reijerse, W. R. Hagen, *J. Am. Chem. Soc.* **1993**, *115*, 5651–5656.
4. D. A. Rodionov, P. Hebbeln, M. S. Gelfand, T. Eitinger, *J. Bacteriol.* **2006**, *188*, 317–327.

5. S. Quiroz, J. K. Kim, S. B. Mulrooney, R. P. Hausinger, in *Nickel and Its Surprising Impact in Nature*, Vol. 2 of *Metal Ions in Life Sciences*, Eds A. Sigel, H. Sigel, R. K. O. Sigel, John Wiley and Sons, Chichester, UK, 2007, pp 519–544.
6. C. M. Phillips, E. R. Schreiter, Y. Guo, S. C. Wang, D. B. Zamble, C. L. Drennan, *Biochemistry* **2008**, *47*, 1938–1946.
7. S. K. Atreya, P. R. Mahaffy, A. S. Wong, *Planet. Space Sci.* **2007**, *55*, 358–369.
8. U. Deppenmeier, *Prog. Nucleic Acid Res. Mol. Biol.* **2002**, *71*, 223–283.
9. R. K. Thauer, *Microbiology* **1998**, *144*, 2377–2406.
10. C. R. Woese, L. J. Magrum, G. E. Fox, *J. Mol. Evol.* **1978**, *11*, 245–251.
11. C. R. Woese, O. Kandler, M. L. Wheelis, *Proc. Natl. Acad. Sci. USA* **1990**, *87*, 4576–4579.
12. G. E. Fox, E. Stackebrandt, R. B. Hespell, J. Gibson, J. Maniloff, T. A. Dyer, R. S. Wolfe, W. E. Balch, R. S. Tanner, L. J. Magrum, L. B. Zablen, R. Blakemore, R. Gupta, L. Bonen, B. J. Lewis, D. A. Stahl, K. R. Luehrsen, K. N. Chen, C. R. Woese, *Science* **1980**, *209*, 457–463.
13. R. K. Thauer, S. Shima, *Ann. N. Y. Acad. Sci.* **2008**, *1125*, 158–170.
14. J. L. Garcia, B. K. Patel, B. Ollivier, *Anaerobe*, **2000** 6, 205–226.
15. R. S. Hanson, T. E. Hanson, *Microbiol. Rev.* **1996**, *60*, 439–471.
16. S. J. Hallam, N. Putnam, C. M. Preston, J. C. Detter, D. Rokhsar, P. M. Richardson, E. F. DeLong, *Science* **2004**, *305*, 1457–1462.
17. M. Kruger, A. Meyerdierks, F. O. Glockner, R. Amann, F. Widdel, M. Kube, R. Reinhardt, J. Kahnt, R. Bocher, R. K. Thauer, S. Shima, *Nature* **2003**, *426*, 878–881.
18. W. Michaelis, R. Seifert, K. Nauhaus, T. Treude, V. Thiel, M. Blumenberg, K. Knittel, A. Gieseke, K. Peterknecht, T. Pape, A. Boetius, R. Amann, B. B. Jorgensen, F. Widdel, J. Peckmann, N. V. Pimenov, M. B. Gulin, *Science* **2002**, *297*, 1013–1015.
19. K. Nauhaus, A. Boetius, M. Kruger, F. Widdel, *Environ. Microbiol.* **2002**, *4*, 296–305.
20. A. A. Raghoebarsing, A. Pol, K. T. van de Pas-Schoonen, A. J. Smolders, K. F. Ettwig, W. I. Rijpstra, S. Schouten, J. S. Damste, H. J. Op den Camp, M. S. Jetten, M. Strous, *Nature* **2006**, *440*, 918–921.
21. S. J. Hallam, P. R. Girguis, C. M. Preston, P. M. Richardson, E. F. DeLong, *Appl. Environ. Microbiol.* **2003**, *69*, 5483–5491.
22. E. F. DeLong, *Nature* **2000**, *407*, 577–579.
23. D. R. Blake, F. Sherwood Rowland, *Science* **1988**, *239*, 1129–1131.
24. D. A. Lashoff, D. Ahuja, *Nature* **1990**, *344*, 213–242.
25. J. Schimel, *Nature* **2000**, *403*, 375, 377.
26. R. S. Wolfe, *Am. Soc. Microbiol. News* **1996**, *62*, 529–534.
27. H. A. Barker, *Bacterial Fermentations*, Wiley, New York, 1956.
28. N. L. Söhngen, *Recl. Trav. Chim. Pays Bas.* **1910**, *29*, 238–250.
29. R. E. Hungate, in *Methods in Microbiology*, Eds J. R. Norris, D. W. Ribbons, Academic Press, Inc., New York, 1969, pp 117–132.
30. W. E. Balch, G. E. Fox, L. J. Magrum, C. R. Woese, R. S. Wolfe, *Microbiol. Rev.* **1979**, *43*, 260–296.
31. R. S. Wolfe, *Methods Enzymol.* **2011**, *494*, 1–22.
32. J. G. Zeikus, R. S. Wolfe, *J. Bacteriol.* **1972**, *109*, 707–715.
33. A. C. Rosenzweig, S. W. Ragsdale, Eds, *Methods in Methane Metabolism*, Vol. 494, Elsevier, Academic Press, Amsterdam, 2011.
34. A. A. DiMarco, T. A. Bobik, R. S. Wolfe, *Ann. Rev. Biochem.* **1990**, *59*, 355–394.
35. G. Diekert, B. Klee, R. K. Thauer, *Arch. Microbiol.* **1980**, *124*, 103–106.
36. G. Diekert, R. Jaenchen, R. K. Thauer, *FEBS Letters* **1980**, *119*, 118–120.
37. W. B. Whitman, R. S. Wolfe, *Biochem. Biophys. Res. Comm.* **1980**, *92*, 1196–1201.
38. G. Diekert, U. Konheiser, K. Piechulla, R. K. Thauer, *J. Bacteriol.* **1981**, *148*, 459–464.
39. J. Telser, in *Structure and Bonding*, Ed R. J. P. Williams, Springer Verlag, Heidelberg, 1998, pp 32–63.
40. B. Jaun, *Met. Ions Biol. Syst.* **1993**, *29*, 287–337.

41. J. Telser, *J. Braz. Chem. Soc.* **2010**, *21*, 1139–1157.

42. A. Pfaltz, B. Juan, A. Fassler, A. Eschenmoser, R. Jaenchen, H. H. Gilles, G. Diekert, R. K. Thauer, *Helv. Chim. Acta* **1982**, *65*, 828–865.

43. G. Färber, W. Keller, C. Kratky, B. Jaun, A. Pfaltz, C. Spinner, A. Kobelt, A. Eschenmoser, *Helv. Chim. Acta* **1991**, *74*, 697–716.

44. H. Won, M. F. Summers, K. Olson, R. S. Wolfe, *J. Am. Chem. Soc.* **1990**, *112*, 2178–2184.

45. W. L. Ellefson, W. B. Whitman, R. S. Wolfe, *Proc. Natl. Acad. Sci. USA* **1982**, *79*, 3707–3710.

46. B. Jaun, *Helv. Chim. Acta* **1990**, *73*, 2209–2216.

47. B. Jaun, A. Pfaltz, *J. Chem. Soc., Chem. Comm.* **1986**, 1327–1329.

48. U. Ermler, W. Grabarse, S. Shima, M. Goubeaud, R. K. Thauer, *Science* **1997**, *278*, 1457–1462.

49. W. G. Grabarse, F. Mahlert, S. Shima, R. K. Thauer, U. Ermler, *J. Mol. Biol.* **2000**, *303*, 329–344.

50. S. Shima, M. Krueger, T. Weinert, U. Demmer, J. Kahnt, R. K. Thauer, U. Ermler, *Nature* **2012**, *481*, 98–101.

51. P. E. Cedervall, M. Dey, X. Li, R. Sarangi, B. Hedman, S. W. Ragsdale, C. M. Wilmot, *J. Am. Chem. Soc.* **2011**, *133*, 5626–5628.

52. W. G. Grabarse, F. Mahlert, E. C. Duin, M. Goubeaud, S. Shima, R. K. Thauer, V. Lamzin, U. Ermler, *J. Mol. Biol.* **2001**, *309*, 315–330.

53. P. E. Cedervall, M. Dey, A. R. Pearson, S. W. Ragsdale, C. M. Wilmot, *Biochemistry* **2010**, *49*, 7683–7693.

54. S. Shima, E. Warkentin, R. K. Thauer, U. Ermler, *J. Biosci. Bioeng.* **2002**, *93*, 519–530.

55. Y.-C. Horng, D. F. Becker, S. W. Ragsdale, *Biochemistry* **2001**, *40*, 12875–12885.

56. S. Ebner, B. Jaun, M. Goenrich, R. K. Thauer, J. Harmer, *J. Am. Chem. Soc.* **2010**, *132*, 567–575.

57. W. Grabarse, F. Mahlert, S. Shima, R. K. Thauer, U. Ermler, *J. Mol. Biol.* **2000**, *303*, 329–344.

58. T. Selmer, J. Kahnt, M. Goubeaud, S. Shima, W. Grabarse, U. Ermler, R. K. Thauer, *J. Biol. Chem.* **2000**, *275*, 3755–3760.

59. J. Kahnt, B. Buchenau, F. Mahlert, M. Kruger, S. Shima, R. K. Thauer, *FEBS J.* **2007**, *274*, 4913–4921.

60. H. Won, K. D. Olson, D. R. Hare, R. S. Wolfe, C. Kratky, M. F. Summers, *J. Am. Chem. Soc.* **1992**, *114*, 6880–6892.

61. C. L. Hamilton, R. A. Scott, M. K. Johnson, *J. Biol. Chem.* **1989**, *264*, 11605–11613.

62. B. Jaun, A. Pfaltz, *J. Chem. Soc. Chem. Commun.* **1988**, *1327*, 293–294.

63. S.-K. Lin, B. Jaun, *Helv. Chim. Acta* **1992**, *75*, 1478–1490.

64. S.-K. Lin, B. Jaun. *Helv. Chim. Acta* , **1991** *74*, 1725–1738.

65. Y. Zhou, D. A. Sliwa, S. W. Ragsdale, in *Handbook of Porphyrin Science*, Eds K. M. Kadish, K. M. Smith, R. Guilard, World Scientific Publishing, 2012, pp 1–42,

66. J. Telser, R. Davydov, Y. C. Horng, S. W. Ragsdale, B. M. Hoffman, *J. Am. Chem. Soc.* **2001**, *123*, 5853–5860.

67. F. Mahlert, C. Bauer, B. Jaun, R. K. Thauer, E. C. Duin, *J. Biol. Inorg. Chem.* **2002**, *7*, 500–513.

68. S. Rospert, R. Böcher, S. P. J. Albracht, R. K. Thauer, *FEBS Lett.* **1991**, *291*, 371–375.

69. E. C. Duin, N. J. Cosper, F. Mahlert, R. K. Thauer, R. A. Scott, *J. Biol. Inorg. Chem.* **2003**, *8*, 141–148.

70. Y. Zhou, A. E. Dorchak, S. W. Ragsdale, *Front. Microbiol. Chem.* **2013**, *4*, 69.

71. D. F. Becker, S. W. Ragsdale, *Biochemistry* **1998**, *37*, 2639–2647.

72. M. Goubeaud, G. Schreiner, R. K. Thauer, *Eur. J. Biochem.* **1997** , *243*, 110–114.

73. D. Hinderberger, S. Ebner, S. Mayr, B. Jaun, M. Reiher, M. Goenrich, R. K. Thauer, J. Harmer, *J. Biol. Inorg. Chem.* **2008**, *13*, 1275–1289.

74. R. Sarangi, M. Dey, S. W. Ragsdale. *Biochemistry* **2009**, *48*, 3146–3156.

75. D. I. Kern, M. Goenrich, B. Jaun, R. K. Thauer, J. Harmer, D. Hinderberger, *J. Biol. Inorg. Chem.* **2007**, *12*, 1097–1105.
76. J. Harmer, C. Finazzo, R. Piskorski, S. Ebner, E. C. Duin, M. Goenrich, R. K. Thauer, M. Reiher, A. Schweiger, D. Hinderberger, B. Jaun, *J. Am. Chem. Soc.* **2008**, *130*, 10907–10920.
77. C. Finazzo, J. Harmer, C. Bauer, B. Jaun, E. C. Duin, F. Mahlert, M. Goenrich, R. K. Thauer, S. Van Doorslaer, A. Schweiger, *J. Am. Chem. Soc.* **2003**, *125*, 4988–4989.
78. E. C. Duin, L. Signor, R. Piskorski, F. Mahlert, M. D. Clay, M. Goenrich, R. K. Thauer, B. Jaun, M. K. Johnson, *J. Biol. Inorg. Chem.* **2004**, *9*, 563–576.
79. J. L. Craft, Y.-C. Horng, S. W. Ragsdale, T. C. Brunold, *J. Am. Chem. Soc.* **2004**, *126*, 4068–4069.
80. M. Dey, X. Li, Y. Zhou, S. W. Ragsdale, *Met. Ions Life Sci.* **2010**, *7*, 71–110.
81. G. K. Lahiri, A. M. Stolzenberg, *Inorg. Chem.* **1993**, *32*, 4409–4413.
82. M. Dey, J. Telser, R. C. Kunz, N. S. Lees, S. W. Ragsdale, B. M. Hoffman, *J. Am. Chem. Soc.* **2007**, *129*, 11030–11032.
83. N. Yang, M. Reiher, M. Wang, J. Harmer, E. C. Duin, *J. Am. Chem. Soc.* **2007**, *129*, 11028–11029.
84. R. C. Kunz, Y.-C. Horng, S. W. Ragsdale, *J. Biol. Chem.* **2006**, *281*, 34663–34676.
85. X. Li, J. Telser, B. M. Hoffman, G. Gerfen, S. W. Ragsdale, *Biochemistry* **2010**, *49*, 6866–6876.
86. B. Jaun, R. K. Thauer, *Met. Ions Life Sci.* **2009**, *6*, 115–132.
87. R. C. Kunz, M. Dey, S. W. Ragsdale, *Biochemistry* **2008**, *47*, 2661–2667.
88. J. Ellermann, A. Kobelt, A. Pfaltz, R. K. Thauer, *FEBS Lett.* **1987**, *220*, 358–362.
89. Y. Ahn, J. A. Krzycki, H. G. Floss, *J. Am. Chem. Soc.* **1991**, *113*, 4700–4701.
90. S. Scheller, M. Goenrich, R. K. Thauer, B. Jaun, *J. Am. Chem. Soc.* **2013**, *135*, 14985–14995.
91. S. Scheller, M. Goenrich, R. K. Thauer, B. Jaun, *J. Am. Chem. Soc.* **2013**, *135*, 14975–14984.
92. V. Pelmenschikov, M. R. A. Blomberg, P. E. M. Siegbahn, R. H. Crabtree, *J. Am. Chem. Soc.* **2002**, *124*, 4039–4049.
93. M. Dey, X. Li, R. C. Kunz, S. W. Ragsdale, *Biochemistry* **2010**, *49*, 10902–10911.
94. M. Goenrich, F. Mahlert, E. C. Duin, C. Bauer, B. Jaun, R. K. Thauer, *J. Biol. Inorg. Chem.* **2004**, *9*, 691–705.
95. R. G. Matthews, *Acc. Chem. Res.* **2001**, *34*, 681–689.
96. S. W. Ragsdale, in *The Porphyrin Handbook*, Eds K. M. Kadish, K. M. Smith, R. Guilard, Academic Press, New York, 2003, pp 205–228.
97. D. Hinderberger, R. P. Piskorski, M. Goenrich, R. K. Thauer, A. Schweiger, J. Harmer, B. Jaun, *Angew. Chem. Int. Ed. Engl.* **2006**, *45*, 3602–3607.
98. M. Dey, R. C. Kunz, D. M. Lyons, S. W. Ragsdale, *Biochemistry* **2007**, *46*, 11969–11978.
99. A. Ghosh, T. Wondimagegn, H. Ryeng, *Curr. Opin. Chem. Biol.* **2001**, *5*, 744–750.
100. V. Pelmenschikov, P. E. Siegbahn, *J. Biol. Inorg. Chem.* **2003**, *8*, 653–662.
101. C. Finazzo, J. Harmer, B. Jaun, E. C. Duin, F. Mahlert, R. K. Thauer, S. Van Doorslaer, A. Schweiger, *J. Biol. Inorg. Chem.* **2003**, *8*, 586–593.
102. S.-L. Chen, M. R. A. Blomberg, P. E. M. Siegbahn, *Chem. Eur. J.* **2012**, *18*, 6309–6315.
103. S. Scheller, M. Goenrich, R. Boecher, R. K. Thauer, B. Jaun, *Nature* **2010**, *465*, 606–608.
104. M. Dey, X. Li, R. C. Kunz, S. W. Ragsdale, *Biochemistry* **2010**, *49*, 10902–10911.
105. M. Goenrich, E. C. Duin, F. Mahlert, R. K. Thauer, *J. Biol. Inorg. Chem.* **2005**, *10*, 333–342.
106. C. A. Haynes, R. Gonzalez, *Nat. Chem. Biol.* **2014**, *10*, 331–339.
107. R. J. Conrado, R. Gonzalez, *Science* **2014**, *343*, 621–623.

Chapter 7
Cleaving the N,N Triple Bond: The Transformation of Dinitrogen to Ammonia by Nitrogenases

Chi Chung Lee, Markus W. Ribbe, and Yilin Hu

Contents

ABSTRACT 148
1 INTRODUCTION 148
2 THE STRUCTURAL AND BIOCHEMICAL PROPERTIES
  OF Mo-NITROGENASE 150
  2.1 The Fe Protein and Its Associated Metal Clusters 151
      2.1.1 The Polypeptide 151
      2.1.2 The [Fe$_4$S$_4$] Cluster 152
  2.2 The MoFe Protein and Its Associated Metal Clusters 154
      2.2.1 The Polypeptide 154
      2.2.2 The P-cluster 156
      2.2.3 The FeMoco 156
3 THE CATALYTIC MECHANISM OF Mo-NITROGENASE 157
  3.1 The Thorneley–Lowe Model 158
      3.1.1 The Fe Protein Cycle 158
      3.1.2 The MoFe Protein Cycle 160
  3.2 Further Development and Modifications of the Thorneley–Lowe Model 162
      3.2.1 Intermediates of the MoFe Protein Cycle 162
      3.2.2 The Reductive Dihydrogen Elimination Mechanism 164
      3.2.3 The Alternating Dinitrogen Reduction Pathway 165

C.C. Lee • Y. Hu
Department of Molecular Biology and Biochemistry, 2236 McGaugh Hall,
University of California, Irvine, CA 92697-3900, USA
e-mail: chichul@uci.edu; yilinh@uci.edu

M.W. Ribbe (✉)
Department of Molecular Biology and Biochemistry, 2236 McGaugh Hall,
University of California, Irvine, CA 92697-3900, USA

Department of Chemistry, University of California, Irvine, CA 92697-2025, USA
e-mail: mribbe@uci.edu

© Springer Science+Business Media Dordrecht 2014
P.M.H. Kroneck, M.E. Sosa Torres (eds.), *The Metal-Driven Biogeochemistry of Gaseous Compounds in the Environment*, Metal Ions in Life Sciences 14, DOI 10.1007/978-94-017-9269-1_7

4   THE DISTINCT STRUCTURAL AND CATALYTIC FEATURES
    OF V-NITROGENASE 166
    4.1   The Structural Features 166
          4.1.1   The Fe Protein 166
          4.1.2   The VFe Protein 167
    4.2   The Catalytic Features 169
          4.2.1   The Reduction of Dinitrogen 169
          4.2.2   The Reduction of Carbon Monoxide 171
5   CONCLUSIONS 172
ABBREVIATIONS 173
ACKNOWLEDGMENT 173
REFERENCES 173

Abstract Biological nitrogen fixation is a natural process that converts atmospheric nitrogen (N_2) to bioavailable ammonia (NH_3). This reaction not only plays a key role in supplying bio-accessible nitrogen to all life forms on Earth, but also embodies the powerful chemistry of cleaving the inert N,N triple bond under ambient conditions. The group of enzymes that carry out this reaction are called nitrogenases and typically consist of two redox active protein components, each containing metal cluster(s) that are crucial for catalysis. In the past decade, a number of crystal structures, including several at high resolutions, have been solved. However, the catalytic mechanism of nitrogenase, namely, how the N,N triple bond is cleaved by this enzyme under ambient conditions, has remained elusive. Nevertheless, recent biochemical and spectroscopic studies have led to a better understanding of the potential intermediates of N_2 reduction by the molybdenum (Mo)-nitrogenase. In addition, it has been demonstrated that carbon monoxide (CO), which was thought to be an inhibitor of N_2 reduction, could also be reduced by the vanadium (V)-nitrogenase to small alkanes and alkenes. This chapter will begin with an introduction to biological nitrogen fixation and Mo-nitrogenase, continue with a discussion of the catalytic mechanism of N_2 reduction by Mo-nitrogenase, and conclude with a survey of the current knowledge of N_2- and CO-reduction by V-nitrogenase and how V-nitrogenase compares to its Mo-counterpart in these catalytic activities.

Keywords ammonia • ATP hydrolysis • carbon monoxide • crystallography • dinitrogen • ENDOR spectroscopy • FeMoco • Fe protein • [Fe_4S_4] cluster • MoFe protein • P-cluster • Thorneley–Lowe model • VFe protein

Please cite as: *Met. Ions Life Sci.* 14 (2014) 147–176

1   Introduction

Biological nitrogen fixation refers to the natural process in which the atmospheric nitrogen (N_2) is converted to a bio-accessible form, ammonia (NH_3) [1–16]. This process represents a critical step in the global nitrogen cycle. Since nitrogen, one of the major constituents of bio-macromolecules (e.g., proteins and nucleic acids), would eventually be converted to N_2 through denitrification, the conversion of N_2 to NH_3 provides the necessary re-entry point for the inorganic nitrogen back into the biosphere and, therefore, is essential for all life forms on Earth [17]. Biological nitrogen fixation has been estimated to reduce (or fix) approximately 90 million tons of nitrogen per year, which supports roughly half of the human population [18, 19]. The other half of the population relies on the Haber-Bosch process, an industrial counterpart of biological nitrogen fixation, for the fixed nitrogen supplies. Developed in 1908 by Fritz Haber, this process combines N_2 with hydrogen gas (H_2) into NH_3 at high temperature (300–400 °C) and pressure (~300 atm) [8]. These conditions require approximately 1–2 % of annual power production worldwide [20]. Such a high energy demand does not come as a surprise, though, because the N, N triple bond has a bond dissociation energy of 225 kcal (945 kJ) per mole and, therefore, is one of the most stable bonds found in Nature [21]. Given the energy crisis and population growth on our planet, it is highly desirable to develop an energy-efficient alternative to the Haber-Bosch process for NH_3 production.

Biological nitrogen fixation may just provide a suitable blueprint for future development of such an alternative approach, as this process converts N_2 to NH_3 at ambient temperature and pressure while producing H_2, a clean energy source, as an abundant side product. The enzymes responsible for such a conversion are nitrogenases, which can be found in a select group of microbes called diazotrophs. These metal-containing enzymes couple the reduction of N_2 with the hydrolysis of adenosine 5′-triphosphate (ATP), thereby overcoming the energy barrier of the reaction and facilitating the cleavage of the N,N triple bond under ambient conditions. Recently, it has been demonstrated that nitrogenases are also capable of converting carbon monoxide (CO) to hydrocarbons [22]. As such, the reactions catalyzed by nitrogenases can be appreciated not only from the perspective of energy conservation, but also from the perspective of the useful products they generate.

Four classes of nitrogenases have been identified to date: the molybdenum (Mo)-nitrogenase, the vanadium (V)-nitrogenase, the iron (Fe)-only nitrogenase and the superoxide dismutase (SOD)-dependent nitrogenase from *Streptomyces thermoautotrophicus* [5, 8, 10, 23, 24]. The first three nitrogenases are highly homologous in primary sequences and are believed to be structurally similar to one another, namely, they all consist of two protein components, each of which houses metal cluster(s) crucial for catalysis [1–16]. The three nitrogenases also function similarly in that the two protein components act as redox partners, allowing electrons to flow sequentially through their respective metal clusters for the reduction of N_2. The Mo-nitrogenase is the best characterized among these nitrogenases, with more than 35 crystal structures deposited in the Protein Data Bank (PDB) and the largest body of spectroscopic and catalytic data collected on this protein. Consequently, most of our current knowledge regarding the mechanism of biological nitrogen fixation centers on the Mo-nitrogenase. The V-nitrogenase, on the other hand, was much less studied than its Mo-counterpart. This nitrogenase displays structural and catalytic features that are distinct from those of the Mo-nitrogenase [5, 22]. Most notably, it reduces carbon monoxide to short-chain hydrocarbons, such as ethane, ethylene, propane, and propylene, at much higher efficiencies than its Mo-counterpart [22]. Given the isoelectronic properties of CO and N_2, it is conceivable that knowledge of the mechanism of CO reduction may contribute, if indirectly, to our understanding of the mechanism of N_2 reduction.

The objective of this article is to review the most recent advances toward the understanding of the reaction mechanism of nitrogenase, with sections introducing the structural properties of Mo-nitrogenase (Section 2), presenting the classical model of nitrogenase mechanism and the recent development of catalysis by Mo-nitrogenase (Section 3), comparing the structural and catalytic features of Mo- and V-nitrogenases (Section 4) and briefly summarizing the current status of this research topic (Section 5).

2 The Structural and Biochemical Properties of Mo-Nitrogenase

Mo-nitrogenase is the best studied member of this enzyme family and it has a wide prevalence in Nature. This nitrogenase is found in diazotrophic (or nitrogen-fixing) microorganisms, such as *Azotobacter chroococcum, Azotobacter vinelandii, Klebsiella pneumoniae*, and *Clostridium pasteurianum* [9, 10, 25]. Biochemical analysis has long established Mo-nitrogenase as a two-component enzyme system, with one component referred to as the iron (Fe) protein and the other referred to as the molybdenum-iron (MoFe) protein (Figure 1). Both component proteins contain FeS clusters: the Fe protein carries a ferrodoxin-type [Fe_4S_4] cluster; whereas the MoFe protein carries two types of complex FeS clusters: the P-cluster ([Fe_8S_7]) and the iron-molybdenum cofactor (or FeMoco; [$MoFe_7S_9C$-homocitrate]) (Figure 1).

The three FeS clusters of nitrogenase are essential for catalysis, as they can be aligned into an electron transfer pathway upon ATP-dependent docking of the Fe protein on the MoFe protein, which then allows electrons to be transferred from the [Fe_4S_4] cluster of the former protein, via the P-cluster, to the FeMoco of the latter protein, where substrate reduction eventually occurs (Figure 1) [1–16]. To better understand this ATP-dependent, inter-protein electron transfer process, it is crucial for us to take a look at the structural and chemical properties of the two component proteins of Mo-nitrogenase and their associated metal clusters.

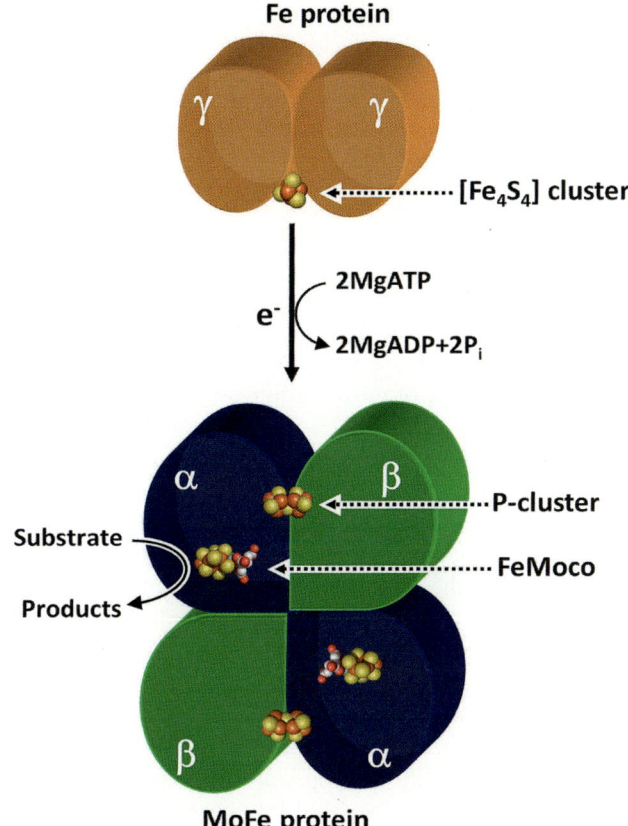

Figure 1 Schematic presentation showing the interaction between the two component proteins of Mo-nitrogenase during catalysis.

2.1 The Fe Protein and Its Associated Metal Clusters

The Fe protein of Mo-nitrogenase couples the energy derived from ATP hydrolysis to the transfer of electrons to its redox partner, the MoFe protein. This protein and its associated $[Fe_4S_4]$ cluster display a number of interesting structural and spectroscopic features, which are discussed in detail below.

2.1.1 The Polypeptide

The Fe protein of *A. vinelandii* is encoded by the *nifH* gene. It is a γ_2-homodimer of ~60 kDa, and its two subunits are bridged by a $[Fe_4S_4]$ cluster [1–4, 6–9, 26–28]. Each subunit of the Fe protein contains a Walker A motif (residues 9–16), which provides the site for the binding and hydrolysis of MgATP [26–28]. While the

X-ray crystal structures of both the nucleotide-free and the MgADP-bound forms of the Fe protein have been solved (Figure 2a) [29], the structure of the MgATP-bound form of this protein is still unavailable. Nevertheless, biochemical and mutagenetic studies have revealed the potential involvement of a so-called P-loop at the Walker A motif in the binding of MgATP to the Fe protein [6, 10, 27, 30–32]. The P-loop features Ser^{16}, which coordinates a Mg^{2+} ion and a conserved Lys^{10} residue through its hydroxyl group, thereby forming a salt bridge with the β-phosphate of the nucleotide and consequently locking the nucleotide in its binding site (Figure 2a, b).

Upon ATP binding, the Fe protein undergoes a conformational rearrangement that renders its $[Fe_4S_4]$ cluster more exposed to solvent and, consequently, there is an increase in the reactivity of the $[Fe_4S_4]$ cluster toward metal chelators [33]. Consistent with this observation, electron paramagnetic resonance (EPR) analysis also suggests a potential change of the environment of the $[Fe_4S_4]$ cluster in the Fe protein upon incubation with MgATP [33–36]. Interestingly, the crystal structures of the Fe protein show that the nucleotide-binding sites are 15–19 Å away from the $[Fe_4S_4]$ cluster [4, 6, 9, 29]. The lack of direct contact between the nucleotide- and cluster-binding sites provides further support for the proposed conformational change of the Fe protein upon MgATP binding that alters the location of the $[Fe_4S_4]$ cluster [4]. While the structural information of such a conformational rearrangement has not been acquired so far, there is indirect evidence from mutagenic studies that certain site-directed variants of the Fe protein cannot undergo conformational changes due to the perturbation of the signal transduction between the nucleotide- and cluster-binding sites [31, 32]. Further investigations of these variants suggest that the Walker A motif, the so-called Switch I (residues 38–44) and Switch II (residues 125–129) regions, and a conserved region around the residue Ala^{157} are all crucial for the MgATP-dependent protein conformational changes of the Fe protein (Figure 2b) [4, 6, 10, 27]. Based on these results, as well as the crystal structure of the MgADP-bound Fe protein, it has been proposed that (i) Switch I acts as a direct link that communicates between the nucleotide-binding site and the Fe protein/MoFe protein docking interface; and (ii) Switch II mediates signal transduction between the nucleotide-binding site and the $[Fe_4S_4]$ cluster [29, 37]. The purpose of this MgATP binding-induced change in protein conformation is still under debate; however, this conformational change is clearly essential for catalysis, as any mutation that prevents it from occurring renders the Fe protein inactive in substrate reduction.

2.1.2   The [Fe₄S₄] Cluster

The $[Fe_4S_4]$ cluster is symmetrically ligated by two cysteine residues, Cys^{97} and Cys^{132}, from each subunit of the Fe protein (Figure 2c). It has a +1 oxidization state when the protein is isolated in excess dithionite. In this state, the cluster displays a mixed electronic spin state of $S = 1/2$ and $S = 3/2$ [38, 39]. The distribution of

Figure 2 Crystal structures of the MgADP-bound Fe protein of Mo-nitrogenase (**a**), the catalytically important domains (**b**), and the [Fe$_4$S$_4$] cluster (**c**) of the Fe protein. The subunits of the Fe protein are colored orange, and the cluster is shown in ball-and-stick presentation. The P-loop is colored green, the Switch I region is colored blue, and the Switch II region is colored red. Atoms are colored as follows: C, white; O, red; N, blue; P, green; Fe, orange; S, yellow. The cysteinyl ligands from the two subunits are labeled Cys and Cys*, respectively. PYMOL was used to generate this figure (PDB entry 1FP6).

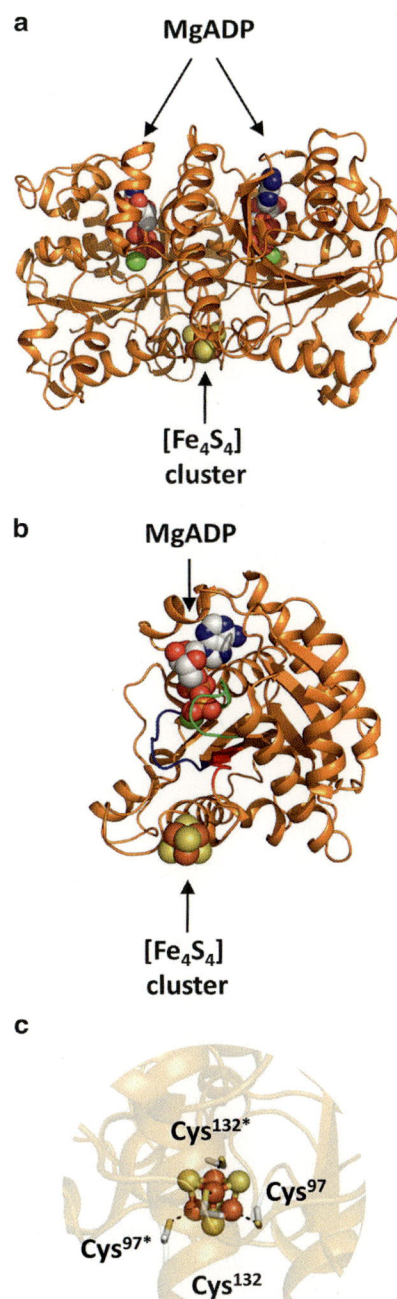

these two spin states is thought to be influenced by a number of factors, such as temperature, viscosity, and ionic strength of the medium [4, 38, 39]. The $[Fe_4S_4]^{1+}$ cluster can be oxidized by one electron to the diamagnetic +2 state, either enzymatically by its redox partner (i.e., MoFe protein), or chemically by oxidizing dyes [e.g., indigodisulfonate (IDS)]. The $[Fe_4S_4]^{1+/2+}$ couple is believed to be physiologically relevant, with a midpoint potential of approximately −290 mV versus NHE for the Fe protein of *A. vinelandii* [40]. Interestingly, the midpoint potential of the $[Fe_4S_4]^{1+/2+}$ couple drops notably (by ~120 mV) to approximately −430 mV versus NHE [40]. While it is generally assumed that such a significant decrease in redox potential is important for the electron transfer between the Fe protein and the MoFe protein, the exact relevance of this change to nitrogenase catalysis remains elusive.

The $[Fe_4S_4]^{1+}$ cluster can be further reduced to the all-ferrous, $[Fe_4S_4]^0$ state by a strong reductant like Ti(III) citrate, which has a redox potential of −800 mV versus NHE or lower [41, 42]. This observation implies that the Fe protein may be able to transfer two electrons per catalytic cycle via a $[Fe_4S_4]^{0/2+}$ redox couple, although there is currently no evidence supporting the theory that the all-ferrous state is a physiologically or catalytically relevant redox state of the $[Fe_4S_4]$ cluster of the Fe protein.

2.2 The MoFe Protein and Its Associated Metal Clusters

The MoFe protein is the catalytic component of Mo-nitrogenase. The structural and spectroscopic properties of this protein and its associated metal clusters, which are central to the understanding of N_2 activation and reduction, are discussed in detail below.

2.2.1 The Polypeptide

The MoFe protein of *A. vinelandii* (Figure 3) is an $\alpha_2\beta_2$-heterotetramer of ~220 kDa, and its α- and β-subunits are encoded by the *nifD* and *nifK* genes, respectively. The two αβ-dimers are related by a two-fold rotational axis (Figure 3a). Each αβ-dimer of the MoFe protein contains a pair of P-cluster and FeMoco and, therefore, can be viewed as a functional unit on its own. Each of the α- and β-subunits are divided into three structural domains: αI, αII, αIII, and βI, βII, βIII, respectively, and all of these domains consist of α-helical- and parallel-β-sheet-type polypeptide folds [43–49]. The P-cluster is located at the interfaces of the αI- and βI-domains, whereas the FeMoco is buried within a cavity between the αI-, αII-, and αIII-subunits, 14 Å away from the P-cluster [43–49].

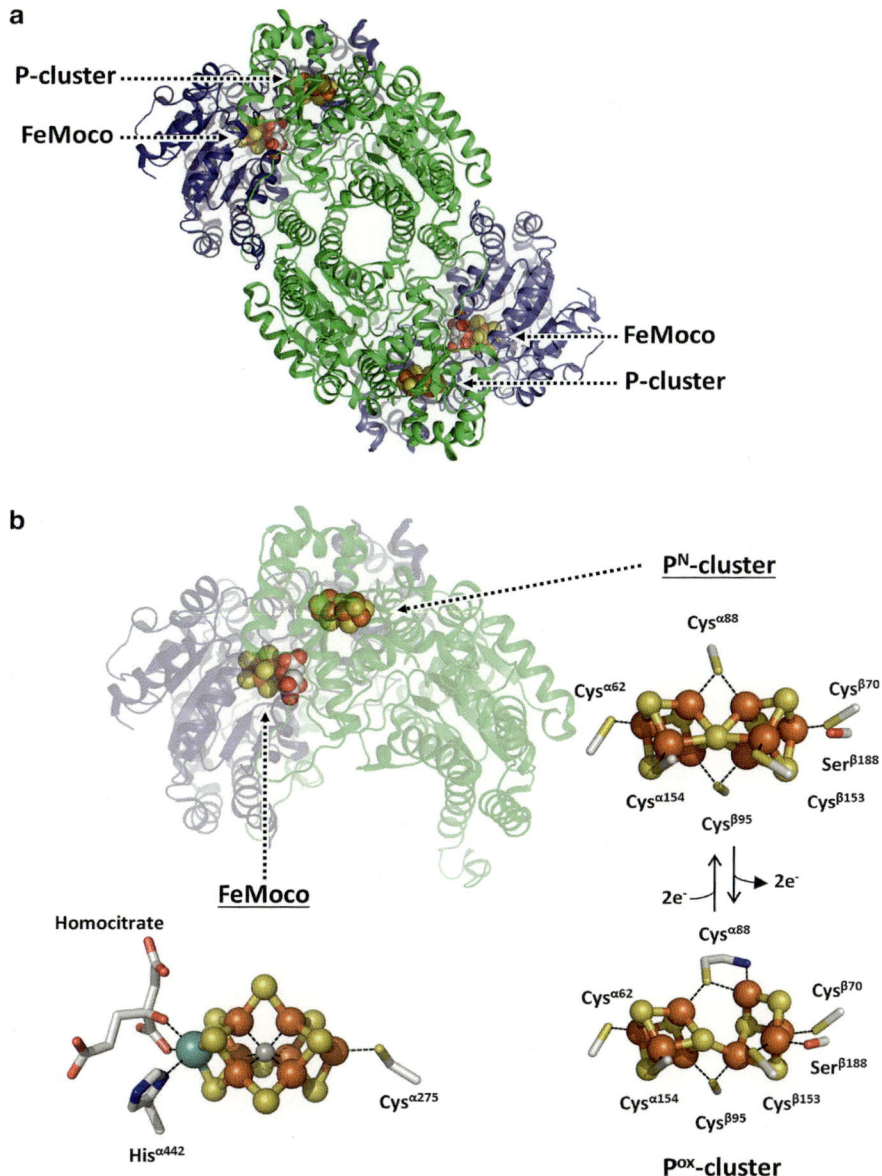

Figure 3 Crystal structures of the MoFe protein from *A. vinelandii* (**a**), the αβ-dimer of the MoFe protein (**b**, *upper left*), the FeMoco (b, *lower left*), and the P-cluster in the P^N (**b**, *upper right*) and P^{ox} state (**b**, *lower right*). The α- and β-subunits are colored blue and green, respectively, and the clusters are shown in ball-and-stick presentations. Atoms are colored as follows: Fe, orange; S, yellow; Mo, cyan; C, white; O, red; N, blue. The interstitial C^{4-} atom in the FeMoco is colored light gray. PYMOL was used to generate this figure (PDB entry 1M1N).

2.2.2 The P-cluster

The P-cluster is a $[Fe_8S_7]$ cluster that is bridged between each $\alpha\beta$-dimer of the MoFe protein by six cysteinyl ligands: three from the α-subunit (Cys$^{\alpha62}$, Cys$^{\alpha88}$, Cys$^{\alpha154}$) and three from the β-subunit (Cys$^{\beta70}$, Cys $^{\beta95}$, Cys $^{\beta153}$) (Figure 3) [7, 10]. In the presence of excess dithionite, the P-cluster adopts a so-called P^N state, which can be structurally viewed as two $[Fe_4S_4]$ sub-cubanes sharing a μ_6-sulfide (Figure 3b). EPR and Mössbauer spectroscopy further demonstrates that the dithionite-reduced P^N-cluster is essentially diamagnetic, which leads to the general belief that all Fe atoms in this cluster are present in the all-ferrous state [4, 50, 51].

   While there is no evidence that the P^N-cluster can undergo further reduction, it has been shown that this cluster can be oxidized stepwise into the so-called P^{1+}- and P^{2+}-oxidation states. The P^{1+}-state can be generated by an enzymatic one-electron oxidation process, and it displays an interesting, mixed spin state of $S = 1/2$, $S = 5/2$ [52]. The P^{2+}-state, also known as the P^{OX}-state, can be generated by a two-electron oxidation process in the presence of IDS. This cluster is far better characterized than the P^{1+}-cluster. It displays a $S = 3$ or 4 integer spin state, which can be best recognized by a parallel-mode EPR signal at $g = 11.8$ [50, 52]. The crystal structure of the oxidized MoFe protein also reveals that the P^{2+}-cluster is structurally different than the P^N-cluster in that it adopts a more open confirmation than the latter (Figure 3b). In particular, the sub-cubane coordinated by the three cysteines from the β-subunit undergoes ligands exchange, with the coordination of two of the μ_6-sulfide-ligated Fe atoms replaced by the coordination of these atoms to two protein ligands: the Oγ atom from the side chain of Ser$^{\beta188}$ and the N atom from the backbone amide of Cys$^{\alpha88}$ (Figure 3b) [10, 46]. Interestingly, the P^N- and P^{OX}-states are readily reversible despite the significant structural reorganization in this process, although it is not clear if such a redox/conformational change of the P-cluster is relevant to nitrogenase catalysis.

2.2.3 The FeMoco

The FeMoco is located within the α-subunit of the MoFe protein, and it is coordinated in the protein by only two residues: His$^{\alpha442}$ at the Mo end and Cys$^{\alpha275}$ at the opposite Fe end of the cluster (Figure 3b) [7, 10, 26]. This cluster consists of a $[MoFe_7S_9]$ core, which contains a carbide (C^{4-}) atom in the center and a homocitrate entity at the Mo end [53–55]. The metal-sulfur core of the FeMoco can be viewed as $[Fe_4S_3]$ and $[MoFe_3S_3]$ sub-cubanes bridged by three μ_2-sulfides and one μ_6-carbide in between. This arrangement renders all Fe atoms μ_4-coordinated by three inorganic sulfides and one additional ligand that is either provided by the cysteinyl thio group (in the case of the terminal Fe atom) or the carbide (in the cases of all other Fe atoms). The Mo atom, on the other hand, has a μ_6-coordination, with homocitrate acting as a bidentate ligand through two oxygen atoms (one carboxyl group and one hydroxyl group) in addition to the ligation provided by three sulfides and one nitrogen atom of $His^{\alpha442}$ (Figure 3). The homocitrate entity forms an extensive hydrogen-bonding network with the sulfides of the metal-sulfur core [4], and it is thought to be responsible for the overall negative charge of the FeMoco, which in turn is important for the insertion of FeMoco along a positively charged path into its binding site within the MoFe protein [56, 57].

The protein-bound FeMoco exhibits a well-characterized $S = 3/2$ EPR signal at $g = 4.7$, 3.7 and 2.0 in the presence of excess dithionite [4]. Electrochemical analysis shows that FeMoco can be either reduced or oxidized by one electron to a diamagnetic state [4, 52]. The EPR signal of the resting-state FeMoco disappears when the MoFe protein is enzymatically reduced by the Fe protein, which is interpreted as FeMoco being further reduced to facilitate substrate turnover. The FeMoco can be extracted as an intact entity into organic solvents, such as N-methylformamide (NMF) [3, 4, 58–60]. Upon extraction, the FeMoco exhibits an $S = 3/2$ EPR signal at $g = 5.94$, 4.66, and 3.50, which is similar to, yet much broader in shape, than that of its protein-bound counterpart [58], suggesting that the polypeptide surroundings of FeMoco have a significant impact on its electronic properties. The solvent-extracted FeMoco can be used to reconstitute and activate the FeMoco-deficient, apo MoFe protein. More excitingly, it can be used directly to reduce certain substrates, such as protons, cyanide, and CO, in the absence of ATP and both protein components of the Mo-nitrogenase [61]. While it is yet to be demonstrated that the extracted FeMoco can reduce N_2, it is conceivable that conditions can be established that allow this cofactor to serve as a catalyst for N_2 reduction in the isolated state.

3   The Catalytic Mechanism of Mo-Nitrogenase

The reaction of N_2 reduction by Mo-nitrogenase is generally depicted as follows:

$$N_2 + 8\,H^+ + 8\,e^- + 16\,ATP \rightarrow 2\,NH_3 + H_2 + 16\,ADP + 16\,P_i \quad (1)$$

In the case of the V-nitrogenase, however, the ratio of H_2 and NH_3 formation was shown to differ from what has been proposed for its Mo-counterpart (see Section 4 below). Structural and biochemical analyses have provided significant insights into the catalytic mechanism of Mo-nitrogenase. The crystal structure of the $MgADP \cdot AlF_4^-$-stabilized Mo-nitrogenase complex has effectively captured a "transition state" of this nitrogenase and provided compelling evidence for the formation of an electron transfer pathway during catalysis [62] (Figure 4).

The large body of biochemical experiments conducted by Thorneley and Lowe, on the other hand, has allowed the successful construction of a classical catalytic scheme of Mo-nitrogenase (see Section 3.1 below). This model provided a great framework for recent spectroscopic studies, which led to the characterization of a number of potential intermediates of N_2 reduction (see Section 3.2 below).

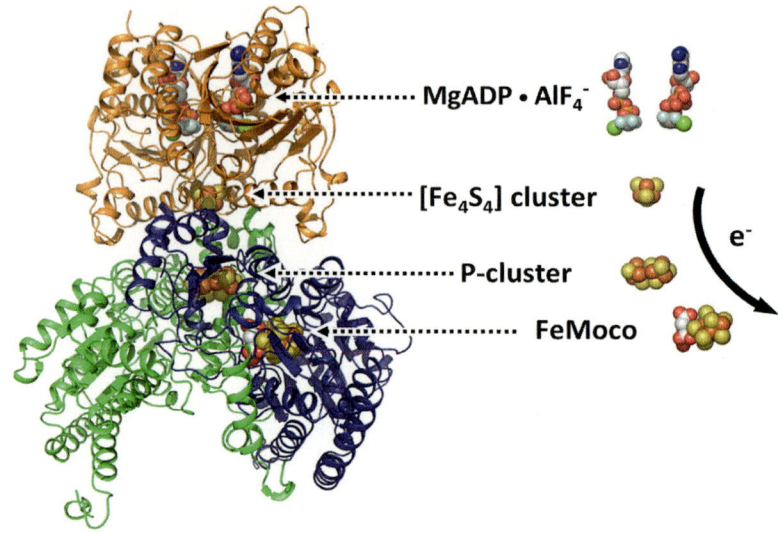

Figure 4 Crystal structure of half of the ADP·AlF$_4^-$-stabilized Fe protein/MoFe protein complex (*left*) and the relative positions of components involved in the electron flow (*right*). The two subunits of the Fe protein are colored orange, and the α- and β-subunits of the MoFe protein are colored blue and green, respectively. All clusters and ADP·AlF$_4^-$ are shown as space-filling models. Atoms are colored as follows: Fe, orange; S, yellow; Mo, cyan; O, red; C, white; N, dark blue; Mg, green; Al, beige; F, light blue. PYMOL was used to create the figure (PDB entry 1 M34).

3.1   The Thorneley–Lowe Model

The classical scheme of N$_2$ reduction, known as the Thorneley–Lowe model, was first introduced in the 1980s to summarize the kinetic data of Mo-nitrogenase [28]. It consists of a so-called Fe protein cycle and a so-called MoFe protein cycle, each representing the series of events or reactions that occur on each protein component for the successful completion of one catalytic cycle by Mo-nitrogenase. It should be noted that, while the MoFe protein cycle describes events required for the complete reduction of one N$_2$ molecule to two NH$_3$ molecules, the Fe protein cycle only depicts events involved in the donation of one electron to the MoFe protein upon the hydrolysis of two ATP molecules. Thus, based on the proposed stoichiometry of N$_2$ reduction (see eqn. 1 above), the completion of one MoFe protein cycle requires the completion of eight Fe protein cycles.

3.1.1   The Fe Protein Cycle

The Fe protein cycle starts with the reduction of Fe protein either by dithionite *in vitro* or ferredoxins *in vivo*, which renders its [Fe$_4$S$_4$] cluster in the +1 oxidation state (Scheme 1). Subsequently, the reduced Fe protein binds two MgATP molecules and undergoes a conformational change, which allows it to complex with the MoFe protein (i.e., stage (i) in Scheme 1). The formation of the Mo-nitrogenase complex then facilities the inter-protein, one-electron transfer from the $[Fe_4S_4]^{1+}$ cluster of the Fe protein to the P-cluster of the MoFe protein, which occurs concomitantly with the hydrolysis of two MgATP molecules. Although the exact order of ATP hydrolysis and electron transfer is not clear [4], it has been proposed that, upon the transfer of electrons, the inorganic phosphates are released from the $FeP^{ox}\{MgADP\text{-}Pi\}_2MoFeP^{red}$ complex (where FeP^{ox} refers to the oxidized Fe protein and $MoFeP^{red}$ refers to the reduced MoFe protein), followed by the dissociation of the Fe protein from the MoFe protein (Scheme 1). The release of the inorganic phosphates signifies that the energy derived from the Fe protein is relayed to the MoFe protein; whereas the dissociation of the Fe protein allows the re-reduction of its $[Fe_4S_4]^{1+}$ cluster following the release of MgADP and the start of another round of electron transfer (Scheme 1).

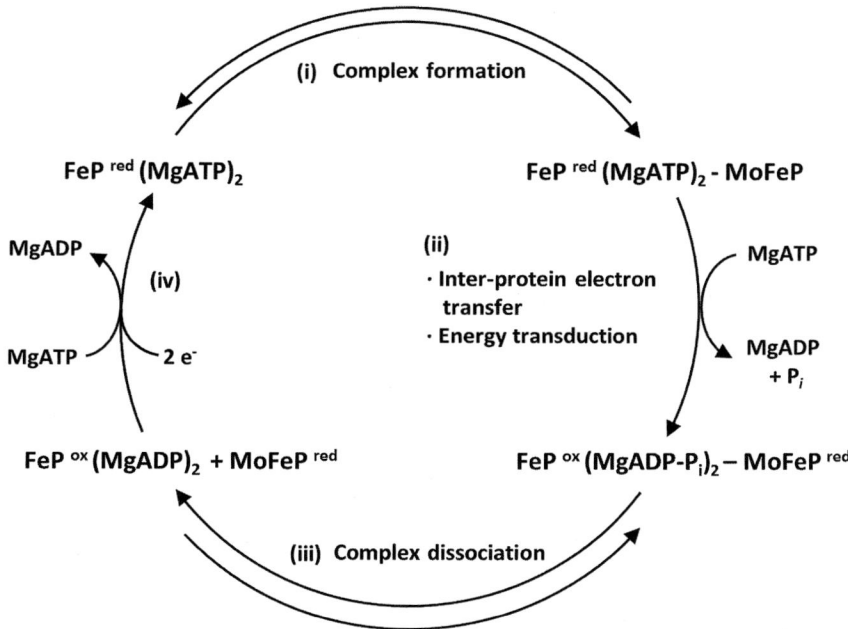

Scheme 1 The Fe protein cycle of Mo-nitrogenase. This cycle describes (**i**) the complex formation of the MgATP-bound Fe protein (FeP) with the MoFe protein (MoFeP); (**ii**) the ATP hydrolysis-dependent electron transfer from FeP to MoFeP; (**iii**) the dissociation of the complex upon the release of the inorganic phosphate (P_i); and (**iv**) the re-reduction of FeP and the replacement of MgADP by MgATP. Superscripts: red, reduced; ox, oxidized.

Excitingly, structural insights into the Fe protein cycle have been provided by a number of Fe protein-MoFe protein complexes generated by (i) chemical crosslinking; (ii) addition of the non-hydrolyzable MgATP analog, MgAMPPCP; (iii) addition of the non-hydrolyzable "MgADP-P_i" analog, MgADP-AlF$_4^-$; and

(iv) addition of MgADP [37, 62, 63]. Studies of these structures reveal that, upon complex formation, the Fe protein adopts different docking conformations, while the MoFe protein remains entirely unchanged in conformation. Remarkably, the [Fe$_4$S$_4$] cluster–P-cluster distance is shortened by ~20 % when MgAMPPCP or MgADP $\cdot$ AlF$_4^-$ is bound. This observation supports the proposal that the binding of MgATP induces a conformational rearrangement of the Fe protein, which may in turn facilitate the transfer of electrons from the Fe protein to the MoFe protein (Table 1).

Table 1 Parameters describing the docking interactions between the Fe protein and the MoFe protein under different conditions.[a]

Complex conditions	ϕ (°)[b]	r (Å)[c]
Chemically cross-linked	30	23.2
MgAMPPCP-stabilized	21	17.8
MgADP-stabilized	26–33	22.6–23.7
MgADP $\cdot$ AlF$_4^-$-stabilized	12–13	17.5–17.6

[a]This table was adapted from [37].
[b]ϕ is the angle between the helices from residue γ98 to γ112 of each of the Fe protein subunits. It is defined that $\phi = 0°$ when the helices are in co-planar arrangement, where the Fe protein docking interface is maximally flattened.
[c]r is the [Fe$_4$S$_4$]–P-cluster distance and it is defined between the centroids of these clusters in each complex.

Additionally, comparison of the MgAMPPCP- or MgADP $\cdot$ AlF$_4^-$-bound structures, which mimic the stage II and III, respectively, of the Fe protein cycle, shows that the binding surface of the Fe protein is substantially flattened, thereby allowing the Fe protein to dock more snugly onto the MoFe protein (Table 1). These structural insights highlight the role of the Fe protein in energetically coupling different nucleotide-binding states to nitrogenase catalysis, which may have some general relevance to other nucleotide-dependent molecular motor systems [63].

3.1.2 The MoFe Protein Cycle

The MoFe protein cycle consists of eight distinct steps, each representing one of the eight electron/proton transfer steps that are required to reduce one molecule of N$_2$ to two molecules of NH$_3$ while evolving one molecule of H$_2$ as the side product (Scheme 2). In this cycle, the resting state of one functional $\alpha\beta$-dimer of the MoFe protein is designated M$_0$, which contains a FeMoco in the dithionite-reduced state. Upon receipt of electrons from the Fe protein, the M$_0$ state is converted stepwise to a series of turnover states. These turnover states are designated M$_n$, where n represents the number of proton/electron pairs added on M$_0$.

In the absence of any substrate, the two-electron-reduced state, M$_2$, is cycled back to M$_0$ through the release of H$_2$ (Scheme 2a). As such, in the absence of N$_2$, the MoFe protein only cycles between M$_0$ and M$_2$, and the enzyme can be viewed as an ATP-dependent hydrogenase. Kinetic data [4, 28] suggest that the binding of N$_2$ to the MoFe protein occurs at a state that is a 3- or 4-electrons more reduced state than

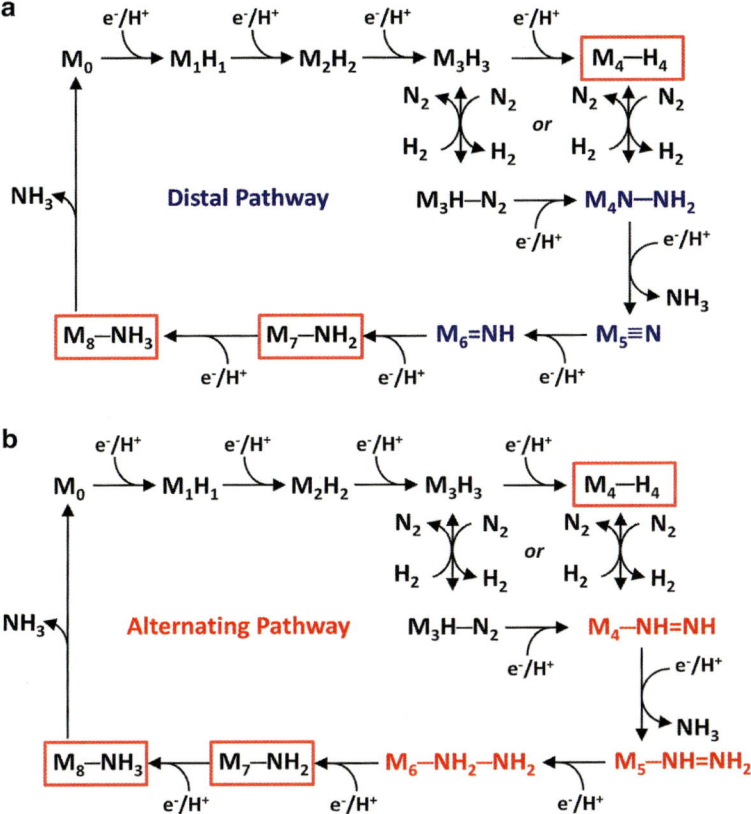

Scheme 2 The MoFe protein cycle of Mo-nitrogenase, featuring the distal (**a**) and alternating (**b**) pathway of N_2 reduction. (**a**) Subscripts 0–8 indicate the number of electrons transferred to the FeMoco of the MoFe protein via the Fe protein cycle (see Scheme 1). The distal pathway describes one plausible sequence of events during N_2 reduction, in which the distal N is reduced first and released as NH_3 before the M_5 stage, followed by the subsequent reduction of the proximal N and the release of a second NH_3 at the M_8 stage. (**b**) The alternating pathway shares most of the steps with the distal pathway, but differs at stages M_4, M_5, and M_6 (highlighted in red). It features the alternating addition of electrons and protons at both the distal and proximal N atoms. Consequently, the two NH_3 molecules are released at stage M_4 and M_8, respectively, late in the sequence. The stages unique to each pathway are colored blue and red, respectively, for the distal pathway and the alternating pathway. Intermediates that have been identified so far are highlighted in red boxes.

M_0 (i.e., M_3 or M_4) (Scheme 2a). Thus, other substrates that bind to a relatively more oxidized state of the MoFe protein, such as acetylene and cyanide, would become non-competitive inhibitors of N_2 and, consequently, they will be preferentially reduced when there is a low electron flux through the MoFe protein [4, 28, 64]. Upon binding of N_2, however, the MoFe protein would undergo four (if N_2 binds at M_4) or five (if N_2 binds at M_3) more proton/electron addition steps to yield two molecules of NH_3 (Scheme 2a). The successive reduction of M_0 is carried out through the stepwise reduction of the two N atoms of N_2, where one N atom is reduced to NH_3 in stage M_4 and M_5, and the other N atom is reduced to NH_3 between stage M_6 and M_8. Thus, this model predicts the formation of $M=N-NH_2$ and $M=NH$ as intermediates at stage M_4 and M_6, respectively, as well as the stepwise release of two NH_3 molecules at stage M_5 and M_8, respectively. Following the release of the second NH_3, MoFe protein is cycled back to the resting state, M_0, and ready for another round of electron/proton addition and substrate reduction (Scheme 2a).

The N_2 reduction pathway depicted in the Thorneley–Lowe model is later designated the distal pathway (Scheme 2a), as it involves the reduction of the distal nitrogen prior to that of the proximal one. A different N_2 reduction pathway was proposed recently [65]. This pathway, termed the alternating pathway, describes the reduction of N_2 as an event occurring alternately on the two N atoms (Scheme 2b). This alternating pathway shares the first five (i.e., the M_0-M_4 states) and the last two stages (i.e., the M_7 and M_8 states) with the distal pathway, but predicts formation of a M–NH=NH intermediate at stage M_4 and a M–NH_2–NH_2 intermediate at stage M_6. It also suggests that the two NH_3 molecules are released at stage M_4 and M_8, respectively (see Section 3.2.3 for details).

3.2 Further Development and Modifications of the Thorneley–Lowe Model

Based on the Thorneley–Lowe model, efforts have been made in the past decade to trap and characterize the reaction intermediates, as well as the individual stages depicted in the MoFe protein cycle, for a better understanding of the mechanism of nitrogenase. One approach employed a combination of spectroscopic methods, such as electron nuclear double resonance (ENDOR), electron spin-echo envelope modulation (ESEEM), and hyperfine sublevel correlation (HYSCORE) spectroscopy, with the mutagenic manipulations of the MoFe protein. This approach has led to the identification of potential reaction intermediates [14, 16] and the proposal of plausible structures of the M_4, M_7 and M_8 states of the MoFe protein cycle [16, 65–68].

3.2.1 Intermediates of the MoFe Protein Cycle

3.2.1.1 The Substrate-Free M_4 Intermediate

The presence of the M_4 state intermediate was first noticed in the $Ile^{\alpha70}$-substituted MoFe protein, which had limited activities in the reduction of all substrates except protons, and later identified in the wild-type MoFe protein, which was present at low concentrations in the reaction mixture [69, 70]. Kinetic and ^{57}Fe ENDOR spectroscopic studies suggest that the M_4 state intermediate is likely four electrons more reduced than the dithionite-reduced MoFe protein [67]. In the absence of any substrate, the MoFe protein in the M_4 state exhibits a novel $S = 1/2$ EPR signal when freeze-quenched under slow turnover conditions [71]. Annealing of this frozen species, on the other hand, results in the relaxation to the resting state via the successive loss of two H_2 molecules, thereby providing additional proof that this species is an M_4 state intermediate [71].

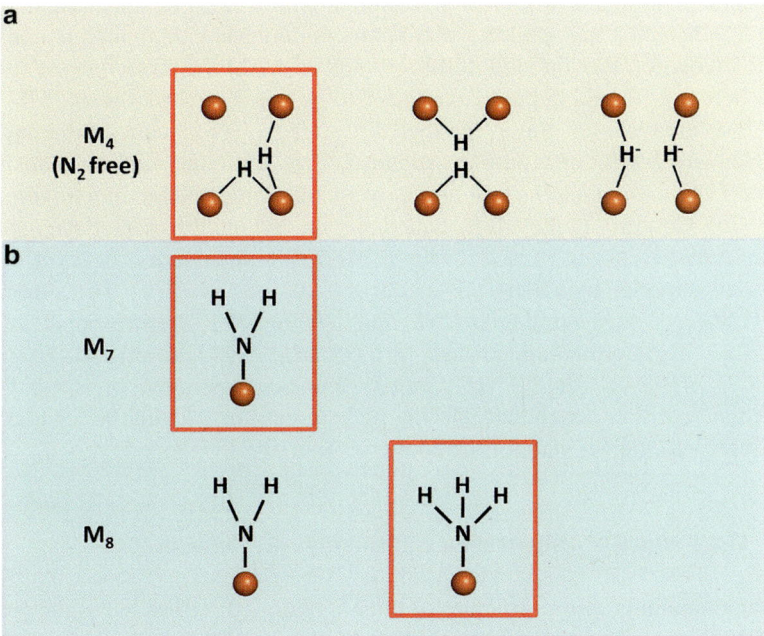

Figure 5 Possible structures of intermediates of N_2 reduction at stage M_4 (**a**), M_7 and M_8 (**b**). All possible structures of intermediates were constructed based on the available spectroscopic data. The most favored candidates of intermediates are highlighted in red boxes based on chemical considerations. The orange spheres represent the Fe atoms of FeMoco.

Results of ^{1,2}H and ^{95}Mo ENDOR spectroscopic analyses suggest that the novel $S = 1/2$ EPR signal displayed by this intermediate may correspond to two M–H–M fragments, where M represents an Fe atom (Figure 5) [69, 72]. The proposal that the M_4 state contains two Fe atom-bridging hydrides on the FeMoco would further imply that the four reducing equivalents are partially, if not completely, carried by hydrides. Such a proposal [4, 28, 64, 73] is somewhat appealing in that (i) it offers a potential rationale for the FeMoco to be constructed with the elaborate trigonal prismatic cage; (ii) it removes the requirement for the FeMoco to acquire multiple redox states for the conversion of M_0 to M_4; and (iii) the recycling step back to M_0 can be easily understood as the protonation of the bound hydrides. This plausible intermediate structure, depicted in Figure 5, also forms the basis for the hypothesis that N_2 may be reduced through a reductive elimination mechanism (see Section 3.2.3 for details).

3.2.1.2   The M_7 and M_8 Intermediates

The M_7 and M_8 intermediates were identified by EPR spectroscopy as the two common intermediate states upon turnover of the $Ile^{\alpha70}/Gln^{\alpha195}$ double-substituted MoFe protein by various substrates, such as N_2, diazene (N_2H_2), hydrazine (N_2H_4), and methyldiazene ($NH=N–CH_3$) [74, 75]. The M_7 state intermediate, designated **H** at the time of discovery, showed a broad EPR signal at low field in the Q-band EPR spectra. This signal was later determined to arise from an integer-spin system (where $S \geq 2$, with a ground-state non-Kramers doublet) [75], which is consistent with the prediction that the odd-number states of the MoFe protein cycle (i.e., M_n states where n is an odd number) are likely to be non-Kramer systems [57, 76].

ESEEM analysis of the ^{15}N-labeled diazene, ^{15}N-labeled hydrazine, and ^2H-labeled methyldiazene further suggests that this intermediate consists of only one FeMoco-bound N atom and is most likely a $[-NH_2]$ moiety (Figure 5) [67, 68, 75]. Contrary to the M_7 intermediate, the M_8 intermediate, designated **I** at the time of discovery, displays a strong $S = 1/2$ EPR signal that is easily accessible for further analysis by advanced spectroscopic methods [74]. In combination with HYSCORE measurements, $^{1,2}H$, and ^{15}N ENDOR spectroscopic analyses deduce that this intermediate consists of a bound $[-NH_x]$ moiety, where x = 2 or 3 [65, 67, 74]. Since the $[-NH_2]$ moiety had been assigned to **H** (or the M_7 intermediate), it was concluded that the NH_x moiety in **I** had an x = 3 and that **I** would represent the M_8 state of the MoFe protein cycle (Figure 5).

3.2.2   The Reductive Dihydrogen Elimination Mechanism

The characterization of the structure of the M_4 intermediate has led to the proposal of a mechanism that involves the activation of N_2 upon the reductive elimination of H_2 (Figure 6). This mechanism presumably begins with a transient conversion of bridging hydrides to terminal hydrides that renders an elevated reactivity of the Fe atoms of FeMoco toward N_2. Previous studies involving the synthesis of N_2-bound Fe, Co, and other transition metal complexes have demonstrated that N_2 binding was promoted upon the reductive elimination of H_2, likely through the reduction of the metal center that lowered the energy barrier for N_2 activation [77–80]. In the case of nitrogenase, the eliminated H_2 is thought to carry away only two of the four reducing equivalents, allowing the remaining reducing equivalents to be used for the reduction of N_2. A bound N_2H_2 species may be generated in this process through the coupling of N_2 with the two protons that are possibly bound to the sulfides (Figure 6). Since the elimination of hydrides as H_2 would be the key to the binding and reduction of N_2, this proposed mechanism could provide an adequate explanation for the obligate formation of H_2 during N_2 reduction (see equation 1).

Efforts to substantiate this proposed mechanism are currently underway, with one attempted approach focusing on the micro-reversibility of the H_2 elimination step. This approach is based on the hypothesis that the reductive elimination of hydrides can be reversed by the oxidative addition of H_2 and, in one such

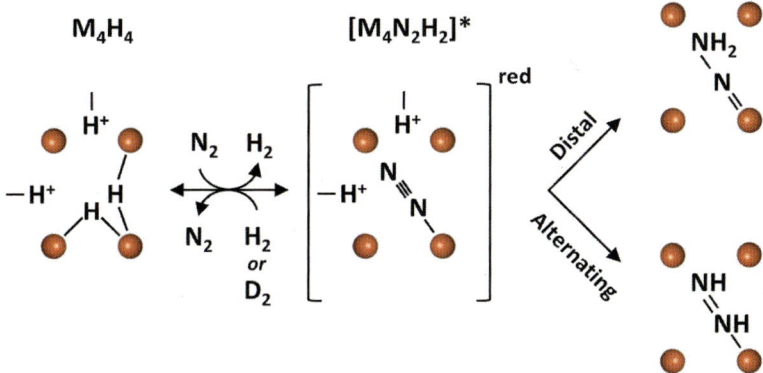

Figure 6 The proposed reductive elimination mechanism for N_2 activation at stage M_4. The four reducing equivalences accumulated via the four electron transfer events are stored on the FeMoco as two bridging hydrides. Upon N_2 binding, one molecule of H_2 is eliminated, thereby carrying away two of the four reducing equivalences. The remaining two reducing equivalences transiently render the FeMoco more reduced and, consequently, more reactive. The elevated reactivity then facilitates the activation of the substrate, N_2, and the subsequent cleavage and reduction of the N,N triple bond. This reduction can take place either on the distant N atom first (the distal pathway) or back and forth on both N atoms (the alternating pathway). The orange spheres represent the Fe atoms of FeMoco.

experiment, D_2 was added to the MoFe protein variant upon N_2 turnover in hopes of achieving the oxidative addition of D_2 on the N_2-bound M_4 intermediate and the subsequent formation of an N_2-free M_4 intermediate with bound deuteride (D^-) (Figure 6). Acetylene (C_2H_2), which can be reduced by FeMoco at more oxidized states than that required for N_2 reduction, was then added to the turnover mixture to "outcompete" N_2 for the deuterides bound to FeMoco [81]. The detection of $C_2H_2D_2$ and C_2H_3D by GC-MS appeared to be in line with the argument that deuteride was added on the FeMoco upon addition of D_2 and incorporated into C_2H_2 during substrate turnover [81]. Nevertheless, direct evidence is yet to be acquired to demonstrate the formation of FeMoco-bound deuterides upon addition of D_2, which will provide the definitive proof for the proposed mechanism of N_2 activation via the reductive elimination of H_2.

3.2.3   The Alternating Dinitrogen Reduction Pathway

Support for the distal N_2 reduction pathway (Scheme 2a) came mainly from the work on the synthetic, NH_3-forming, mononuclear Mo complexes [82–84]. The strongest experimental support for the proposed distal pathway was the detection of hydrazine formation upon acid- and base-quenching of the pre-steady-state turn-over, which was interpreted as the release of the two-electron reduced hydrazido- or bound $=N–NH_2$ intermediate at the M_4 stage [4, 28, 64] (Scheme 2a). Support for the alternating reduction pathway (Scheme 2b), on the other hand, came mainly from the recent spectroscopic studies of a number of MoFe protein variants [65, 67].

The observation that N_2H_4, N_2H_2 and N_2 are all substrates of nitrogenase suggests that N_2H_2 and N_2H_4 may represent intermediates generated along the alternating pathway [65, 67, 68]; however, there is currently no evidence demonstrating that N_2H_2 or N_2H_4 is indeed formed during the reduction of N_2 by Mo-nitrogenase [85, 86]. The observation of two common intermediates during the turnover of N_2H_4, N_2H_2, and N_2 was also used to argue in favor of the alternating pathway; yet, these common intermediates – identified lately as the M_7 and M_8 states by spectroscopic characterization – are shared by both the distal and alternating pathways and, therefore, cannot be used to exclude one pathway from another (Figure 5, Scheme 2b). Clearly, further efforts are required to distinguish the two proposed pathways; in particular, the capture and characterization of the nitrogenous interme- diates bound at stages M_4-M_6 will likely provide definitive evidence in this regard.

4   The Distinct Structural and Catalytic Features of V-Nitrogenase

Discovered some 30 years ago, the V-nitrogenase is usually expressed in the absence of Mo as an "alternative" form of its more efficient Mo-counterpart [5]. Like the Mo-nitrogenase, the V-nitrogenase is a binary system comprising an iron (Fe) protein component and a vanadium iron (VFe) protein component, each containing metallocluster(s) analogous to those found in their respective Mo-counterpart. Catalysis by V-nitrogenase also involves the association/dissociation between the Fe protein and the VFe protein, which facilitates the ATP-dependent, inter-protein electron transfer from the former to the latter and the subsequent substrate reduction at the active cofactor site [5, 22, 87]. Despite their similarities, the V-nitrogenase displays some unique structural and catalytic features that are clearly distinct from its Mo-counterpart. These features will be discussed in detail below.

4.1   The Structural Features

4.1.1   The Fe Protein

The Fe protein (encoded by *vnfH*) of the *A. vinelandii* V-nitrogenase shares ~91 % of sequence identity with its counterpart (encoded by *nifH*) in the Mo-nitrogenase [5, 22, 87]. Like NifH, VnfH is a ~ 60 kDa homodimer that contains 4 Fe atoms and 4 acid-labile S atoms per protein. The cysteinyl ligands of the [Fe_4S_4] cluster, as well as the Walker A motif, are all perfectly conserved in the primary sequences of VnfH and NifH [88]. In the dithionite-reduced state, both VnfH and NifH display the same, mixed $S = 3/2 : S = 1/2$ EPR signals, although the $S = 3/2$ and $S = 1/2$ signals of VnfH are 10 % and 27 %, respectively, more intense than those of NifH [89, 90]. In addition, VnfH displays a midpoint potential comparable to that of NifH in the MgADP-bound state, and it also utilizes the $[Fe_4S_4]^{1+/2+}$ couple for the one-electron transfer event during catalysis.

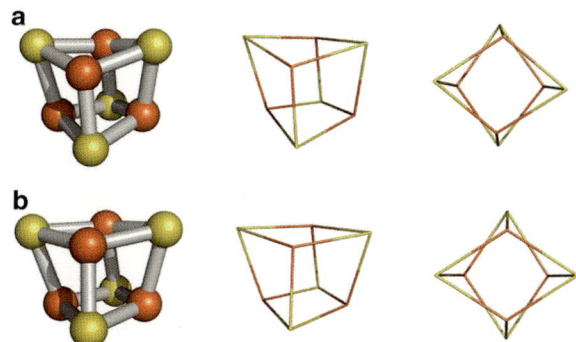

Figure 7 EXAFS-derived models of $[Fe_4S_4]$ clusters in NifH (**a**) and VnfH (**b**). The $[Fe_4S_4]$ clusters are shown in ball-and-stick (*left*) and line (*middle* and *right*) presentations. Both clusters have been rotated to show the overlapping of the $[Fe_2S_2]$ rhomboids (*right*). In the case of NifH, the two rhomboids are unequal and bent to a greater degree out of plane relative to each other (**a**); whereas in VnfH, two identical $[Fe_2S_2]$ rhomboids are stacked and offset by 90° (**b**). Atoms are colored as follows: Fe, orange; S, yellow. PYMOL was used to generate this figure, with coordinates of the crystal structure of the $[Fe_4S_4]$ cluster in NifH (PDB entry 1NIP) modified on the basis of the EXAFS fits given in [95].

Surprisingly, despite the high homology between the two Fe proteins, Fe K-edge XAS/EXAFS analysis shows distinct differences between the geometric arrangements of their associated $[Fe_4S_4]$ clusters [91] (Figure 7). The $[Fe_4S_4]$ cluster in VnfH can be best modeled by two identical $[Fe_2S_2]$ rhomboids, which are stacked on and offset by 90° from each other (Figure 7b); whereas the $[Fe_4S_4]$ cluster in NifH can be modeled alternatively by two unequal $[Fe_2S_2]$ rhomboids, which are bent to a greater degree out of plane relative to each other (Figure 7a). The exact effect of these subtle differences on the catalytic capacities of the V- and Mo-nitrogenases is still unclear, and further biochemical or structural studies need to be performed on these proteins to gain a better understanding in this regard.

4.1.2 The VFe Protein

The α-, β-, and δ-subunits of the VFe protein are encoded by *vnfD*, *vnfK*, and *vnfG* genes, respectively. Contrary to the $\alpha_2\beta_2$-tetrameric MoFe protein, the VFe protein from *A. vinelandii* (Figure 8a, b) has the composition of an $\alpha_2\beta_2\delta_4$-octamer, with its α- and β-subunits sharing ∼ 33 % and ∼ 32 % of sequence identity with their respective counterparts in the MoFe protein and its δ-subunit being a unique feature of the VFe protein [5, 22]. In addition, there is a high degree of sequence identity between the MoFe protein and the VFe protein, which centers around the residues that are either ligated to or in the surroundings of the P-cluster and the cofactor [5, 22].

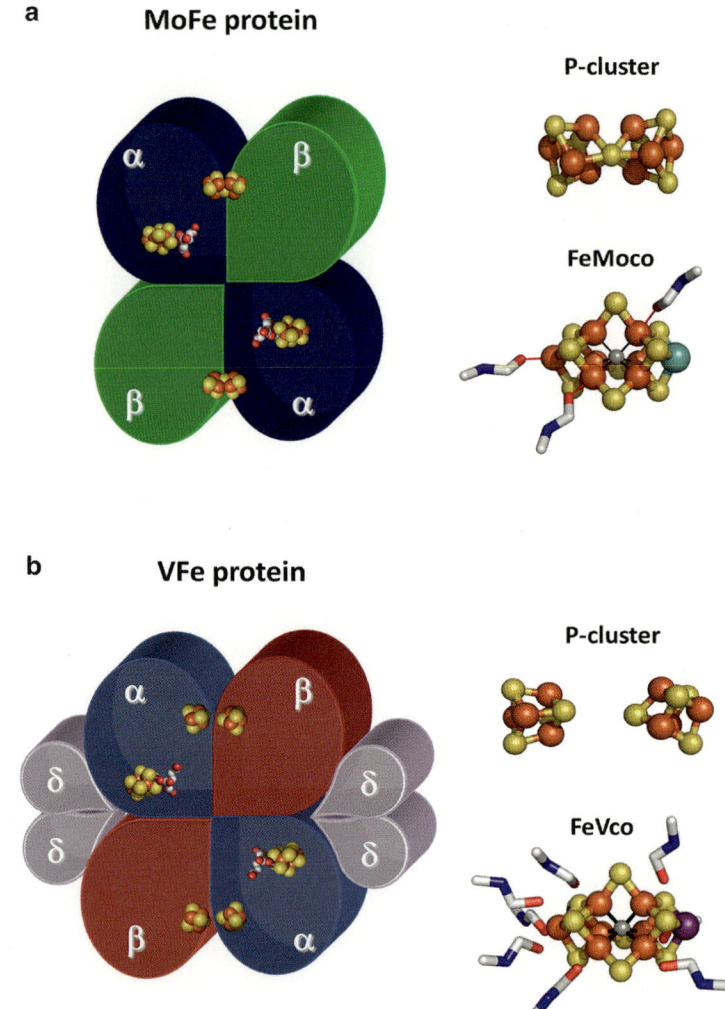

Figure 8 Schematic representations of the MoFe (**a**, *left*) and VFe (**b**, *left*) proteins, EXAFS-derived models of the P-clusters in the MoFe (**a**, *top right*) and VFe (**b**, *top right*) proteins, and EXAFS-derived models of the NMF-extracted FeMoco (**a**, *bottom right*) and FeVco (**b**, *bottom right*). The MoFe protein is an $\alpha_2\beta_2$ heterotetramer (**a**, *left*); whereas the VFe protein is an $\alpha_2\beta_2\delta_4$ heterooctamer (**b**, *left*). The P-cluster of the MoFe protein has an [Fe$_8$S$_7$] topology (**a**, *top right*); whereas the P-cluster of the VFe protein consists of paired [Fe$_4$S$_4$] clusters (**b**, *top right*). The isolated FeVco (**b**, *bottom right*) resembles the isolated FeMoco (**a**, *bottom right*) in core structure, although it is more extended in structure and bound with more NMF molecules. Clusters are shown in ball-and-stick presentations; NMF molecules are shown as sticks. Atoms are colored as follows: Fe, orange; S, yellow; V, dark purple; Mo, cyan; C, white; N, blue; O, red. PYMOL was used to generate this figure, with coordinates of the P-cluster and the FeMoco in the crystal structure of the MoFe protein (PDB entry 1M1N) modified on the basis of the Fe-edge EXAFS fits.

The P-cluster of the VFe protein has long been thought to closely resemble the P-cluster of the MoFe protein in structure, but the recent characterization of the *A. vinelandii* VFe protein suggested otherwise [22, 92]. The EPR data of the dithionite-reduced and IDS-oxidized forms of the VFe protein strongly indicate that the electronic properties of the P-cluster in the VFe protein are quite different than those of its counterpart in the MoFe protein [92]. This observation is consistent with the outcome of the Fe K-edge EXAFS analysis of a cofactor-deficient form of the VFe protein, which suggests that this protein carries a P-cluster species that consists of a pair of [Fe_4S_4]-like clusters (Figure 8b) [93]. Interestingly, a highly similar conformation has also been observed earlier in the case of the MoFe protein, where a precursor form to the P-cluster was identified as a pair of [Fe_4S_4]-like clusters [94].

The iron-vanadium cofactor (FeVco) of the VFe protein has been characterized in the protein-bound and solvent-extracted states by a number of spectroscopic methods, including EPR, XAS/EXAFS, and Mössbauer spectroscopy [95–101]. Results from the Fe K-edge XAS/EXAFS studies suggest that the FeVco is structurally similar to the FeMoco except for the presence of a different heterometal in the cofactor [95–98]. Further, these studies reveal that the isolated FeVco adopts an overall octahedral geometry, whereas the isolated FeMoco assumes an overall tetrahedral/trigonal pyramidal structure. This observation suggests that the FeS core of the FeVco, albeit structurally similar to the FeS core of the FeMoco, is slightly elongated compared to the latter (Figure 8b). This argument is also supported by the observation of a longer Fe–V backscattering path in FeVco (2.88 Å) than the Fe–Mo backscattering path in FeMoco (2.68 Å) [101].

Consistent with the outcome of the XAS/EXAFS studies, EPR analysis reveals differences in the electronic properties of the two cofactors. The isolated FeVco displays an $S = 3/2$ signal that is similar to, yet significantly broader in line-shape, than that of the isolated FeMoco, suggesting a modulating effect of the electronic properties of these cofactors by the presence of different heterometals [101]. Other than the information gathered on the core structure of the FeVco, there is currently no direct evidence proving homocitrate as the organic moiety in the FeVco, nor is there conclusive data demonstrating the presence of an interstitial carbide in this cofactor. However, given the similarity between the structural and biochemical properties of the two cofactors, it is assumed that a homocitrate entity (or one of its analogs) can be found along with a central carbide in the structure of the FeVco.

4.2   The Catalytic Features

4.2.1   The Reduction of Dinitrogen

While the V-nitrogenase resembles the Mo-nitrogenase in catalysis, it exhibits some catalytic features that are notably distinct from those of its Mo-counterpart.

Compared to the Mo-nitrogenase, the V-nitrogenase reduces substrates at lower efficiencies despite having an electron flux similar to that through its Mo-counterpart during substrate turnover (Table 2) [5, 92]. In the case of C_2H_2 reduction, the V- and Mo-nitrogenases use ~60 % and <10 %, respectively, of the total electron flux for the formation of H_2, a side product of the reaction (Table 2).

Table 2 Specific substrate-reducing activities of Mo- and V-nitrogenases of *A. vinelandii*.[a,b]

	H_2 Formation under Ar	H_2 Formation under N_2	NH_3 Formation under N_2	Electron Partitioning Ratio of NH_3/H_2[c]
Mo-Nitrogenase	489	133	205	2.3
V-Nitrogenase	419	192	111	1.1

[a]This table was adapted from [96].
[b]All activities are expressed in nmol product per nmol protein per min.
[c]Calculations are based on the assumption that the formation of each H_2 requires two electrons, while the formation of each NH_3 requires three electrons.

Similarly, in the case of N_2 reduction, the V- and Mo-nitrogenases produce H_2 and NH_3 at ratios of 1:1 and 1:2, respectively, again showing a clear favoritism of the V-nitrogenase toward H_2 formation (Table 2) [92, 99, 102]. As such, the stoichiometry of N_2 reduction by the V-nitrogenase could be tentatively depicted as follows:

$$N_2 + 10\,H^+ + 10\,e^- \rightarrow 2\,NH_3 + 2\,H_2$$

Such a stoichiometry could be physiologically relevant, since the *in vivo* measurements of N_2 reduction by V-nitrogenase also yielded similar ratios to those obtained in the *in vitro* assay [5]. The elevated level of H_2 formation by the V-nitrogenase suggests the possible presence of additional sites in the nitrogenase that are dedicated to H_2 evolution, which may or may not be related to the reduction of N_2. Alternatively, the increased $H_2:NH_3$ ratio in the case of the V-nitrogenase may reflect the utilization of a specific "VFe protein cycle" by this nitrogenase, in which two molecules of H_2 (instead of one in the case of the Mo-nitrogenase) need to be released in order to make ready the FeVco for the activation/reduction of N_2. Finally, it was reported earlier that a sub-stoichiometric amount of N_2H_4 could be generated at elevated temperatures by the V-nitrogenase of *A. chroococcum* during the turnover of N_2 [85, 86]. The formation of N_2H_4 by this V-nitrogenase correlated positively with the decrease of NH_3 formation between 40 °C and 50 °C, which was interpreted by a looser association of the reaction intermediate(s) to the V-nitrogenase, rendering this nitrogenase more "leaky" than its Mo-counterpart during substrate turnover [85].

4.2.2 The Reduction of Carbon Monoxide

The fact that CO – an isoelectronic molecule to N_2 – is also a substrate of nitrogenase was first discovered in the case of the V-nitrogenase. The observation that the formation of H_2 by the V-nitrogenase was inhibited by CO led to the hypothesis that a portion of the electrons flowing through V-nitrogenase were "re-routed" for the reduction of CO. Subsequently, this hypothesis was proven by GC-MS analyses, which demonstrated the ability of the V-nitrogenase to convert CO to C1-C4 hydrocarbons (i.e., CH_4, C_2H_4, C_2H_6, C_3H_6, C_3H_8, 1-C_4H_8, and n-C_4H_{10}) in a reaction analogous to the industrial Fischer-Tropsch process [22, 103]. There are notable differences, however, between this enzymatic reaction and its industrial counterpart, particularly with regard to the reducing power and energy source in these reactions. The enzymatic reaction uses protons and electrons to reduce CO to hydrocarbons, and it is driven by the chemical energy released from ATP hydrolysis, whereas the industrial process uses H_2 to reduce CO, and it occurs at high temperature and pressure [22, 104]. Our current knowledge of CO reduction by V-nitrogenase can be summarized by a somewhat "abstract" equation as follows:

$$CO + H^+ + e^- \rightarrow CH_4 + C_2H_4 + C_2H_6 + C_3H_6 + C_3H_8 + 1\text{-}C_4H_8$$
$$+ \ n\text{-}C_4H_{10} + H_2$$

Apparently, much work needs to be done to establish the stoichiometry of the reaction, trace the fate of the oxygen atom, and elucidate the mechanistic details of this reaction.

Following the discovery of CO reduction by V-nitrogenase, the reactivity of Mo-nitrogenase toward CO was re-examined, which revealed the ability of the Mo-nitrogenase to reduce CO to C2-C3 hydrocarbons. However, the activity of CO reduction by Mo-nitrogenase is less than 0.5 % compared to that of its V-counterpart. The gap in reactivity closes significantly, though, between the two nitrogenases when H_2O is replaced by D_2O (Figure 9), which enhances the CO-reducing activity of the Mo-nitrogenase by more than 20-fold while hardly impacting the activity of the V-nitrogenase (Figure 9) [22, 105]. The reduction of CO by Mo-nitrogenase also differs from that by V-nitrogenase in terms of the product distribution profile. While the V-nitrogenase produces almost exclusively C_2H_4 (~94 % of total products) from CO, the Mo-nitrogenase has a more evenly distributed product profile despite the fact that it still produces C_2H_4 (~54 % of total products) as the major product of CO reduction (Figure 9). These differences suggest that, even if the two nitrogenases share a common mechanism for CO reduction, there is still room to fine tune the enzymatic reactivity and product profile through the modulation of the structural and electronic properties of these homologous systems.

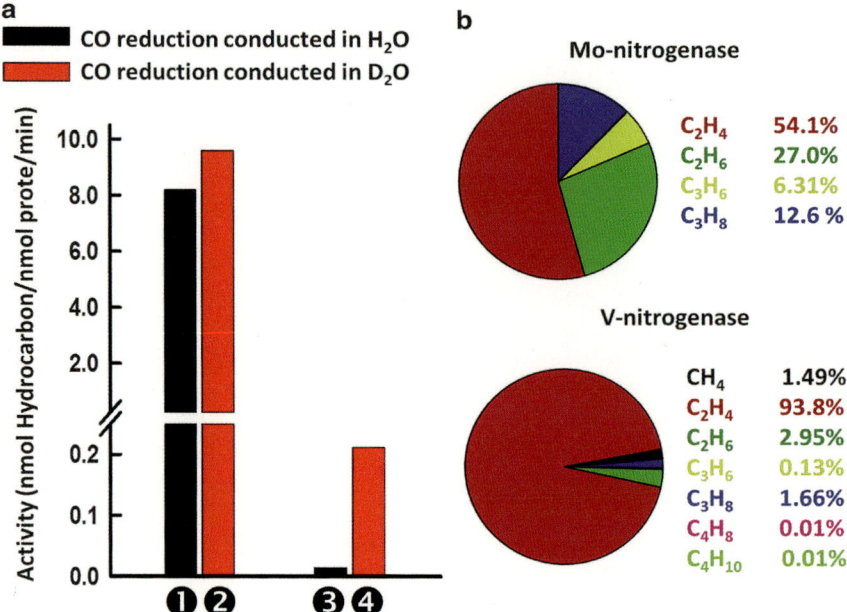

Figure 9 Comparison of total specific activities of CO reduction by Mo- and V-nitrogenases in the presence of H_2O and D_2O (**a**) and the product distributions of hydrocarbons in the reactions catalyzed by the two nitrogenases (**b**). Shown are the total activities of hydrocarbon formation from CO by Mo- and V-nitrogenases in H_2O- (*black*) and D_2O- (*red*) based reactions. In H_2O, the V-nitrogenase is 680-fold more active than the Mo-nitrogenase (❸ *versus* ❶); whereas in D_2O, V-nitrogenase is only 50-fold more active than the Mo-nitrogenase (❷ *versus* ❹). The activity of V-nitrogenase is minimally impacted upon D_2O substitution (❷ *versus* ❶); whereas the activity of Mo-nitrogenase is significantly increased upon D_2O substitution (❸ *versus* ❹). For calculations of product distributions, the total amounts of hydrocarbons formed in V- and Mo-nitrogenase-catalyzed reactions were set as 100 %, and the percentages of individual products were determined accordingly.

5   Conclusions

For decades, nitrogenase has remained one of the most enigmatic topics in the field of metalloprotein biochemistry. From the unveiling of its structure, to the quest of its reaction mechanism, there are surprises in almost every footstep toward a better understanding of this complex metalloenzyme system.

   While significant strides have been made in elucidating the structure of the Mo-nitrogenase and the key steps in the reaction of N_2 reduction by this enzyme, the exact mechanism of nitrogenase is yet to be defined. Much work needs to be done to address the question of how nitrogenase overcomes the great energy barrier to cleave the N,N triple bond, as well as the question of how nitrogenase breaks the isoelectronic C,O triple bond, which may provide valuable, albeit indirect, insights into the mechanism of N_2 reduction by nitrogenase.

Abbreviations

ADP	adenosine 5′-diphosphate
AMPPCP	5′-adenylyl (β,γ-methylene) diphosphonate
ATP	adenosine 5′-triphosphate
ENDOR	electron nuclear double resonance
EPR	electron paramagnetic resonance
ESEEM	electron spin-echo envelope modulation
EXAFS	extended X-ray fine structure
FeMoco	iron molybdenum cofactor
FeVCo	iron vanadium cofactor
GC-MS	gas chromatography-mass spectrometry
HYSCORE	hyperfine sublevel correlation
IDS	indigodisulfonate
NHE	normal hydrogen electrode
NMF	N-methylformamide
PDB	Protein Data Bank
P_i	inorganic phosphate
SOD	superoxide dismutase
XAS	X-ray absorption spectroscopy

Acknowledgment Work in our laboratory is supported by National Institute of Health Grant GM-67626 (MWR) and Herman Frasch Foundation Grant 617-HF07 (MWR).

References

1. R. M. Allen, R. Chatterjee, M. S. Madden, P. W. Ludden, V. K. Shah, *Crit. Rev. Biotechnol.* **1994**, *14*, 225–249.
2. J. W. Peters, K. Fisher, D. R. Dean, *Annu. Rev. Microbiol.* **1995**, *49*, 335–366.
3. J. B. Howard, D. C. Rees, *Chem. Rev.* **1996**, *96*, 2965–2982.
4. B. K. Burgess, D. J. Lowe, *Chem. Rev.* **1996**, *96*, 2983–3011.
5. R. R. Eady, *Chem. Rev.* **1996**, *96*, 3013–3030.
6. L. C. Seefeldt, D. R. Dean, *Acc. Chem. Res.* **1997**, *30*, 260–266.
7. D. C. Rees, J. B. Howard, *Curr. Opin. Chem. Biol.* **2000**, *4*, 559–566.
8. B. E. Smith, *Adv. Inorg. Chem.* **1999**, *47*, 159–218.
9. J. Christiansen, D. R. Dean, L. C. Seefeldt, *Annu. Rev. Plant Physiol. Plant Mol. Biol.* **2001**, *52*, 269–295.
10. D. M. Lawson, B. E. Smith, *Met. Ions Biol. Syst.* **2002**, *39*, 75–119.
11. J. Frazzon, D. R. Dean, *Met. Ions Biol. Syst.* **2002**, *39*, 163–186.
12. R. R. Eady, *Coord. Chem. Rev.* **2003**, *237*, 23–30.
13. D. C. Rees, F. A. Tezcan, C. A. Haynes, M. Y. Walton, S. L. Andrade, O. Einsle, J. B. Howard, *Philos. Trans. Roy. Soc. Lond. A* **2005**, *363*, 971–984.
14. B. M. Barney, H. I. Lee, P. C. Dos Santos, B. M. Hoffman, D. R. Dean, L. C. Seefeldt, *Dalton Trans.* **2006**, *19*, 2277–2284.
15. J. B. Howard, D. C. Rees, *Proc. Natl. Acad. Sci. USA* **2006** *103*, 17088–17093.

16. L. C. Seefeldt, B. M. Hoffman, D. R. Dean, *Annu. Rev. Biochem.* **2009**, *78*, 701–722.
17. P. M. Vitousek, S. Hattenschwiler, L. Olander, S. Allison, *Ambio* **2002**, *31*, 97–101.
18. N. Gruber, J. N. Galloway, *Nature* **2008**, *45*, 293–296.
19. J. N. Galloway, E. B. Cowling, *Ambio* **2002**, *31*, 64–71.
20. B. E. Smith, *Science* **2002**, *297*, 1654–1655.
21. T. A. Bazhenova, A. E. Shilov, *Coord. Chem. Rev.* **1995**, *144*, 69–145.
22. Y. Hu, C. C. Lee, M. W. Ribbe, *Dalton Trans.* **2012**, *41*, 1118–1127.
23. D. Gadkari, G. Mörsdorf, O. Meyer, *J. Bacteriol.* **1992**, *174*, 6840–6843.
24. M. W. Ribbe, D. Gadkari, O. Meyer, *J. Biol. Chem.* **1997**, *272*, 26627–26633.
25. R. R. Eady, *Met. Ions Biol. Syst.* **1995**, *31*, 363–405.
26. D. C. Rees, *Annu. Rev. Biochem.* **2002**, *71*, 221–246.
27. J. B. Howard, D. C. Rees, *Annu. Rev. Biochem.* **1994**, *63*, 235–264.
28. R. N. F. Thorneley, D. J. Lowe, *J. Biol. Inorg. Chem.* **1996**, *1*, 576–580.
29. S. B. Jang, L. C. Seefeldt, J. W. Peters, *Biochemistry* **2000**, *39*, 14745–14752.
30. N. Gavini, B. K. Burgess, *J. Biol. Chem.* **1992**, *267*, 21179–21186.
31. E. H. Bursey, B. K. Burgess, *J. Biol. Chem.* **1998**, *273*, 16927–16934.
32. E. H. Bursey, B. K. Burgess, *J. Biol. Chem.* **1998**, *273*, 29678–29685.
33. T. L. Deits, J. B. Howard, *J. Biol. Chem.* **1989**, *264*, 6619–6628.
34. T. Ljones, R. H. Burris, *Biochemistry* **1978**, *17*, 1866–1872.
35. M. J. Ryle, W. N. Lanzilotta, L. E. Mortenson, G. D. Watt, L. C. Seefeldt, *J. Biol. Chem.* **1995**, *270*, 13112–13117.
36. W. H. Orme-Johnson, W. D. Hamilton, T. L. Jones, M. Y. Tso, R. H. Burris, V. K. Shah, W. J. Brill, *Proc. Natl. Acad. Sci. USA* **1972**, *69*, 3142–3145.
37. F. A. Tezcan, J. T. Kaiser, D. Mustafi, M. Y. Walton, J. B. Howard, D. C. Rees, *Science* **2005**, *309*, 1377–1380.
38. P. A. Lindahl, E. P. Day, T. A. Kent, W. H. Orme-Johnson, E. Münck, *J. Biol. Chem.* **1985**, *260*, 1160–1173.
39. G. D. Watt, J. W. McDonald, *Biochemistry* **1985**, *24*, 7226–7231.
40. W. N. Lanzilotta, M. J. Ryle, L. C. Seefeldt, *Biochemistry* **1995**, *34*, 10713–10723.
41. G. D. Watt, K. R. N. Reddy, *J. Inorg. Biochem.* **1994**, *53*, 281–294.
42. M. Guo, F. Sulc, M. W. Ribbe, P. J. Farmer, B. K. Burgess, *J. Am. Chem. Soc.* **2002**, *124*, 12100–12101.
43. J. Kim, D. C. Rees, *Science* **1992**, *257*, 1677–1682.
44. J. Kim, D. C. Rees, *Nature* **1992**, *360*, 553–560.
45. M. K. Chan, J. Kim, D. C. Rees, *Science* **1993**, *260*, 792–794.
46. J. W. Peters, M. H. Stowell, S. M. Soltis, M. G. Finnegan, M. K. Johnson, D. C. Rees, *Biochemistry* **1997**, *36*, 1181–1187.
47. M. Sorlie, J. Christiansen, B. J. Lemon, J. W. Peters, D. R. Dean, B. J. Hales, *Biochemistry* **2001**, *40*, 1540–1549.
48. O. Einsle, F. A. Tezcan, S. L. A. Andrade, B. Schmid, M. Yoshida, J. B. Howard, D. C. Rees, *Science* **2002**, *297*, 1696–1700.
49. B. Schmid, M. W. Ribbe, O. Einsle, M. Yoshida, L. M. Thomas, D. R. Dean, D. C. Rees, B. K. Burgess, *Science* **2002**, *296*, 352–356.
50. K. K. Surerus, M. P. Hendrich, P. D. Christie, D. Rottgardt, W. H. Orme-Johnson, E. Münck, *J. Am. Chem. Soc.* **1992**, *114*, 8579–8590.
51. R. Zimmermann, W. H. Orme-Johnson, E. Münck, V. K. Shah, W. I. Brill, M. T. Henzl, J. Rawlings, *Biochim. Biophys. Acta* **1978**, *537*, 185–207.
52. A. J. Pierik, H. Wassink, H. Haaker, W. R. Hagen, *Eur. J. Biochem.* **1993**, *212*, 51–61.
53. A. Wiig, Y. Hu, C. C. Lee, M. W. Ribbe, *Science* **2012**, *337*, 1672–1675.
54. K. M. Lancaster, M. Roemelt, P. Ettenhuber, Y. Hu, M. W. Ribbe, F. Neese, U. Bergmann, S. DeBeer, *Science* **2011**, *334*, 974–977.
55. T. Spatzal, M. Aksoyoglu, L. Zhang, S. L. Andrade, E. Schleicher, S. Weber, D. C. Rees, O. Einsle, *Science* **2011**, *334*, 940.

56. H. I. Lee, B. J. Hales, B. M. Hoffman, *J. Am. Chem. Soc.* **1997**, *119*, 11395–11400.
57. S. J. Yoo, H. C. Angove, V. Papaefthymiou, B. K. Burgess, E. Münck, *J. Am. Chem. Soc.* **2000**, *122*, 4926–4936.
58. B. K. Burgess, *Chem. Rev.* **1990**, *90*, 1377–1406.
59. V. K. Shah, W. J. Brill, *Proc. Natl. Acad. Sci. USA* **1977**, *74*, 3249–3253.
60. A. W. Fay, C. C. Lee, J. A. Wiig, Y. Hu, M. W. Ribbe, *Methods Mol. Biol.* **2011**, *766*, 239–248.
61. C. C. Lee, Y. Hu, M. W. Ribbe, *Angew. Chem. Int. Ed. Engl.* **2012**, *51*, 1947–1949.
62. H. Schindelin, C. Kisker, J. L. Schlessman, J. B. Howard, D. C. Rees, *Nature* **1997**, *387*, 370–376.
63. B. Schmid, O. Einsle, H. J. Chiu, A. Willing, M. Yoshida, J. B. Howard, D. C. Rees, *Biochemistry* **2002**, *41*, 15557–15565.
64. R. N. F. Thorneley, D. J. Lowe, in *Molybdenum Enzymes*, Ed T. G. Spiro, Wiley-Interscience, New York, 1985, pp. 221–284.
65. B. M. Hoffman, D. Lukoyanov, D. R. Dean, L. C. Seefeldt, *Acc. Chem. Res.* **2013**, *46*, 587–595.
66. K. Danyal, Z. Y. Yang, L. C. Seefeldt, *Methods Mol. Biol.* **2011**, *766*, 191–205.
67. L. C. Seefeldt, B. M. Hoffman, D. R. Dean, *Curr. Opin. Chem. Biol.* **2012**, *16*, 19–25.
68. B. M. Hoffman, D. Lukoyanov, Z. Y. Yang, D. R. Dean, L. C. Seefeldt, *Chem. Rev.* **2014**, *114*, 4041–4062.
69. R. Y. Igarashi, P. Laryukhin, P. C. Dos Santos, H.-I. Lee, D. R. Dean, L. C. Seefeldt, B. M. Hoffman, *J. Am. Chem. Soc.* **2005**, *127*, 6231–6241.
70. M. G. Yates, D. J. Lowe, *FEBS Lett.* **1976**, *72*, 127–130.
71. D. Lukoyanov, B. M. Barney, D. R. Dean, L. C. Seefeldt, B. M. Hoffman, *Proc. Natl. Acad. Sci. USA* **2007**, *104*, 1451–1455.
72. D. Lukoyanov, Z.-Y. Yang, D. R. Dean, L. C. Seefeldt, B. M. Hoffman, *J. Am. Chem. Soc.* **2010**, *132*, 2526–2527.
73. J. H. Guth, R. H. Burris, *Biochemistry* **1983**, *22*, 5111–5122.
74. D. Lukoyanov, S. A. Dikanov, Z.-Y. Yang, B. M. Barney, R. I. Samoilova, K. V. Narasimhulu, D. R. Dean, L. C. Seefeldt, B. M. Hoffman, *J. Am. Chem. Soc.* **2011**, *133*, 11655–11664.
75. D. Lukoyanov, Z.-Y. Yang, B. M. Barney, D. R. Dean, L. C. Seefeldt, B. M. Hoffman, *Proc. Natl. Acad. Sci. USA* **2012**, *109*, 5583–5587.
76. B. H. Huynh, M. T. Henzl, J. A. Christner, R. Zimmermann, W. H. Orme-Johnson, E. Münck, *Biochim. Biophys. Acta* **1980**, *623*, 124–138.
77. G. M. Bancroft, M. J. Mays, B. E. Prater, F. P. Stefanini, *J. Chem. Soc. A* **1970**, 2146–2149.
78. A. Sacco, M. Rossi, *Chem. Commun.* **1967**, 316.
79. M. L. H. Green, W. E. Silverthorn, *J. Chem. Soc. D* **1971**, 557–558.
80. J. Ballmann, R. F. Munhá, M. D. Fryzuk, *Chem. Commun.* **2010**, *46*, 1013–1025.
81. Z. Y. Yang, N. Khadka, D. Lukoyanov, B. M. Hoffman, D. R. Dean, L. C. Seefeldt, *Proc. Natl. Acad. Sci. USA* **2013** *110*, 16327–16332.
82. J. Chatt, J. R. Dilworth, R. L. Richards, *Chem. Rev.* **1978**, *78*, 589–625.
83. R. R. Schrock, *Acc. Chem. Res.* **2005**, *38*, 955–962.
84. R. R. Schrock, *Angew. Chem., Int. Ed. Engl.* **2008**, *47*, 5512–5522.
85. M. J. Dilworth, M. E. Eldridge, R. R. Eady, *Biochem. J.* **1993**, *289*, 395–400.
86. M. J. Dilworth, R. R. Eady, *Biochem. J.* **1991**, *277*, 465–468.
87. B. J. Hales, *Adv. Inorg. Biochem.* **1990**, *8*, 165–198.
88. P. E. Bishop, R. Premakumar, in *Biological Nitrogen Fixation*, Eds G. Stacey, R. H. Burris, H. J. Evans, Chapman and Hall, New York, 1992, pp. 736–762.
89. B. J. Hales, D. J. Langosch, E. E. Case, *J. Biol. Chem.* **1986**, *261*, 5301–5306.
90. J. Bergström, R. R. Eady, R. N. F. Thorneley, *Biochem. J.* **1988**, *251*, 165–169.
91. M. A. Blank, C. C. Lee, Y. Hu, K. O. Hodgson, B. Hedman, M. W. Ribbe, *Inorg. Chem.* **2011**, *50*, 7123–7128.

92. C. C. Lee, Y. Hu, M. W. Ribbe, *Proc. Natl. Acad. Sci. USA* **2009**, *106*, 9209–9214.
93. Y. Hu, M. C. Corbett, A. W. Fay, J. A. Webber, B. Hedman, K. O. Hodgson, M. W. Ribbe, *Proc. Natl. Acad. Sci. USA* **2005**, *102*, 13825–13830.
94. M. C. Corbett, Y. Hu, F. Naderi, M. W. Ribbe, B. Hedman, K. O. Hodgson, *J. Biol. Chem.* **2004**, *279*, 28276–28282.
95. N. Ravi, V. Moore, S. G. Lloyd, B. J. Hales, B. H. Huynh, *J. Biol. Chem.* **1994**, *269*, 20920–20921.
96. I. Harvey, J. M. Arber, R. R. Eady, B. E. Smith, C. D. Garner, S. S. Hasnain, *Biochem. J.* **1990**, *266*, 929–931.
97. J. M. Arber, B. R. Dobson, R. R. Eady, P. Stevens, S. S. Hasnain, C. D. Garner, B. E. Smith, *Nature* **1987**, *325*, 372–374.
98. G. N. George, C. L. Coyle, B. J. Hales, S. P. Cramer, *J. Am. Chem. Soc.* **1988**, *110*, 4057–4059.
99. B. J. Hales, E. E. Case, J. E. Morningstar, M. F. Dzeda, L. A. Mauterer, *Biochemistry* **1986**, *25*, 7251–7255.
100. D. J. Lowe, K. Fisher, R. N. F. Thorneley, *Biochem. J.* **1993**, *292*, 93–98.
101. A. W. Fay, M. A. Blank, C. C. Lee, Y. Hu, K. O. Hodgson, B. Hedman, M. W. Ribbe, *J. Am. Chem. Soc.* **2010**, *132*, 12612–12618.
102. R. L. Robson, R. R. Eady, T. H. Richardson, R. W. Miller, M. Hawkins, J. R. Postgate, *Nature* **1986**, *322*, 388–390.
103. C. C. Lee, Y. Hu, M. W. Ribbe, *Science* **2010**, *329*, 642.
104. C. C. Lee, Y. Hu, M. W. Ribbe, *Angew. Chem. Int. Ed. Engl.* **2011**, *50*, 5545–5547.
105. Y. Hu, C. C. Lee, M. W. Ribbe, *Science* **2011**, *333*, 753–755.

Chapter 8

No Laughing Matter: The Unmaking of the Greenhouse Gas Dinitrogen Monoxide by Nitrous Oxide Reductase

Lisa K. Schneider, Anja Wüst, Anja Pomowski,
Lin Zhang, and Oliver Einsle

Contents

ABSTRACT 178
1 INTRODUCTION: THE BIOGEOCHEMICAL NITROGEN CYCLE 179
2 NITROUS OXIDE: ENVIRONMENTAL EFFECTS
  AND ATMOSPHERIC CHEMISTRY 181
  2.1 Chemical Properties of Dinitrogen Monoxide 181
  2.2 Nitrous Oxide, the Greenhouse Effect, and Ozone Depletion 182
  2.3 Abiotic and Biotic Sources 183
  2.4 Bacterial Denitrification 183
3 NITROUS OXIDE REDUCTASE 184
  3.1 Anatomy of an Unusual Copper Enzyme 185
    3.1.1 Distinct Forms of the Enzyme 185
    3.1.2 Three-Dimensional Structures 189
  3.2 Cu$_A$: More than an Electron Transfer Center 190
    3.2.1 Spectroscopic Properties of Cu$_A$ 190
    3.2.2 Three-Dimensional Structure(s) of Cu$_A$ 191
    3.2.3 Unexpected Flexibility: Cu$_A$ in *P. stutzeri* Nitrous Oxide Reductase 194
  3.3 The Tetranuclear Cu$_Z$ Center 195
    3.3.1 Structural Data on Cu$_Z$ 195
    3.3.2 States of Cu$_Z$ and Catalytic Properties 198
4 BIOGENESIS AND ASSEMBLY OF NITROUS OXIDE REDUCTASE 200
  4.1 The *nos* Operon 200
  4.2 Protein Maturation and Cu$_A$ Insertion 202
  4.3 Assembly of Cu$_Z$ 202

L.K. Schneider • A. Wüst • A. Pomowski • L. Zhang • O. Einsle (✉)
Institute for Biochemistry, Albert-Ludwigs-Universität Freiburg,
D-79104 Freiburg im Breisgau, Germany
e-mail: einsle@biochemie.uni-freiburg.de

© Springer Science+Business Media Dordrecht 2014
P.M.H. Kroneck, M.E. Sosa Torres (eds.), *The Metal-Driven Biogeochemistry
of Gaseous Compounds in the Environment*, Metal Ions in Life Sciences 14,
DOI 10.1007/978-94-017-9269-1_8

5   ACTIVATION OF NITROUS OXIDE: THE WORKINGS
    OF NITROUS OXIDE REDUCTASE 203
    5.1   Substrate Access 203
    5.2   Gated Electron Transfer 205
    5.3   Activation of Nitrous Oxide 205
    5.4   The Fate of the Products 206
6   GENERAL CONCLUSIONS 206
ABBREVIATIONS AND DEFINITIONS 207
ACKNOWLEDGMENT 207
REFERENCES 208

Abstract The gas nitrous oxide (N_2O) is generated in a variety of abiotic, biotic, and anthropogenic processes and it has recently been under scrutiny for its role as a greenhouse gas. A single enzyme, nitrous oxide reductase, is known to reduce N_2O to uncritical N_2, in a two-electron reduction process that is catalyzed at two unusual metal centers containing copper. Nitrous oxide reductase is a bacterial metalloprotein from the metabolic pathway of denitrification, and it forms a 130 kDa homodimer in which the two metal sites Cu_A and Cu_Z from opposing monomers are brought into close contact to form the active site of the enzyme. Cu_A is a binuclear, valence-delocalized cluster that accepts and transfers a single electron. The Cu_A site of nitrous oxide reductase is highly similar to that of respiratory heme-copper oxidases, but in the denitrification enzyme the site additionally undergoes a conformational change on a ligand that is suggested to function as a gate for electron transfer from an external donor protein. Cu_Z, the tetranuclear active center of nitrous oxide reductase, is isolated under mild and anoxic conditions as a unique [4Cu:2S] cluster. It is easily desulfurylated to yield a [4Cu:S] state termed Cu_Z^* that is functionally distinct. The Cu_Z form of the cluster is catalytically active, while Cu_Z^* is inactive as isolated in the [$3Cu^{1+}$:$1Cu^{2+}$] state. However, only Cu_Z^* can be reduced to an all-cuprous state by sodium dithionite, yielding a form that shows higher activities than Cu_Z. As the possibility of a similar reductive activation in the periplasm is unconfirmed, the mechanism and the actual functional state of the enzyme remain under debate. Using enzyme from anoxic preparations with Cu_Z in the [4Cu:2S] state, N_2O was shown to bind between the Cu_A and Cu_Z sites, suggesting direct electron transfer from Cu_A to the substrate after its activation by Cu_Z.

Keywords copper enzymes • global warming • nitrogen cycle • nitrous oxide • X-ray crystallography

Please cite as: *Met. Ions Life Sci.* 14 (2014) 177–210

1   Introduction: The Biogeochemical Nitrogen Cycle

The bulk elements carbon, oxygen, hydrogen, nitrogen, and sulfur are the fundamental building blocks for all classes of biomolecules. They are abundant in our environment and were readily available during evolution to be combined into molecules of increasing complexity that form the basic classes of biological macro-molecules: carbohydrates, lipids, amino acids, and nucleic acids. Water is an obvious and omnipresent source both for oxygen and hydrogen, and nature has devised various different pathways for the assimilation of carbon, the most prominent being the light-driven fixation of CO_2 in the Calvin cycle during photosynthesis. The remaining elements, nitrogen and sulfur, are present in a series of different modifi-cations and oxidation states that are chemically or enzymatically interconverted in global, biogeochemical cycles [1]. For nitrogen, this cycle spans eight oxidation levels ranging from $+V$ in nitrate, NO_3^-, to $-III$ in ammonia, NH_3 [2, 3], and only the latter can be incorporated into amino acids *via* the reactions of glutamine synthase or glutamate dehydrogenase. Nitrate undergoes a two-electron reduction to nitrite, NO_2^-, catalyzed by a molybdenum-containing nitrate reductase, and nitrite in turn constitutes the central metabolic hub of the nitrogen cycle (Figure 1).

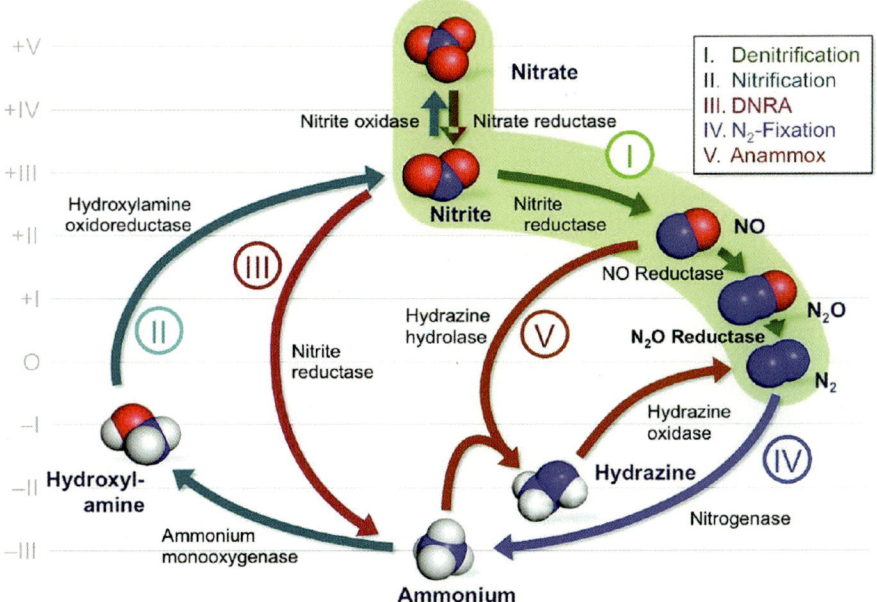

Figure 1 The biogeochemical nitrogen cycle. A network of reactions catalyzed by metal-containing proteins connects the different modifications of the element. The pathway of denitrification (highlighted in green) is a four-step metabolism to reduce nitrate (NO_3^-) *via* nitrite (NO_2^-), nitric oxide (NO), and nitrous oxide (N_2O) to dinitrogen (N_2). Nitrous oxide reductase catalyzes the final step of denitrification, and although the reaction is thermodynamically highly favored, a high activation energy barrier conveys substantial kinetic stability to N_2O.

From here, the reductive *ammonification* pathway bridges a step of six electrons with a single enzyme that exists in two distinct versions. Assimilatory nitrite reductase is a siroheme-dependent metalloprotein that in bacteria and plants is employed for the production of ammonium to be incorporated into biomolecules, consequently using NADPH as an electron donor. In contrast, the dissimilatory variant of nitrite reductase is a multiheme *c* enzyme that obtains electrons from the pool of menaquinol in the membrane, coupling its oxidation to the generation of a proton motive force for energy conservation. The enzyme, NrfA, contains five covalently attached heme groups per monomer and forms functional dimers [4]. The reversal of nitrate ammonification is realized in the pathway of *nitrification*, the sole oxygen-dependent process in the nitrogen cycle. Here, ammonium is oxidized to hydroxylamine by ammonia monooxygenase, in a process with strong parallels to methane oxidation to methanol by methane monooxygenase. Hydroxylamine is then converted to nitrite by an octaheme hydroxylamine dehydrogenase, followed by the two-electron oxidation to nitrate by a molybdenum-dependent nitrite oxidase. An alternative route for ammonium oxidation exists as an anaerobic process, in which ammonium is comproportionated with nitric oxide, NO, to yield hydrazine, N_2H_4, by hydrazine hydrolase. The product is then oxidized to N_2 by an octaheme hydrazine dehydrogenase, and this *anammox* process – while being the latest discovery in the nitrogen cycle – is now known to predominate in many habitats, giving it a significant influence on the global nitrogen balance [5]. Nevertheless, the highest metabolic flux through the nitrogen cycle occurs along the remaining reductive pathway, leading in four enzymatic steps from nitrate *via* nitrite, nitric oxide and nitrous oxide, N_2O, to the stable end product dinitrogen, N_2. Each step is energy-conserving through the generation of a proton motive force, and in summary this *denitrification* pathway is the most energetically favorable process known to operate in the absence of molecular oxygen [2, 6, 7]. Its final two intermediates, N_2O and N_2, play a particular role in the nitrogen cycle. Both are stable, inert gases, and for each nature seems to have evolved only a single enzyme able to catalyze a reductive conversion. For N_2, this is the assimilatory enzyme nitrogenase, a complex iron-sulfur enzyme and the agent of biological nitrogen fixation [8]. N_2O, in contrast, is converted to N_2 and H_2O by the copper enzyme nitrous oxide reductase (N_2OR) (equation 1), the topic of the present chapter.

$$N_2O + 2\,H^+ + 2\,e^- \rightarrow N_2 + H_2O \tag{1}$$

N_2OR is unusual in various ways, as it contains two non-canonical metal centers that allow reacting a highly inert substrate molecule. Due to the limited stability of this enzyme, N_2O is a common byproduct of this microbial metabolism and gives rise to substantial environmental problems (see Section 2.4).

2   Nitrous Oxide: Environmental Effects and Atmospheric Chemistry

The colorless, slightly sweet-smelling gas nitrous oxide was first described by Joseph Priestley in 1774 as 'dephlogisticated nitrous air' in his *"Experiments and Observations on Different Kinds of Air"* that also encompassed the first description of O_2, NH_3, and NO [9]. It was found to be a useful anesthetic and also to cause a certain light-headedness that was soon popularized through laughing gas parties and traveling shows [10]. Today the gas is still in use as a mild sedative in dental surgery and in obstetrics, where the application in medical-grade purity does no longer give rise to fits of uncontrolled happiness. For major surgical procedures, nitrous oxide is used as a carrier gas for more potent agents such as halothane.

2.1   Chemical Properties of Dinitrogen Monoxide

The N_2O molecule is isoelectronic to CO_2 and shares several of its physicochemical properties. From rotational spectroscopy measurements, the bond lengths in N_2O gas were determined to be 1.1282 Å for the N–N bond and 1.1842 Å for the N–O bond. These bonds are notably shorter than the respective average values of 1.25 Å and 1.21 Å for the corresponding double bonds, and this can be ascribed to the existence of two resonance structures (Figure 2). The central nitrogen carries a positive partial charge in both structures, resulting in an orbital contraction that helps to rationalize the shortened bond lengths.

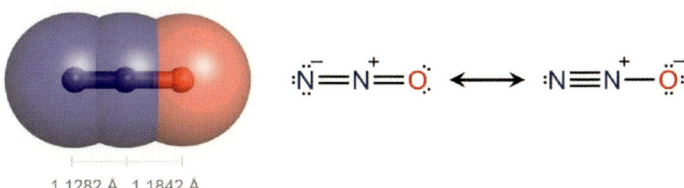

1.1282 Å   1.1842 Å

Figure 2   The N_2O molecule. The linear molecule shows N–N and N–O bond lengths that are shorter than those expected for double bonds. This is due to resonance stabilization that also conveys a slightly positive partial charge to the central nitrogen atom, while both terminal atoms carry a slightly negative partial charge. The resulting small dipole moment of N_2O substantially contributes to its solubility in polar solvents such as water.

The two resonance structures carry opposing charges on their terminal atoms, resulting in a low dipole moment of 0.161 D in spite of the terminal oxygen atom. N_2O has an electronic configuration of $(1\sigma)^2(2\sigma)^2(3\sigma)^2(4\sigma)^2(5\sigma)^2(6\sigma)^2$ $(1\pi)^4(7\sigma)^2(2\pi)^4$, so that the highest occupied π orbitals, 1π and 2π, each are a degenerate pair of allyl-type orbitals. Calculations have suggested the 2π orbital to be slightly N–N bonding and N–O antibonding, in agreement with the shorter

N–N distance. The LUMO is an antibonding 3π molecular orbital, whose π character is strengthened by the fact that oxygen has a higher electronegativity than nitrogen [10]. The N_2O molecule exhibits three fundamental absorptions with superimposed rotational fine structure in infrared spectroscopy, an asymmetric stretch (Σ^+) at $\nu_1 = 1284.91$ cm^{-1}, a bend (Π) at $\nu_2 = 588.77$ cm^{-1}, and a symmetric stretch (Σ^+) at $\nu_3 = 2223.76$ cm^{-1}. Both stretches (ν_1 and ν_3) also appear in the Raman spectrum, and these properties provided initial proof that the linear structure of the molecule was asymmetric. Nitrous oxide liquefies at 183.7 K and solidifies at 170.8 K. It is well soluble in water with a mole fraction of $4.4 \cdot 10^{-4}$ at 298 K, so that solvated nitrous oxide is readily accessible as an enzyme substrate [11]. Hydration of N_2O in water could potentially yield hyponitrous acid ($H_2N_2O_2$), a weak acid with $pK_1 = 7.2$ and $pK_2 = 11.5$, but the equilibrium for this reaction is far on the side of N_2O, which is consequently the dominant species in aqueous solution.

2.2 Nitrous Oxide, the Greenhouse Effect, and Ozone Depletion

In the atmosphere, nitrous oxide appears as a trace gas with a current concentration of about 325 ppb, and it can be traced through climate history by analyzing air entrapped in polar ice cores. Interestingly, the present concentration is higher than at any time during the last 45,000 years [12]. From low levels during the last glacial period (the Würm ice age in the European Alpine region) the atmospheric concentration of N_2O increased to about 275 ppb and remained largely constant for the last 10,000 years. It started to rise in the 19th century, concomitant with the onset of industrialization, and is increasing since, at an average annual rate of 0.8 ppbv [13]. Due to its chemical stability, the atmospheric half-life of N_2O is estimated to be 120 years, so that any increase in production will lead to an atmospheric accumulation that persists for centuries to come [14].

The problematic aspect of this increase is that N_2O is a potent greenhouse gas that absorbs sunlight reflected from the Earth's surface with approximately the 300-fold efficiency of CO_2 [14, 15]. The control and reduction of N_2O emissions thus should be an imminent aspect of global climate policies, but while attempts are undertaken to regulate and limit CO_2 emissions, the impact of nitrous oxide has long gone unnoticed. Only in 2009, the gas has been designated the dominant anthropogenic emission of the 21st century [16], and public awareness is rising. Most critically, nitrous oxide not only acts as a greenhouse gas, but is also involved in complex atmospheric chemistry.

N_2O is the predominant source of stratospheric NO_x that react with and deplete the reservoir of ozone, O_3. However, in conjunction with the notorious ozone-depleting chlorofluorocarbons (CFC) the effect of NO_x may even be beneficial, as chlorine oxide radicals generated through the decomposition of CFCs react with

NO and NO_2 in the lower stratosphere to form stable compounds such as $ClONO_2$ that do not react with ozone [10, 17].

Today the international treaties to limit the production of CFCs have shown substantial effects, but at the same time this brings back N_2O as a major sink for ozone that will play a key role in the foreseeable future [16]. Nitrous oxide is among the greenhouse gases considered in the second phase of the Kyoto Protocol to the United Nations Framework Convention on Climate Change, together with CO_2, CH_4, SF_6, and the CFCs. This treaty covers the commitment period 2013–2020, but remains to be ratified by most countries, with a decision by the European Parliament projected for 2015 [18].

2.3   Abiotic and Biotic Sources

Nitrous oxide is primarily produced in the course of a variety of biological processes, in particular bacterial denitrification (see Section 2.4), but in recent times a number of abiotic routes for the generation of substantial quantities of the gas were found. The nitrate-dependent oxidation of Fe(II) was recognized as a central geochemical process, but while for most minerals the reduced nitrogen species generated here are nitrite or ammonium, the mineral siderite ($FeCO_3$) was described to yield N_2O as a main product [19]. More recently, a study investigating a brine pond in Antarctica revealed that several Fe(II) components of the mineral dolerite could also generate N_2O in large quantities [20].

On a global scale, such processes nevertheless are dwarfed by the biological generation of nitrous oxide through the pathways of nitrification and denitrification in the nitrogen cycle [6, 21]. Approximately 40 % of the global nitrous oxide emissions originate directly or indirectly from human activities (US Environmental Protection Agency, http://epa.gov/climatechange/ghgemissions/gases/n2o.html). Herein, 70 % are related to agricultural soil management, with a minor fraction of approximately 5 % being due to the breakdown of nitrogen compounds in the manure and urine of livestock, while the bulk is due to the use of nitrogen fertilizers in industrial crop production (see Section 2.4). About 10 % of anthropogenic N_2O release are attributed to combustion engines in industry and transportation, and another 10 % to industrial and chemical production processes. A major factor hereby is the release of N_2O as a side product in the synthesis of adipic acid, one of the two components required for the production of nylon [22].

2.4   Bacterial Denitrification

In the metabolic pathway of denitrification, N_2O is generated through the reductive coupling of two molecules of nitric oxide (NO) by the membrane-integral nitric oxide reductase. It is released to the periplasmic space in Gram-negative bacteria, where a soluble nitrous oxide reductase catalyzes the two-electron reduction to N_2 (see equation 1 in Section 1), using electrons that originate from menaquinol. Energetically, denitrification is the most efficient respiratory pathway in the absence of molecular oxygen as a terminal electron acceptor, and consequently denitrifiers thrive in environments with low oxygen levels and abundant supply of nitrate [7]. Coincidentally, these conditions are met quite precisely in modern industrial agriculture, where the growth-limiting element nitrogen is provided as a fertilizer, commonly in the form of nitrate salts. Soil denitrifiers compete with the assimilating crops for this nutrient, to the effect that approximately half of the fertilizer dispersed on the fields is returned to the N_2 from which it was produced in the Haber-Bosch process [3].

With a current global fertilizer production of approximately 160 Mt per year, the overall metabolic flux through the denitrification pathway is therefore substantial. The second condition for denitrification, an anoxic or microoxic environment, is met quite readily as well, as the availability of O_2 diminishes drastically already millimeters or centimeters into the rather stratified soil of a regularly watered field. Nitrate reduction increases sharply at this oxycline, where traces of O_2 may be encountered, but can generally be tolerated. The weakest link in the chain of denitrificatory enzymes, however, is the one catalyzing the final step from N_2O to N_2, nitrous oxide reductase. Its copper centers show the highest sensitivity, and while the reaction of the enzyme is presumably coupled to the generation of a proton motive force [23], the organism can tolerate its failure, in particular because the gaseous substrate N_2O can be released and will not accumulate to inhibit the preceding NO reductase. The result is a direct correlation of the increasing use of nitrogen fertilizers in agriculture and the rise of atmospheric N_2O released through incomplete denitrification. While agriculture thus is just one of the factors contributing to the rise of atmospheric levels of nitrous oxide, it is the one most directly related to human population growth with a direct link to the ongoing industrialization of highly populated countries such as India or China [16].

3   Nitrous Oxide Reductase

The enzymatic conversion of N_2O into the stable products N_2 and water (equation 1) is carried out by copper-containing nitrous oxide reductase. After noticing a general requirement for copper for N_2O reduction by *Alcaligenes faecalis* [24], the enzyme itself was identified in a dedicated search for copper-containing enzymes in Pseudomonads [25], with the first ortholog to be characterized being the one from *Pseudomonas stutzeri* strain ZoBell (originally *P. perfectomarina*) [26].

A variety of other orthologs was characterized over the years and detailed insight was gained concerning the two metal sites, but it was only in the year 2000 that a first crystal structure became available for the enzyme from *Pseudomonas nautica* [27, 28] (now *Marinobacter hydrocarbonoclasticus*) [29] and in 2003 for *Paracoccus denitrificans* [27, 30]. Structural data was later presented for N_2O reductase from *Achromobacter cycloclastes* [31], and more recently for the originally studied enzyme from *P. stutzeri* [32]. Gaining an unambiguous picture of the function and properties of N_2O reductase has been complicated by the fact that the enzyme has been isolated in various forms that differ in content and state of the active centers, as well as in their activity.

3.1   Anatomy of an Unusual Copper Enzyme

The enzyme N_2O reductase is the product of the *nosZ* gene, an open reading frame encoding a protein of approximately 640 amino acid residues that is located in the bacterial periplasmic space. Variations to this organization were reported for the denitrifying hyperthermophilic crenarchaeon *Pyrobaculum aerophilum* [33], and for few genera of bacteria including *Thiomicrospira denitrificans* and *Wolinella succinogenes* that contain an additional cytochrome *c* domain in their *nosZ* gene [23]. The two domains of the canonical NosZ protein fold into an N-terminal, seven-bladed β-propeller, and a C-terminal β-sandwich that attains the cupredoxin fold commonly observed in mononuclear type-I copper proteins [34]. Accordingly, each individual domain has a distinct metal center embedded, and the interaction of two chains in the dimeric enzyme assures that they are located in an appropriate orientation to afford a functional interaction (Figure 3).

NosZ proteins form stable homodimers with a strict head-to-tail arrangement, where the N-terminal domain of one monomer interacts with the C-terminal domain of the other and *vice versa*. In consequence, the metal centers within a single protein chain are more than 40 Å apart in the NosZ dimer, while the distance between the centers in different chains amounts to a mere 10 Å, placing it well within the range required for efficient electron transfer [35]. Both metal centers of N_2O reductase exclusively employ copper, and both exhibit a degree of specialization that is only observed here. The cupredoxin domain is the site of the binuclear Cu_A center, an electron transfer site with a characteristic spectroscopic signature that is also found in many respiratory oxidases. Its function in N_2O reductase is to mediate single-electron transfer from an external electron donor, commonly a *c*-type cytochrome or a cupredoxin, to the second metal site where substrate reduction occurs. This second site is a tetranuclear Cu center termed Cu_Z that so far has only been found in N_2O reductase. Its architecture and exact composition have been a matter of debate for many years (see Section 3.3).

3.1.1   Distinct Forms of the Enzyme

The observed differences in operon architecture are in at least one case directly related to the structure of the NosZ protein itself. *Nos* operons that lack the *nosR* gene and instead contain *nosG*, *nosH*, and further periplasmic cytochromes, typically contain a larger open reading frame for *nosZ* that includes a third domain with the α-helical secondary structure and the typical CXXCH heme binding motif of

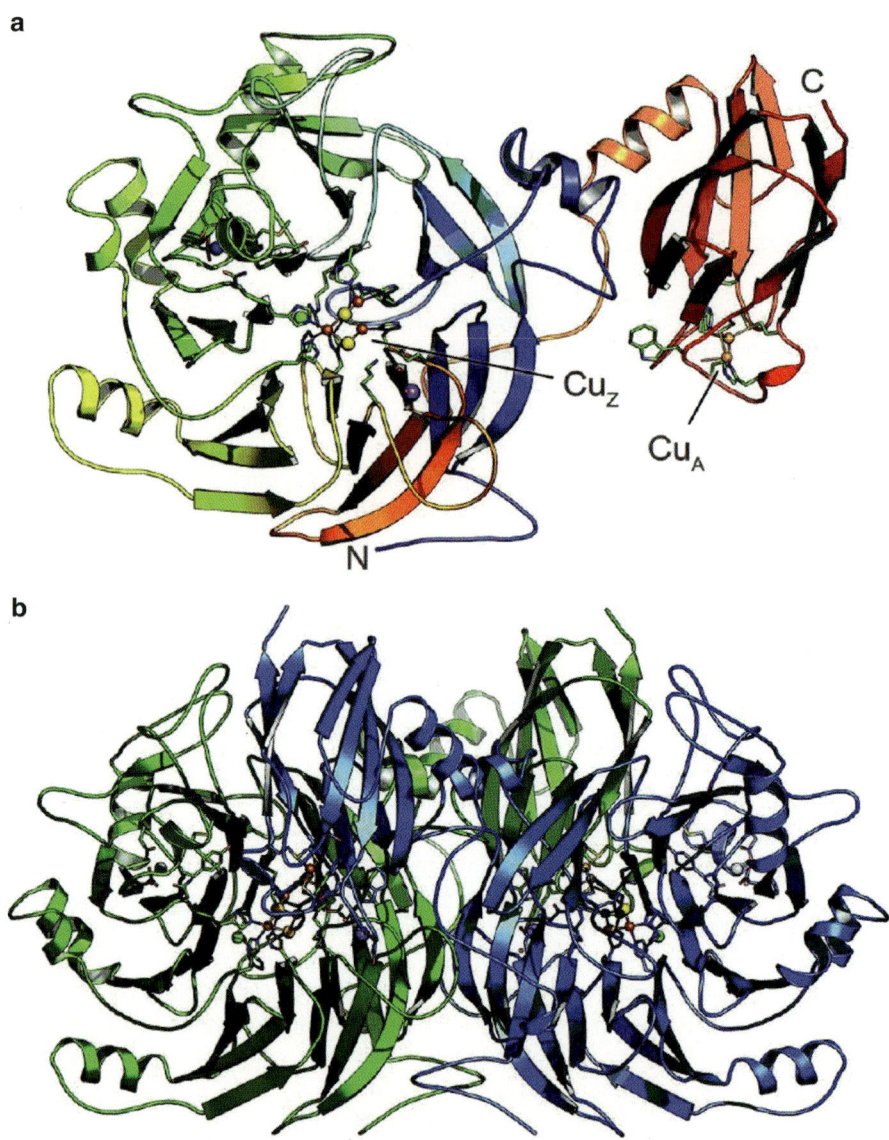

Figure 3 Structure of the copper enzyme N_2O reductase (PDB ID 3SBR). (**a**) N_2OR is the product of the *nosZ* gene, and the peptide chain is organized into two distinct domains that harbor one of the metal centers each. The N-terminal domain attains a seven-bladed β-propeller fold that coordinates the tetranuclear active site Cu_Z through seven histidine residues at its hub. The C-terminal domain shows a cupredoxin fold similar to type-I copper proteins and binds the dinuclear mixed-valent Cu_A center. (**b**) The distance between the two metals centers within a single polypeptide chain is above 40 Å, but the enzyme forms stable homodimers in a head-to-tail arrangements that brings the Cu_A site of one monomer within a distance of 10 Å from the Cu_Z site of the other and *vice versa*.

c-type cytochromes. This has been analyzed in *W. succinogenes* [23, 36], and it is commonly interpreted to originate from a fusion event of the reductase enzyme with a soluble electron donor. A cytochrome c_{552} of similar architecture serves as electron donor to the enzyme of *M. hydrocarbonoclasticus* [37], while for other NosZ orthologs the type-I copper protein pseudoazurin was described to also fulfill this function [38]. This functional variability in the architecture of electron transfer chains is not unprecedented within the pathway of denitrification and a very similar case was described for cytochrome cd_1 nitrite reductase [39].

As an additional layer of complexity, NosZ itself was found to occur in a series of forms that are distinct in terms of spectroscopy and reactivity [40]. When isolated in the absence of dioxygen, preparations had a deep purple color, with high catalytic activity and approximately 8 mol of Cu per mol of dimeric enzyme of *P. stutzeri* [41, 42]. This was designated form I of N_2O reductase, and it is considered to be the physiologically relevant active form [43], although the copper content at the time was significantly underestimated. When isolated in the presence of dioxygen, the enzyme showed reduced catalytic activity and a different optical spectrum.

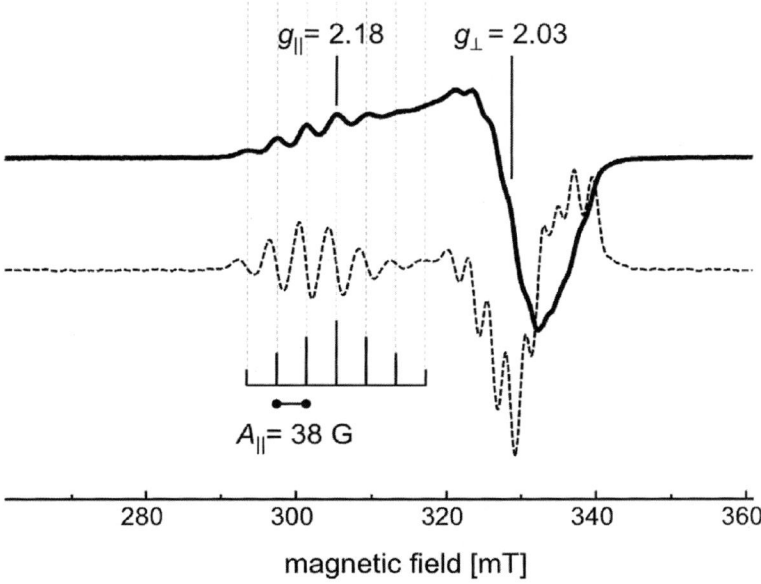

Figure 4 Electron paramagnetic resonance spectrum of N_2O reductase. In a continuous-wave EPR experiment at X-band ($\nu = 9.4$ GHz) recorded at 10 K, N_2OR shows a characteristic axial signal dominated by the contribution from the Cu_A site. Both the $g_{\parallel}$ and the $g_{\perp}$ region of the EPR spectrum are split into a complex hyperfine pattern that was best resolved in the purple form I of the enzyme where the second metal site, Cu_Z, was in the [4Cu:2S] configuration. The seven-line pattern is well resolved in the first harmonic (solid), and the second harmonic (dashed) shows that both g regions are split with an intensity ratio of 1:2:3:4:3:2:1 that is explained by two coupled copper nuclei ($^{63,65}Cu$ nuclear spin $I = 3/2$) in a fully valence-delocalized arrangement.

In this form II, the 7-line hyperfine pattern in the EPR spectrum of Cu_A (Figure 4) was generally less well resolved in the $g_\parallel$ region, and at the time this was ascribed to an underlying signal of a distinct form of Cu_Z, termed Cu_Z^* [40, 43]. As such, Cu_Z was thus associated with an active state, Cu_Z^* with an inactive state of the center. Anoxic reduction of Cu_A with a 10-fold molar excess of dithionite changed the color of the protein to blue (form III), leading to a catalytically inactive state dominated by a single charge-transfer band with a maximum at 650 nm, and an axial $S = \frac{1}{2}$ EPR signal without a resolved hyperfine structure in the $g_\parallel$ region [40]. With Cu_A in the reduced state, the spectroscopic features of form III solely originate from the Cu_Z center, but while a distinction was made for the EPR data, it was not initially recognized that the UV/vis signatures reveal the identity and ratio of Cu_Z *versus* Cu_Z^* that reflect on the structure/function relationship of the enzyme's active site (Figure 5) [44].

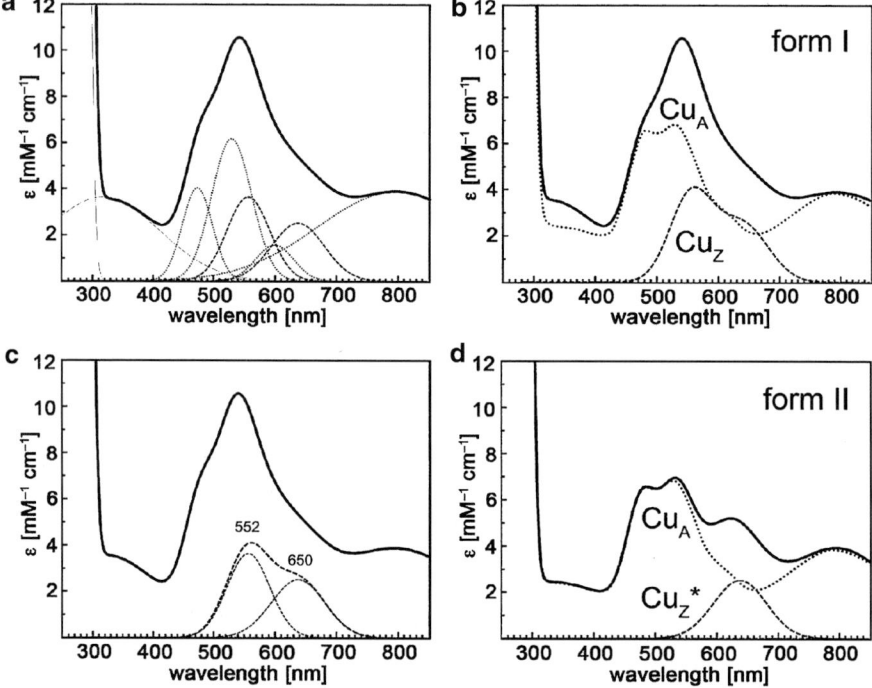

Figure 5 Electron excitation spectra of N_2O reductase. (**a**) The spectrum of purple N_2OR (form I) shows maxima at 538 nm and 795 nm and can be deconvoluted into individual bands by a multi-Gaussian fit. (**b**) The individual contributions of the Cu_A and Cu_Z centers can be reconstructed from the fit, so that the properties of the two sites can be studied in isolation. In addition, the reduction with sodium ascorbate will only reduce Cu_A, and the Cu_Z contribution is readily accessible experimentally. (**c**) In N_2OR form I, the contribution of the Cu_Z site can be modelled by two bands with maxima at 552 nm and 650 nm. Further reduction with sodium dithionite will only deplete the transition at 552 nm, while the remaining band persists. (**d**) In form II N_2OR, Cu_Z is present in the [4Cu:S] Cu_Z^* state, with only a single band with a maximum at 650 nm from the tetranuclear cluster. Contrary to form I, this state can be further reduced with sodium dithionite, yielding an enzyme that is all-Cu(I) and thus colorless.

A fourth form of N_2O reductase was described as a 'reconstituted' enzyme, where Cu was first removed to generate the apo-protein that could subsequently be replenished with copper. This was achieved by anoxic dialysis against KCN, followed by the addition of $Cu(en)_2SO_4$ [40]. This procedure led to an enzyme with a lower copper content of approximately 4 Cu per homodimer of N_2OR that had the spectroscopic features of an intact Cu_A site, but with Cu_Z fully absent. Besides providing evidence that the assembly of Cu_Z required further maturation factors, this procedure also gave access to a form of the enzyme in which one of the metal centers could be studied without interference from the other. The same effect was reached through a variant strain, *P. stutzeri* MK402, in which the biogenesis of Cu_Z was prevented through a Tn5 insertion, leading to an enzyme containing exclusively Cu_A, termed form V [40].

The Cu content of different preparations of nitrous oxide reductase was generally between 7 and 8 Cu per dimer, but the exogenous addition of copper salts to the growth medium yielded a highly active form of the enzyme with an increased Cu content of 10–12/dimer [45, 46]. This could be rationalized from the crystal structures that became available at the same time, and showed the Cu_Z site to be an entirely novel, tetranuclear copper center, with four metal ions arranged around a central ligand and coordinated to a total of seven histidine residues at the hub of the N-terminal β-propeller domain [27, 28, 30]. Both relevant forms of Cu_Z, the Cu_Z site of N_2OR form I and the Cu_Z^* site of form II, contain the same number of copper ions within the cluster, and even the original assumption that Cu_Z^* represents an oxidatively damaged form of Cu_Z was recently questioned by the finding that an enzyme with the typical features of N_2OR form I could be isolated under oxic conditions from *M. hydrocarbonoclasticus* [47].

3.1.2 Three-Dimensional Structures

Initial advances towards structural data for the metal sites of nitrous oxide reductase was made for the Cu_A site in respiratory cytochrome *c* oxidase or in engineered cupredoxins as detailed below, but no information was available for the active site of the denitrificatory enzyme, the Cu_Z center. In the initial description of the ortholog from *P. nautica* (*M. hydrocarbonoclasticus*) [29, 48], Cu_Z was identified as a novel tetranuclear site coordinated by seven histidines, with a central oxygen ligand [28, 49, 50]. This structure was obtained from a preparation corresponding to form III of N_2O reductase, with Cu_A in the reduced state and the distinct blue color of an absorption band with a maximum at 600 nm that consequently had to originate from Cu_Z. This assignment raised immediate criticism from the spectroscopic community, as an environment consisting exclusively of imidazole nitrogens and an oxygen lacked the soft, donating ligand required to yield the prominent ligand → metal charge-transfer transition (LMCT) represented by the blue color of the preparation. Indeed, a re-determination of the *M. hydrocarbonoclasticus* N_2OR structure, combined with the structure of the enzyme from *Paracoccus denitrificans* at 1.6 Å resolution soon thereafter clarified that the central oxygen in Cu_Z was indeed a sulfido species that likely was misinterpreted due to partial occupancy [27, 30].

Both structures were obtained as a blue form III of the enzyme and interpreted to show the Cu_Z^* form of the active center. In 2006, the enzyme from *A. cycloclastes* was crystallized and refined to 1.7 Å resolution, showing for the first time an enzyme in form II, which differed from the previous structures in having the Cu_A site in the oxidized state [31]. Note, however, that the copper sites of nitrous oxide reductase are likely prone to swift photoreduction upon exposure to X-rays, so that spectroscopic monitoring of crystals will be required before unambiguous assignments of oxidation state can be made. The most recent structural data on N_2OR came from *P. stutzeri*, the organism where the enzyme was originally discovered. Using anoxically isolated protein, purple crystals of form I N_2OR were obtained [51], and the resulting structure showed several relevant features, including the Cu_Z state of the active site cluster [32].

3.2 Cu_A: More than an Electron Transfer Center

Beside its role as an electron transfer site in nitrous oxide reductase, further interest in the Cu_A site arose from the fact that its spectroscopic features indicated a possible structural and electronic similarity to one of the copper sites in respiratory cytochrome *c* oxidase [52]. Due to a better resolution of the observed spectroscopic features, nitrous oxide reductase eventually proved to be highly instrumental to understand the structure and properties of this site, but agreement was only reached after lengthy debates with the advent of several crystal structures [53]. Yet, only the most recent analyses of the denitrificatory enzyme have revealed intriguing properties that after all do set the Cu_A site in N_2OR apart from the one in cytochrome *c* oxidase and that help to explain the improved resolution in EPR spectra of the latter [32, 44].

3.2.1   Spectroscopic Properties of Cu_A

Copper-containing proteins are frequently classified within a scheme of three basic types according to a proposal by Malkin and Malmström [54]. When the enzyme nitrous oxide reductase was initially isolated from *P. stutzeri* in different form as described above [43], it was readily apparent that this novel protein would not fit into the established scheme. For the purple form I of the enzyme, EPR studies at X-band showed a distinct axial signal with $g_{\parallel} = 2.18$ and $g_{\perp} = 2.03$, with characteristic hyperfine splittings with seven lines in a 1:2:3:4:3:2:1 intensity ratio that were not consistent with a mononuclear copper site ($^{63/65}Cu$ nuclear spin $I = 3/2$) (Figure 4) [43]. The $g_{\parallel}$ region featured seven equidistant lines with $A_{\parallel} = 3.83$ mT, and a similar pattern was observed in the $g_\perp$ region with $A_\perp = 2.8$ mT. This hyperfine structure disappeared upon reduction of the protein in the absence of dioxygen to yield to a broad featureless signal with g values at 2.18 and 2.06. Notably, the pink form II of the enzyme that is characterized by an intact Cu_A site, but a Cu_Z site in the Cu_Z^* state, showed a similar spectrum to form I, but with an inferior resolution for the hyperfine pattern in both the $g_{||}$ and $g_\perp$ regions.

Based on these and further studies, Kroneck and coworkers concluded that Cu_A in nitrous oxide reductase was a valence-delocalized $[Cu^{1.5+}:Cu^{1.5+}]$ binuclear center, and that the same was true for the corresponding site in cytochrome c oxidase [52]. This claim was soon contested by the proponents of a mononuclear Cu_A site in the respiratory complex [55], largely under the impression of the available spectroscopic data for the respiratory enzyme. While the additional copper center there was found to show an unusual $g_{||}$ value of 2.18 and lacked the four-line hyperfine pattern typical for type-I and type-II copper sites unless it was chemically denatured, the characteristic seven-line pattern (Figure 4) was never observed clearly. In retrospect this was ascribed to the additional presence of heme groups in cytochrome c oxidase [53, 56], but more recent data has provided an alternative explanation (see Section 3.2.3).

In the absence of a crystal structure, Cu K-edge EXAFS was highly valuable for obtaining initial information on the architecture of Cu_A. A study on the solubilized cupredoxin domain of subunit II in cytochrome c oxidase from *Bacillus subtilis* indicated the presence of two coppers, each of which was coordinated to a single histidine and two cysteines [57]. In addition, a direct metal-metal interaction was observed at a distance of 2.5 Å. In a follow-up, the authors compared the Cu_A domains of the oxidases of *B. subtilis* and *Thermus thermophilus* in different redox states, and obtained a Cu-Cu distance of 2.43 Å for the valence-delocalized oxidized state, *versus* 2.51 Å for the fully-reduced $[Cu^{1+}:Cu^{1+}]$ state [58].

3.2.2   Three-Dimensional Structure(s) of Cu_A

While the Cu_A site in N_2O reductase had proved valuable for understanding the electronic properties of the center, structural information was first provided along different lines. Extensive efforts to solve the three-dimensional structure of respiratory cytochrome c oxidase were undertaken by several groups, until the prokaryotic ortholog from *Paracoccus denitrificans* was solved to 2.8 Å resolution in 1995 [59], followed shortly thereafter by the eukaryotic enzyme from bovine heart mitochondria [60, 61]. The same arrangement was also found in a solubilized form of subunit II of cytochrome bo_3 oxidase from *Escherichia coli* [62]. These structures unequivocally settled the debate concerning the number of metal ions in Cu_A, and they underlined that the ligand environment in Cu_A is highly conserved.

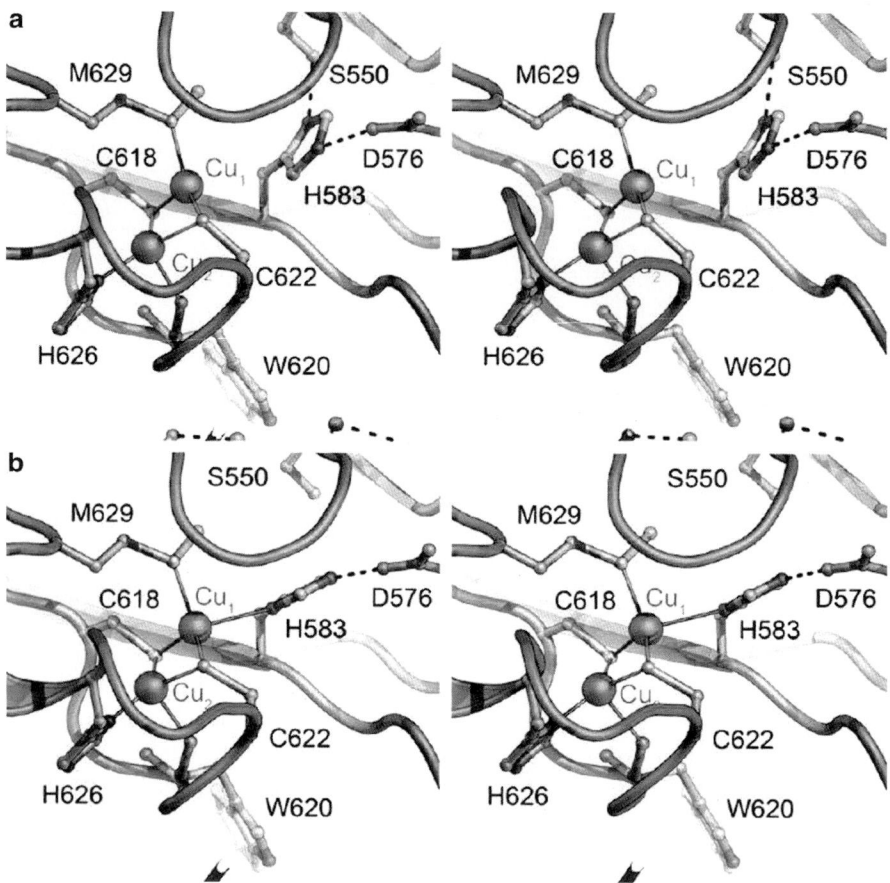

Figure 6 The electron transfer site Cu_A of N_2O reductase. Stereo representations of the two conformations observed in structures of N_2OR. (**a**) In purple N_2OR from *P. stutzeri*, as isolated, residue His583 was in most cases not a ligand to Cu_1 of the Cu_A site resulting in a shift of the copper ion into the plane formed by the two cysteine residues, Cys618 and Cys622, and Met629 (PDB ID 3SBQ). The experimental data is consistent with a sharpening of the EPR hyperfine structure in this conformation. (**b**) In all instances with substrate bound to Cu_Z, the side chain of His583 was flipped back to coordinate Cu_1 of the Cu_A center, linking the metal cluster to the protein surface *via* residue Asp576 to allow for electron transfer from an external donor (PDB ID 3SBR).

The two Cu ions of Cu_A show a distorted tetrahedral coordination environment that represents the classical 'entatic' state between a square-planar geometry with hard ligands (preferred by Cu(II)) and a tetrahedral geometry with soft ligands (preferred by Cu(I)) [63]. In Cu_A, the two cysteine ligands are shared in a μ^2-bridging fashion, and each ion in addition has a histidine ligand (Figure 6). The coordination environment is complemented in each case by a further ligand that differs between the two metals. For Cu_1, this ligand is the oxygen atom of a backbone carbonyl group, while for Cu_2 the first coordination sphere is completed by the sulfur of a methionine, at a slightly longer distance. The site thus is of high – if not perfect – symmetry, and shows the features of electronic delocalization, in particular the characteristic seven-line hyperfine pattern.

Based on the crystal structures, an excellent biomimetic model complex was presented soon thereafter [64]. While the complex was centrosymmetric and therefore had perfectly identical geometries for both metals, the observed seven-line hyperfine pattern in EPR spectroscopy showed a hyperfine splitting of 4.99 mT for $A_{\parallel}$ and 3.63 mT for $A_{\perp}$, notably larger than for the protein. This was concomitant with a Cu-Cu distance in the model complex of 2.92 Å that was significantly longer than the 2.5 Å determined by EXAFS spectroscopy for Cu_A [57, 65]. The short distance between both metals in this center is commonly interpreted as one of the rare cases of a direct metal-metal interaction in Biology [57]. In the structure of *P. denitrificans* N_2OR determined at 1.6 Å resolution, the Cu-Cu distance was 2.51 Å [30], but was significantly longer at 2.63 Å in the purple form I structure from *P. stutzeri* [32]. For this enzyme, ^{63}Cu- or ^{65}Cu-enrichment as well as ^{15}N-labeling had allowed for a significantly improved resolution for EPR spectroscopy that then formed the basis for molecular orbital calculations on a structural core unit consisting of Cu ions with two μ^2-bridging sulfides and a terminal amine ligand on each metal [66]. The obtained values agreed very well with the experimental data, and subsequent ENDOR studies helped to refine the picture further (see Section 5.1) [67, 68].

The Cu_A center is located in a domain that shows a conserved tertiary structure commonly termed the cupredoxin fold [69, 70]. Cupredoxins are mononuclear copper proteins ('type-I copper proteins') that coordinate a single copper ion in a rigid binding site at the periphery of a 100–140 aa peptide with a characteristic β-barrel forming a Greek-key motif [34]. In cupredoxins, the metal is coordinated by a cysteine that allows for a LMCT, giving rise to an intense absorption maximum around 600 nm that conveys the typical color of these 'blue copper proteins'. Further ligands are two histidines and a fourth ligand that commonly is a methionine, but can also be a different amino acid that modulates the midpoint potential of the Cu^+/Cu^{2+} redox pair [71]. Within the protein chain, one of the histidine ligands is found in the first third of the sequence, while the other three ligands cluster in a single loop near the C-terminal that connects the last two β strands.

Interestingly, the very same architecture is found both in cytochrome *c* oxidase and nitrous oxide reductase, where a cupredoxin domain holds the Cu_A site. In *P. stutzeri* N_2O reductase, the ligands to Cu_A are His583 in the N-terminal part of the cupredoxin domain, and Cys618, Trp620, Cys622, His626, and Met629 that all form part of the terminal loop between β strands 8 and 9 (Figures 3 and 6). This loop is longer than in a typical type-I copper protein, and in fact Sanders-Loehr, Canters, and coworkers could show that a simple insertion of a Cu_A-binding loop into a cupredoxin such as amicyanin led to the assembly of an intact Cu_A center in a valence-delocalized state [72, 73]. The copper-binding loop of the Cu_A site in nitrous oxide reductase faces the second metal site, Cu_Z, and two residues, Phe621 and Met627 in *P. stutzeri*, form part of the substrate binding pocket, as detailed below.

3.2.3 Unexpected Flexibility: Cu$_A$ in *P. stutzeri* Nitrous Oxide Reductase

In the structure of purple N$_2$O reductase from *P. stutzeri*, the Cu$_A$ site differed not only in the consistently longer metal-metal bond when compared to earlier structure determinations, but also in the conformation of one of the coordinating ligands to a copper, His583 [32]. While all earlier structure determinations of proteins containing a Cu$_A$ site – natively or through engineering of the Cu$_A$-binding loop – invariably showed the conserved ligand environment described above, His583 of *P. stutzeri* N$_2$O reductase was found in two different conformations, with its N$_\delta$ atom either coordinating Cu$_2$ of the Cu$_A$ site, or forming a hydrogen bond to the hydroxyl group of nearby residue Ser550 (Figure 6). The two different conformations are related through a rotation of the imidazole moiety of the side chain of His583 by approximately 135° [32]. With its N$_\varepsilon$ nitrogen, His583 retains a hydrogen bond to the β-carboxy group of Asp576, an aspartate residue located at the surface of the enzyme, in an area that was postulated as a possible docking site for an electron transfer partner [37]. The arrangement observed in the crystal structures thus gave rise to the hypothesis that His583 can act as a gate for electron transfer from an external electron donor to the Cu$_A$ site of N$_2$O reductase. If a direct coordination of His583 to Cu$_2$ was indeed a prerequisite for reduction of Cu$_A$, then the rotation of the histidine side chain towards Ser550 would serve this conduit and prevent the enzyme from executing its catalytic function [32, 44].

However, in order to function as an efficient gate the flip of the imidazole moiety should be linked to a triggering event within the catalytic cycle. In various structures of N$_2$OR form I, His583 was observed in either conformation, flipped in or out with respect to Cu$_2$, indicating a significant degree of structural flexibility on the part of the reductase. Notably, in several crystal structure determinations the conformation of His583 differed between the two monomers of the same enzyme dimer. This is not straightforward to rationalize in a crystal structure that always represents the macroscopic conformational average over all unit cells, and it may imply a slight asymmetry of the dimeric arrangement that affects the crystallization process. No such asymmetry has been described to date, and it will be interesting to conduct an in-depth analysis based on high-quality structural data.

To date His583 (or the corresponding residue in other Cu$_A$ domains) was so far only found in a flipped-out state in the purple form I of N$_2$OR (Figure 6). While Ser550 is a conserved residue in N$_2$O reductase, it is absent in heme-copper oxidases, where no flip of a histidine ligand to Cu$_A$ seems to occur. In N$_2$O reductase, the purple form I has the second metal site, Cu$_Z$, in a [4Cu:2S] conformation, so that there must be structural changes in the vicinity of both metal sites that enable His583 to abandon its coordination to Cu$_2$ of Cu$_A$. The exact determinants of these changes remain to be identified, but their functional relevance is underlined by the finding that the one case where His583 was always found flipped towards Cu$_A$ – thus enabling electron transfer from an external partner – was when substrate was bound to the nearby Cu$_Z$ site (see Section 5.3).

3.3   The Tetranuclear Cu_Z Center

In contrast to Cu_A, the second metal site in N_2O reductase, Cu_Z, has no precedence in any other known enzyme, and its exact architecture and relevant states remain to be fully understood. As the addition of the strong reductant sodium dithionite, $Na_2S_2O_4$, produced a blue species with an absorption maximum around 640 nm and a distinct EPR spectrum, the presence of a second copper site in N_2OR was suspected early on [26]. A copper site that was not reduced by dithionite seemed highly unusual, and as copper determinations at the time yielded on average 8 Cu per enzyme dimer [41, 42], the site – now designated Cu_Z – was considered to be a second dinuclear center with a sulfide ligand that gave rise to the characteristic charge-transfer band [74]. The EPR signature of Cu_Z, with $g_{\parallel} = 2.18$ and $g_{\perp} = 2.06$, was distinct from the signature of Cu_A (Figure 4) and remained after the latter was reduced to an all-Cu(I), diamagnetic state [43]. This form of Cu_Z was described as an $S = \frac{1}{2}$ resting state of the cluster [75]. In retrospect, many studies on Cu_Z suffered from the intrinsic structural and electronic heterogeneity found in Cu_Z, and only with the advent of three-dimensional structures for different forms of nitrous oxide reductase the picture has become clearer.

A recurring theme in literature on N_2OR is on the different states of the Cu_Z site. After the initial description of two forms, a catalytically active Cu_Z and a catalytically inactive Cu_Z^* [74, 76, 77], more recent data have led to a more differentiated (and complicated) picture. Ascorbate reduces the Cu_A site to a colorless and EPR-silent [Cu^+:Cu^+] state, so that Cu_Z can be straightforwardly analyzed without influence from the other cluster. Here, the difference between both forms of Cu_Z is readily apparent. In electron excitation spectroscopy, Cu_Z^* shows the single LMCT band at 640 nm described above, while the Cu_Z center is characterized by two distinct bands at 550 nm and 650 nm. Further reduction of Cu_Z with sodium dithionite depletes the absorption at 550 nm, but notably a complete reduction of the site is not achieved and the band at 650 nm is retained even after prolonged incubation with strong reductants. In contrast, dithionite in combination with the redox mediator methyl viologen can reduce Cu_Z^* to an all-cuprous, colorless state. After reduction of Cu_A, both Cu_Z and Cu_Z^* appear blue in color, and according to the original assignment both species would be classified as form III N_2OR. Nevertheless the key to understanding the reactivity of nitrous oxide reductase most likely lies exactly within the distinction of these states.

3.3.1   Structural Data on Cu_Z

After the initial clarification of the central ligand of Cu_Z to be a partially occupied sulfide rather than on oxide [27, 28], the Cu_Z centers of *M. hydrocarbonoclasticus* [27], *P. denitrificans* [30], and *A. cycloclastes* [31] were consistently described as [4Cu:μ-S] centers [44, 78]. The picture was amended with the structure of purple form I N_2OR from *P. stutzeri* that showed a [4Cu:2S] configuration (Figure 7) and

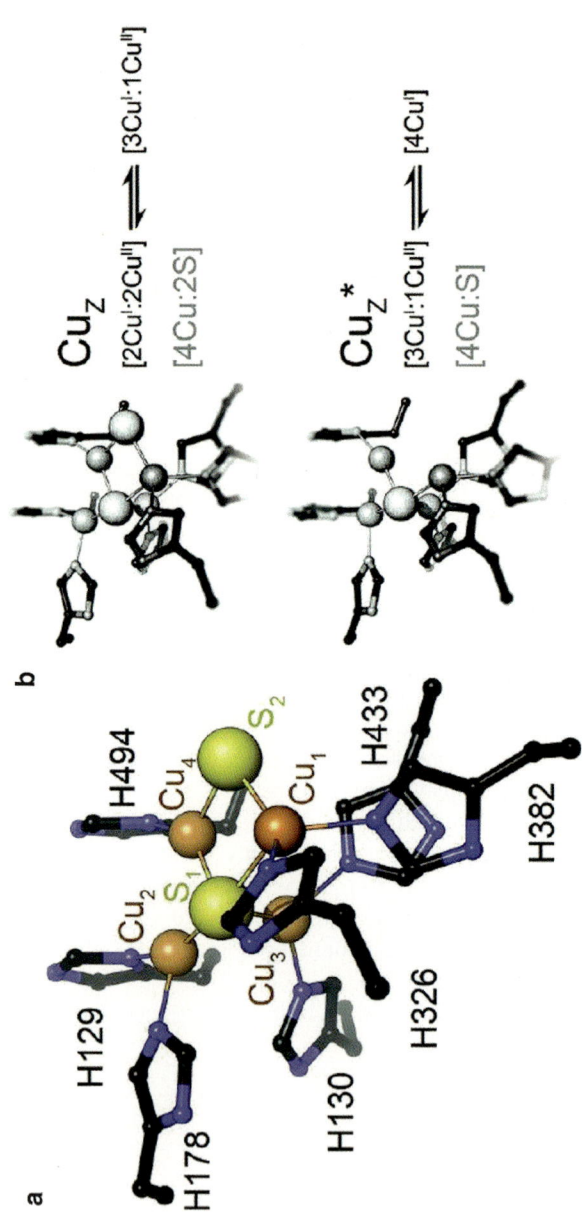

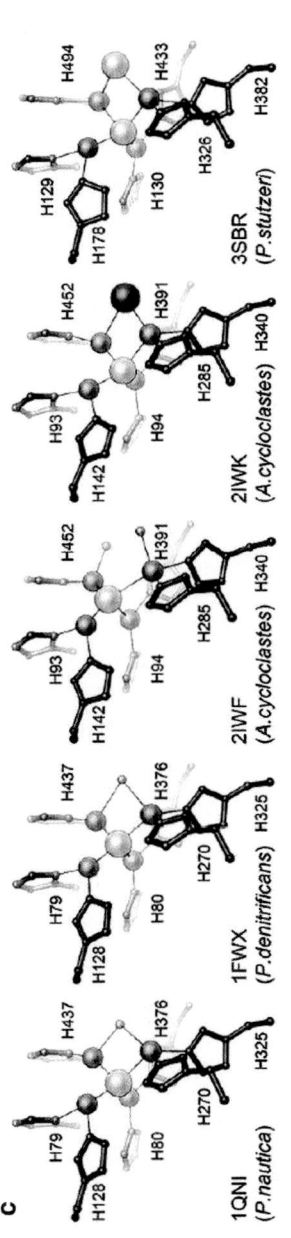

Figure 7 Cu_Z, the catalytic center of N_2O reductase. (**a**) The tetranuclear Cu_Z site is located at the hub of the N-terminal β-propeller domain. It constitutes a unique [4Cu:2S] cluster coordinated to seven histidine residues. (**b**) The spectroscopically and functionally distinct forms Cu_Z (above) and Cu_Z^* (below) differ in the presence of the second sulfur atom in the cluster. While Cu_Z undergoes a redox transition from a [2Cu$^+$:2Cu^{2+}] state to a [3Cu$^+$:1Cu^{2+}] and cannot be further reduced, Cu_Z^* is isolated as [3Cu$^+$:1Cu^{2+}] and can be further reduced to a [4Cu$^+$] state by sodium dithionite and methyl viologen. (**c**) The variation of Cu_Z structures from different crystal structures of nitrous oxide reductases. The first structures (PDB ID 1QNI, 1FWX) showed mixtures of Cu_Z and Cu_Z^*, where a partially occupied sulfur atom between Cu_1 and Cu_4 was interpreted as a bridging water or hydroxo species. A form II structure (PDB ID 2IWF) then had two water ligands at the Cu_1–Cu_4 edge, but would bind the inhibitor iodide in a bridging fashion (PDB ID 1IWK). Only in the purple form I structure (PDB 3SBR), Cu_Z was in the complete [4Cu:2S] state.

revealed that this was associated to the Cu_Z state, while the single-sulfide species [4Cu:μ-S] represented Cu_Z^* [32, 44]. The tetranuclear site seems to be rather susceptible to partial degradation during the isolation procedure of the enzyme, and while the four copper ions are stably retained, it is in the presence of the additional sulfide ion bridging atoms Cu_1 and Cu_4 where the differences manifest.

Degradation of the site in the Cu_Z state leads to formation of Cu_Z^*, and the relative abundance of both states tends to vary from one protein batch to another. In addition, the reverse process, a conversion of Cu_Z^* to Cu_Z, had not been achieved *in vitro*, and it is assumed that this is only possible with the help of additional maturation factors during the biogenesis of the enzyme [11]. As a consequence, the electron density maps obtained from X-ray diffraction data are prone to showing an average of different forms of the cluster, and this refers in particular to mixtures of the Cu_Z and Cu_Z^* states. The magnitude of a peak in an electron density map primarily reflects the number of electrons at a given position, and in addition the value obtained represents an average of all unit cells in the entire crystal. For a 1:1 mixture of Cu_Z and Cu_Z^* within a crystal, this means that the electron density maximum at the position of atom S_{Z2}, where both forms of the cluster differ, will have half the magnitude of a fully occupied Cu_Z state. In number of electrons this corresponds to only 8 of the 16 electrons expected for the element sulfur, but it also corresponds to a fully occupied oxygen atom, and in the original structures it was indeed modeled as such. An oxygen atom in the model will satisfy the observed electron density so that no conspicuous residual peaks will be visible in F_o–F_c difference electron density maps. Oxygen was modeled in this position in the structures of *M. hydrocarbonoclasticus* and *P. denitrificans*, but in both cases the Cu–'O' bond lengths refined to 2.3 Å (PDB ID 1QNI, 1FWX) [27, 30], falling far more in the expected range of a Cu–S bond [44]. The current interpretation of the existing Cu_Z structures (Figure 7) thus suggests that CuZ exists as a [4Cu:2S] cluster, while Cu_Z^* only lacks the bridging S_{Z2}, but does not contain a water ligand connecting Cu_1 and Cu_4. The situation differed slightly in the structure of the *A. cycloclastes* enzyme, where two distinct electron density maxima were interpreted as water ligands at reasonable bond distances of 2.0–2.1 Å (PDB ID 2IWF), or where iodide as an inhibitor bound between Cu_1 and Cu_4 in a bridging manner (PDB ID 2IWK) [31]. This enzyme had the spectroscopic properties of a form II N_2OR with Cu_Z in the Cu_Z^* state, but it did unambiguously show exogenous ligands binding to the center for the first time.

3.3.2   States of Cu_Z and Catalytic Properties

Detailed activity assays for N_2O reductase were first presented for the *P. stutzeri* enzyme [43]. As detailed above, the observed activities were highest for the anoxically isolated form I of the enzyme that we now understand to contain the intact [4Cu:2S] Cu_Z cluster (Figure 7). Subsequently, Snyder and Hollocher published activities for a 'purple' form of *P. denitrificans* N_2OR that were two orders of magnitude higher, but the precise state of the enzyme preparation was not clearly stated [79]. More recent activity data on the form II preparation of *M. hydrocarbonoclasticus* N_2OR provided further insights [50]. Here, the protein was isolated with the tetranuclear site in the Cu_Z^* state, and the observed activity of 0.01 U N_2O was far below the one reported for other orthologs. However, the enzyme could be activated by complete reduction of Cu_Z^* to an all-Cu(I) form using sodium dithionite in conjunction with methyl viologen as a redox mediator (see Section 3.3). In this state, the activity of the enzyme was determined to 160 U N_2O, and the conclusion drawn was that Cu_Z^* must be the active form of the Cu_Z cluster and that reductive activation of the enzyme is a necessary requirement for catalytic activity [47, 78]. In this mechanism, the activation of the enzyme still is far slower than the observed catalytic turnover, and it is not straightforward to reconcile fast oxidation of Cu_Z^* during N_2O reduction with the observed, slow activation of the cluster with dithionite [44].

A possible solution for the issue came with the identification of a further state of the Cu_Z site that was proposed to be an additional catalytic intermediate. This species was characterized by an additional absorption maximum at 680 nm and termed Cu_Z^0 [80]. Solomon and coworkers identified Cu_Z^0 as a $[3Cu^+:1Cu^{2+}]$ form of the cluster that was, however, distinct from the isoelectronic Cu_Z^* and that can be rapidly reduced back to the all-Cu(I) state of the site [81]. For the mechanism of nitrous oxide reduction, this suggested that within the catalytic cycle, the fully reduced state was required to bind and activate the kinetically inert N_2O molecule, but that the catalytic one-electron oxidation of the tetranuclear cluster would lead to Cu_Z^0 rather than the Cu_Z^* resting state, allowing for a quick reductive reactivation of the site to keep up with the observed rate of catalysis. At the same time, a single-turnover experiment, where activated and reduced N_2OR was incubated with its substrate in the absence of further reductant, yielded the enzyme in the Cu_Z^* resting state, while N_2O had obtained one electron each from Cu_A and Cu_Z to yield the products N_2 and H_2O [81]. Note that this mechanistic proposal strictly depends on the availability of the all-Cu(I) state of Cu_Z, which only seems to be accessible with the [4Cu:S] Cu_Z^* site, while in the [4Cu:2S] form, Cu_Z, even extended incubation with sodium dithionite and viologens will not reduce all four copper ions [44, 47]. This may be explained with the presence of the additional sulfur donor ligand that forms highly covalent bonds with Cu_1 and Cu_4 of the cluster and stabilizes the more oxidized state [34, 81]. The activity reported for the purple form I N_2OR from *P. stutzeri* would thus either be due to a residual fraction of Cu_Z^* that was reductively activated, or it would reflect the catalytic potential of the Cu_Z itself that, however, was too low to explain the activities observed *in vivo* [34]. In this mechanism, N_2O was still suggested to bind to the exposed edge of Cu_Z^* formed by Cu_1 and Cu_4 in the absence of a bridging sulfur [81–83], and the binding mode observed experimentally [32] was proposed to reflect the lower activity of the purple form I of N_2OR [81].

From a physiological point of view, however, this mechanistic proposal encounters two major problems. First, the reductive activation of N_2OR to generate the all-Cu(I) state of Cu_Z^* that is a prerequisite for catalysis can only be achieved at very low redox potentials *in vitro*, and it additionally required the presence of an abiological redox mediator. Both components will not be readily available in the oxidizing environment of the bacterial periplasm, and to date there is no mechanism known to drive such low-potential redox chemistry outside the cytoplasm in an *in vivo* situation [44]. Second, the most careful isolations of N_2OR, in particular those in the absence of dioxygen, commonly yield a high Cu_Z/Cu_Z^* ratio, i.e., a high proportion of the tetranuclear site in the [4Cu:2S] state [32, 40, 47]. It thus seems safe to assume that the Cu_Z^* center is only generated from Cu_Z through the loss of a sulfur. N_2OR is isolated from its natural host and from cells that have turned over N_2O in a physiological context. If indeed the [4Cu:2S] state was an inactive one, this would imply that the bacteria have a way to remove the S_2 sulfide anion from Cu_Z for reductive activation and catalysis, but to subsequently replenish the center to its [4Cu:2S] form that then can be isolated. The biogenesis of CuZ is not fully understood to date (see Section 4.3), but it has long been known that there is no straightforward way to re-introduce sulfur into a defective Cu_Z center.

4   Biogenesis and Assembly of Nitrous Oxide Reductase

Denitrification is a respiratory metabolism that relies on the generation of a proton motive force across the cytoplasmic membrane. All its enzymatic steps are located in the periplasm (in the case of Gram-negative denitrifiers), and the required metalloenzymes therefore must be assembled in this compartment that in several aspects differs from the 'inside' of the cell, the cytoplasm. First, the assembly of metal centers in biomolecules is never an unspecific process. This is particularly true for the relatively rare and highly toxic copper that is tightly chaperoned at all times in a physiological context [84]. Denitrifiers thus require not only the different enzymes, but in most cases also a complex biogenesis pathway that assembles the active sites post-translationally. Second, the general currency of energy for the cell, ATP, is not available in the periplasm, so that any energy-consuming step in said pathways must be fueled in a different way.

4.1   *The* nos *Operon*

The structural gene for N_2O reductase is termed *nosZ* [85] (Figure 8). In most instances, the derived NosZ protein shows the two-domain architecture mentioned above (Figure 3), but in some cases an additional domain encoded for a *c*-type cytochrome that presumably represents a fused electron transfer partner (see Section 3.1) [23, 78]. This may not represent a fundamental difference in the functionality of the enzymes, but it has substantial consequences for the required machinery for N_2OR biogenesis. Two-domain NosZ proteins are exported *via* the twin-arginine translocation (Tat) pathway, where the entire protein domain can cross the cytoplasmic membrane in a folded or partially folded state [86]. However, if a c-type cytochrome domain is included, as in *W. succinogenes* NosZ, its maturation, i.e., the covalent attachment of the heme group to the protein chain [87], is strictly dependent on the Sec translocon that transports proteins in their unfolded state [88]. In the latter case, all assembly steps of the two copper sites of N_2O reductase thus have to take place in the periplasm, *after* the attachment of the heme group to the cytochrome domain, while Tat-dependent export of the more canonical two-domain NosZ would in principle allow to assemble the Cu_A and Cu_Z fully or in part in the cytoplasm. Although it is presently assumed that the key steps of metal center biogenesis do occur in the periplasm in both variants of the *nos* operon, the operon architecture itself is different [23].

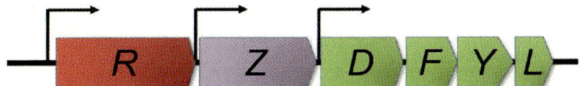

Figure 8 The *nos* operon of *Pseudomonas stutzeri*. The *nosZ* gene (Z) is the structural gene for N_2O reductase, while the integral membrane protein NosR, encoded by the preceding gene (R), is required both for the biogenesis of NosZ and for electron transfer to the reductase. NosF (F) and NosY (Y) form an ABC transporter that is presumed to shuttle sulfur into the periplasm, in a form that can be incorporated into Cu_Z with the help of the *nosD* gene (D) product. NosL (L) is a putative copper chaperone that provides Cu(I) for the assembly of both metal sites.

The *W. succinogenes nos* operon consists of twelve open reading frames, the first of which encodes the larger version of NosZ with the additional cytochrome domain. The operon in addition encodes for two periplasmic assembly factors, NosD and NosL, and ABC Transporter, NosFY, and contains three further open reading frames (*orfs*) of unknown function. It also holds a putative menaquinol oxidase, NosGH, and two periplasmic cytochromes, NosC1 and NosC2, with a putative role in electron transfer to the enzyme [23]. In contrast, *P. stutzeri* possesses a smaller *nos* cluster that is organized into three transcriptional units (Figure 8) [89]. It commences with the monocistronic *nosR* gene, encoding a complex, cofactor-containing membrane protein required both for NosZ maturation and for N_2O reduction [90].

Following *nosR*, the *nosZ* gene encoding the actual enzyme is a second monocistronic transcription unit that includes an N-terminal leader peptide with the typical twin-arginine motif for Tat-dependent translocation. The following *nosD* operon is a four-gene transcription unit that holds the factors required for periplasmic maturation of Cu_Z (and possibly Cu_A) [89]. All four genes, *nosDFYL*, are also present in the *W. succinogenes* operon, indicating that this is a general maturation machinery, while NosGH, NosC1 and NosC2 may combine to play the complex role of the NosR protein in *P. stutzeri*. As the main focus of the present text is on *P. stutzeri* N_2OR, we will restrict the discussion of maturation factors to this version of the operon (Figure 8).

4.2   Protein Maturation and Cu$_A$ Insertion

In *P. stutzeri*, the *nosZ* gene is translated in the cytoplasm, so that its folded – or partially folded – protein product is exported *via* the Tat pathway. It is currently not known whether NosZ is exported as a monomer or a dimer, and whether it reaches the periplasm as a metal-free apo protein or already has pre-assembled copper sites included. The presence and location of the known maturation factors, however, strongly indicates that the major part of the assembly process occurs in the extracellular compartment. Here, the less intricate Cu$_A$ site is assembled along a different route, as deletions of any of the genes *nosDFYL* will leave the binuclear site unaffected, while the resulting protein is fully deplete of NosZ [11]. In a metal-depleted NosZ protein, Cu$_A$ can be readily reconstituted by the addition of Cu(II) salts, while Cu$_Z$ cannot [43], and it is assumed that *in vivo* the Cu$_A$ site employs a copper chaperone, such as ScoA, that is also required for the assembly of heme-copper oxidases [11]. This is in line with the concept of rack-induced metal binding by cupredoxin domains and the facile reconstitution of the binuclear CuA site has also been observed in model proteins with an engineered CuA-binding loop [72, 73].

4.3   Assembly of Cu$_Z$

The ABC transporter NosFY and the periplasmic proteins NosD and NosL that form part of all known gene clusters encoding a functional N$_2$O reductase system are required for assembly of the Cu$_Z$ active site. This site is substantially more complex than Cu$_A$ (Figures 6 and 7), and consequently its biogenesis is more intricate, to the point that *in vitro* reconstruction without the help of additional factors has not been achieved [7]. The Cu$_Z$ center consists of two essential building blocks, the four copper ions and the two sulfide ions, that must be provided along different routes.

As mentioned previously, Cu trafficking is a highly regulated process in all organisms, and specialized chaperones are required to provide the metal, commonly in the Cu(I) state [84]. For N$_2$OR, the NosL protein was suggested for this role [11, 78, 91]. An NMR structure of NosL is available that shows the apo-state of the protein, but contains cysteine residues that may well serve as ligands for Cu(I) [92]. NosFY is an ABC transporter that shuttles a necessary assembly intermediate from the cytoplasm to the periplasm, but its actual cargo is unknown to date [89]. It is, however, homologous to the Atm1 type of ABC transporters that are involved in iron-sulfur cluster biogenesis. Two structures of Atm1 orthologs have become available most recently and imply that the transported species may be a glutathione persulfide [93, 94]. In conjunction with NosFY, NosD is thought to serve as an assembly chaperone or a periplasmic binding protein for the species transported by NosFY. The emerging picture thus is that NosL is the copper donor at least for the assembly of Cu_Z in N_2OR, while NosD obtains an activated sulfide from the cytoplasm, transported *via* NosFY, and shuttles it to the enzyme. The process is undoubtedly more complex than this, as the addition of any combination of sulfides or thiols and Cu(I) or Cu(II) salts *in vitro* is not sufficient to achieve even a partial reconstitution of Cu_Z.

5 Activation of Nitrous Oxide: The Workings of Nitrous Oxide Reductase

The reaction of nitrous oxide reductase requires two electrons per N_2O molecule (equation 1) that are provided by a soluble single-electron donor such as cytochrome c_{552} and reach the enzyme *via* an electron input site close to the Cu_A center. The structural analysis of purple N_2OR I from *P. stutzeri* not only revealed novel structural features of the metal sites, but it also allowed to demonstrate the mode of substrate binding by pressurization of crystals with N_2O gas prior to flash-cooling and diffraction data collection [32]. In contrast, earlier studies based on the $[4Cu:\mu\text{-}S]$ structure of $Cu_Z{}^*$ and the iodide-inhibited state of the site observed in *A. cycloclastes* N_2OR [31] (Figure 7) led to a model where nitrous oxide bound to the cluster edge formed by atoms Cu_1 and Cu_4 that in the Cu_Z state was occupied by the second sulfide [78, 82, 83]. While we have discussed the electronic and functional implications above, this distinction will be relevant for mechanistic considerations in the context of the architecture of the enzyme's active site.

5.1 Substrate Access

While the substrate N_2O is gaseous at ambient temperatures, it is well soluble in water and overall slightly polar, albeit far less so than the surrounding solvent, water. The structure of nitrous oxide reductase shows a feature that is present in an overwhelming majority of gas-converting enzymes, even for gas molecules as small as H_2 [95]. In N_2OR, a hydrophobic (i.e., devoid of ordered solvent molecules) channel leads from the protein surface to the substrate binding site observed between Cu_A and Cu_Z in the *P. stutzeri* enzyme [32].

The channel is formed by the interface of the two monomers that form the tight N_2OR dimer, and it terminates in a small vestibule located right at the tetranuclear site (Figure 9). This is the only visible direct connection from the protein surface to the metal sites, and although, at present, an unambiguous conformation of the functionality of this channel by obstruction through site-directed mutagenesis has not been made, the overall arrangement is highly suggestive. In this position, however, the substrate channel does not help to make a distinction concerning the relevance of the substrate binding modes predicted by X-ray crystallography *versus*

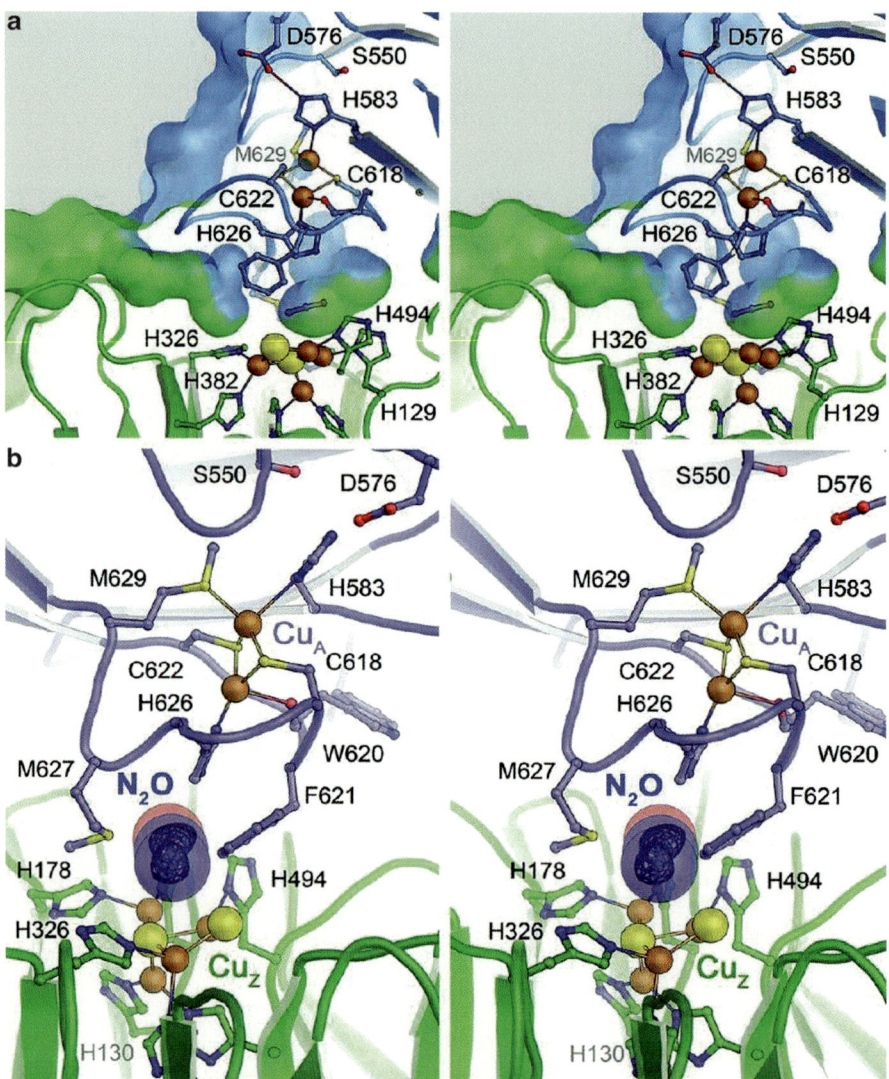

Figure 9 Substrate access and N_2O binding at the active site of N_2O reductase. (**a**) Stereo representation of the molecular surface at the interface between the cupredoxin domain (blue) and the β-propeller domain with the active site situated below. Substrate access is provided *via* a hydrophic channel along the domain interface, leading to a vestibule close to Cu_Z. The access to the N_2O binding site on Cu_Z is controlled by residues Phe621 and Met627. (**b**) Observed binding of N_2O between Cu_A and Cu_Z in gas-pressurized crystals of *P. stutzeri* N_2OR form I (PDB ID 3SBR). A difference electron density map for the substrate is shown below a transparent van-der-Waals surface. Upon N_2O binding, residue His583 at Cu_A was invariably found to coordinate to the nearby copper ion (Figure 6). The positioning of the substrate between the two metal sites suggests that electron transfer might occur directly from Cu_A to N_2O, rather than involving a reduction of Cu_Z.

theory [34, 44]. While the hydrophobic channel indeed leads right up to a gate formed by two bulky amino acid side chains (Phe621 and Met627 in *P. stutzeri*) on the face of Cu_Z that is directed towards Cu_A, the additional space gained through the loss of the second sulfide, S_2, in the Cu_Z^* state would allow the substrate to access the Cu_1-Cu_4 edge of the cluster as well.

5.2 Gated Electron Transfer

An unexpected finding in the structure of *P. stutzeri* N_2OR was the flexibility of His583, a ligand to Cu_2 of the Cu_A site (Figure 6) [32]. Based on the available structural data the 135° flip of the imidazole side chain of this histidine residue cannot be related to the redox state of the site, and both conformations seem accessible. However, all structures of gas-pressurized crystals of the *P. stutzeri* enzyme showed the ligand to be coordinated to the metal ion, as did all earlier observations, where the tetranuclear site was in the Cu_Z^* state. The flipped-out state of His583 was thus only observed in form I of N_2OR, i.e., with an intact [4Cu:2S] cluster, and in the absence of the substrate N_2O. As His583 in both its conformations remains hydrogen-bonded to residue Asp576 that is presumed to be the electron entry site from an external electron transfer protein [37], the on-off coordination to Cu_A was suggested to form a conformational gate for intra-molecular electron transfer, with the consequence that electron flow will only be permitted *after* the substrate is bound at Cu_Z [32, 44]. Obviously, this intricate coupling mechanism is tightly linked to the physiological activity of the enzyme, based on a reduced electron transfer protein as a redox partner. Using small-molecule reductants such as ascorbate or sodium dithionite, complete reduction of Cu_A is swift, but there is no way of controlling at which points the reductants will interact with the enzyme. The question of gated electron transfer in nitrous oxide reductase thus touches the distinction between reactivities observed *in vitro* and those occurring in the actual physiological environment of the bacterial periplasm.

5.3 Activation of Nitrous Oxide

The reduction of N_2O to N_2 by nitrous oxide reductase (equation 1) takes place at the tetranuclear site, Cu_Z. The N_2O molecule observed in the *P. stutzeri* X-ray structure was not directly coordinated to any of the metal ions of the cluster, but rather it was fixed by hydrogen bonds to the oxygen atom, originating from His626, a ligand to Cu_A, and a structural water molecule (Figure 9) [32]. This structure did not represent a coordination complex in the classical sense, but rather revealed a substrate-binding site in close proximity to the metal centers. This binding event is presumed to be the trigger for His583 to flip back towards the Cu_A center and enable electron transfer to the substrate. Nevertheless, the suggestive implication that derives from this mode of N_2O binding is that the substrate is located *in between* the two metal sites, making it unlikely that in this arrangement electron transfer from Cu_A would proceed first into Cu_Z and then to the substrate, but rather directly from Cu_A to an activated state of N_2O that is coordinated to Cu_Z.

5.4   The Fate of the Products

The substrate N_2O is a weakly polar molecule with a positive partial charge on the central nitrogen atom and according negative partial charges on the terminal N and O (Figure 2). It is well suited to access the active site of N_2O reductase *via* a largely hydrophobic access channel. After reduction by two electrons, the gaseous, apolar product N_2 will be able to exit the enzyme along the same route. However, the oxygen atom of N_2O remains as a polar water molecule, and the hydrophobic substrate channel seems to disfavor the passage of water.

In the structures of N_2O reductase, water molecules are found in close proximity to the metal sites, and indeed the position held by the oxygen atom of N_2O in the substrate complex structure (Figure 9) is commonly occupied by water if substrate is absent. From here, a network of coordinated water molecules, interacting through hydrogen bonds, stretches through the protein matrix. This arrangement suggests that after the reduction of N_2O, the fate of the two product molecules differs, and while N_2 will exit through the original substrate channel, the water molecule will leave the active site in a different way, to move through a hydrophilic cavity and exit the protein at a remote site. In many enzymes, such optimizations serve to prevent product inhibition and accelerate the reaction rate, but in the present case it seems more likely that this is a mechanism to accommodate the very different physicochemical properties of the two reaction products.

6   General Conclusions

Nitrous oxide reductase is a complex and fascinating metalloenzyme that catalyzes the conversion of an inert gas of high chemical interest and ecological relevance. Its copper sites have features that so far are unique in bioinorganic chemistry, and the seemingly straightforward two-electron redox process it catalyzes is of sufficient complexity to defy a final elucidation.

At present, the model of the action of N_2O reductase brought forward by Solomon, Moura, and coworkers integrates the majority of the available experimental data and presents the outlines of a mechanism in which the simultaneous coordination of N_2O to two copper ions within $Cu_Z{}^*$ is required to overcome the activation energy barrier for substrate reduction [34, 81]. It places the purple form I of the enzyme on a catalytic sideline, but this in particular raises the question why all biochemical evidence then points to $Cu_Z{}^*$ being a degradation product of the intact [4Cu:2S] Cu$_Z$ site, and why this doubly-sulfurated form I of N$_2$OR can readily be isolated from bacteria that have actively turned over N$_2$O during denitrificatory growth, using the exact protein that was isolated after harvesting of the biomass. It therefore is imminent to address the issue of assembly and maturation of the Cu$_Z$ center and generate structural pictures – ideally verified by single-crystal spectroscopy – that align the different states of the enzyme with the spectroscopy data at the highest resolution possible.

Moreover, it will also be essential to determine whether a full-reduced, all-Cu(I) state of Cu$_Z$ can be generated *in vivo* and which mechanisms might be in place to deliver the required low-potential electrons to a periplasmic enzyme. No such process was observed to date and it seems quite difficult to envision indeed, but one of the great aspects of science surely is that very few things, really, seem to be impossible in nature. Astounding progress has been made in recent years, and the field seems all set to eventually provide some of those famed 'final answers' regarding the biological reduction of nitrous oxide.

Abbreviations and Definitions

aa	amino acid
anammox	anaerobic ammonia oxidation
ATP	adenosine 5′-triphosphate
CFC	chlorofluorocarbon
DNRA	dissimilatory nitrate reduction to ammonia
en	ethylenediamine (= 1,2-diaminoethane)
EPR	electron paramagnetic resonance
ENDOR	electron nuclear double resonance
EXAFS	extended X-ray absorption fine structure
LMCT	ligand-to-metal charge transfer
LUMO	lowest unoccupied molecular orbital
NADPH	nicotinamide adenine dinucleotide phosphate (reduced)
NMR	nuclear magnetic resonance
N$_2$OR	nitrous oxide reductase
NO$_x$	atmospheric nitrogen oxides (NO$_x$=NO+NO$_2$)
orf	open reading frame
ppbv	parts per billion by volume
Tat	twin-arginine translocation
U	unit of enzymatic activity (μmol (substrate)$\cdot$min$^{-1}\cdot$mg^{-1} (protein))

Acknowledgment The authors thank Peter Kroneck, Walter Zumft, Jörg Simon, Sofia Pauleta, and Isabel Moura for stimulating discussions. This work was supported by Deutsche Forschungsgemeinschaft, Deutscher Akademischer Austauschdienst, the BIOSS Centre for Biological Signalling Studies, and the European Research Council.

References

1. P. M. H. Kroneck, in *Biogeochemical Cycles of Elements*, Vol. 43 of *Metal Ions in Biological Systems*, Eds A. Sigel, H. Sigel, R. K. O. Sigel, Taylor & Francis, Boca Raton, USA, 2005, pp. 1–7.
2. O. Einsle, P. M. H. Kroneck, *Biol. Chem.* **2004**, *385*, 875–883.
3. D. E. Canfield, A. N. Glazer, P. G. Falkowski, *Science* **2010**, *330*, 192–196.
4. O. Einsle, *Methods Enzymol.* **2011**, *496*, 399–422.
5. B. Kartal, W. J. Maalcke, N. M. de Almeida, I. Cirpus, J. Gloerich, W. Geerts, H. J. M. O. den Camp, H. R. Harhangi, E. M. Janssen-Megens, K. J. Francoijs, H. G. Stunnenberg, J. T. Keltjens, M. S. M. Jetten, M. Strous, *Nature* **2011**, *479*, 127–132.
6. R. Knowles, *Microbiol. Rev.* **1982**, *46*, 43–70.
7. W. G. Zumft, *Microbiol. Mol. Biol. Rev.* **1997**, *61*, 533–616.
8. D. C. Rees, F. A. Tezcan, C. A. Haynes, M. Y. Walton, S. Andrade, O. Einsle, J. B. Howard, *Philos. Trans. R. Soc. A* **2005**, *363*, 971–984.
9. J. Priestley, *Experiments and Observations on Different Kinds of Air*, J. Johnson, London, 1774.
10. W. C. Trogler, *Coord. Chem. Rev.* **1999**, *187*, 303–327.
11. W. G. Zumft, P. M. H. Kroneck, *Adv. Microb. Physiol.* **2007**, *52*, 107–225.
12. M. Leuenberger, U. Siegenthaler, *Nature* **1992**, *360*, 449–451.
13. W. C. Trogler, *J. Chem. Educ.* **1995**, *72*, 973–976.
14. J. T. Houghton, *Climate Change 1996: The Science of Climate Change*, Cambridge University Press, Cambridge, 1996.
15. O. Badr, S. D. Probert, *Appl. Energ.* **1993**, *44*, 197–231.
16. A. R. Ravishankara, J. S. Daniel, R. W. Portmann, *Science* **2009**, *326*, 123–125.
17. P. O. Wennberg, R. C. Cohen, R. M. Stimpfle, J. P. Koplow, J. G. Anderson, R. J. Salawitch, D. W. Fahey, E. L. Woodbridge, E. R. Keim, R. S. Gao, C. R. Webster, R. D. May, D. W. Toohey, L. M. Avallone, M. H. Proffitt, M. Loewenstein, J. R. Podolske, K. R. Chan, S. C. Wofsy, *Science* **1994**, *266*, 398–404.
18. European Commission, in "Climate action: Commission proposes ratification of second phase of Kyoto Protocol", 2013.
19. S. Rakshit, C. J. Matocha, M. S. Coyne, *Soil Sci. Soc. Am. J.* **2008**, *72*, 1070–1077.
20. V. A. Samarkin, M. T. Madigan, M. W. Bowles, K. L. Casciotti, J. C. Priscu, C. P. Mckay, S. B. Joye, *Nat. Geosci.* **2010**, *3*, 341–344.
21. G. A. Kowalchuk, J. R. Stephen, *Annu. Rev. Microbiol.* **2001**, *55*, 485–529.
22. R. A. Reimer, C. S. Slaten, M. Seapan, M. W. Lower, P. E. Tomlinson, *Environ. Prog.* **1994**, *13*, 134–137.
23. J. Simon, O. Einsle, P. M. H. Kroneck, W. G. Zumft, *FEBS Lett.* **2004**, *569*, 7–12.
24. H. Iwasaki, T. Saigo, T. Matsubara, *Plant Cell Physiol* **1980**, *21*, 1573–1584.
25. T. Matsubara, W. G. Zumft, *Arch. Microbiol.* **1982**, *132*, 322–328.
26. W. G. Zumft, T. Matsubara, *FEBS Lett.* **1982**, *148*, 107–112.
27. K. Brown, K. Djinovic-Carugo, T. Haltia, I. Cabrito, M. Saraste, J. J. G. Moura, I. Moura, M. Tegoni, C. Cambillau, *J. Biol. Chem.* **2000**, *275*, 41133–41136.
28. K. Brown, M. Tegoni, M. Prudêncio, A. S. Pereira, S. Besson, J. J. G. Moura, I. Moura, C. Cambillau, *Nat. Struct. Biol.* **2000**, *7*, 191–195.
29. C. Sproer, E. Lang, P. Hobeck, J. Burghardt, E. Stackebrandt, B. J. Tindall, *Int. J. Syst. Bacteriol.* **1998**, *48*, 1445–1448.
30. T. Haltia, K. Brown, M. Tegoni, C. Cambillau, M. Saraste, K. Mattila, K. Djinovic-Carugo, *Biochem. J.* **2003**, *369*, 77–88.
31. K. Paraskevopoulos, S. V. Antonyuk, R. G. Sawers, R. R. Eady, S. S. Hasnain, *J. Mol. Biol.* **2006**, *362*, 55–65.
32. A. Pomowski, W. G. Zumft, P. M. H. Kroneck, O. Einsle, *Nature* **2011**, *477*, 234–237.

33. P. Völkl, R. Huber, E. Drobner, R. Rachel, S. Burggraf, A. Trincone, K. O. Stetter, *Appl. Environ. Microbiol.* **1993**, *59*, 2918–2926.
34. E. I. Solomon, D. E. Heppner, E. M. Johnston, J. W. Ginsbach, J. Cirera, M. Qayyum, M. T. Kieber-Emmons, C. H. Kjaergaard, R. G. Hadt, L. Tian, *Chem. Rev.* **2014**, *114*, 3659–3853.
35. C. C. Page, C. C. Moser, X. X. Chen, P. L. Dutton, *Nature* **1999**, *402*, 47–52.
36. S. Teraguchi, T. C. Hollocher, *J. Biol. Chem.* **1989**, *264*, 1972–1979.
37. S. Dell'Acqua, S. R. Pauleta, E. Monzani, A. S. Pereira, L. Casella, J. J. G. Moura, I. Moura, *Biochemistry* **2008**, *47*, 10852–10862.
38. K. Mattila, T. Haltia, *Proteins: Struct. Funct. Bioinform.* **2005**, *59*, 708–722.
39. I. V. Pearson, M. D. Page, R. J. M. van Spanning, S. J. Ferguson, *J. Bacteriol.* **2003**, *185*, 6308–6315.
40. J. Riester, W. G. Zumft, P. M. H. Kroneck, *Eur. J. Biochem.* **1989**, *178*, 751–762.
41. W. G. Zumft, P. M. H. Kroneck, *Adv. Inorg. Chem.* **1996**, *11*, 193–221.
42. P. Wunsch, H. Körner, F. Neese, R. J. M. van Spanning, P. M. H. Kroneck, W. G. Zumft, *FEBS Lett.* **2005**, *579*, 4605–4609.
43. C. L. Coyle, W. G. Zumft, P. M. H. Kroneck, H. Körner, W. Jakob, *Eur. J. Biochem.* **1985**, *153*, 459–467.
44. A. Wüst, L. Schneider, A. Pomowski, W. G. Zumft, P. M. H. Kroneck, O. Einsle, *Biol. Chem.* **2012**, *393*, 1067–1077.
45. J. M. Charnock, A. Dreusch, H. Körner, F. Neese, J. Nelson, A. Kannt, H. Michel, C. D. Garner, P. M. H. Kroneck, W. G. Zumft, *Eur. J. Biochem.* **2000**, *267*, 1368–1381.
46. T. Rasmussen, B. C. Berks, J. Sanders-Loehr, D. M. Dooley, W. G. Zumft, A. J. Thomson, *Biochemistry* **2000**, *39*, 12753–12756.
47. S. Dell'Acqua, S. R. Pauleta, J. J. Moura, I. Moura, *Philos. Trans. R. Soc. B* **2012**, *367*, 1204–1212.
48. M. J. Gauthier, B. Lafay, R. Christen, L. Fernandez, M. Acquaviva, P. Bonin, J. C. Bertrand, *Int. J. Syst. Bacteriol.* **1992**, *42*, 568–576.
49. M. Prudêncio, A. S. Pereira, P. Tavares, S. Besson, I. Cabrito, K. Brown, B. Samyn, B. Devreese, J. Van Beeumen, F. Rusnak, G. Fauque, J. J. G. Moura, M. Tegoni, C. Cambillau, I. Moura, *Biochemistry* **2000**, *39*, 3899–3907.
50. M. Prudêncio, A. S. Pereira, P. Tavares, S. Besson, I. Moura, *J. Inorg. Biochem.* **1999**, *74*, 267–267.
51. A. Pomowski, W. G. Zumft, P. M. H. Kroneck, O. Einsle, *Acta Crystallogr. Sect. F Cryst. Comm.* **2010**, *66*, 1541–1543.
52. P. M. H. Kroneck, W. A. Antholine, J. Riester, W. G. Zumft, *FEBS Lett.* **1988**, *242*, 70–74.
53. H. Beinert, *Eur. J. Biochem.* **1997**, *245*, 521–532.
54. R. Malkin, B. G. Malmström, *Adv. Enzymol. Relat. Subj. Biochem.* **1970**, *33*, 177–244.
55. P. M. Li, B. G. Malmström, S. I. Chan, *FEBS Lett.* **1989**, *248*, 210–211.
56. P. M. H. Kroneck, W. A. Antholine, J. Riester, W. G. Zumft, *FEBS Lett.* **1989**, *248*, 212–213.
57. N. J. Blackburn, M. E. Barr, W. H. Woodruff, J. Vanderooost, S. de Vries, *Biochemistry* **1994**, *33*, 10401–10407.
58. N. J. Blackburn, S. deVries, M. E. Barr, R. P. Houser, W. B. Tolman, D. Sanders, J. A. Fee, *J. Am. Chem. Soc.* **1997**, *119*, 6135–6143.
59. S. Iwata, C. Ostermeier, B. Ludwig, H. Michel, *Nature* **1995**, *376*, 660–669.
60. T. Tsukihara, H. Aoyama, E. Yamashita, T. Tomizaki, H. Yamaguchi, K. Shinzawa-Itoh, R. Nakashima, R. Yaono, S. Yoshikawa, *Science* **1995**, *269*, 1069–1074.
61. T. Tsukihara, H. Aoyama, E. Yamashita, T. Tomizaki, H. Yamaguchi, K. Shinzawa-Itoh, R. Nakashima, R. Yaono, S. Yoshikawa, *Science* **1996**, *272*, 1136–1144.
62. M. Wilmanns, P. Lappalainen, M. Kelly, E. Sauer-Eriksson, M. Saraste, *Proc. Natl. Acad. Sci. USA* **1995**, *92*, 11955–11959.
63. B. L. Vallee, R. J. P. Williams, *Proc. Natl. Acad. Sci. USA* **1968**, *59*, 498–505.
64. R. P. Houser, V. G. Young, W. B. Tolman, *J. Am. Chem. Soc.* **1996**, *118*, 2101–2102.

65. G. Henkel, A. Müller, S. Weissgräber, G. Buse, T. Soulimane, G. C. M. Steffens, H. F. Nolting, *Angew. Chem. Int. Ed.* **1995**, *34*, 1488–1492.
66. F. Neese, W. G. Zumft, W. E. Antholine, P. M. H. Kroneck, *J. Am. Chem. Soc.* **1996**, *118*, 8692–8699.
67. F. Neese, R. Kappl, J. Hüttermann, W. G. Zumft, P. M. H. Kroneck, *J. Biol. Inorg. Chem.* **1998**, *3*, 53–67.
68. B. Epel, C. S. Slutter, F. Neese, P. M. H. Kroneck, W. G. Zumft, I. Pecht, O. Farver, Y. Lu, D. Goldfarb, *J. Am. Chem. Soc.* **2002**, *124*, 8152–8162.
69. P. Wittung, B. Kallebring, B. G. Malmström, *FEBS Lett.* **1994**, *349*, 286–288.
70. M. Saraste, *Q. Rev. Biophys.* **1990**, *23*, 331–366.
71. E. T. Adman, *Adv. Protein Chem.* **1991**, *42*, 145–197.
72. C. R. Andrew, P. Lappalainen, M. Saraste, M. T. Hay, Y. Lu, C. Dennison, G. W. Canters, J. A. Fee, C. E. Slutter, N. Nakamura, J. Sanders-Loehr, *J. Am. Chem. Soc.* **1995**, *117*, 10759–10760.
73. C. Dennison, E. Vijgenboom, S. de Vries, J. Vanderoost, G. W. Canters, *FEBS Lett.* **1995**, *365*, 92–94.
74. J. A. Farrar, A. J. Thomson, M. R. Cheesman, D. M. Dooley, W. G. Zumft, *FEBS Lett.* **1991**, *294*, 11–15.
75. P. Chen, S. I. Gorelsky, S. Ghosh, E. I. Solomon, *Angew. Chem. Int. Edit.* **2004**, *43*, 4132–4140.
76. M. L. Alvarez, J. Y. Ai, W. G. Zumft, J. Sanders-Loehr, D. M. Dooley, *J. Am. Chem. Soc.* **2001**, *123*, 576–587.
77. T. Rasmussen, B. C. Berks, J. N. Butt, A. J. Thomson, *Biochem. J.* **2002**, *364*, 807–815.
78. S. R. Pauleta, S. Dell'Acqua, I. Moura, *Coord. Chem. Rev.* **2013**, *257*, 332–349.
79. S. W. Snyder, T. C. Hollocher, *J. Biol. Chem.* **1987**, *262*, 6515–6525.
80. S. Dell'Acqua, S. R. Pauleta, P. M. Paes de Sousa, E. Monzani, L. Casella, J. J. Moura, I. Moura, *J. Biol. Inorg. Chem.* **2010**, *15*, 967–976.
81. E. M. Johnston, S. Dell'Acqua, S. Ramos, S. R. Pauleta, I. Moura, E. I. Solomon, *J. Am. Chem. Soc.* **2014**, *136*, 614–617.
82. S. Ghosh, S. I. Gorelsky, P. Chen, I. Cabrito, J. J. G. Moura, I. Moura, E. I. Solomon, *J. Am. Chem. Soc.* **2003**, *125*, 15708–15709.
83. S. I. Gorelsky, S. Ghosh, E. I. Solomon, *J. Am. Chem. Soc.* **2006**, *128*, 278–290.
84. T. V. O'Halloran, R. A. Pufhal, G. Munson, D. Huffman, *Biochemistry* **1996**, *35*, 110–110.
85. A. Viebrock, W. G. Zumft, *J. Bacteriol.* **1988**, *170*, 4658–4668.
86. T. Palmer, B. C. Berks, *Nat. Rev. Microbiol.* **2012**, *10*, 483–496.
87. R. Kranz, R. Lill, B. Goldman, G. Bonnard, S. Merchant, *Mol. Microbiol.* **1998**, *29*, 383–396.
88. V. A. M. Gold, F. Duong, I. Collinson, *Mol. Membr. Biol.* **2007**, *24*, 387–394.
89. U. Honisch, W. G. Zumft, *J. Bacteriol.* **2003**, *185*, 1895–1902.
90. P. Wunsch, W. G. Zumft, *J. Bacteriol.* **2005**, *187*, 1992–2001.
91. M. A. McGuirl, J. A. Bollinger, N. Cosper, R. A. Scott, D. M. Dooley, *J. Biol. Inorg. Chem.* **2001**, *6*, 189–195.
92. L. M. Taubner, M. A. McGuirl, D. M. Dooley, V. Copie, *Biochemistry* **2006**, *45*, 12240–12252.
93. J. Y. Lee, J. G. Yang, D. Zhitnitsky, O. Lewinson, D. C. Rees, *Science* **2014**, *343*, 1133–1136.
94. V. Srinivasan, A. J. Pierik, R. Lill, *Science* **2014**, *343*, 1137–1140.
95. F. Leroux, S. Dementin, B. Burlatt, L. Cournac, A. Volbeda, S. Champ, L. Martin, B. Guigliarelli, P. Bertrand, J. Fontecilla-Camps, M. Rousset, C. Leger, *Proc. Natl. Acad. Sci. USA* **2008**, *105*, 11188–11193.

Chapter 9
The Production of Ammonia by Multiheme Cytochromes *c*

Jörg Simon and Peter M.H. Kroneck

Contents

ABSTRACT 211
1 INTRODUCTION 212
2 AMMONIA AND ITS ROLE IN THE ENVIRONMENT 214
3 ENZYMES INVOLVED IN AMMONIA TURNOVER 216
4 CYTOCHROME *c* NITRITE REDUCTASE AS PARADIGM 217
    4.1   Cytochrome *c* Nitrite Reductase-Containing Organisms and Their Physiology .. 221
    4.2   Biochemistry and Structure of Cytochrome *c* Nitrite Reductase 221
    4.3   Electron Transfer Routes to Cytochrome *c* Nitrite Reductase 227
    4.4   Role of Cytochrome *c* Nitrite Reductase in Stress Defense 230
5 OTHER MULTIHEME CYTOCHROMES *c* 231
6 ENVIRONMENTAL ISSUES AND CONCLUSIONS 231
ABBREVIATIONS 232
ACKNOWLEDGMENTS 233
REFERENCES 233

Abstract The global biogeochemical nitrogen cycle is essential for life on Earth. Many of the underlying biotic reactions are catalyzed by a multitude of prokaryotic and eukaryotic life forms whereas others are exclusively carried out by microorganisms. The last century has seen the rise of a dramatic imbalance in the global nitrogen cycle due to human behavior that was mainly caused by the invention of the Haber-Bosch process. Its main product, ammonia, is a chemically reactive and

J. Simon (✉)
Microbial Energy Conversion and Biotechnology, Department of Biology, Technische Universität Darmstadt, Schnittspahnstrasse 10, D-64287 Darmstadt, Germany
e-mail: simon@bio.tu-darmstadt.de

P.M.H. Kroneck
Fachbereich Biologie, Universität Konstanz, Universitätsstrasse 10, D-78457 Konstanz, Germany
e-mail: peter.kroneck@uni-konstanz.de

© Springer Science+Business Media Dordrecht 2014   211
P.M.H. Kroneck, M.E. Sosa Torres (eds.), *The Metal-Driven Biogeochemistry of Gaseous Compounds in the Environment*, Metal Ions in Life Sciences 14,
DOI 10.1007/978-94-017-9269-1_9

biotically favorable form of bound nitrogen. The anthropogenic supply of reduced nitrogen to the biosphere in the form of ammonia, for example during environmental fertilization, livestock farming, and industrial processes, is mandatory in feeding an increasing world population. In this chapter, environmental ammonia pollution is linked to the activity of microbial metalloenzymes involved in respiratory energy metabolism and bioenergetics. Ammonia-producing multiheme cytochromes c are discussed as paradigm enzymes.

Keywords biogeochemical nitrogen cycle • climate change • cytochrome c nitrite reductase • multiheme cytochrome c family • NrfA

Please cite as: *Met. Ions Life Sci.* 14 (2014) 211–236

1   Introduction

The biogeochemical nitrogen cycle has received considerable attention over the past decades because of its global importance for life on Earth (see also Chapters 7 and 8 in this book) [1–7]. Nitrogen is a basic element for life because it is a component of essential biomolecules, such as amino acids, proteins, and nucleic acids. In the biosphere, nitrogen exists in several oxidation states, ranging from +5 as in the nitrate anion (NO_3^-) to −3 as in ammonia (NH_3; IUPAC name azane), and the ammonium cation, NH_4^+. Interconversions of the various nitrogen species constitute the global biogeochemical nitrogen cycle, which is sustained by biological processes with microorganisms playing a predominant role (Figure 1) [8]. Most remarkably, microbes perform these many chemical reactions in every nook and cranny from the near surface to the depths, including even the most extreme environments [9].

In the context of this article, the most notable nitrogen compound conversions are (i) nitrification ($NH_3 \rightarrow NO_3^-$), equation (1), (ii) denitrification ($NO_3^- \rightarrow N_2$), equation (2), whereby nitrate is successively transformed to nitrite (NO_2^-), nitrogen monoxide (NO), dinitrogen monoxide (N_2O), and dinitrogen (N_2), and (iii) nitrate ammonification ($NO_3^- \rightarrow NH_3$), equation (3), the latter using nitrite as intermediate (Figure 1).

$$NH_3 + 3\,H_2O \rightarrow NO_3^- + 8\,e^- + 9\,H^+ \tag{1}$$

$$2\,NO_3^- + 10\,e^- + 12\,H^+ \rightarrow N_2 + 6\,H_2O \tag{2}$$

$$NO_3^- + 8\,e^- + 9\,H^+ \rightarrow NH_3 + 3\,H_2O \tag{3}$$

In addition, it was shown more recently that large scale conversion of fixed inorganic nitrogen (i.e., NH_3/NH_4^+ and NO_2^-) to N_2 can occur through another

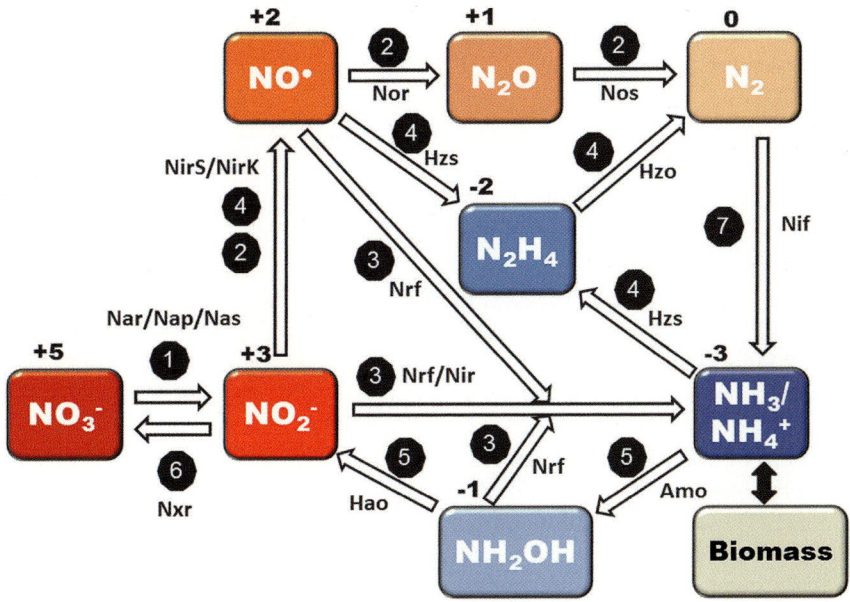

Figure 1 Conversion of nitrogen compounds serving as substrates in nitrogen cycle processes relevant to this article. The oxidation state of nitrogen atoms is indicated above the boxed compound. Numbers in black decagons refer to the following processes: 1, respiratory nitrate reduction to nitrite; 2, denitrification of nitrite to N_2; 3, Nrf-dependent ammonification; 4, anaerobic ammonium oxidation (anammox, i.e., comproportionation of nitrite and ammonium to form dinitrogen); 5, ammonium oxidation to nitrite; 6, nitrite oxidation to nitrate; 7 nitrogen fixation. These metabolic pathways are catalyzed by distinct respiratory enzyme systems that are designated by the following abbreviations: Amo, ammonium monooxygenase; Hao, hydroxylamine oxidoreductase; Hzo, hydrazine oxidoreductase; Hzs, hydrazine synthase; Nap, periplasmic nitrate reductase; Nar, membrane-bound nitrate reductase; Nas, assimilatory nitrate reductase; Nif, nitrogenase; Nir, assimilatory nitrite reductase; NirK, copper nitrite reductase; NirS, cytochrome cd_1 nitrite reductase; Nor, nitric oxide reductase; Nos, nitrous oxide reductase; Nrf, cytochrome *c* nitrite reductase; Nxr, nitrite oxidoreductase. Modified from [24].

important route, the so-called anammox process (anaerobic ammonium oxidation). This process occurs in several Planctomycetes and depends on a specialized cell compartment, the anammoxosome, in which ammonium is converted to N_2 via NO and hydrazine (N_2H_4) intermediates (see Chapter 7 in [7]). It produces twice the amount of N_2 per molecule of nitrite consumed as denitrification and does not necessarily require an external reductant (equation 4) [10–14].

$$NH_4^+ + NO_2^- \rightarrow N_2 + 2\,H_2O \qquad (4)$$

Naturally occurring dinitrogen gas makes up 78 % (by volume) of the Earth's atmosphere. However, it usually has to be converted to its biologically useful form, ammonia, because of its unavailability for most living organisms (equation 5).

In biology, this process is known as nitrogen fixation catalyzed by the enzyme nitrogenase, which is followed by the incorporation of ammonia into organic molecules (Chapter 7 in this book).

$$N_2 + 8\,e^- + 8\,H^+ \rightarrow 2\,NH_3 + H_2 \tag{5}$$

The flow of nitrogen compounds between the oceans and the atmosphere is central to life. Much is known about the nitrogen cycle of the oceans, however, two important questions remain unanswered: (i) is the marine nitrogen budget currently in balance, and (ii) are the processes that add and remove nitrogen to and from the seas closely linked? According to Deutsch and coworkers [15] the primary process responsible for putting nitrogen compounds into the sea, biological nitrogen fixation (equation 5), is intimately associated, both geographically and temporally, with marine nitrogen removal. Further work also implies that the ratio of nitrogen to phosphorus in seawater may be the central factor that regulates nitrogen fixation [16]. Two main processes are responsible for nitrogen loss in the ocean, denitrification and anammox. These processes rely on fundamentally different organisms and metabolic pathways (Figure 1). Denitrification and anammox together, occurring in the oxygen minimum zones and ocean sediments, effectively consume the organic material sinking out of the surface waters and account for all the oceanic nitrogen loss. The dependence of nitrogen loss rates on organic matter supply implies a tightly coupled oceanic nitrogen cycle, which controls the fertility of surface waters [17].

In addition to nitrogen fixation, many organisms are able to produce ammonia from nitrate reduction according to equation (3) using nitrate reduction to nitrite, equation (6), as the initial reaction.

$$NO_3^- + 2\,e^- + 2\,H^+ \rightarrow NO_2^- + H_2O \tag{6}$$

Thus, nitrate plays a key role in the biogeochemical nitrogen cycle, which is contributed to by both prokaryotic and eukaryotic organisms. It is a source of nitrogen for assimilation into organic nitrogen compounds by nitrate (and nitrite) ammonification according to equation (3) followed by incorporation of ammonia into biomass (Figure 1). The same series of reactions, though usually catalyzed by distinct enzymes, can also occur as part of an anaerobic nitrate/nitrite respiration process in various bacteria, whereby nitrate and nitrite serve as terminal electron acceptors [1, 3, 18].

In this chapter we will briefly discuss the role of ammonia in the environment (Section 2) and subsequently summarize recent advances in the field of microbial ammonia production and consumption in relation to energy metabolism (Sections 3, 4, and 5). The main focus is on multiheme cytochromes c and their role in the biogeochemical nitrogen cycle, with emphasis on the production of ammonia. Physiological, structural, and mechanistic aspects of such enzymes will be discussed with respect to electron and proton transfer reactions at unique heme iron centers with unusual chemical and electronic properties. Finally, the emerging picture of the role of multiheme cytochromes c in natural habitats and in microbial ecology is presented (Section 6). In view of the huge amount of data accumulating in the field, we recommend comprehensive reviews as primary references [3, 19–25].

2 Ammonia and Its Role in the Environment

NH_3 is a colorless gas with a typical pungent odor. It has a pyramidal shape, with an H-N-H angle of 106.7° (compared to 93.3° in phosphane, PH_3), and contains a single lone pair of electrons. Ammonia has a dipole moment of 1.47 D (compared to 0.57 D for PH_3). The solubility of ammonia in water is greater than that of any other gas because of hydrogen bond formation between NH_3 and H_2O. In aqueous solution, ammonia equilibrates with the ammonium ion (NH_4^+) which is slightly acidic (equations 7 and 8). The equilibrium depends on pH and temperature.

$$NH_3 + H_2O \rightarrow NH_4^+ + OH^- \quad K_B = 1.8 \times 10^{-5} \quad pK_B(NH_3) = 4.75$$

$$(7)$$

$$NH_4^+ + H_2O \rightarrow H_3O^+ + NH_3 \quad K_A = 5.6 \times 10^{-10} \quad pK_A(NH_4^+) = 9.25$$

$$(8)$$

Ammonia can be obtained by the action of water on nitrides, e.g., lithium nitride (Li_3N), by heating of ammonium chloride (NH_4Cl) with calcium hydroxide [$Ca(OH)_2$], or the reduction of nitrate with metallic zinc in basic solution. In the so-called Ostwald process, ammonia is converted to nitric acid (HNO_3) in three stages. It is oxidized by heating with oxygen in the presence of a platinum-rhodium catalyst to form NO and water, a strongly exothermic reaction (equation 9). NO is further oxidized to yield nitrogen dioxide (NO_2) which is then readily absorbed by the water, yielding the desired product HNO_3 (equations 10 and 11). Thus, the Ostwald process provides the main raw material for the most common type of fertilizer production.

$$4\,NH_3 + 5\,O_2 \rightarrow 4\,NO + 6\,H_2O \quad (9)$$

$$2\,NO + O_2 \rightarrow 2\,NO_2 \quad (10)$$

$$3\,NO_2 + H_2O \rightarrow 2\,HNO_3 + NO \quad (11)$$

The coordination chemistry of ammonia is well established. In metal ammine complexes, NH_3 acts as σ-donor and is located in the middle of the *Spectrochemical Series*: $I^- < Br^- < SCN^- < Cl^- < NO_3^- < N_3^- < F^- < OH^- < H_2O < NCS^- < CH_3CN < $ pyridine $< NH_3 < $ ethylenediamine $< 2,2'$-bipyridine $< 1,10$-phenanthroline $< NO_2^- < CN^- \approx CO$. Most metal ions will bind ammonia as a ligand, e.g., the Cu(II) tetraammine complex $[Cu(NH_3)_4]^{2+}$, or the Ag(I) diammine complex $[Ag(NH_3)_2]^+$. Ammine complexes of Cr(III) were key for Alfred

Werner's theory on the nature of coordination compounds, awarded 1913 with the Nobel Prize in Chemistry [26, 27]. Another prominent example of a metal ammine complex is Cisplatin, $PtCl_2(NH_3)_2$, currently one of the most widely used anticancer drugs in the world [28].

Just over one century has passed since Fritz Haber's Nobel-prize-winning work on ammonia synthesis, in which he showed how ammonia could be synthesized from its constituent parts, under high pressure, high temperature conditions in the presence of an iron-containing catalyst (equation 12).

$$N_2 + 3H_2 \rightarrow 2NH_3 \tag{12}$$

This discovery led to the production of ammonia from atmospheric N_2 on an industrial scale, thereby enabling agricultural intensification across the globe. The equipment for the industrial production of ammonia was developed by the engineer, Carl Bosch, so the process is called the Haber-Bosch process. Historically and practically, it is closely associated with the Ostwald process in providing its requisite raw material, NH_3.

Unfortunately, the amount of reactive nitrogen entering the environment has significantly increased as a result, leading to a host of ecological problems [29, 30]. Ammonia is a gas readily released into the air from a variety of biological sources, as well as from industrial and combustion processes. It is the most prevalent alkaline gas in the atmosphere. While NH_3 has many beneficial uses, it can detrimentally affect the quality of the environment through acidification and eutrophication of natural ecosystems, the associated loss of biodiversity, and the formation of secondary particles in the atmosphere, which can reduce visibility [31]. In the lower atmosphere, ammonia reacts readily with compounds such as nitric acid (HNO_3) and sulfuric acid (H_2SO_4) to form ammonium particulates and aerosols. Fine fraction particulate matter, including ammonium sulfate [$(NH_4)_2SO_4$] and ammonium nitrate (NH_4NO_3), has been linked to medical conditions such as asthma, and ammonium aerosols may influence the global climate by altering the transmission of atmospheric and terrestrial radiation [32, 33]. Anthropogenic perturbations of the nitrogen cycle originate from production of energy and food to sustain human populations, which cause the release of reactive nitrogen compounds, principally as nitrogen oxides (NO, NO_2), nitrous oxide (N_2O), nitrate, and ammonia. Human activities have more than doubled the annual production of reactive nitrogen to satisfy human needs for food, energy, fiber, and other products [34–36]. In consequence, the biogeochemical nitrogen cycle has become severely unbalanced in recent decades and one main reason for this fact is the anthropogenic supply of reduced nitrogen to the biosphere in the form of NH_3/NH_4^+, for example during environmental fertilization and livestock farming.

3 Enzymes Involved in Ammonia Turnover

In addition to the interconversion of NH_3/NH_4^+ and biomass, ammonia is either produced or consumed by various microbial enzymes (Figure 1; see Table 1 for a comprehensive summary of some important enzymes and their properties). Accordingly, ammonia is produced from dinitrogen (by nitrogenase; equation 5) or from nitrite, nitric oxide or hydroxylamine. The latter three reactions (equations 13, 14, and 15) are catalyzed by members of the respiratory multiheme cytochrome *c* (MCC) family (see Sections 4 and 5 for details; see Figure 2 in Section 4 for the structure of heme *c*), of which the cytochrome *c* nitrite reductase (NrfA) is the best characterized enzyme.

$$NO_2^- + 6\,e^- + 8\,H^+ \rightarrow NH_4^+ + 2\,H_2O \quad E^{0'}\left(NO_2^-/NH_4^+\right) = +340\,mV\;[37]$$
$$(13)$$

$$NO + 5\,e^- + 5\,H^+ \rightarrow NH_3 + H_2O \tag{14}$$
$$NH_2OH + 2\,e^- + 2\,H^+ \rightarrow NH_3 + H_2O \tag{15}$$

Conversely, ammonia is used as a substrate by the enzymes hydrazine synthase (involved in the anammox process, equation 16) and ammonia monooxygenase, which catalyzes the first step in the nitrification process, i.e., oxidation of ammonia to nitrite (equation 17).

$$NO + NH_3 + 3\,e^- + 3\,H^+ \rightarrow N_2H_4 + H_2O \tag{16}$$
$$NH_3 + O_2 + 2e^- + 2\,H^+ \rightarrow NH_2OH + H_2O \tag{17}$$

Hydrazine synthase has been recently purified from *Candidatus* Kuenenia stuttgartiensis and characterized as an apparently anammox-specific heterotrimeric complex (HzsABC) that contains two diheme cytochromes (gene products of kuste2860 and kuste2861) [14, 38]. In contrast, ammonia monooxygenase is a membrane-integral copper-iron enzyme (see Chapter 6 in [7]).

4 Cytochrome *c* Nitrite Reductase as Paradigm

In principle, nitrite-reducing enzymes can be divided into two major classes. The first class includes those enzymes reducing NO_2^- to NO in the denitrification pathway, the second class comprises those enzymes which reduce NO_2^- directly to NH_3 (Figure 1). In bacterial denitrification, two groups of dissimilatory nitrite reductases are known, either a cytochrome cd_1 or a copper enzyme carrying a type-1 and a type-2 Cu center. In nitrite-ammonifying bacteria there may exist two different processes of nitrite reduction to ammonia, with different physiological

Table 1 Selected enzymes and enzyme complexes involved in respiratory ammonia production or consumption. Modified from [24].

Enzyme designation[a]	Physiological function	Redox partner	Selected model organism(s)[b]
AMMONIA-PRODUCING ENZYMES			
1. NITROGENASE			
1.1 (NifDK)$_2$ complex, the heterotetra-meric MoFe protein (containing the FeMo cofactor, a [Mo:7Fe:9S:C]; homocitrate moiety)	Cytoplasmic nitrogen reduction to ammonia	ATP-hydrolyzing NifH homodimer (Fe protein)	*Azotobacter vinelandii* (γ), *Klebsiella pneumoniae* (γ)
2. NITRITE REDUCTASES			
2.1 NrfHA complex (NrfA: 5 heme c; one CX$_2$CK motif; NrfH: four heme c; Dimers of NrfHA$_2$ assembly in crystal structure)	Periplasmic nitrite ammonification as well as reduction of NO and NH$_2$OH to ammonia	Quinol (oxidized by NrfH)	*Wolinella succinogenes* (ε), *Desulfovibrio vulgaris* (δ), *Desulfovibrio desulfuricans* (δ), *Bacillus vireti* (Firmicutes)
2.2 NrfA (5 heme c; five CH$_2$CH motifs)	ditto	ditto	*Campylobacter jejuni* (ε)
2.3 NrfA (5 heme c; one CX$_2$CK motif)	ditto	Tetraheme cytochrome c quinol dehydrogenase CymA	*Shewanella oneidensis* (γ)
2.4 NrfA (5 heme c; one CX$_2$CK motif; dimeric)	ditto	Pentaheme cytochrome c NrfB	*Escherichia coli* (γ); *Neisseria gonorrhoeae*
2.5 Otr (8 heme c)	Not known. Otr reduces nitrite and hydroxylamine to ammonia. It also interconverts tetrathionate and thiosulfate.	Not known	*Shewanella oneidensis* (γ), *Geobacter* species (δ)
2.6 Onr (8 heme c; hexameric)	Not known. Onr reduces nitrite and hydroxylamine to ammonia. It also reduces sulfite and hydrogen peroxide.	Not known	*Thioalkalivibrio nitratireducens* (γ), *Thioalkalivibrio paradoxus* (γ), *Geobacter* species (δ)
2.7 eHao (8 heme c)	Most likely nitrite reduction. The reaction product has not been reported.	Presumably a cyto-chrome c of the NapC/NrfH family	*Campylobacter concisus*, *Campylobacter curvus*, *Campylobacter fetus*, *Nautilia profundicola* (all ε)

AMMONIA-CONSUMING ENZYMES

3. HYDRAZINE SYNTHASE

3.1 HzsABC (4 heme *c*)	Comproportionation of nitric oxide and ammonium in anammoxosome	Cytochromes *c* in anammoxosome	*Candidatus* Kuenenia stuttgartiensis (Planctomycetes)

4. AMMONIA MONOOXYGENASE

4.1 AmoABC (Cu, Fe; trimeric)	Periplasmic ammonia oxidation	Quinone pool (assumed)	*Nitrosomonas europaea* (β)

[a]The metal/cofactor content of individual enzymes or enzyme complexes as well as their typical multimerization status is provided if reported.
[b]Names in bold face indicate that a high-resolution enzyme structure from the respective organism is available. Phyla or proteobacterial classes (Greek letters) are given in parentheses.

functions and enzymes equipped with different catalytic centers: (i) assimilatory enzymes which host a special porphyrin cofactor in close neighborhood to a [4Fe-4S] cluster, the so-called siroheme-[4Fe-4S] catalytic center, and (ii) respiratory multiheme c-type cytochromes which contain multiple heme c groups (Figure 2).

Both metal enzymes show highly complex EPR spectra, with characteristic resonances ranging from $g \approx 18$ to 1.5 (siroheme) to $g \approx 10$ and 3.8 (multiheme c-type cytochromes), respectively. In short, sirohemes are Fe complexes of isobacteriochlorin, a class of hydroporphyrins with eight carboxylic acid side chains. They have been detected in assimilatory nitrite ($NO_2^- \rightarrow NH_3$) and sulfite ($SO_3^{2-} \rightarrow H_2S$) reductases as well as in dissimilatory sulfite ($SO_3^{2-} \rightarrow H_2S$) reductases. A salient feature of the isobacteriochlorin skeleton is its ease of oxidation and difficulty of reduction compared with those of porphyrins and chlorins. This facile oxidation led to the proposal that the siroheme macrocycle itself might be involved in the multi-electron, multi-proton transfer reactions (13) and (18) [20, 21, 25].

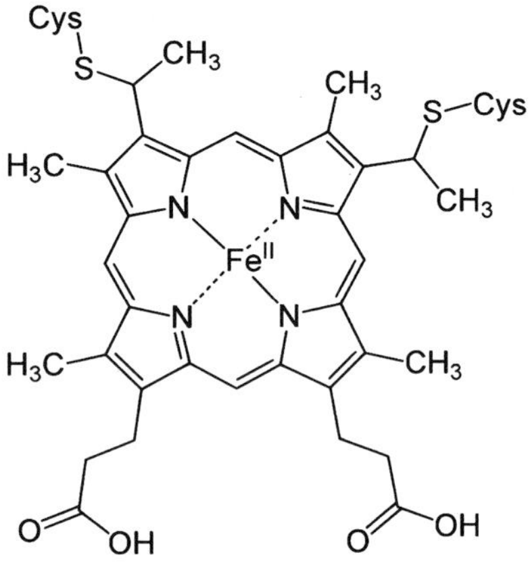

Figure 2 Structure of heme c, the covalently bound heme group found in cytochromes c. During cytochrome c biosynthesis, two thioether bonds are formed between two vinyl groups of heme b and the two sulfhydryl ($-SH$) residues of the apocytochrome heme c binding motif (usually Cys-X-X-Cys-His).

Since the early studies by Fujita on soluble cytochromes in Enterobacteriaceae and the first characterization of the then cytochrome $c552$ enzyme [39], research on multiheme nitrite reductases, especially in view of their key role in the biogeochemical nitrogen cycle (Figure 1) (transformation $NO_2^- \rightarrow NH_3$), has attracted a major interest as documented by the large number of publications and comprehensive reviews in biology, ecology, and chemistry [19, 23, 24, 40–42]. Nowadays, the enzyme is known as pentaheme cytochrome c nitrite reductase (or NrfA) and is *the* prototypic enzyme of respiratory nitrite reduction to ammonium (equation 13) [19, 43].

Originally discovered in enteric bacteria, the range of NrfA-producing organisms is ever increasing in many habitats suggesting that NrfA contributes significantly to global nitrogen turnover. Part of the NH_4^+ is released as NH_3 leading to loss of nitrogen, similar to the pathway of denitrification which generates the gaseous compounds NO, N_2O, and/or N_2. On the other hand, since NH_4^+ can be taken up for assimilatory purposes, the cytochrome c NrfA enzyme is regarded to contribute to nitrogen retention in the habitat.

4.1 Cytochrome c Nitrite Reductase-Containing Organisms and Their Physiology

The NrfA enzyme is encoded on various bacterial genomes (see [44] for a comprehensive list of bacteria coding for a NrfA homolog). On the other hand, *nrfA* genes have not been reported from archaeal or eukaryotic species. Organisms that use cytochrome c nitrite reductase as a nitrite-ammonifying enzyme are abundantly present in soil, water, and host-associated habitats such as the intestine or the rumen of animals. NrfA-containing microbes are assigned to a huge variety of bacterial phyla (including both Gram-negative and Gram-positive organisms) and vary tremendously in their physiological lifestyles. Prominent examples are (i) growth by anaerobic respiration using nitrate, nitrite and/or nitrous oxide as terminal electron acceptor, (ii) growth by anaerobic respiration using non-nitrogen compounds as electron acceptor such as sulfate, sulfite or fumarate, (iii) growth by (micro)aerobic respiration, and (iv) growth by fermentation of carbohydrates. The best characterized organisms with respect to structure and function of NrfA are host-associated bacteria such as *Escherichia coli* and *Wolinella succinogenes* as well as free-living species such as *Shewanella oneidensis*, *Sulfurospirillum deleyianum*, *Desulfovibrio desulfuricans*, and *Desulfovibrio vulgaris* [24]. More recently, *nrfA* genes were also reported to be present in the genomes of several Gram-positive bacteria (both Firmicutes and Actinobacteria), for example in *Bacillus selenatireducens*, *Bacillus azotoformans*, *Bacillus bataviensis*, *Bacillus vireti*, and *Arcanobacterium haemolyticum* [44–46].

4.2 Biochemistry and Structure of Cytochrome c Nitrite Reductase

Purified NrfA enzymes exhibit high specific activities of up to $>1,000$ µmol NO_2^- min^{-1} mg^{-1} at optimum pH 7.0, which led to the development of an electrochemical nitrite sensor. Nitric oxide (NO) and hydroxylamine (NH_2OH) as well as its O-methyl derivative are also transformed to ammonia, however at reduced activities. Similar findings were reported for the assimilatory sulfite reductase from *D. vulgaris*, which catalyzes another important multi-electron, multi-proton transfer reaction in biology, the six-electron reduction of sulfite (SO_3^{2-}) to hydrogen sulfide (H_2S) (equation 18) [47–49].

$$HSO_3^- + 6\,e^- + 6\,H^+ \rightarrow HS^- + 3\,H_2O \quad E^{0'}\left(HSO_3^-/HS^-\right) = -116\,mV\ [37]$$

$$(18)$$

NH_2OH was converted ten times faster than its O-methyl derivative by the assimilatory sulfite reductase which brought the authors to the conclusion that the substrate interacted through its nitrogen moiety with the iron center, as confirmed later by X-ray crystallography. Notably, ammonia-forming NrfA does also catalyze reaction (18) as discussed recently [25].

Approximately two decades ago, Cole and coworkers located the structural gene (nrfA) for cytochrome c552, the terminal reductase of the formate-dependent pathway for nitrite reduction to ammonia (nrf, reduction of nitrite by formate) on the E. coli chromosome [50]. This enzyme was also called ammonia-forming cytochrome c nitrite reductase, or cytochrome c nitrite reductase. The corresponding DNA sequence was described to encode a tetraheme c-type cytochrome, based on the presence of four Cys-X-X-Cys-His heme-binding motifs, in contrast to the hexaheme cytochrome c reported earlier on the basis of biochemical and spectroscopic data. Within that sequence, the authors also found a Cys-X-X-Cys-Lys motif, which they did not assign to a heme-binding site at that time, in agreement with analytical data (heme and Fe content) for the cytochrome c nitrite reductase from S. deleyianum and W. succinogenes [40, 50–54].

NrfA displays typical c-type cytochrome absorption spectra, with absorption maxima at 280, 409, and 534 nm, and a shoulder at 615 nm in the oxidized (as isolated) state [55]. The shoulder at 615 nm can be assigned to a high-spin Fe(III) heme center [40]. Upon reduction with sodium dithionite, new maxima appear at 420, 523, and 553 nm. In the second derivative mode, the spectra reveal minima at 419, 427, 523, 530, 550, and 555 nm. The maxima at 419 and 550 nm can be assigned to low-spin Fe(II) heme centers, whereas those at 427 and 555 nm are consistent with spectral properties of high-spin Fe(II) heme [56, 57].

The first two crystal structures were reported for the enzymes from S. deleyianum [58] and W. succinogenes [59] (enzyme class 2.1 in Table 1; Figure 3) and revealed a homodimeric architecture for NrfA, with five tightly packed c-type heme centers in each subunit [42]. Hereafter, more three-dimensional structures were published, such as for NrfA from E. coli (enzyme class 2.4 in Table 1) [60] and D. desulfuricans ATCC27774 [61], and for the structure of the NrfA from D. vulgaris Hildenborough in complex with a second multiheme protein, designated NrfH.

Note that the cytochrome c nitrite reductases from sulfate-reducing bacteria D. desulfuricans and D. vulgaris Hildenborough were usually co-purified with their physiological electron donor, the tetraheme NrfH, as a complex with a well-defined NrfA:NrfH 2:1 stoichiometry [62]. This organization was confirmed by the X-ray structure of the $(NrfA_2H)_2$ nitrite reductase complex from D. vulgaris Hildenborough, which represented a major breakthrough in our understanding of the dissimilatory reduction of nitrite to ammonia [63]. Recently, the structure of S. oneidensis NrfA became available (enzyme class 2.3 in Table 1). This enzyme is

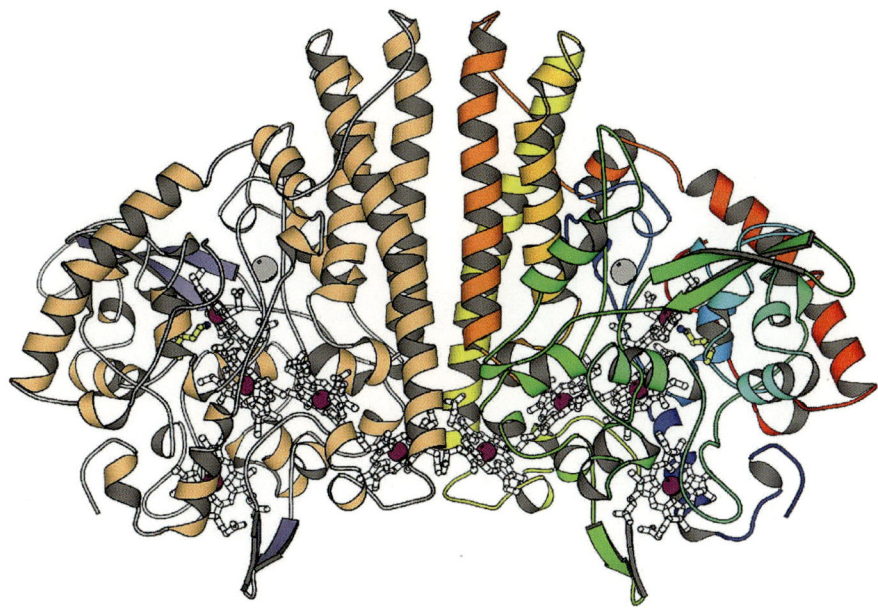

Figure 3 Three-dimensional structure of the cytochrome *c* nitrite reductase (NrfA) dimer from *Sulfurospirillum deleyianum*. A front view with the dimer axis oriented vertically, the five hemes in each monomer (white), the Ca^{2+} ions (grey), and residue Lys133 which coordinates the active site iron atom (yellow). In the right monomer, the protein chain is colored blue from the amino-terminal end to red at the carboxy-terminal end, in the left mono-mer according to secondary structure. The dimer interface is dominated by three long α-helices per monomer. All hemes in the dimer are covalently attached to the protein. Taken by permission from [58]; copyright 1999 Nature Publishing Group. PDB code: 1QDB.

thought to interact with the membrane-bound tetraheme cytochrome *c* CymA (see Section 4.3) [64].

The NrfA protein binds five *c*-type heme groups via thioether bonds to the cysteines of conserved heme *c* attachment motifs (Figures 2 and 4), four of them with the *classical* Cys-X-X-Cys-His and one with the Cys-X-X-Cys-Lys sequence. Thus, in most known NrfA proteins, all heme irons are bis-histidinyl-coordinated (hemes 2–4, sp^2-$N_{imidazole}$) except for the catalytic center (heme 1), which is bound to a lysine as the 5th coordinate ligand (sp^3-NH_2) of an unprecedented structural motif. A minority of known NrfA sequences, however, contained five Cys-X-X-Cys-His heme *c* binding motifs, such as the enzyme from *Campylobacter jejuni* which was shown to catalyze nitrite reduction at a high specific activity (enzyme class 2.2 in Table 1) [65].

A recent phylogenetic analysis of 272 full-length NrfA protein sequences distinguished 18 NrfA clades with robust statistical support [44]. Three clades possessed a Cys-X-X-Cys-His motif in the first heme-binding domain with representative organisms being classified in a surprisingly diverse range of bacterial phyla such as Proteobacteria (delta and epsilon classes), Planctomycetes,

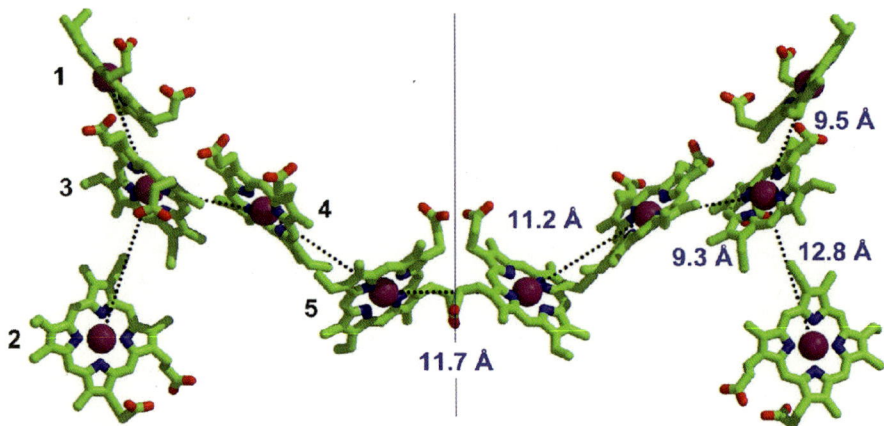

Figure 4 Arrangement of the ten heme c centers in the cytochrome c nitrite reductase (NrfA) dimer from *Sulfurospirillum deleyianum*. The overall orientation of the hemes corresponds to that in Figure 3, with the active site located at heme 1 and the line indicating the NrfA dimer interface. Hemes in the left monomer are numbered according to their attachment to the protein chain. In the right monomer, the Fe-Fe distances (Å) between the hemes are given. Taken by permission from [58]; copyright 1999 Nature Publishing Group. PDB code: 1QDB.

Actinobacteria, Verrucomicrobia, and Acidobacteria. All other clades had a Cys-X-X-Cys-Lys motif in this location. In all NrfA protein hemes 2–4 serve as electron transfer modules, whereas heme center 1 is the catalytic site which binds nitrite and also sulfite [42, 59] (Figures 5, 6, and 7).

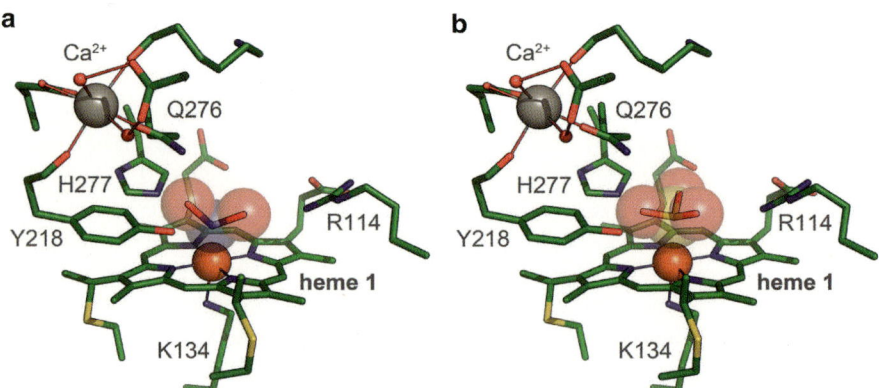

Figure 5 Structures of substrate complexes of cytochrome c nitrite reductase (NrfA) from *Wolinella succinogenes*. (**a**) Nitrite adduct, PDB code: 2E80, with the NO_2^- anion binding to the heme iron through the electron pair at the nitrogen atom. (**b**) Sulfite adduct, PDB code: 3BNF. The binding mode is highly similar to nitrite but the bond distance to iron is slightly longer due to the larger van der Waals radius of sulfur [72]. Taken by permission from [25]; copyright 2013 Elsevier.

NO_2^- is reduced to NH_4^+ in a consecutive series of electron and proton transfer steps. On the basis of the crystallographic observation of reaction intermediates and of density functional theory (DFT) calculations, it was proposed that nitrite reduction started with a heterolytic cleavage of the N—O bond, which is facilitated by a pronounced back-bonding interaction of nitrite coordinated through nitrogen to the reduced [Fe(II)] but not the oxidized [Fe(III)] active site iron (Figures 5 and 6) [66–69].

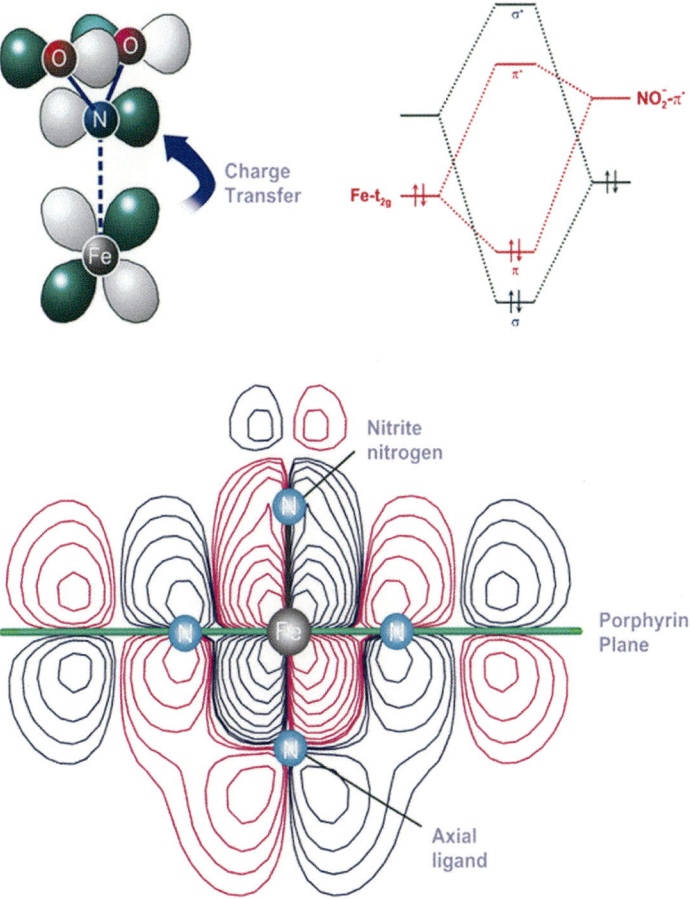

Figure 6 Activation of nitrite at heme center 1 of cytochrome *c* nitrite reductase (NrfA). Upper part: Back-bonding interaction in the porphyrin-nitrite complex transfers charge from the occupied iron-derived t_{2g}-like orbitals into the nitrite π^* orbital. Lower part: Back-bonding interaction in $[Fe(II)(porphyrin)(Lys-NH_3)(NO_2^-)]^-$. The contour shows a cut through the iron d_{yz}-based HOMO that has a constructive overlap with the nitrite π^* LUMO, which indicates the back-bonding interaction. The oxygen atoms of nitrite are below and above the plane of the paper, respectively. The reduction of nitrite starts with a heterolytic cleavage of the N–O bond which is facilitated by a pronounced back-bonding interaction of nitrite coordinated through nitrogen to the reduced Fe(II) but not the oxidized Fe(III) active site iron. This step leads to the formation of an $\{FeNO\}^6$ species according to the Enemark-Feltham notation [66] and a water molecule and is further facilitated by a hydrogen bonding network that induces an electronic asymmetry in the nitrite molecule that weakens one N–O bond and strengthens the other. Taken by permission from [66]; copyright 2002 American Chemical Society.

No intermediates become liberated in the course of this multi-electron, multi-proton reduction process, such as nitric oxide (NO) or hydroxylamine (H_2NOH). This requires a remarkable flexibility of the active site combined with a finely tuned proton and electron delivery system. However, NrfA will convert NO and H_2NOH to ammonia, and it will react with N_2O to a so far unidentified product [70]. The possible role of second-sphere active-site amino acids as proton donors and the role of the Ca^{2+} ion close to the active site heme was investigated by computational chemistry [67]. Note that the active site residues His277, Tyr218, and Arg114 (*W. succinogenes* NrfA numbering) as well as the Ca^{2+} ion are strictly conserved in NrfA enzymes and provide an environment of positive electrostatic potential around the active site [71]. His277 was suggested as the most probable proton donor, whose spatial orientation and fine-tuned acidity led to energetically feasible, low-barrier protonation reactions. An alternative candidate was Arg114 according to the theoretical studies by Bykov and Neese [67]. Interestingly, Tyr218 did not appear to participate in the reaction during the initial stage of the reduction process. However, exchange of Tyr218 by phenylalanine led to a more or less complete loss of nitrite reductase activity, whereas the sulfite reductase activity of the Tyr218Phe variant remained unaffected [72].

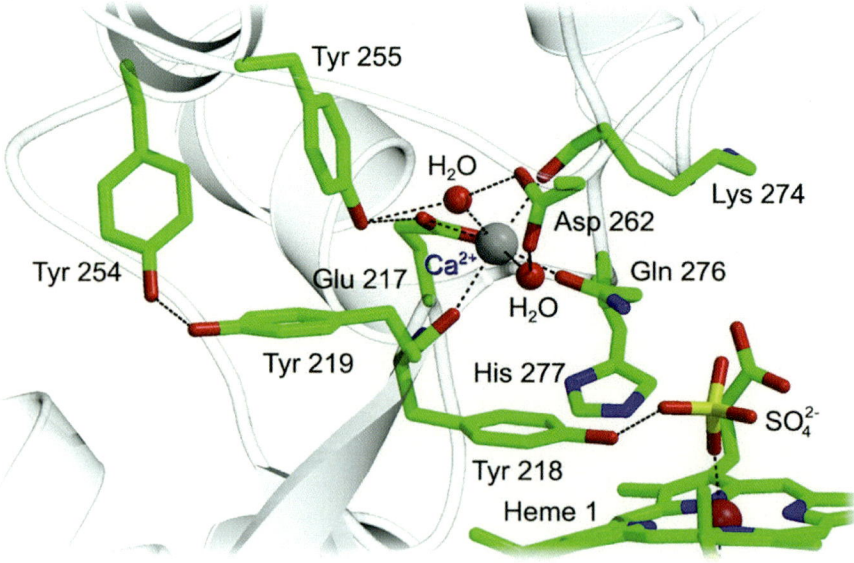

Figure 7 The active site of cytochrome *c* nitrite reductase (NrfA) from *Wolinella succinogenes*. The sulfate molecule (SO_4^{2-}, from crystallization buffer) occupies the substrate binding site at heme 1 (see also Figure 5). The structure emphasizes the set of conserved amino acid residues including four tyrosine residues Tyr255, Tyr254, Tyr219, and Tyr218 that might play a role in radical stabilization during catalysis [69]. PDB code: 1FS8. This research was originally published in [59] and taken by permission; copyright 2000 American Society for Biochemistry and Molecular Biology.

In a consecutive study, DFT theory was employed to investigate the recharging of the active Fe heme center 1 (Figure 6) with protons and electrons via a series of reaction intermediates, Fe(II)–NO$^+$, Fe(II)–NO$^•$, Fe(II)–NO$^-$, and Fe(II)–HNO [68]. The activation barriers for the various proton and electron transfer steps were estimated in the framework of Marcus theory. A radical transfer role for the active-site Tyr218 could not be found in the calculations, whereas the important role of the highly conserved Ca^{2+} ion located in the direct proximity of the active site in proton delivery was confirmed, in agreement with earlier experimental findings [70]. Most recently, the second half-cycle of the six-electron NO$_2^-$ → NH$_3$ reduction mechanism was analyzed by Neese and coworkers [69]. In total, three electrons and four protons have to be delivered to obtain the final product, NH$_3$, starting from the HNO intermediate. Two isomeric radical intermediates, HNOH$^•$ and H$_2$NO$^•$, are postulated which are readily converted to H$_2$NOH, most likely through intramolecular proton transfer from residues Arg114 or His277. After N–O bond cleavage a radical intermediate H$_2$N$^•$ is formed which finally reacts with Tyr218, assigning for the first time a specific role of this amino acid residue in the final step of the nitrite reduction process, as expected from the earlier mutational investigations [72].

Obviously, the active site heme 1 accommodates anions and uncharged molecules, such as NO or hydroxylamine, and releases the NH$_4^+$ cation only after the full six-electron reduction of NO$_2^-$. The preference for anions is reflected by a positive electrostatic potential around and inside of the active site cavity (Figure 8). Considering the good accessibility of the active site for water molecules and the presumably lowered pH on the periplasmic side of the cytoplasmic membrane where nitrite reductase is located (Section 4.3.), the product of nitrite reduction will be the positively charged NH$_4^+$ ion rather than uncharged NH$_3$. The cationic product can use a second exit channel leading to the protein surface opposite to the entry channel. It branches before reaching the protein surface and ends in areas with a significantly negative electrostatic surface potential. The presence of separate pathways for substrate and product with matched electrostatic potential will contribute to the high specific activity of nitrite reductase (Figure 8).

4.3 Electron Transfer Routes to Cytochrome c Nitrite Reductase

As described in the previous section, NrfA was found to form a stable complex with the tetraheme cytochrome *c* NrfH in species such as *W. succinogenes* and *Desulfovibrio vulgaris*. In this arrangement, NrfH anchors the complex in the membrane and catalyzes menaquinol oxidation as well as electron transport to NrfA (Figure 9, top and Table 1, enzyme classes 2.1 and 2.2) [63, 73, 74]. NrfH is a member of the widespread NapC/NrfH family [24, 75, 76]. Such proteins are membrane-bound tetra- or pentaheme cytochromes *c* that comprise an N-terminal

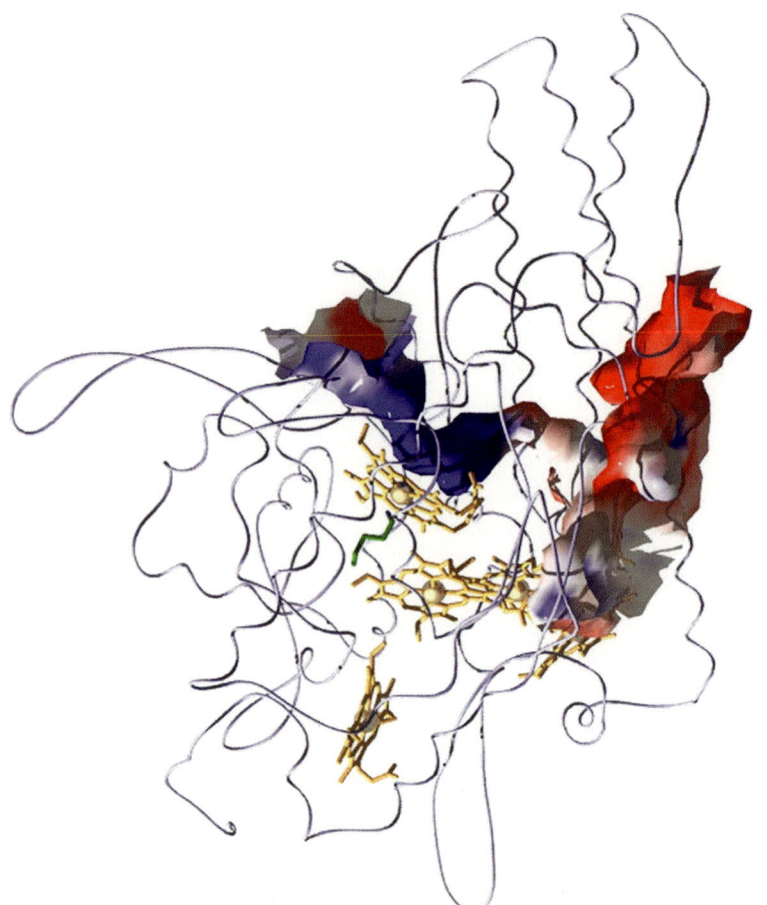

Figure 8 The substrate entry (blue) and product exit (red) channels of cytochrome c nitrite reductase (NrfA). Apart from the entry channel leading from the protein surface to the catalytic site heme 1, a second exit channel exists that reaches the protein surface on the opposite side of the molecule. The whole channel is colored according to the electrostatic surface potential; blue for a positive and red for a negative potential. Modified from [58].

membrane-spanning helix and a globular cytochrome c domain situated at the outside of the bacterial membrane.

Within their enzymic contexts, NapC and NrfH donate electrons, either directly or indirectly, to periplasmic nitrate reductase (NapA) or NrfA, respectively. Another member of this family (CymA) from *Shewanella* species such as *S. oneidensis* was shown to serve as an electron hub within several respiratory chains that terminate with, for example, periplasmic nitrate reductase (NapA), cytochrome c nitrite reductase (NrfA) or cytochrome c fumarate reductase (Table 1, enzyme class 2.3). From the crystal structure of the *D. vulgaris*

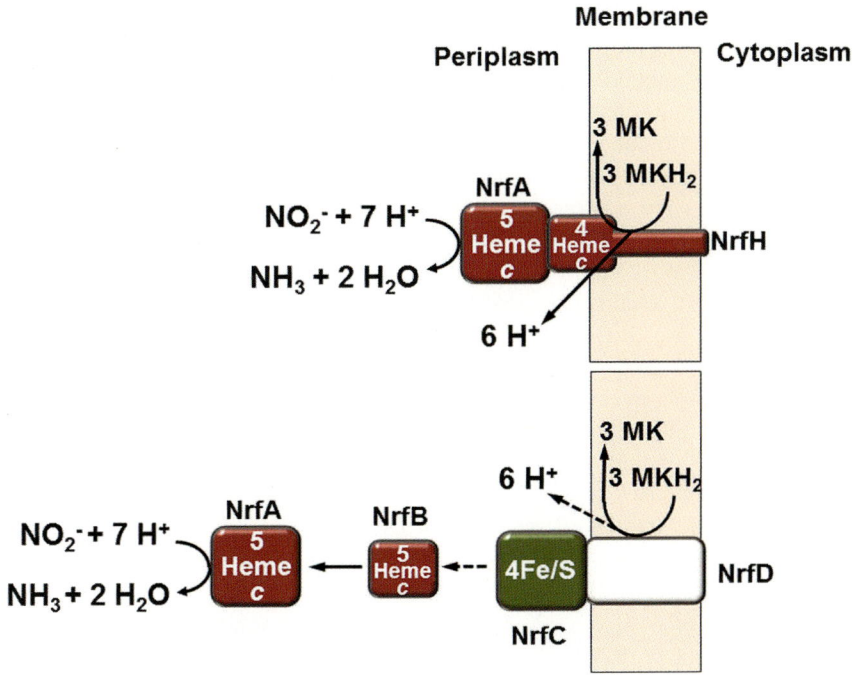

Figure 9 Electron transport chain models catalyzing the oxidation of menaquinol by nitrite in representative respiratory Nrf systems. **Top:** Nrf system of *Wolinella succinogenes*. **Bottom:** Nrf system of *Escherichia coli*. See text and Table 1 for details. The dashed arrows denote that proton release to the periplasmic side of the membrane by NrfD as well as direct electron transfer between NrfC and NrfB are speculative. For simplicity, only monomeric enzyme forms are shown. MK, menaquinone; MKH_2, menaquinol; Fe/S, iron-sulfur center. Modified from [24].

$(NrfHA_2)_2$ complex, it is inferred that menaquinol binds at the periplasmic side of the membrane in the vicinity of heme 1. Therefore, it is conceivable that protons are released to the periplasmic space upon menaquinol oxidation. This would make the catalysis of menaquinol oxidation by nitrite an electroneutral, i.e., non proton motive force-generating, process [24, 73, 77].

Enteric bacteria such as *E. coli* do not encode a NrfH homolog in their *nrf* gene clusters. Instead, *E. coli* was reported to employ a protein assembly consisting of the proteins NrfB, NrfC, and NrfD in order to transfer electrons from menaquinol to NrfA, which appears to be a soluble protein present in the periplasmic space (Figure 9, bottom and Table 1, enzyme class 2.4). Mainly concluded from genetic studies, a membrane-bound NrfCD complex was postulated to oxidize menaquinol near the periplasmic side of the membrane [78–80]. This hypothesis is in line with the experimentally proven location of a quinone binding site in the structurally similar PsrCB subcomplex of a potential polysulfide reductase from *Thermus thermophilus* [81]. As for the NrfHA complex, such a topology of the reactive sites for menaquinol and nitrite in the NrfABCD system would make electrogenic nitrite reduction by menaquinol unlikely [24]. Electron transport between NrfCD and NrfA in *E. coli* is thought to be mediated by the pentaheme cytochrome *c* NrfB, which is distantly related to NrfH [24, 82]. Taken together, it seems that different electron transfer routes have been established during evolution in order to transport electrons from the reduced quinone pool to NrfA, albeit without any influence on the overall bioenergetics of the process.

These findings are a prominent example of the emerging picture of electron transport module families that are involved in the build-up of functionally diverse prokaryotic respiratory chains [24, 83]. Furthermore, differences in the described electron transport chains are also reflected in the fact that *E. coli* and *W. succinogenes* use different so-called cytochrome *c* biogenesis systems for the maturation of NrfA and other cytochromes *c* in the periplasmic space [84]. In both organisms, the unconventional Cys-X-X-Cys-Lys heme *c* attachment motif is processed by a dedicated cytochrome *c* synthase isoenzyme that recognizes its cognate NrfA by an unknown mechanism [65, 85–87]. The corresponding cytochrome *c* synthases of *E. coli* and *W. succinogenes* are largely unrelated in their primary structures and it seems that they have been derived from the respective general cytochrome *c* biogenesis system (named System I or Ccm system in *E. coli* and System II or Ccs system in *W. succinogenes*) in order to enable NrfA maturation [84, 88, 89]. Cys-X-X-Cys-Lys-specific cytochrome *c* synthases are apparently missing in organisms that contain NrfA proteins carrying five conventional Cys-X-X-Cys-His motifs [65].

4.4 Role of Cytochrome c Nitrite Reductase in Stress Defense

Although originally described as a key enzyme of anaerobic nitrite respiration, a prominent function in nitrosative and peroxidative stress defense has been elucidated for NrfA enzymes more recently due to the fact that NrfA reduces NO, hydroxylamine, and hydrogen peroxide efficiently to ammonia and water, respectively, in addition to nitrite, which is also a cytotoxic substance [90–92]. This widens the physiological function of NrfA considerably as it appears that the enzyme represents a unique periplasmic means to combat nitrosative stress, thus supporting the many known cytoplasmic NO-reactive proteins present in aerobic as well as anaerobic microorganisms such as (flavo)hemoglobins, flavorubredoxins, and other flavodiiron proteins [93]. Furthermore, NrfA might functionally replace catalase in some microaerobic bacteria [92, 94].

Taken together, NrfA appears to carry out a versatile role in handling the above mentioned stressors as well as in nitrite tolerance, which has been shown either by physiological experiments using intact cells or with purified enzymes [92]. It might also explain why NrfA is constitutively present in many bacteria irrespective of the presence of nitrite.

5 Other Multiheme Cytochromes *c*

Several interesting variations on the theme of multiheme cytochromes *c* were recently discovered, as reviewed by Simon and colleagues [23] (see also [95] for a phylogenetic model of relevant enzymes within the MCC family). Besides the pentaheme cytochrome *c* NrfA, ammonia production was also reported for purified (i) octaheme cytochrome *c* tetrathionate reductase (Otr enzyme 2.5 in Table 1), (ii) octaheme cytochrome *c* nitrite reductase (Onr, enzyme 2.6 in Table 1), and (iii) hydroxylamine oxidoreductase (Hao) from nitrifiers. In addition, ammonia production from nitrite is also conceivable to occur in some ammonifying Epsilonproteobacteria that usually lack a *nrfA* gene but encode an Hao-type octaheme cytochrome *c*, tentatively named εHao (enzyme 2.7 in Table 1) [22, 23].

The octaheme enzyme from *S. oneidensis*, originally described as tetrathionate reductase, was structurally characterized and shown to reduce NO_2^- and H_2NOH to NH_4^+ [96, 97]. The structures of two Onr enzymes from *Thioalkalivibrio nitratireducens* and *Thioalkalivibrio paradoxus* were reported which also reduced NO_2^- and H_2NOH to NH_4^+, as well as SO_3^{2-} to H_2S [94, 98–100]. Both Onr enzymes showed specific structural features distinguishing them from pentaheme NrfA enzymes: (i) the covalent Tyr-Cys bond in the active site, (ii) the hexameric architecture resulting in the formation of a void space inside the hexamer, and (iii) the product channel that opens into the void interior space of the hexamer. These structural features might explain the higher nitrite reductase activity and the greater preference for nitrite than for sulfite as a substrate compared to NrfA [99, 100]. The heme arrangement within each Onr monomer is highly reminiscent of the octaheme Hao from *Nitrosomonas europaea* which oxidizes H_2NOH to NO_2^- [101] but also reduces either nitrite or NO to ammonia in the presence of a strong reductant [102, 103]. Notably, the siroheme-[4Fe-4S]-dependent sulfite reductase from *Mycobacterium tuberculosis* also contains a covalent Tyr-Cys bond in the active site, in close proximity to the siroheme cofactor. Removal of this covalent bond by site-directed mutagenesis impaired catalytic activity, suggesting that it is important for the enzymatic reaction [104].

Taken together, it appears that the multiheme cytochrome classes designated 2.1 to 2.7 in Table 1 share the property of extracytoplasmic ammonia production.

6 Environmental Issues and Conclusions

As outlined above, ammonia-producing cytochromes *c* of the MCC family are abundant enzymes in various habitats on Earth. However, the detection (not to mention the quantification) of the corresponding genes by PCR-based methods is hampered by primer design, which generally relies on suitable primary structure alignments of representative enzymes. In the case of *nrfA* gene amplification, it appears that genes encoding NrfA enzymes containing a Cys-X-X-Cys-His at the active site have been overlooked by these methods until recently [44] since previous primer pairs were designed to target a nucleotide stretch encoding the NrfA Cys-X-X-Cys-Lys signature [105].

Likewise, metagenomic data have not been rigorously investigated with respect to the presence of ammonia-producing multiheme cytochromes c. Given that these enzymes are generally present at the outside of the cytoplasmic membrane, for example, in the periplasm of Gram-negative bacteria, any produced ammonia would not be readily available to assimilatory cytoplasmic enzymes and had to be taken up by suitable ammonium transporters. Consequently, multiheme cytochromes c might contribute to the release of ammonia into the atmosphere, depending on environmental parameters such as dioxygen availability, temperature, and pH. To our knowledge, this is an as yet largely unaccounted feature in global-scale models of ammonia production in habitats such as anaerobic soil or marine oxygen minimum zones [29, 30, 106–108]. The situation is reminiscent to the discovery that bacterial ammonifiers of nitrate and nitrite, i.e., NrfA-containing microorganisms, might significantly contribute to atmospheric N_2O production, in addition to the denitrifying microbial community [109–112]. In this case, any NO formed either enzymatically or abiotically from nitrite is apparently detoxified under anaerobic conditions to N_2O by a variety of NO-reactive enzymes. Taken together, it seems that the biogeochemical nitrogen cycle still offers a lot of surprising and fascinating discoveries, possibly with many more to come.

Abbreviations

anammox	anaerobic ammonium oxidation (= comproportionation of nitrite and ammonium to form dinitrogen)
DFT	density functional theory
DNA	deoxyribonucleic acid
EPR	electron paramagnetic resonance
Fe/S	iron-sulfur center
Hao	hydroxylamine oxidoreductase
HOMO	highest occupied molecular orbital
LUMO	lowest unoccupied molecular orbital
MCC	multiheme cytochrome c family
MK	menaquinone
MKH$_2$	menaquinol/menahydroquinone
NapA	periplasmic nitrate reductase
Nrf	nitrite reduction by formate
NrfA	pentaheme cytochrome c nitrite reductase
Onr	octaheme cytochrome c nitrite reductase
Otr	octaheme cytochrome c tetrathionate reductase
PCR	polymerase chain reaction

Acknowledgments The authors are grateful to Sascha Hein and Melanie Kern (Technische Universität Darmstadt) for providing unpublished data on NrfA phylogeny, and to Oliver Einsle (Albert-Ludwigs-Universität Freiburg) for stimulating discussions. Cited own work was supported by grants from the Deutsche Forschungsgemeinschaft (DFG) (JS, PK) and the Volkswagen-Stiftung (PK).

References

1. M. Rudolf, P. M. H. Kroneck, *Met. Ions Biol. Syst.* **2005**, *43*, 75–103.
2. J. Rockström, W. Steffen, K. Noone, Å. Persson, F. S. Chapin, E. F. Lambin, T. M. Lenton, M. Scheffer, C. Folke, H. J. Schellnhuber, B. Nykvist, C. A. de Wit, T. Hughes, S. van der Leeuw, H. Rodhe, S. Sörlin, P. K. Snyder, R. Costanza, U. Svedin, M. Falkenmark, L. Karlberg, R. W. Corell, V. J. Fabry, J. Hansen, B. Walker, D. Liverman, K. Richardson, P. Crutzen, J. A. Foley, *Nature* **2009**, *461*, 472 –475.
3. L. B. Maia, J. J. G. Moura, *Chem. Rev.* **2014**, *114*, 5273–5357.
4. J. A. Brandes, A. H. Devol, C. Deutsch, *Chem. Rev.* **2007**, *107*, 577–589.
5. P. G. Falkowsky, *Nature* **2007**, *387*, 272–275.
6. J. N. Galloway, F. J. Dentener, D. G. Capone, E. W. Boyer, R. W. Howarth, S. P. Seitzinger, G. P. Asner, C. C. Cleveland, P. A. Green, E. A. Holland, D. M. Karl, A. F. Michaels, J. H. Porter, A. R. Townsend, C. J. Vörösmarty, *Biogeochemistry* **2004**, *70*,153–226.
7. *Sustaining Life on Planet Earth: Metalloenzymes Mastering Dioxygen and Other Chewy Gases*, Eds P. M. H. Kroneck, M. E. Sosa Torres; Vol. 15 of *Metal Ions in Life Sciences*; Eds A. Sigel, H. Sigel, R. K. O. Sigel; Springer International Publishing AG, Cham, Switzerland, 2015.
8. D. K. Newman, J. F. Banfield, *Science* **2002**, *296*, 1071–1077.
9. A. L. Reysenbach, E. Shock, *Science* **2002**, *296*, 1077–1082.
10. M. Strous, J. A. Fuerst, E. H. Kramer, S. Logemann, G. Muyzer, K. T. van de Pas-Schoonen, R. Webb, J. G. Kuenen, M. S. Jetten, *Nature* **1999**, *400*, 446–449.
11. A. H. Devol, *Nature* **2003**, *422*, 575–576.
12. C. R. Penton, A. H. Devol, J. M. Tiedje, *Appl. Environ. Microbiol.* **2006**, *72*, 6829–6832.
13. M. Ali, L.-Y. Chai, C.-J. Tang, P. Zheng, X.-B. Min, Z.-H. Yang, L. X., Y.-X. Song, *Biomed. Res. Int.* **2013**, doi: 10.1155/2013/134914
14. B. Kartal, W. J. Maalcke, N. M. de Almeida, I. Cirpus, J. Gloerich, W. Geerts, H. J. M. Op den Camp, H. R. Harhangi, E. M. Janssen-Megens, K.-J. Francoijs, H. G. Stunnenberg, J. T. Keltjens, M. S. M. Jetten, M. Strous, *Nature* **2011**, *479*,127–130.
15. C. Deutsch, J. L. Sarmiento, D. M. Sigman, N. Gruber, J. P. Dunne, *Nature* **2007**, *445*, 163–167.
16. D. G. Capone, A. N. Knapp, *Nature* **2007**, *445*, 159–160.
17. B. B. Ward, *Science* **2013**, *341*, 352–353.
18. D. J. Richardson, *Cell. Mol. Life Sci.* **2001**, *58*, 165–178.
19. J. Simon, *FEMS Microbiol. Rev.* **2002**, *26*, 285–309.
20. O. Einsle, P. M. H. Kroneck, *Biol. Chem.* **2004**, *385*, 875–883.
21. G. Fritz, O. Einsle, M. Rudolf, A. Schiffer, P. M. H. Kroneck, *J. Mol. Microbiol. Biotechnol.* **2005**, *10*, 223–233.
22. M. Kern, J. Simon, *Biochim. Biophys. Acta* **2009**, *1787*, 646–656.
23. J. Simon, M. Kern, B. Hermann, O. Einsle, J. N. Butt, *Biochem. Soc. Trans.* **2011**, *39*, 1864–1870.
24. J. Simon, M. G. Klotz, *Biochim. Biophys. Acta* **2013**, *1827*, 114–135.
25. J. Simon, P. M. H. Kroneck, *Adv. Microbial Physiol.* **2013**, *62*, 45–117.
26. J. E. Huheey, E. A. Keiter, R. L. Keiter, *Inorganic Chemistry: Principles of Structure and Reactivity*, 4th edn, HarperCollins College Publishers, 1993, pp. 405–408.

27. E. Housecroft, A. G. Sharpe, *Inorganic Chemistry*, 3rd edn, Pearson, Edinburgh Gate, Harlow, UK, 2008, pp. 433–455.
28. R. A. Alderden, M. D. Hall, T. W. Hambley, *J. Chem. Ed.* **2006**, *83*, 728–734.
29. S.N. Behera, M. Sharma, V. P. Aneja, R. Balasubramanian, *Environ. Sci. Pollut. Res.* **2013**, *20*, 8092–8131.
30. M. A. Sutton, S. Reis, S. N. Riddick, U. Dragosits, E. Nemitz, M. R.Theobald, Y. S. Tang, C. F. Braban, M. Vieno, A. J. Dore, R. F. Mitchell, S. Wanless, F. Daunt, D. Fowler, T. D. Blackall, C. Milford, C. R. Flechard, B. Loubet, R. Massad, P. Cellier, E. Personne, P. F. Coheur, L. Clarisse, M. Van Damme, Y. Ngadi, C. Clerbaux, C. A. Skjøth, C. Geels, O. Hertel, R.J. Wichink Kruit, R. W. Pinder, J. O. Bash, J. T. Walker, D. Simpson, L. Horváth, T. H. Misselbrook, A. Bleeker, F. Dentener, W. de Vries, *Phil. Trans. R. Soc. B* **2013**, *368*, 20130166; doi: 10.1098/rstb.2013.0166.
31. Ammonia Gas Monitoring Network (AMoN), within the US National Atmospheric Deposition Program (http://nadp.sws.uiuc.edu/AMoN/).
32. L. Myles, *Nat. Geosci.* **2009**, *2*, 461–462.
33. S. Singh, B. R. Bakshi, *Environ. Sci. Technol.* **2013**, *47*, 9388–9396.
34. M. Van Damme, L. Clarisse, C. L. Heald, D. Hurtmans, Y. Ngadi, C. Clerbaux, A. J. Dolman, J. W. Erisman, P. F. Coheur, *Atmos. Chem. Phys. Discuss.* **2013**, *13*, 24301–24342.
35. A. Bytnerowicz, P. E. Padgett, S. D. Parry, M. E. Fenn, M. J. Arbaugh, *The Scientific World* **2001**, *1(S2)*, 304–311.
36. B. Gu, J. Chang, Y. Min, Y. Ge, Q. Zhu, J. N. Galloway, C. Peng, *Scientific Reports* **2013**, *3*, 2579, 1–7, doi: 10.1038/srep02579.
37. R. K. Thauer, K. Jungermann, K. Decker, *Bacteriol. Rev.* **1977**, *41*, 100–180.
38. B. Kartal, N. M. de Almeida, W. J. Maalcke, H. J.M. Op den Camp, M. S. M. Jetten, J. T. Keltjens, *FEMS Microbiol. Rev.* **2013**, *37*, 428–461.
39. T. Fujita, *J. Biochem. (Tokyo)* **1966**, *60*, 204–215.
40. T. Brittain, R. Blackmore, C. Greenwood, A. J. Thomson, *Eur. J. Biochem.* **1992**, *209*, 793–802.
41. W. Schumacher, F. Neese, U. H. Hole, P. M. H. Kroneck, in *Transition Metals in Microbial Metabolism*, Eds G. Winkelmann, C. J. Carrano, Harwood Academic, Amsterdam, NL, 1997, pp. 329–356.
42. O. Einsle, *Meth. Enzymol.* **2011**, *496*, 399–422.
43. J. A. Cole, *FEMS Microbiol. Lett.* **1996**, *136*, 1–11.
44. A. Welsh, J. C. Chee-Sanford, L. M. Connor, F. E. Löffler, R. A. Sanford, *Appl. Environ. Microbiol.* **2014**, *80*, 2110–2119.
45. K. Heylen, J. Keltjens, *Front. Microbiol.* **2012**, *3*, article 371, 1–27, doi: 10.3389/fmicb.2012.00371.
46. D. Mania, K. Heylen, R. J. M. van Spanning, Å. Frostegard, *Environ. Microbiol.* **2014**, in press, doi: 10.1111/1462-2920.12478.
47. B. Strehlitz, B. Gründig, W. Schumacher, P. M. H. Kroneck, K.-D. Vorlop, H. Kotte, *Anal. Chem.* **1996**, *68*, 807–816.
48. J. Tan, J. A. Cowan, *Biochemistry* **1991**, *30*, 8910–8917.
49. M. Rudolf, O. Einsle, F. Neese, P. M. H. Kroneck, *Biochem. Soc. Trans.* **2002**, *30*, 649–653.
50. A. Darwin, H. Hussain, L. Griffiths, J. Grove, Y. Sambongi, S. Busby, J. Cole, *Mol. Microbiol.* **1993**, *9*, 1255–1265.
51. W. Schumacher, P. M. H. Kroneck, *Arch. Microbiol.* **1991**, *156*, 70–74.
52. M.-C. Liu, H. D. Peck, Jr., *J. Biol. Chem.* **1981**, *256*, 13159–13164.
53. M.-C. Liu, M.-Y. Liu, W. J. Payne, H. D. Peck, Jr., D. V. DerVartanian, *FEBS Lett.* **1987**, *218*, 227–230.
54. W. Schumacher, U. H. Hole, P. M. H. Kroneck, *Biochem. Biophys. Res. Commun.* **1994**, *205*, 911–916.
55. G. W. Pettigrew, G. R. Moore, *Cytochromes c. Biological Aspects*, Springer-Verlag, Berlin, Heidelberg, New York, London, Paris, Tokyo, 1987.
56. P. M. Wood, *Biochim. Biophys. Acta* **1984**, *768*, 293–317.

57. S. I. Adachi, S. Nagano, K. Ishimori, Y. Watanabe, I. Morishima, T. Egawa, T. Kitagawa, R. Makino, *Biochemistry* **1993**, *32*, 241–252.
58. O. Einsle, A. Messerschmidt, P. Stach, G. P. Bourenkov, H. D. Bartunik, R. Huber, P. M. H. Kroneck, *Nature* **1999**, *400*, 476–480.
59. O. Einsle, P. Stach, A. Messerschmidt, J. Simon, A. Kröger, R. Huber, P. M. H. Kroneck, *J. Biol. Chem.* **2000**, *275*, 39608–39616.
60. V. A. Bamford, H. C. Angove, H. E. Seward, A. J. Thomson, J. Cole, J. N. Butt, A. M. Hemmings, D. J. Richardson, *Biochemistry* **2002**, *41*, 2921–2931.
61. C. A. Cunha, S. Macieira, J. M. Dias, G. Almeida, L. L. Goncalves, C. Costa, J. Lampreia, R. Huber, J. J. G. Moura, I. Moura, M. J. Romao, *J. Biol. Chem.* **2003**, *278*, 17455–17465.
62. M. G. Almeida, S. Macieira, L. L. Goncalves, R. Huber, C. A. Cunha, M. J. Romao, C. Costa, J. Lampreia, J. J. G. Moura, I. Moura, *Eur. J. Biochem.* **1993**, *270*, 3904–3915.
63. M. L. Rodrigues, T. F. Oliveira, I. A. Pereira, M. Archer, *EMBO J.* **2006**, *25*, 5951–5960.
64. M. Youngblut, E. T. Judd, V. Srajer, B. Sayyed, T. Goelzer, S. J. Elliot, M. Schmidt, A. A. Pacheco, *J. Biol. Inorg. Chem.* **2012**, *17*, 647–662.
65. M. Kern, F. Eisel, J. Scheithauer, R. G. Kranz, J. Simon, *Mol. Microbiol.* **2010**, *75*, 122–137.
66. O. Einsle, A. Messerschmidt, R. Huber, P.M.H. Kroneck, F. Neese, *J. Am. Chem. Soc.* **2002**, *124*, 11737–11745.
67. D. Bykov, F. Neese, *J. Biol. Inorg. Chem.* **2011**, *16*, 417–430.
68. D. Bykov, F. Neese, *J. Biol. Inorg. Chem.* **2012**, *17*, 741–760.
69. D. Bykov, M. Plog, F. Neese, *J. Biol. Inorg. Chem.* **2014**, *19*, 97–112.
70. P. Stach, O. Einsle, W. Schumacher, E. Kurun, P. M. H. Kroneck, *J. Inorg. Biochem.* **2000**, *79*, 381–385.
71. T. A. Clarke, A. Hemmings, B. Burlat, J. N. Butt, J. A. Cole, D. J. Richardson, *Biochem. Soc. Trans.* **2006**, *34*,143– 145.
72. P. Lukat, R. Rudolf, P. Stach, A. Messerschmidt, P. M. H. Kroneck, J. Simon, O. Einsle, *Biochemistry* **2008**, *47*, 2080–2086.
73. J. Simon, R. Gross, O. Einsle, P. M. H. Kroneck, A. Kröger, O. Klimmek, *Mol. Microbiol.* **2000**, *35*, 686–696.
74. J. Simon, R. Pisa, T. Stein, R. Eichler, O. Klimmek, R. Gross, *Eur. J. Biochem.* **2001**, *268*, 5776–5782.
75. R. Gross, R. Eichler, J. Simon, *Biochem. J.* **2005**, *390*, 689–693.
76. J. Simon, in *Nitrogen Cycling in Bacteria. Molecular Analysis*, Ed J. W. B. Moir, Caister Academic Press, Norfolk, UK, 2011, pp. 39–58
77. J. Simon, R. J. M. van Spanning, D. J. Richardson, *Biochim. Biophys. Acta* **2008**, *1777*, 1480–1490.
78. H. Hussain, J. Grove, L. Griffiths, S. Busby, J. Cole, *Mol. Microbiol.* **1994**, *12*, 153–163.
79. B. C. Berks, S. J. Ferguson, J. W. B. Moir, D. J. Richardson, *Biochim. Biophys. Acta* **1995**, *1232*, 97–173.
80. J. Simon, M. Kern, *Biochem. Soc. Trans.* **2008**, *36*, 1011–1016.
81. M. Jormakka, K. Yokoyama, T. Yano, M. Tamakoshi, S. Akimoto, T. Shimamura, P. Curmi, S. Iwata, *Nat. Struct. Mol. Biol.* **2008**, *15*, 730–737.
82. T. A. Clarke, J. A. Cole, D. J. Richardson, A. M. Hemmings, *Biochem. J.* **2007**, *406*, 19–30.
83. F. Grein, A. R. Ramos, S. S. Venceslau, I. A. C. Pereira, *Biochim. Biophys. Acta* **2013**, *1827*, 145–160.
84. R. G. Kranz, C. Richard-Fogal, J. S. Taylor, E. R. Frawley, *Microbiol. Mol. Biol. Rev.* **2009**, *73*, 510–528.
85. D. J. Eaves, J. Grove, W. Staudenmann, P. James, R. K. Poole, S. A. White, I. Griffiths, J. A. Cole, *Mol. Microbiol.* **1998**, *28*, 205–216.
86. R. Pisa, T. Stein, R. Eichler, R. Gross, J. Simon, *Mol. Microbiol.* **2002**, *43*, 763–770.
87. M. Kern, J. Scheithauer, R. G. Kranz, J. Simon, *Microbiology* **2010**, *156*, 3773–3781.
88. J. M. Stevens, D. A. Mavridou, R. Hamer, P. Kritsiligkou, A. D. Goddard, S. J. Ferguson, *FEBS J.* **2011**, *278*, 4170–4178.
89. J. Simon, L. Hederstedt, *FEBS J.* **2011**, *278*, 4179–4188.

90. S. R. Poock, E. R. Leach, J. W. B. Moir, J. A. Cole, D. J. Richardson, *J. Biol. Chem.* **2002**, *277*, 23664–23669.
91. P. C. Mills, G. Rowley, S. Spiro, J. C. D. Hinton, D. J. Richardson, *Microbiology* **2008**, *154*, 1218–1228.
92. M. Kern, J. Volz, J. Simon, *Environ. Microbiol.* **2011**, *13*, 2478–2494.
93. R. K. Poole, *Biochem. Soc. Trans.* **2005**, *33*, 176–180.
94. T. V. Tikhonova, A. Slutsky, A. N. Antipov, K. M. Boyko, K. M. Polyakov, D. Y. Sorokin, R. A. Zvyagilskaya, V. O. Popov, *Biochim. Biophys. Acta* **2006**, *1764*, 715–723.
95. M. Kern, M.G. Klotz, J. Simon, *Mol. Microbiol.* **2011**, *82*,1515–1530.
96. C. G. Mowat, E. Rothery, C. S. Miles, L. McIver, M. K. Doherty, K. Drewette, P. Taylor, M. D. Walkinshaw, S. K. Chapman, G. A. Reid, *Nat. Struct. Mol. Biol.* **2004**, *11*, 1023–1024.
97. S. J. Atkinson, C. G. Mowat, G. A. Reid, S. K. Chapman, *FEBS Lett.* **2007**, *581*, 3805–3808.
98. K. M. Polyakov, K. M. Boyko, T. V. Tikhonova, A. Slutsky, A. N. Antipov, R. A. Zvyagilskaya, A. N. Popov, G. P Bourenkov, V. S. Lamzin, V. O. Popov, *J. Mol. Biol.* **2009**, *389*, 846–862
99. T. V. Tikhonova, A. A. Trofimov, V. O. Popov, *Biochemistry (Moscow)* **2012**, *77*, 1129–1138.
100. T. V. Tikhonova, A. Tikhonov, A. Trofimov, K. M. Polyakov, K. M. Boyko, E. Cherkashin, T. Rakitina, D. Y. Sorokin, V. O. Popov, *FEBS J.* **2012**, *279*, 4052–4061.
101. N. Igarashi, H. Moriyama, T. Fujiwara, Y. Fukumori, N. Tanaka, *Nat. Struct. Biol.* **1997**, *4*, 276–284.
102. J. Kostera, M. D. Youngblut, J. M. Slosarczyk, A. A. Pacheco, *J. Biol. Inorg. Chem.* **2008**, *13*, 1073–1083
103. J. Kostera, J. McGarry, A. A. Pacheco, *Biochemistry* **2010**, *49*, 8546–8553.
104. R. Schnell, T. Sandalova, U. Hellman, Y. Lindqvist, G. Schneider, *J. Biol. Chem.* **2005**, *280*, 27319–27328.
105. S. B. Mohan, M. Schmid, M. S. M. Jetten, J. Cole, *FEMS Microbiol. Ecol.* **2004**, *49*, 433–443.
106. J. W. Erisman, A. Bleeker, J. Galloway, M. S. Sutton, *Environ. Pollut.* **2007**, *150*, 140–149.
107. D. Fowler, M. Coyle, U. Skiba, M. A. Sutton, J. N. Cape, S. Reis, L. J. Sheppard, A. Jenkins, B. Grizzetti, J. N. Galloway, P. Vitousek, A. Leach, A. F. Bouwman, K. Butterbach–Bahl, F. Dentener, D. Stevenson, M. Amann, M. Voss, *Phil. Trans. R. Soc. B* **2013**, *368*, 20130164; doi: 10.1098/rstb.2013.0164.
108. M. Voss, H. W. Bange, J. W. Dippner, J. J. Middelburg, J. P. Montoya, B. Ward, *Phil. Trans. R. Soc. B* **2013**, *368*, 20130121; doi: 10.1098/rstb.2013.0121.
109. M. Giles, N. Morley, E. M. Baggs, T. J. Daniell, *Front. Microbiol.* **2012**, *3*, article 407, 1–16.
110. G. Rowley, D. Hensen, H. Felgate, A. Arkenberg, C. Appia-Ayme, K. Prior, C. Harrington, S. Field, J. N. Butt, D. J. Richardson, *Biochem. J.* **2012**, *441*, 755–762.
111. M. A. Streminska, H. Felgate, G. Rowley, D. J. Richardson, E. M. Baggs, *Environ. Microbiol. Rep.* **2012**, *4*, 66–71.
112. M. Luckmann, D. Mania, M. Kern, L. R. Bakken, Å. Frostegård, J. Simon, *Microbiology* **2014**, *160*, 1749–1759.

Chapter 10
Hydrogen Sulfide: A Toxic Gas Produced by Dissimilatory Sulfate and Sulfur Reduction and Consumed by Microbial Oxidation

Larry L. Barton, Marie-Laure Fardeau, and Guy D. Fauque

Contents

ABSTRACT 238
1 INTRODUCTION 238
   1.1 Overview of Bacteria and Archaea Associated
        with Hydrogen Sulfide Metabolism 240
        1.1.1 Sulfate-Reducing Bacteria and Archaea 240
        1.1.2 Sulfur-Reducing Bacteria and Archaea 242
        1.1.3 Sulfur-, Sulfite-, and Thiosulfate-Disproportionating Bacteria 245
        1.1.4 Sulfide-, Sulfur-, Sulfite-, and Thiosulfate-Oxidizing Bacteria 245
   1.2 Properties and Toxicity of Hydrogen Sulfide 249
   1.3 Effects of Hydrogen Sulfide on Gene Expression
        and Physiology of Desulfovibrio vulgaris Hildenborough 250
2 ENZYMOLOGY OF HYDROGEN SULFIDE PRODUCTION
   FROM SULFATE 250
   2.1 Enzymology of Dissimilatory Sulfate Reduction 250
   2.2 ATP Sulfurylase 251
   2.3 Dissimilatory Adenylylsulfate Reductase 253
   2.4 Sulfite Reductases 254
        2.4.1 Dissimilatory High-Spin Sulfite Reductase 254
        2.4.2 Oxy-Sulfur Reductases in Non-sulfate Reducers 260
        2.4.3 Low-Molecular-Mass and Low-Spin Assimilatory-Type Sulfite
               Reductase from Desulfovibrio vulgaris H, Desulfuromonas
               acetoxidans, and Methanosarcina barkeri 261
3 ENZYMOLOGY OF HYDROGEN SULFIDE PRODUCTION
   FROM ELEMENTAL SULFUR 262

L.L. Barton
Department of Biology, University of New Mexico, MSCO3 2020, Albuquerque, NM, USA
e-mail: lbarton@unm.edu

M.-L. Fardeau • G.D. Fauque (✉)
Institut Méditerranéen d'Océanologie (MIO), Aix-Marseille Université, USTV, UMR CNRS
7294/IRD 235, Campus de Luminy, Case 901, F-13288 Marseille Cedex 09, France
e-mail: marie-laure.fardeau@univ-amu.fr; guy.fauque@univ-amu.fr

© Springer Science+Business Media Dordrecht 2014
P.M.H. Kroneck, M.E. Sosa Torres (eds.), *The Metal-Driven Biogeochemistry
of Gaseous Compounds in the Environment*, Metal Ions in Life Sciences 14,
DOI 10.1007/978-94-017-9269-1_10

   3.1   Eubacterial Sulfur Reductase 262
         3.1.1   Sulfur Reductase in *Desulfovibrio* and *Desulfomicrobium* Species 262
         3.1.2   Polysulfide Reductase from *Wolinella succinogenes* 263
         3.1.3   Polysulfide Reductase from *Desulfuromonas acetoxidans* 263
         3.1.4   Sulfur Oxidoreductase from *Sulfurospirillum deleyianum* 265
   3.2   Archaebacterial Sulfur Reductase 265
         3.2.1   Membraneous Sulfur Reductase Complex from *Acidianus ambivalens* .. 265
         3.2.2   Sulfur-Reducing Complex from *Pyrodictium abyssi* 265
         3.2.3   Sulfur Reductase from *Pyrococcus furiosus* 266
4   MICROBIAL OXIDATION OF HYDROGEN SULFIDE TO SULFATE 266
   4.1   Archaebacterial Inorganic Sulfur Compound Oxidation 266
   4.2   Eubacterial Inorganic Sulfur Compound Oxidation 267
         4.2.1   Oxidation of Sulfide 268
         4.2.2   Oxidation of Polysulfides 268
         4.2.3   Oxidation of Stored Sulfur to Sulfite 269
         4.2.4   Oxidation of Sulfite to Sulfate 269
         4.2.5   Oxidation of Thiosulfate 269
5   CONCLUSIONS 270
ABBREVIATIONS AND DEFINITIONS 271
ACKNOWLEDGMENTS 272
REFERENCES 272

Abstract Sulfur is an essential element for the synthesis of cysteine, methionine, and other organo-sulfur compounds needed by living organisms. Additionally, some prokaryotes are capable of exploiting oxidation or reduction of inorganic sulfur compounds to energize cellular growth. Several anaerobic genera of Bacteria and Archaea produce hydrogen sulfide (H_2S), as a result of using sulfate (SO_4^{2-}), elemental sulfur (S^0), thiosulfate ($S_2O_3^{2-}$), and tetrathionate ($S_4O_6^{2-}$) as terminal electron acceptors. Some phototrophic and aerobic sulfur bacteria are capable of using electrons from oxidation of sulfide to support chemolithotrophic growth. For the most part, biosulfur reduction or oxidation requires unique enzymatic activities with metal cofactors participating in electron transfer. This review provides an examination of cytochromes, iron-sulfur proteins, and sirohemes participating in electron movement in diverse groups of sulfate-reducing, sulfur-reducing, and sulfide-oxidizing Bacteria and Archaea.

Keywords hydrogen sulfide production • sulfate reduction • sulfide oxidation • sulfite reduction • sulfur cycle

Please cite as: *Met. Ions Life Sci.* 14 (2014) 237–277

1   Introduction

Sulfur is one of the most versatile elements in life due to its reactivity in different reduction and oxidation states. Sulfur is the element with the highest number of allotropes (about 30), but only a few are found in nature and occur in biological

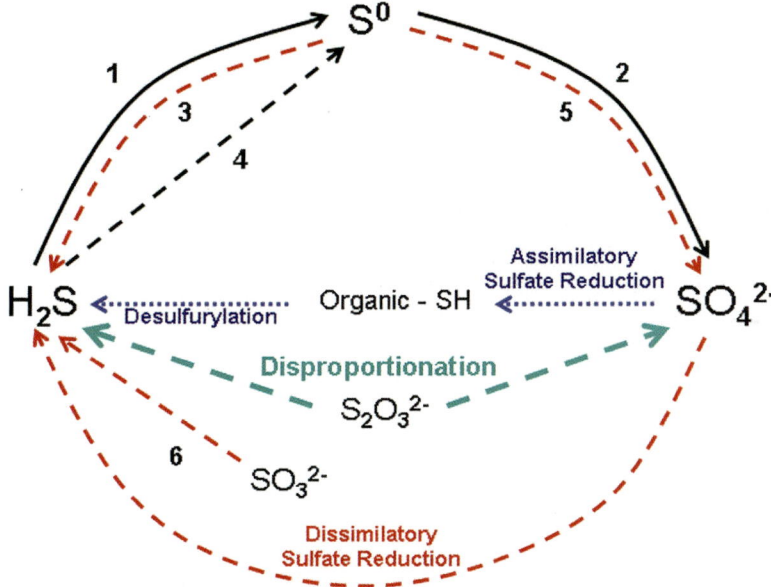

Figure 1 The biological sulfur cycle with roles of bacteria identified. Solid lines indicate aerobic reactions, dashed lines indicate anaerobic reactions, and dotted lines indicate both aerobic and anaerobic activity. **Desulfurylation** by many aerobic and anaerobic prokaryotes, **assimilatory sulfate reduction** by many aerobic and anaerobic microorganisms, **dissimilatory sulfate reduction** by anaerobic organisms listed in Table 1 of this chapter and in Table 1 of Ref. [25], and **disproportionation** of thiosulfate by *Desulfovibrio* and *Desulfocapsa*. **1** and **2**: Sulfide and sulfur oxidation by colorless sulfur bacteria. **3**: Sulfur reduction by the anaerobic microorganisms listed in Table 1 of this chapter. **4** and **5**: Anaerobic sulfide and sulfur oxidation by purple sulfur bacteria and green sulfur bacteria. **6**: Sulfite-reducing bacteria.

systems. The inorganic sulfur compounds of biological relevance which occur in the biological sulfur cycle are elemental sulfur, sulfate, sulfite, thiosulfate, polythionates, sulfide, and polysulfides (Figure 1).

Sulfur can adopt many oxidation states, ranging from −2 to +6. Inorganic sulfur compounds of intermediate oxidation states can serve as electron acceptors or donors in redox processes. In contrast, sulfate and sulfide cannot be further oxidized or reduced, respectively, and they are therefore the final products of most sulfur oxidation or reduction pathways. The biological roles of inorganic sulfur compounds are rather restricted: either they serve as acceptors or donors of electrons for dissimilatory energy-generating electron transport (almost exclusively among prokaryotes), or they are employed as sources for sulfur assimilation, very common in prokaryotes as well as in algae, fungi, and plants.

Despite its toxicity (5-fold higher than CO), H_2S is a fundamental molecule in both anaerobic and aerobic organisms. Since the first description of hydrogen sulfide toxicity by Ramazzini in 1713, most studies about H_2S have been devoted to its toxic effects with little attention paid to its physiological function [1]. The liberation of H_2S is controlled not only by the rate of its production by sulfate- and sulfur-reducing prokaryotes, but also by its tendency to rapidly precipitate as metal sulfides, its pH-dependent speciation, and its fast biological and chemical oxidation. Only the protonated compound hydrogen sulfide (H_2S) is volatile, whereas sulfide (S^{2-}) dominates under alkaline conditions, and at neutral pH, most of inorganic sulfide is present as bisulfide anion (systematically named sulfanide, HS^-).

1.1 Overview of Bacteria and Archaea Associated with Hydrogen Sulfide Metabolism

1.1.1 Sulfate-Reducing Bacteria and Archaea

Dissimilatory sulfate reduction (also known as anaerobic sulfate respiration) is an essential step in the global sulfur cycle and is exclusively mediated by the sulfate-reducing prokaryotes (SRP), a physiologically and phylogenetically versatile group of microorganisms [2, 3]. SRP are of major functional and numerical importance in many ecosystems and they can grow under different physico-chemical conditions. SRP are found in almost all ordinary environments on this planet: they are present in geothermal areas and hot springs, soils, fresh, marine, brackish, and artesian waters, estuarine muds, cyanobacterial microbial mats, oil and natural gas wells, anaerobic sludge, digestive tracts of humans and animals [3–7]. Dissimilatory sulfate reduction has evolved approximately 3.47 billion years ago [8] and sulfate-reducing bacteria (SRB) should be considered as ancestral microorganisms, which have contributed to the primordial biogeochemical cycle for sulfur as soon as life emerged on the planet [9]. SRB contribute to the complete oxidation of organic matter and participate through metal reduction and sulfide production to the overall biogeochemistry of these extreme environments [2, 10].

As of 2012, 65 genera containing 250 species of SRP have been isolated and characterized [2]. They belong to five phyla within the Bacteria [the *Deltaproteobacteria* (the most frequently represented lineage among SRB), the spore-forming *Desulfovirgula*, *Desulfotomaculum*, *Desulfurispora*, *Desulfosporomusa*, and *Desulfosporosinus* species within the phylum *Firmicutes*, the *Thermodesulfovibrio* species within the phylum *Nitrospirae* and two phyla represented by *Thermodesulfobium narugense* and *Thermodesulfobacterium/Thermodesulfatator* species], and two divisions within the Archaea [the euryarchaeotal genus *Archaeoglobus* and the three crenarchaeotal genera *Vulcanisaeta*, *Thermocladium*, and *Caldivirga*, affiliated with the *Thermoproteales*] (Table 1) [3, 6, 7, 11–13].

As of June 2013, a total of 101 genomes of SRP were available at the Integrated Microbial Genomes website including 36 *Desulfovibrio* species, 10 *Desulfotomaculum* species, 8 *Desulfobulbus* species, 5 *Desulfosporosinus* species and 4 *Archaeoglobus* species.

SRP may have a heterotrophic, autotrophic, lithoautotrophic, or respiration-type of life under anaerobiosis and their possible microaerophilic nature has also been reported [2, 3, 10, 14]. More than one hundred compounds including H_2, sugars, amino acids, mono- and dicarboxylic acids, alcohols, and aromatic compounds are

Table 1 Genera of prokaryotes displaying dissimilatory sulfate and sulfur reduction (list incomplete).

	Sulfate Reducers		Sulfur Reducers
Archaea		Archaea	
	Archaeoglobus fulgidus		*Acidianus ambivalens*
	Caldivirga		*Acidilobus*
	Vulcanisaeta		*Caldisphaera*
			Caldivirga
Bacteria	*Ammonifex*		*Caldococcus*
	Candidatus desulforudis		*Desulfolobus*
	Desulfacinum		*Desulfurococcus*
	Desulfobacter		*Hyperthermus*
	Desulfobacterium autotrophicum		*Methanobacterium*
	Desulfobulbus		*Methanococcus*
	Desulfocapsa		*Methanosarcina barkeri*
	Desulfococcus		*Pyrobaculum*
	Desulfocurvus		*Pyrococcus furiosus*
	Desulfofustis		*Pyrodictium abyssi*
	Desulfohalobium		*Staphylothermus*
	Desulfoluna		*Stetteria*
	Desulfomicrobium norvegicum		*Stygiolobus*
	Desulfonatronovibrio		*Thermococcus*
	Desulfosarcina		
	Desulfosporosinus	Bacteria	
	Desulfotomaculum acetoxidans		*Campylobacter*
	Desulfovibrio gigas		*Desulfomicrobium*[a]
	Desulfovibrio vulgaris **H**		*Desulfotomaculum*[a]
	Desulfovirga		*Desulfovibrio*[a]
	Syntrophobacter		*Desulfurella*
	Thermodesulfatator		*Desulfurobacterium*
	Thermodesulfobacterium commune		*Desulfuromonas acetoxidans*
	Thermodesulfobium		*Salmonella*
	Thermodesulfovibrio		*Sulfurospirillum deleyianum*
			Shewanella
			Wolinella succinogenes

Model organisms and/or major species studied are written in bold.
[a]Only some thiophilic species are able to reduce elemental sulfur.

potential electron donors for SRP [2, 3, 15]. SRP are the anaerobic microorganisms that reduce the greatest number of different terminal electron acceptors including inorganic sulfur compounds and various other organic and inorganic substrates [3, 4, 6, 10, 15–17]. The contribution of SRP to the total carbon mineralization process in marine sediments, where sulfate is not limiting, was estimated to be up to 50 % [3, 4].

The most extensive biochemical and physiological studies have been done with SRB of the genus *Desulfovibrio* which are the most rapidly and easily cultivated sulfate reducers. Dissimilatory sulfate reduction in *Desulfovibrio* species is linked to electron transport-coupled phosphorylation because substrate level phosphorylation is inadequate to support their growth [18]. The SRB belonging to the genus *Desulfovibrio* possess a number of unique physiological and biochemical characteristics such as the requirement for ATP to reduce sulfate [18], the cytoplasmic localization of two key enzymes [adenosine 5'-phosphosulfate (APS) reductase and dissimilatory-type sulfite reductase] involved in the pathway of dissimilatory sulfate reduction [19, 20], the periplasmic localization of some hydrogenases [21, 22], and the abundance of multiheme *c*-type cytochromes [15, 16, 23].

A set of unique membrane-bound respiratory complexes are involved in sulfate respiration. The Dsr (dissimilatory sulfite reductase) MKJOP and the Qmo (quinone-interacting membrane oxidoreductase) ABC complexes are present in all SRP and are deemed essential for dissimilatory sulfate reduction [24, 25]. A group of other complexes (Hmc, high-molecular weight cytochrome; Nhc, nineheme cytochrome; Ohc, octaheme cytochrome; Qrc, quinone reductase complex; Tmc, tetraheme membrane cytochrome) are present only in sulfate reducers that are characterized by a high content of multiheme cytochromes *c* (mainly the deltaproteobacterial SRB) [24]. A model reflecting organization of membrane-bound respiratory complexes associated with electron transport and cell energetics in *Desulfovibrio* species is given in Figure 2 [25].

1.1.2   Sulfur-Reducing Bacteria and Archaea

Compared with SRP, the environmental distribution of elemental sulfur reducers, and their quantitative role in carbon cycling, is poorly understood. The mechanism of elemental sulfur reduction (characterization of enzymes and electron carriers) has been much less studied than that of dissimilatory sulfate reduction [25–30].

1.1.2.1   Eubacterial Sulfur Reduction

Elemental sulfur is probably the most widespread sulfur species in sediments and geological deposits. Many biological and chemical oxidation processes of H_2S do not directly produce sulfate but rather elemental sulfur, which may accumulate [3]. Elemental sulfur is relatively reactive and in contrast to sulfate, it requires no energy-dependent activation before a reduction can take place. The problem in the utilization of elemental sulfur mainly concerns the low solubility of sulfur flower in water (0.16 μmole per liter at 25 °C) [31]. The so-called "hydrophilic sulfur" is probably the form available in aqueous medium; it consists of elemental sulfur associated with small portions of oxocompounds such as polythionates.

Several genera of the domain Bacteria are able to grow by a dissimilatory reduction of elemental sulfur to sulfide in a respiratory type of metabolism [2, 3, 15, 27, 29, 30, 32–36]. The facultative sulfur-reducing eubacteria, such as the SRB, utilize elemental (or colloidal) sulfur as a respiratory substrate in the absence of

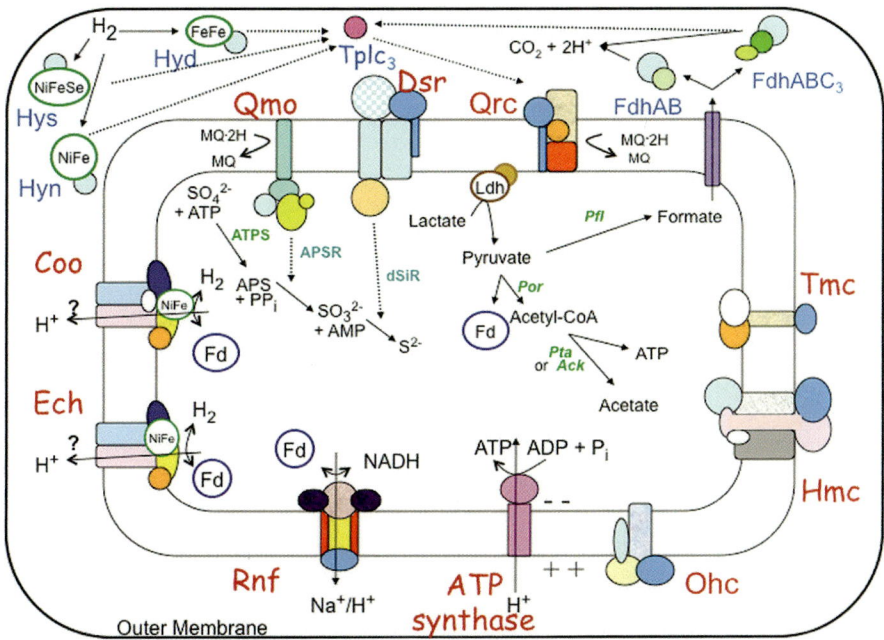

Figure 2 Model reflecting the organization of protein complexes associated with electron transport and cell energetics in sulfate-reducing bacteria. Abbreviations are as follows: Qmo = QmoABC complex, Qrc = QrcABCD complex, Dsr = DsrKMJOP complex, Tmc = TmcABCD complex, Hmc = HmcABCDEF complex, ATP synthase = proton-driven ATP synthase, Coo = carbon monoxide dehydrogenase-hydrogenase membrane complex system, Ech = multi-subunit membrane-bound hydrogenase, Fd = ferredoxin, Fdh = formate dehydrogenase, Hyd = periplasmic [Fe] hydrogenase, Hyn = periplasmic [NiFe] hydrogenase, Hys = periplasmic [NiFeSe] hydrogenase, Ohc = octaheme cytochrome c membrane complex, Rnf = NADH:quinone oxidoreductase membrane complex, Tplc$_3$ = periplasmic type I cytochrome c_3. Cytoplasmic enzymes are as follows: Ack = acetate kinase, APSR = adenylylsulfate reductase, ATPS = ATP sulfurylase, dSiR = dissimilatory sulfite reductase, Ldh = lactate dehydrogenase, Pfl = pyruvate formate lyase, Por = pyruvate:ferredoxin oxidoreductase, and Pta phosphotransacetylase. Reproduced by permission from [25]; copyright 2012, Academic Press.

other possible terminal electron acceptors such as sulfate, sulfite, thiosulfate, nitrate or nitrite. The growth of many species of SRB in the presence of sulfate is inhibited by elemental sulfur, probably because sulfur as an oxidant shifts the potential of redox couples in the medium and cells to unfavorable and positive values [3, 30]. Even if most of the SRB are not able to grow by dissimilatory elemental sulfur reduction, some thiophilic species of SRB, belonging to fifteen genera (such as *Desulfomicrobium*, *Desulfovibrio*, *Desulfohalobium*, *Desulfosporosinus*, *Desulfotomaculum*) can use elemental sulfur as an alternative electron acceptor (Table 1) [25, 37–39]. The eubacterial sulfur reducers comprise both facultative and true (or strict) respiratory microorganims. Sulfur reducers use many different types of metabolic systems for oxidizing organic compounds. Both complete and incomplete oxidation of organic electron donors has been reported for sulfur-reducing eubacteria.

The sulfur reductase (SR) (EC 1.12.98.4-sulfhydrogenase, formerly EC 1.97.1.3-sulfur reductase) is a constitutive enzyme in sulfur-reducing eubacteria: *Desulfuromonas (Drm.) acetoxidans, Wolinella (W.) succinogenes, Sulfuros-pirillum (S.) deleyianum, Desulfomicrobium (Dsm.)*, and *Desulfovibrio (D.)* species [26, 27]. From genome organization, it appears that a multisubunit polysulfide reductase (PSR) (α, β, and γ) is found in 8 Bacteria genera and 3 Archaea genera [25].

1.1.2.2   Archaeal Sulfur Reduction

Many genera of Archaea are able to grow with elemental sulfur as terminal electron acceptor in the energy metabolism [40]. The dissimilatory reduction of elemental sulfur to hydrogen sulfide is linked with energy conservation as evidenced by growth on H_2 and S^0 ($E^{0'}S^0/SH^- = -270$ mV) [41]. SR is also a constitutive enzyme in the Archaea: *Methanosarcina (Ms.) barkeri* 227, *Methanococcus thermolithotrophicus*, and *Methanobacterium thermoautotrophicum* Marburg [30].

All archaeal sulfur reducers are extremely thermophilic, whereas sulfur-reducing eubacteria may be mesophilic or moderately thermophilic. The reduction of S^0 is widespread among members of the Archaea, including deep-branching hyperthermophilic genera. In the *Euryarchaeota*, sulfur reduction is present in the orders *Thermoplasmatales, Thermococcales*, and many methanogens; in the *Crenarchaeota*, sulfur reduction is found in the orders *Desulfurococcales, Sulfolobales*, and *Thermoproteales* [29, 30].

Four mechanisms of S^0 reduction are known in Archaea: (a) The most wide-spread metabolism consists in the facultative or obligate chemolitho-autotrophic reduction of S^0 with H_2, accomplished by many hyperthermophiles from the *Crenarchaeota*, including members of the genera *Thermoproteus, Sulfolobus, Stygiolobus, Pyrobaculum, Ignicoccus, Acidianus, Thermoplasma* (Table 1) [28, 30]. (b) Some members of the Archaea including representatives from the genera *Thermococcus, Thermodiscus, Hyperthermus, Stetteria, Thermocladium, Pyrodictium, Pyrococcus*, and *Desulfurococcus* utilize S^0 reduction as a H_2 sink during fermentative metabolism (Table 1) [29, 30]. (c) Some members of the order *Thermoproteales,* and *Pyrobaculum (Pyb.) islandicum*, can respire heterotrophi-cally with S^0 in an apparently energy-gaining metabolism [30, 42]. (d) Some hydrogen-oxidizing methanogenic Archaea like *Methanococcus, Methanosarcina, Methanobacterium, Methanothermus,* and *Methanopyrus* can also reduce S^0 with H_2 or methanol [43].

1.1.3 Sulfur-, Sulfite-, and Thiosulfate-Disproportionating Bacteria

The disproportionation (or dismutation) of inorganic sulfur intermediates (also called "inorganic sulfur compound fermentation") at moderate temperatures (0–80 °C) consists of a microbiologically catalyzed chemolithotrophic process in which compounds such as sulfite, thiosulfate, and S^0 serve as both electron donor and acceptor to produce sulfate plus sulfide [44]. Reactions involving disproportionation of thiosulfate, sulfite, and S^0 are listed in equations (1), (2), and (3), respectively.

$$S_2O_3^{2-} + H_2O \rightarrow SO_4^{2-} + HS^- + H^+ \qquad \Delta G^{0'} = -21.9 \text{ kJ mol}^{-1} S_2O_3^{2-} [41] \tag{1}$$

$$4 SO_3^{2-} + H^+ \rightarrow 3 SO_4^{2-} + HS^- \qquad \Delta G^{0'} = -58.9 \text{ kJ mol}^{-1} SO_3^{2-} [41] \tag{2}$$

$$4 S^0 + 4 H_2O \rightarrow SO_4^{2-} + 3 HS^- + 5 H^+ \quad \Delta G^{0'} = +10.2 \text{ kJ mol}^{-1} S^0 [41] \tag{3}$$

Desulfovibrio sulfodismutans DSM 3696 was the first bacterium isolated able to carry out the disproportionation of sulfite and thiosulfate to sulfide plus sulfate [45]. *Desulfocapsa sulfoexigens* was the first microorganism able to disproportionate S^0 to sulfate and sulfide [46]. The disproportionation of sulfur compounds is associated with only small free energy changes [47]. Only three *Desulfocapsa* species are able to disproportionate S^0, sulfite, and thiosulfate with growth.

The capacities of dissimilatory sulfate reduction and of sulfite and thiosulfate disproportionation are constitutively present in *D. desulfuricans* CSN (DSM 9104) and *D. sulfodismutans*. In contrast, ATP sulfurylase and sulfite oxidoreductase activities were not detected in these last two strains. During sulfite and thiosulfate dismutation, sulfate is formed via APS reductase and ATP sulfurylase, but not by sulfite oxidoreductase [47]. Elemental sulfur-disproportionating bacteria can be traced back in time as long as 3.5 billion years indicating that elemental sulfur dismutation would be one of the oldest biological processes on Earth [44, 48].

1.1.4 Sulfide-, Sulfur-, Sulfite-, and Thiosulfate-Oxidizing Bacteria

Biological oxidation of hydrogen sulfide to sulfate is one of the major reactions of the biological sulfur cycle (Figure 1). Chemotrophic sulfur-oxidizing bacteria (SOB) are found in four classes of the Proteobacteria (Alphaproteobacteria, Betaproteobacteria, Gammaproteobacteria, and Epsilonproteobacteria) whereas anoxygenic phototrophic sulfur bacteria are only present in Gammaproteobacteria [49].

1.1.4.1   Anoxygenic Phototrophic Bacteria

Reduced inorganic sulfur compounds play an important role as electron donors for photosynthetic carbon dioxide fixation in anoxygenic phototrophic sulfur bacteria. Four major phylogenetic groups of anoxygenic phototrophic bacteria can be distinguished: (a) The green sulfur bacteria (GSB) (family *Chlorobiaceae*), (b) the purple sulfur (PSB) and purple non-sulfur bacteria (PNSB), (c) the Gram-positive Heliobacteria, (d) the filamentous and gliding green bacteria (*Chloroflexaceae*) [50, 51].

Dissimilatory sulfur metabolism (i.e., the use of inorganic sulfur compounds, such as sulfide, elemental sulfur, polysulfides, sulfite, or thiosulfate, as sources or sinks of electrons) has been mainly investigated in the PSB of the families *Ectothiorhodospiraceae* and *Chromatiaceae*. Many of the PNSB can also utilize inorganic sulfur compounds as a source of electrons. The GSB of the family *Chlorobiaceae*, some cyanobacteria, and some members of the filamentous anoxygenic phototrophic bacteria (family *Chloroflexaceae*) are also able to grow phototrophically using reduced sulfur compounds as electron donors [50, 51].

Anoxygenic phototrophic PSB constitute a major group of bacteria widely distributed in nature, primarily in aquatic environments. The two families of PSB, the *Ectothiorhodospiraceae* and *Chromatiaceae*, respectively, produce external and internal sulfur granules. Typical habitats of PSB of the family *Chromatiaceae* are freshwater lakes and intertidal sandflats. The family *Ectothiorhodospiraceae* is found mainly in hypersaline waters. Many species of PSB are "extremophilic", growing at high salt and/or pH. PNSB have been isolated from almost every environment, including freshwater, marine systems, soils, plants, and activated sludge. PSB (more than 30 genera) consist of a variety of morphological types and belong to the Gammaproteobacteria (order *Chromatiales*) (Table 2). GSB are found in various types of aquatic habitats such as the pelagial of lagoons or lakes, bacterial mats in hot springs, or bottom layers of bacterial mats in intertidal sediments.

More than twenty genera of PNSB are now recognized (Table 2). PNSB constitute a physiologically versatile group of purple bacteria that can grow well both in darkness and phototrophically.

1.1.4.2   Colorless Sulfur Bacteria

The name "colorless sulfur bacteria" (CSB) has been utilized since the time of Winogradsky [49] to designate microorganisms able to use reduced sulfur compounds (e.g., sulfide, elemental sulfur, and organic sulfides) as sources of energy for growth. The adjective "colorless" is utilized because of the lack of photopigments in these organisms, although colonies and dense cultures of these bacteria could actually be brown or pink due to their high cytochrome content [49]. CSB play an essential role in the oxidative side of the biological sulfur cycle (Figure 1).

Table 2 Anoxygenic phototrophic sulfur bacteria using reduced inorganic sulfur compounds as electron donors (list incomplete).

Purple Sulfur Bacteria	Green Sulfur Bacteria	Purple Non-Sulfur Bacteria
Allochromatium vinosum	*Ancalochloris*	*Blastochloris*
Amoebobacter	***Chlorobaculum tepidum***	*Phaeovibrio*
Chromatium	*Chlorobium*	*Rhodobaca*
Ectothiorhodosinus	*Chloroherpeton*	***Rhodobacter capsulatus***
Ectothiorhodospira	*Clathrochloris*	*Rhodobium*
Halochromatium	*Pelodictyon*	*Rhodomicrobium*
Halorhodospira halophila	*Prosthecochloris*	*Rhodopila*
Isochromatium		*Rhodoplanes*
Lamprobacter		***Rhodopseudomonas palustris***
Lamprocystis		*Rhodospira*
Marichromatium		***Rhodospirillum rubrum***
Rhabdochromatium		*Rhodothallasium*
Thermochromatium		*Rhodovibrio*
Thioalkalicoccus		*Rhodovivax*
Thiobaca		*Rhodovulum*
Thiocapsa roseopersicina		*Roseospira*
		Roseospirillum
Thiococcus		*Rubrivivax*
Thiodictyon		
Thiocystis		
Thioflavicoccus		
Thiolamprovum		
Thiopedia		
Thiophaeococcus		
Thiorhodococcus		
Thiorhodovibrio		
Thiorhodospira		
Thiospirillum		

Model organisms and/or major species studied are written in bold.

A wide variety of CSB oxidize various inorganic sulfur compounds in nature. Most of these bacteria are chemolithoautotrophs coupling sulfide oxidation with nitrate or oxygen reduction. Based on comparative analysis of 16 rRNA sequences, the known CSB are grouped into four phylogenetic lineages, one within the Archaea and three within the Bacteria [49]. Most of the CSB belong to the phylum Proteobacteria, in particular the class Gammaproteobacteria (*Thiomicrospira, Thioalkalimicrobium, Thioalkalivibrio, Thiothrix, Thiohalospira, Thiohalomonas, Halothiobacillus,* and the *Acidithiobacillaceae*), the class Betaproteobacteria (*5 Thiobacillus* spp. and *Sulfuricella denitrificans*), the class Alphaproteobacteria (*2 Starkeya* spp. and *Thioclava pacifica*), and the class Epsilonproteobacteria (*3 Sulfurimonas* spp., *Sulfurovulum,* and *Thiovulum*) (Table 3).

Table 3 Colorless sulfur bacteria: obligately and facultative chemolithoautotrophic genera able to gain energy from oxidizing inorganic sulfur compounds (list incomplete).

Obligately Chemolithoautotrophic Genera	Facultative Chemolithoautotrophic Genera
*Acidianus**	*Acidianus**
*Acidithiobacillus**	*Acidiphilium*
Aquifex	*Acidithiobacillus**
Arcobacter	*Alkalilimnicola*
*Beggiatoa**	*Aquaspirillum*
Halothiobacillus	*Beggiatoa**
Hydrogenivirga	*Hydrogenobacter*
Hydrogenovibrio	*Magnetospirillum*
*Sulfolobus**	*Paracoccus*
Sulfuricella	*Pseudaminobacter*
Sulfuricurvum	*Sphaerotilus*
*Sulfurihydrogenibium**	*Starkeya*
Sulfurimonas	*Stygiolobus*
Sulfurivirga	*Sulfobacillus*
Sulfurovum	*Sulfolobus**
*Thermithiobacillus**	*Sulfurihydrogenibium**
*Thermothrix**	*Sulfurisphaera*
Thioalkalibacter	*Sulfuritalea*
Thioalkalimicrobium	*Sulfurococcus*
Thioalkalispira	*Tetrathiobacter*
Thioalkalivibrio	*Thermithiobacillus**
*Thiobacillus**	*Thermocrinis*
Thiobacter	*Thermothrix**
Thiofaba	*Thiobacillus**
Thiohalobacter	*Thioclava*
Thiohalomonas	*Thiomargarita*
Thiohalophilus	*Thiomonas*
Thiohalorhabdus	*Thioploca*
Thiohalospira	*Thiosphaera*
Thiomicrospira	*Thiospira*
Thioprofundum	*Thiothrix*
Thiovirga	
Thiovulum	

Archaebacteria are written in bold.
*Genera containing both facultative and obligately chemolithoautotrophic species.

In addition to the Proteobacteria, five *Sulfobacillus* species belong to the phylum *Firmicutes* and five *Sulfurihydrogenibium* species belong to the phylum *Aquificae*. Within the Archaea, *Sulfurisphaera ohwakuensis*, four *Acidianus* species and six *Sulfolobus* species belong to the family *Sulfolobaceae* within the phylum *Crenarchaeota* [49]. Haloalkaliphilic and neutrophilic halophilic chemolithoautotrophic SOB comprise four and seven different groups, respectively, within the

Gammaproteobacteria [52]. CSB can be found wherever reduced sulfur compounds are available (e.g., in soils sediments, at aerobic/anaerobic interfaces in water, and at volcanic sources such as the hydrothermal vents). CSB growing at neutral to slightly alkaline pH values are found in soils, freshwater, and marine sediments [52]. The acidophilic CSB are mainly found in acid mine-drainage water.

1.2   Properties and Toxicity of Hydrogen Sulfide

The toxicological effect of H_2S was first described in 1713 by Ramazzini and Scheele was the first one to synthesize H_2S gas [1]. H_2S is a small gaseous molecule freely permeable through a membrane. Like CO and NO, H_2S fulfills all of the criteria to define a gasotransmitter of clinical relevance [1, 53]. H_2S is produced endogenously by 3 enzymes in mammalian cells and plays important roles in physiological and pathophysiological conditions [1].

H_2S is the only product excreted from the dissimilatory sulfur metabolism of SRB. If sulfate is the energetically stable form of sulfur under aerobic conditions, H_2S is the stable form under anaerobic reduced conditions. Intermediary oxidation states of sulfur (such as sulfite, thiosulfate, and elemental sulfur) may be formed in natural habitats by incomplete biological or chemical oxidation of sulfide, or during anaerobic sulfide oxidation by purple and green sulfur microorganisms.

Under ambient temperature and pressure, H_2S (CAS registry number: 7783-06-4) is a colorless and flammable gas heavier than air (d = 1.19), slightly soluble in water, with a molecular weight of 34.08. The smell of H_2S is characteristic of rotten eggs or the obnoxious odor of a blocked sewer. The melting point of H_2S is -82.3 °C, its boiling point is -60.3 °C, and its freezing point is -86 °C. H_2S is a weak acid in aqueous solution with an acid dissociation constant (pK_a) of 6.76 at 37 °C. H_2S is a highly lipophilic molecule and can diffuse through cell membranes without facilitation of membrane channels. The half-life of H_2S in air varies from 12 to 37 hours.

Ambient air H_2S comes from two different sources. Inorganic H_2S sources include volcanic gases, sulfur deposits, petroleum refinery, natural gas, manure pits, pulp and paper mill industry, and sulfur springs. Organic H_2S sources include bacteria and decomposition of organic matters such as released from sewers, water treatment plants, or septic tanks.

Sulfide is a well-known toxin with the potential to harm organisms through, for example, mitochondrial depolarization [54], decreased hemoglobin oxygen affinity [55], inhibition of twenty enzymes involved in aerobic metabolism [56], and reversible inhibition of cytochrome c oxidase [57]. H_2S is as toxic as hydrogen cyanide (HCN) because of the capacity of their corresponding anions S^{2-} and CN^- to coordinate and precipitate metal cations [58]. H_2S is the primary toxic form of the compound because of its ability to diffuse across cellular membranes.

1.3   Effects of Hydrogen Sulfide on Gene Expression and Physiology of Desulfovibrio vulgaris Hildenborough

H_2S is toxic to most life forms including the SRB themselves for which its presence presents a stress, which these organisms must overcome. The activity of SRB is influenced by the presence of metals such as iron, manganese, copper, cadmium, nickel, lead, and zinc. Inhibitory or even lethal effects are observed at high concentrations of heavy metals, while a low concentration could promote the SRB activity [59]. Most SRB tolerate more than 10 mM sulfide and sulfate reducers growing on aromatic hydrocarbons formed as much as 25 mM sulfide before growth ceased. Some *Desulfotomaculum* species are more sensitive to sulfide, which affects their growth at concentrations of 4–7 mM [59]. Sulfidic sulfur can be present in three different forms (H_2S, HS^-, and S^{2-}) of which the relative fractions depend on the pH of the environment, see reactions (4) and (5) [60].

$$H_2S \rightarrow H^+ + HS^- \qquad pK_a1 = 6.97 \ (25\,°C) \qquad (4)$$
$$HS^- \rightarrow H^+ + S^{2-} \qquad pK_a2 = 12.9 \ (25\,°C) \qquad (5)$$

The response of *Desulfovibrio vulgaris* Hildenborough (hereafter referred to as *D. vulgaris* H) cells to high sulfide stress has been determined [60]. The growth of *D. vulgaris* H cells was compared in an open system, where sulfide was removed as H_2S by continuous gassing (low sulfide, 1 mM), with a closed system, where sulfide was accumulated (high sulfide, 10 mM). High sulfide decreased the final cell density by 33 % and the specific growth rate constant by 52 %, indicating a decrease in bioenergetics fitness. Under high sulfide conditions the transcription of ribosomal protein-encoding genes was decreased, in agreement with the lower *D. vulgaris* H growth rate. The expression of the DsrD gene, located downstream of the Dsr genes was also strongly down-regulated. In contrast, the expression of many genes involved in proteolysis, stress response, and iron accumulation were increased. High sulfide represents a significant stress condition, in which the bioavailability of metals like iron may be lowered [60].

2   Enzymology of Hydrogen Sulfide Production from Sulfate

2.1   Enzymology of Dissimilatory Sulfate Reduction

The hallmark characteristic of microorganisms utilizing dissimilatory sulfate reduction is the production of copious amounts of H_2S. Prior to reduction, sulfate must be activated by ATP sulfurylase to produce adenosine 5'-phosphosulfate (APS) plus pyrophosphate (PP_i), reaction (6).

$$SO_4^{2-} + 2\,H^+ + ATP \rightarrow APS + PP_i \qquad \Delta G^{0'} = +46.0\,kJ/mol\ [41] \qquad (6)$$

The sulfate activation reaction is thermodynamically unfavorable and depletion of end products (APS and PP$_i$) is required to favor APS production. Hydrolysis of PP$_i$ by a cytoplasmic inorganic pyrophosphatase (EC 3.6.1.1) producing inorganic phosphate, enhances the production of APS, reaction (7).

$$PP_i + H_2O \rightarrow 2\,P_i \qquad \Delta G^{0'} = -21.9\,kJ/mol\ [41] \qquad (7)$$

Metal ions (Zn^{2+}, Mn^{2+}, Co^{2+}, Mg^{2+}) are known to stabilize inorganic pyrophosphatases of SRB with greatest activity observed with Mg^{2+} [61, 62].

SRB use APS reductase to catalyze the production of HSO_3^- from APS. The reaction of APS/AMP + HSO_3^- is slightly exergonic with $E^{0'} = -60\,mV$ [41]. Production of H$_2$S from APS is a two step process. APS reductase catalyzes the two electron reduction of APS to hydrogen sulfite (reaction 8) [41] which is followed by the six electron reduction converting hydrogen sulfite to sulfide by dissimilatory sulfite reductase (reaction 9).

$$APS + H_2 \rightarrow HSO_3^- + AMP + H^+ \qquad \Delta G^{0'} = -68.9\,kJ/mol\ [41] \qquad (8)$$

$$HSO_3^- + 3\,H_2 \rightarrow HS^- + 3\,H_2O \qquad \Delta G^{0'} = -171.7\,kJ/mol\ [41] \qquad (9)$$

While dissimilatory sulfate reduction proceeds, SRB also use the assimilatory sulfate reduction pathway to synthesize the amino acid cysteine. In both sulfate reduction pathways, sulfate is activated to APS by ATP sulfurylase.

Initially, it was proposed that a trithionate pathway was used to metabolize intermediates produced by the dissimilatory sulfate reduction and several reports with cell-free systems supported this [63–71]. However, there is no convincing evidence for the presence of trithionate reductases in SRB, either by genome analysis or biochemically, and thiosulfate reductases have been reported only in a few SRB. It is probable that under general growth conditions trithionate and thiosulfate are not produced as intermediates in dissimilatory sulfite reduction [72–74].

2.2   ATP Sulfurylase

ATP sulfurylase (ATPS) is also known as adenylylsulfate pyrophosphorylase or sulfate adenylyltransferase (EC 2.7.7.4) and is a product of the *ppa* gene. The pathway utilizing ATPS is indicated in Figure 3.

For optimal activity, ATPS from *Desulfotomaculum* (*Dst.*) (formerly *Clostridium*) *nigrificans* and *D. desulfuricans* strain 8303 require Mg^{2+} to neutralize the

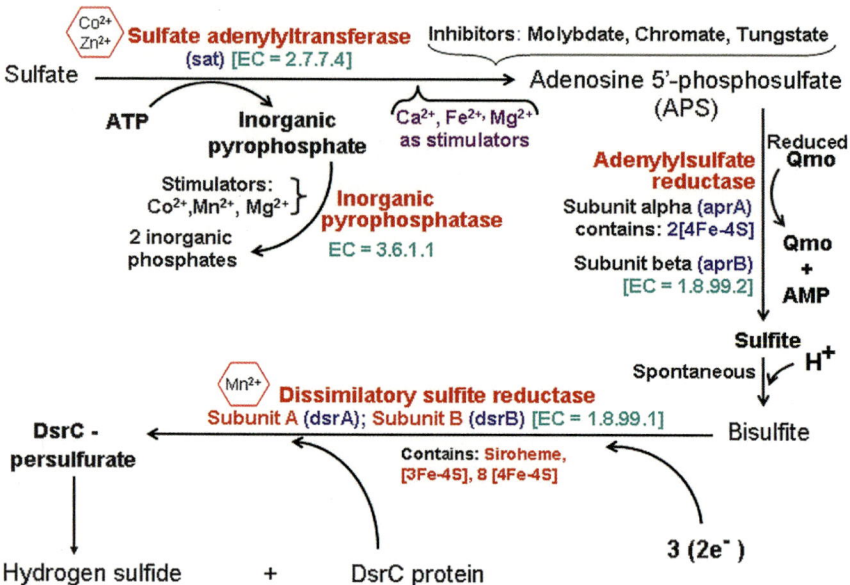

Figure 3 Pathway for sulfide production from dissimilatory sulfate reduction by *Desulfovibrio* species.

charge on ATP and PP$_i$ (reaction 10) [75, 76]. Inhibitors of ATP sulfurylase include CrO_4^{2-}, MoO_4^{2-}, and WO_4^{2-} which produce unstable intermediates.

$$ATP\text{-}Mg + SO_4^{2-} \rightarrow APS + PP_i\text{-}Mg \qquad (10)$$

The ATPS from *Archaeoglobus (Ar.) fulgidus* DSM 4304 has a molecular mass of 150 kDa (53.1 kDa subunits) compared to 141 kDa (46.9 kDa subunits) for the enzyme from *D. desulfuricans* ATCC 27774 [77, 78]. Using extended X-ray absorption fine structure (EXAFS) and electron paramagnetic resonance (EPR) spectroscopies, Co^{2+} and Zn^{2+} were found to bind to three sulfur atoms and one nitrogen atom in a tetrahedral coordination in the enzymes from *D. desulfuricans* and *D. gigas* [78]. Such a tetrahedral Zn^{2+} site was observed in the crystal structure of *Thermus thermophilus* ATPS [79] and analysis of the crystal structure of ATPS from the sulfur-oxidizing purple sulfur bacterium *Alc. vinosum* reveals that three cysteine residues and one histidine are involved in the zinc-binding site [75]. Similarly, four coordinating amino residues are conserved in the ATPS of *Ar. fulgidus, Pyrodictium (P.) abyssi, Sulfolobus solfataricus, D. desulfuricans,* and *D. gigas* [75, 80] and may be the site for Zn^{2+} binding. ATPS occurs in dissimilatory sulfate-reducing microorganisms as homotrimers with a zinc ion bound to each monomer [78] and in dissimilatory SOB, ATPS is a homodimer [75]. From structural and genetic analysis, adjacent monomers of ATPS from *Alc. vinosum* have the GXXKXXD sequence and zinc ion stabilizes the APS and PP$_i$ binding sites

[75]. The binding of cobalt to ATPS is unresolved at this time but may contribute to stability of the polymeric structure [78].

2.3   Dissimilatory Adenylylsulfate Reductase

In the pathway of sulfate reduction, APS is reduced to sulfite by APS reductase (APSR) (adenylylsulfate reductase, EC 1.8.99.2). A $\alpha\beta$ heterodimer forms the basic enzyme unit with the active site FAD located in the AprA subunit and two ferredoxin-like [4Fe-4S] clusters in the AprB subunit. The crystal structure of APSR has been reported for *D. gigas* at 3.1 Å [81] and for *Ar. fulgidus* at 1.6 Å [82]. Using homology modeling of 20 different species, Meyer and Kuever [83] have compared the dissimilatory APSR (AprBA) of SRP and SOB and report that the protein matrix around the [4Fe-4S] clusters and the FAD cofactor show high similarity.

The *D. gigas* APSR has a hexamer structure consisting of six $\alpha\beta$ heterodimers [81]. The α subunit of *D. gigas* APSR has a distinct region that binds FAD; another area is the capping attributed to the specific molecular configuration and the third is the helical domain [81]. FAD is attached to the α subunit domain by six hydrogen bonds and a shallow cleft is formed in the α subunit by the action of the FAD and helical domains. The β subunit of *D. gigas* APSR has a domain for binding the [4Fe-4S] cluster, a segment that contains the β-sheet protein and the C-terminal region. In formation of the dimer, the domain of the β subunit containing the [4Fe-4S] clusters is buried in the shallow cleft occurring in the α subunit. The β-sheet protein and the C-terminal domains stabilize the interaction between the α subunit and β subunit of APSR. Cluster I [4Fe-4S] is buried in the β subunit while cluster II [4Fe-4S] is located on the surface of the β subunit where it presumably acquires electrons from the electron donor [81]. The redox potentials for the cluster I and cluster II are 0 and -400 mV, respectively [84]. The electron donor for the dissimilatory APSR in SRB is the membrane QmoABC complex [85]. Using deletion mutants of *D. vulgaris* H, it has been shown that the Qmo complex (*qmoABC* genes) is not essential for sulfite or thiosulfate reduction [86]. The proximity of *AprBA* genes to *QmoABC* for *D. gigas* and *D. vulgaris* H are given in Figure 4.

A comparison of the *D. gigas* APSR crystal structure with the structure of the *Ar. fulgidus* enzyme reveals considerable similarity [81]. The α and β subunits from *Ar. fulgidus* APSR have the same three domains in each subunit as the *D. gigas* APSR. In the β subunit from *Ar. fulgidus* APSR, the redox potential for cluster I [4Fe-4S] is -60 mV while that for cluster II [4Fe-4S] is -520 mV [82]. The C-terminus of the *D. gigas* APSR β subunit is longer than the C-terminus of *Ar. fulgidus* APSR.

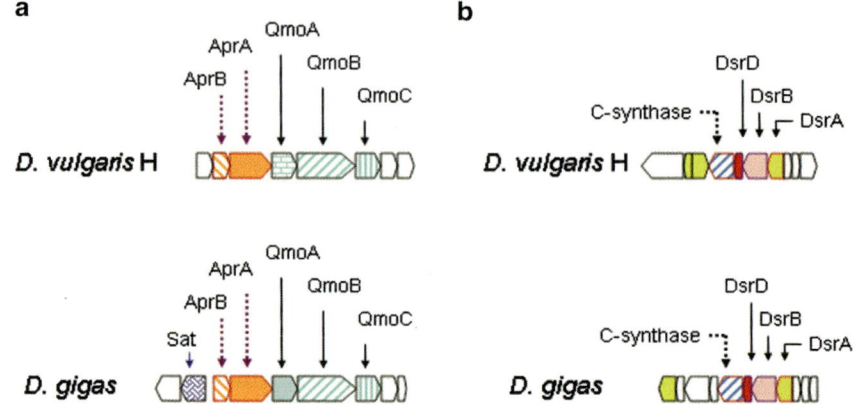

Figure 4 Genes for dissimilatory sulfate reduction in *D. vulgaris* Hildenborough and *D. gigas* DSM 1382. (**a**) Loci of genes for adenylyl sulfate reductase (AprAB), sulfate adenylyl transferase (Sat), and heterodisulfide reductase (QmoABC). (**b**) Loci of genes for dissimilatory sulfite reduction (DsrABD) and cobyrinic acid a,c–diamide synthase (C-synthase). Draft sequence is used for *D. gigas*.

2.4   Sulfite Reductases

Two classes of sulfite reductases can be defined in SRP on the basis of their physiological function. The first class comprises the low-molecular-mass and low-spin sulfite reductases, also called assimilatory-type sulfite reductases (aSiR) (EC 1.8.99.1). They have a molecular mass around 27 kDa, consist of a single polypeptide chain and contain one [4Fe-4S] cluster coupled to a siroheme in a low-spin state [87]. The second class is constituted by the high-spin dissimilatory sulfite reductases (EC 1.8.99.3) dSiR, which possess a molecular mass around 200 kDa and a complex molecular architecture hosting siroheme centers and [4Fe-4S] clusters. While dSiR is preferred it had been customary to refer to dissimilatory sulfite reductase as Dsr and currently the genes for dSiR are designated as *dsr* and their gene products as Dsr. Both aSiR and dSiR enzymes have iron-sulfur clusters and an iron-chelating sirohydrochlorin referred to as siroheme [88].

2.4.1   Dissimilatory High-Spin Sulfite Reductase

High-spin dSiR has either an $\alpha_2\beta_2$ structure, as in the case of *Ar. fulgidus* [89] while some SRB have an $\alpha_2\beta_2\gamma_m\delta_n$ multimeric structure, with α 50 kDa, β 45 kDa, γ 11 kDa, and δ 8 kDa [90, 91]. Catalytic activity of dSiR is attributed to the $\alpha_2\beta_2$ structure and the third protein, γ subunit (DsrC), is an independent protein that functions as a sulfur carrier protein to facilitate the release of H_2S following reduction of sulfite (Figure 3) [92, 93]. The reduced sulfur atom forms a persulfide with a conserved cysteine on DsrC, the γ subunit, and with the release of sulfide, the intramolecular disulfide linkage of DsrC is reformed by interaction with DsrK, a subunit of the DsrMKJOP transmembrane complex [94–96]. The DsrD, δ subunit, lacks cysteine residues [97] and has no function in electron transport.

Five different types of enzymes belonging to the high-spin DSR class (desulfoviridin, desulfofuscidin, desulforubidin, P-582, and Archaeal) have been isolated and characterized from different genera of SRP [2, 15, 16]. These five enzymes differ mainly by the behavior of their siroheme moieties, their major optical absorption, and EPR spectra [2, 15, 16] (Table 4).

Metals as cofactors are important in dSiR with each $\alpha_2\beta_2$ structure having associated with it two sirohemes, a [4Fe-4S] cluster closely associated with each siroheme and four ferredoxin-type [4Fe-4S] clusters at some distance from the sirohemes. The binding of a siroheme and the [4Fe-4S] cluster to the α subunit and β subunit would be attributed to the (Cys-X_5–Cys)–X_n–(Cys-X_3–Cys) arrangement [98]. Binding of the ferredoxin-type [4Fe-4S] clusters to the α and β subunits is predicted to follow the arrangement of Cys-X_2–Cys-X_2–Cys that is preceded or followed by a Cys-Pro sequence [99–101].

2.4.1.1   Desulfoviridin-Type Sulfite Reductase

The green protein, desulfoviridin, is the dSiR characteristic of the genus *Desulfovibrio* but it has also been found in some species of the genera *Desulfococcus, Desulfomonile, Desulforegula,* and *Desulfonema* [15, 90, 102–105]. The structure of the dSiR of *D. vulgaris* H was reported to be a $\alpha_2\beta_2\gamma_2$ structure [67] and similar to the $\alpha_2\beta_2\gamma_2$ structure found in *D. vulgaris oxamicus* (Monticello), *D. gigas,* and *D. desulfuricans* ATCC 27774 [106]. DsrA and DsrB in *D. vulgaris* H are the products of the *dsrAB* operon while *dsrC,* which encodes for DsrC, is located at another site on the chromosome [95, 96].

Associated with the $\alpha_2\beta_2$ structure are two sirohemes and two sirohydrochlorins which are positioned at the interface of DsrA and DsrB. The sirohydrochlorin accounts for the absorption maximum at 628 nm. The sulfur atom from sulfite, the substrate, is bound to the iron atom of siroheme in DsrA and this region is surrounded by basic amino acids. The other side of the siroheme is surrounded by residues from DsrB. Access to this catalytic site is through a positively charged channel; however, a similar channel is lacking in the region where sirohydrochlorin is bound to DsrA [95, 96]. The *D. vulgaris* H DsrC subunit is in close proximity to the cleft formed between DsrA and DsrB. The C-terminus of DsrC contains a cysteine moiety which may interface with the siroheme and may participate in the catalytic reaction. Oliveira et al. [95, 96] propose that the initial reduction of sulfite is a four and not a six electron step with S^0 formed as an intermediate product. S^0 would interact with the terminal cysteine on DsrC to form a persulfide which would be reduced to sulfide. Reduction of DsrC could be achieved by heterodisulfide reductase activity of the membrane-bound DsrMKJOP which has an appropriate iron-sulfur center for this reduction process [107].

Table 4 Characteristics of selected bacterial and archaeal high-spin and low-spin sulfite reductases. Iron compounds are listed for intact enzyme structure.

Protein	λmax (nm)	Organism	M_r(kDa)	Non-heme Iron	Siroheme (sirohydrochlorin)	[4Fe-4S] Cluster	Reference
High-spin "dissimilatory-type" sulfite reductase							
Desulfoviridin	628	D. gigas	200	34	2(2)	8	[64, 92, 93]
	630	D. vulgaris H	200	34	2(2)	8	[95, 96, 165]
P-582	582	Dst. nigrificans	194	16	1.3	4	[65, 111]
	582	Dst. thermocisternum[a]	196	16	2	4	[112]
Desulfofuscidin	576	T. commune[b]	167	20–21	4	4	[69, 120]
	578	T. mobile	190	32	4	8	[119]
Desulforubidin	545	Dsm. novegicum[c]	225	36	4	8	[15, 93, 102, 113]
	545	Ds. variabilis	208	15	2	8	[166]
Archaeal	593	Ar. fulgidus	218	36	4	8	[89, 117]
	NR[d]	Ar. profundus	198	24	2	6	[112]
Low-spin "assimilatory-type" sulfite reductase							
	590	D. vulgaris H	27.2	5	1	1	[67, 143]
	587	Drm. acetoxidans	23.6	5	1	1	[145]
	590	Ms. barkeri	23	5	1	1	[87, 144]

[a]Calculated from gene analysis.
[b]Formerly known as D. thermophilus.
[c]Formerly known as D. desulfuricans Norway 4 and D. baculatus Norway 4.
[d]NR: not reported

Adjacent to the genes for dissimilatory sulfite reduction in *D. vulgaris* H is a gene that encodes for DsvD, a peptide of only 78 amino acids [108]. A similar gene encoding 77 amino acids is also present in the genome of *Ar. fulgidus* and it is downstream of the dsrB gene. The DsvD, also known as DsrD, is not the γ subunit and would not be associated with electron transfer because it lacks cysteine residues [97]. DsvD has structural homology to DNA-binding proteins and may have a role in transcription or translation [109].

D. gigas is a unique sulfate reducer in the sense that it has a multimeric dissimilatory sulfite reductase. Dsr-I and Dsr-II have enzymatic activity while Dsr-III is inactive [92]. The $\alpha_2\beta_2\gamma_2$ structure of Dsr-I type from *D. gigas* contains eight [4Fe-4S] clusters, two planar sirohydrochlorins and two saddle-shaped sirohemes while the Dsr-II type contains two sirohemes, two hydrochlorins, two [3Fe-4S] clusters, and six [4Fe-4S] clusters [92]. Dsr-III from *D. gigas* has iron-sulfur clusters similar to Dsr-II and the inactivity of Dsr-III is attributed to the absence of iron in the siroheme [92].

Analysis of the sirohydrochlorin from several *Desulfovibrio* species reveals that it is different from that isolated from *Escherichia* (*E.*) *coli*. Notably, one of the eight carboxylates of the tetrapyrrole moiety is replaced by an amide group at the 2'-acetate [110]. Additional [4Fe-4S] clusters, referred to as remote iron-sulfur clusters, are located on the surfaces of the α and β subunits. In *D. gigas*, the S atom of sulfite binds to the Fe of the siroheme and a positively charged arginine along with two lysine moieties in the α subunit has an electrostatic interaction with the three oxygen atoms on sulfite. Another arginine forms hydrogen bonding with sulfite and directs the release of sulfide and water molecules through a positively charged water channel.

The γ subunit of *D. gigas* Dsr-I is aligned adjacent to the α/β subunits and the C-terminus of the γ subunit is proposed to have multiple conformations that may influence catalytic function. It has been proposed that activity of the C-terminus of the γ subunit could explain the mechanism for three products (sulfide, thiosulfate, and trithionate) produced by Dsr [92]. The association of dsr genes in *D. gigas* and *D. vulgaris* H is given in Figure 4.

2.4.1.2   P-582-Type Sulfite Reductase

P-582-type sulfite reductase has only been reported in several species of the spore-forming SRB genera *Desulfosporosinus* and *Desulfotomaculum* [15, 25, 65, 111]. The structure of dSiR from *Dst. thermocisternum,* a Gram-positive thermophilic sulfate reducer, has been deduced from gene analysis [112]. From cloning and sequence analysis, it is suggested that the *dsr* operon in *Dst. thermocisternum* is similar to that of *Ar. fulgidus* and *D. vulgaris* H in terms of sequence and gene organization.

The *dsrA* gene encodes for a protein of 54.1 kDa with a 76 % similarity to *Ar. fulgidus* DsrA and an 83 % similarity to DsrA of *D. vulgaris* H. Down-stream of the *dsrA* gene in *Dst. thermocisternum* is the *dsrB* gene and it encodes a protein of

44.2 kDa with an 89 % similarity to the *Ar. fulgidus* DsrB. Down-stream from the *dsrB* is the *dsrD* gene which encodes for a peptide of 11.1 kDa and the DsrD, a hypothetical δ-polypeptide, has a projected sequence identity of 38 % with *Ar. fulgidus* DsrD and 41 % identity to *D. vulgaris* H DsrD. The proposed operon for dissimilatory sulfite reductase in *Dst. thermocisternum* consists of *dsrABD*.

2.4.1.3   Desulforubidin-Type Sulfite Reductase

The red brown protein, desulforubidin, belongs to the genera *Desulfohalobium*, *Desulfosarcina*, *Desulfomicrobium*, *Desulfocurvus*, *Desulfobulbus*, *Desulfofustis*, and *Desulfobacter* [2, 39, 113, 114]. The crystal structure at 2.5 Å resolution of the *Desulfomicrobium* (*Dsm.*) *norvegicum* dSiR was reported by Oliveira et al. [93]. Due to the distinctive spectral characteristics, the dSiR is a desulforubidin-type with the $\alpha_2\beta_2\gamma_2$ structure and the isolated complex contains two DsrC proteins. The *Dsm. norvegicum* dSiR contains four siroheme groups and eight [4Fe-4S] clusters.

One of the cofactors in the dSiR is a siroheme bound to a [4Fe-4S] cluster and each of the peptides making up the $\alpha_2\beta_2$ structure contains a siroheme-[4Fe-4S] moiety. The siroheme-[4Fe-4S] component is at the interface of the DsrA and DsrB structures with the [4Fe-4S] cluster bound into the DsrB subunit. It is suggested by Oliveira et al. [93] that the two siroheme-[4Fe-4S] cofactors bound in the DsrA structures are not involved in the enzymatic sulfite reduction because they are not readily accessible. In comparison, the site where the possibly inactive siroheme-[4Fe-4S] is bound in *Dsm. norvegicum* is occupied by a sirohydrochlorin in *D. vulgaris* H and *D. gigas*. The DsrC of *Dsm. norvegicum* has a helical structure where the C-terminal segment of the protein is flexible and is inserted in the cleft between DsrA and DsrB.

The dSiR produced by *Desulfobacter* (*Dba.*) *vibrioformis* and *Desulfobulbus* (*Dbu.*) *rhabdoformis* is of the desulforubidin-type and the operons for dSiR were characterized by gene analysis [115]. As in the case of dSiR from the genera *Desulfovibrio*, *Desulfotomaculum*, *Desulfomicrobium*, and *Archaeoglobus*, the α and β subunits of dSiR in *Dba. vibrioformis* and *Dbu. rhabdoformis* are encoded on *dsrA* and *dsrB*, respectively. The polypeptide encoded on *dsrA* is 48–50 kDa and for *dsrB* is 42–43 kDa in *Dba. vibrioformis* and *Dbu. rhabdoformis*, respectively, and is comparable to polypeptides reported for other siroheme-containing sulfite reductases. Important characteristics of siroheme-containing dSiR are the homology region of H1-H5 and the binding site for the ferredoxin motif [116, 117] and these are also present in *Dba. vibrioformis* and *Dbu. rhabdoformis*. High sequence identity was reported when *dsrA* and *dsrB* from *Dba. vibrioformis* and *Dbu. rhabdoformis* were compared to respective genes found in *D. vulgaris* H, *Dst. thermocisternum*. *Ar. fulgidus*, *Ar. profundus*, *Pyb. islandicum*, and *Allochromatium* (*Alc.*) *vinosum*.

Also in the *dsr* operons of *Dba. vibrioformis* and *Dbu. rhabdoformis* was *dsr*D which would encode a polypeptide of 9.8–8.7 kDa [115]. While the DsrD function is unknown, its distribution appears to be only in sulfate-reducing microorganisms. Very good sequence similarity of DsrD between *Dba. vibrioformis, Dbu. rhabdoformis, D. vulgaris* H, *Dst. thermocisternum, Ar. fulgidus,* and *Ar. profundus,* was reported. A gene comparable to *dsrD* appears absent in *Pyb. islandicum* and *Alc. vinosum.* The *dsr* operons of *Dba. vibrioformis* and *Dbu. rhabdoformis* contain a gene, *dnrN,* that encodes for a 53 kDa protein of unknown function. The *dsrN* shows considerable homology to *cbiA,* a gene for amination of cobyrinic acid to cobyrinic acid a,c-diamine [118]. Larsen et al. [115] suggest a possible role for DsrN in sulfate reducers as the amidation of siroheme since siroamide is a prosthetic group present in sulfite reductase of *Desulfovibrio* species [110].

2.4.1.4 Desulfofuscidin-Type Sulfite Reductase

The dark brown-colored protein, desulfofuscidin is the dSiR of thermophilic eubacterial sulfate reducers such as *Thermodesulfovibrio* strains *hydrogeniphilus* and *yellowstoni* and *Thermodesulfobacterium (T.) commune* and *T. mobile* [119–121]. The reduction of sulfite by *T. mobile* and *T. commune* is accomplished by a sulfite reductase that has the $\alpha_2\beta_2$ subunit structure.

The tetrameric dSiR from *T. commune* has a molecular mass of 167 kDa and consists of nonidentical subunits of approximately 47 kDa [120], compared to 175 kDa (gel filtration) or 190 kDa (sedimentation equilibrium) found for the dSiR from *T. mobile,* with subunits of 38–44 kDa [119]. Spectral absorption maxima of dSiR from *T. commune* are at 576, 389, and 279 nm while for dSiR *T. mobile* these are at 578, 392, and 281 nm. Four siroheme groups are found in the dSiR of *T. mobile* and *T. commune* with four [4Fe-4S] clusters in *T. commune* [69] and eight [4Fe-4S] clusters in *T. mobile* [119]. The EPR spectrum of desulfofuscidin exhibits resonances assigned to high-spin Fe(III) heme centers, with *g* values of 7.02, 4.81, and 1.91 for the enzyme from *T. commune* [120], and 7.26, 4.78, and 1.92 for the enzyme from *T. mobile* [119].

2.4.1.5 Archaeal-Type Sulfite Reductase

The dissimilatory *Ar. fulgidus* sulfite reductase accounts for 0.5 % of the soluble proteins and the isolated enzyme occurs as a tetramer with an $\alpha_2\beta_2$ structure [117]. The oxidized enzyme has an α-absorption maximum at 593 nm; upon reduction with dithionite, the maximum shifts to 598 nm. In the oxidized protein an absorption band with a maximum at 715 nm is observed; however, this band disappears upon dithionite reduction suggesting that the siroheme is in the high-spin state [117, 122].

Ar. veneficus grows with dissimilatory sulfite or thiosulfate reduction but is unable to couple growth to the transfer of electrons from the electron donor to sulfate [123]. The Dsr gene locus of *Ar. profundus* has *dsrA* contiguous to *dsrB* [112]. The DsrD, as product of *Ar. profundus dsrD,* is projected to be about 9.3 kDa and would be comparable to the γ subunit present in the dSiR of *Desulfovibrio* species where it functions in sulfite-binding [108]. Downstream of *dsrABD* in both *Archaeoglobus* species is a gene for ferredoxin and this is significant because ferredoxin may be the electron donor for enzymatic sulfite reduction.

2.4.2   Oxy-Sulfur Reductases in Non-sulfate Reducers

There are several genera of bacteria that are capable of reducing tetrathionate, thiosulfate, and sulfite to H_2S (see reactions 11, 12, and 13), respectively. Sulfite, an intermediate of dissimilatory sulfate reduction, is readily utilized by SRB and the metabolism of tetrathionate or thiosulfate with the production of sulfide is covered in Section 2.1. In the environment, thiosulfate is found in marine and freshwater sediments [124] and in the lumen of mammalian large intestine [125]. Tetrathionate is present in soils, where SRB are growing [126], and is produced in the gut as a result of inflammatory response [127]. Several microorganisms that are taxonomically unrelated have enzymes that function under anaerobic conditions as facilitating sulfite respiration (reactions 11, 12, and 13) [20].

$$S_4O_6^{2-} + 2\,e^- \qquad\qquad \rightarrow 2\,S_2O_3^{2-} \qquad\qquad E^{0'} = +170\,mV\ [41] \qquad (11)$$

$$S_2O_3^{2-} + 2\,H^+ + 2\,e^- \rightarrow SO_3^{2-} + H_2S \qquad E^{0'} = -402\,mV\ [41] \qquad (12)$$

$$SO_3^{2-} + 8\,H^+ + 6\,e^- \rightarrow H_2S + 3\,H_2O \qquad E^{0'} = -160\,mV\ [41] \qquad (13)$$

The sulfite reductase, Fsr, present in *Methanocaldococcus jannaschii* (M_r 70 kDa) could have arisen from gene fusion since the N-terminal part of the enzyme is a homolog of the β subunit of coenzyme F_{420}-reducing hydrogenase (FpoF or FqoF) while the C-terminal half has homology to DsrA and DsrB of the siroheme-containing dSiR [128]. A homolog of this bifunctional sulfite reductase from *Methanocaldococcus jannaschii* is also present in *Methanopyrus kandleri* and *Methanococcoides burtonii* [129].

The dissimilatory sulfite reductase, dSiR, of *Bilophila wadsworthia* has a $\alpha_2\beta_2\gamma_n$ ($n \geq 2$) multimeric structure. This enzyme has an absorption maximum at 630 nm [130] indicating that it is a desulfoviridin-type enzyme. The α subunit (M_r 49 kDa) contains the siroheme-[4Fe-4S] while the β subunit polypeptide is 54 kDa. The β subunit is a fusion protein resulting from fused *dsrB* and *dsrD* genes. Based on phylogenic analysis, *Bilophila wadsworthia* is closely related to *D. desulfuricans* Essex 6 and at one time may have been a dissimilatory sulfate reducer [130].

An inducible dissimilatory sulfite reductase in *Clostridium pasteurianum* couples growth to the reduction of sulfite [131]. The isolated enzyme (M_r 83.6 kDa), has an absorption maximum at 585 nm, but has no siroheme cofactor. Using reduced methyl viologen as the electron donor, thiosulfate or trithionate could not replace sulfite; however, nitrite and hydroxylamine were slowly reduced by the sulfite reductase.

Respiratory-linked tetrathionate reduction is present in soil bacterial communities [132] and species of *Citrobacter, Proteus, Salmonella,* and *Pseudomonas* [133]. In *Salmonella (Sal.) enterica* Serovar Typhimurium, tetrathionate reductase is encoded on the *ttr* operon with expression controlled by the global regulator OxrA. This membrane-bound tetrathionate reductase is induced by tetrathionate and has a bis-molybdopterin guanine dinucleotide (MGD) cofactor [134].

Dissimilatory thiosulfate reduction is found in numerous environmental bacteria including *Shewanella oneidensis* [135] and enteric bacterial pathogens [125] such as *Sal. enterica*. Anaerobic thiosulfate reductase activity (reaction 12; EC 1.97.1-) in *Sal. enterica* is linked to the plasma membrane and the enzyme is a product of the *phsABC* operon. Subunit phsA contains the active site and the cofactor MGD [136]. The cytochrome *b* subunit, phsC, is an integral membrane protein that contains two heme cofactors and a site to interact with the electron donor, naphthoquinone-8 [125]. Subunit phsB has four iron-sulfur clusters which transfer electrons between the subunits phsA and phsC.

Sulfite is reduced to sulfide by a cytoplasmic sulfite reductase and in *Sal. enterica* the anaerobic sulfite reductase is a product of the chromosomal *asr* (anaerobic sulfite reduction) operon [137, 138] with *asrA, asrB,* and *asrC* encoding for peptides of 40, 31, and 37 kDa, respectively. From genome analysis, it is predicted that a [4Fe-4S] ferredoxin is present in both the asrA and asrC peptides and a siroheme in the asrC subunit. NADH is proposed to be bound into the asrB subunit. This anaerobic sulfite reductase forms a large complex of about 360,000 M_r. An even larger complex has been detected for the assimilatory sulfite reductase (670,000 M_r) which has an $\alpha_8\beta_4$ structure with a flavin moiety in the α subunit and siroheme plus [4Fe-4S] clusters in the β subunit [139].

Dissimilatory reduction of sulfite to sulfide by *W. succinogenes* is attributed to an octaheme cytochrome *c* without the involvement of a coupled siroheme-[4Fe-4S] cofactor [140]. The octaheme cytochrome *c*, MccA, contains a heme bound by the unique motif of $CX_{15}CH$, and is encoded on the *mccABCD* gene cluster. An octaheme cytochrome *c*, SirA, in *Shewanella oneidensis* MR-1 also displays dissimilatory sulfite reductase activity [141].

2.4.3   Low-Molecular-Mass and Low-Spin Assimilatory-Type Sulfite Reductase from *Desulfovibrio vulgaris* H, *Desulfuromonas acetoxidans*, and *Methanosarcina barkeri*

A low-molecular-mass assimilatory-type sulfite reductase has been purified and characterized from *D. vulgaris* H [67, 142, 143]. This enzyme has a molecular mass of 27.2 kDa and its optical spectrum exhibits maxima at 405, 545, and 590 nm (Table 4). This hemoprotein is able to reduce sulfite to sulfide in the presence of reduced methyl viologen. The specific sulfite reductase activity was 900 mU/mg protein [87].

The *D. vulgaris* H assimilatory-type sulfite reductase has been studied by chemical, EPR, and Mössbauer techniques [143]. This protein contains one siroheme and a single [4Fe-4S] cluster. As purified, the siroheme is in a low-spin Fe(III) state (S = 1/2) which exhibits characteristic EPR resonances at g = 2.44, 2.36, and 1.77. Hereby, the iron-sulfur cluster is in the [4Fe-4S]$^{2+}$ state. Similar to the hemoprotein subunit of *E. coli* sulfite reductase, low-temperature Mössbauer spectra of *D. vulgaris* H sulfite reductase also show evidence for an exchange-coupled siroheme-[4Fe-4S] unit [143]. The presence of an assimilatory-type sulfite reductase in *D. vulgaris* H is surprising because this strain produces large amounts of sulfide during normal growth on sulfate and also because the enzymes responsible for dissimilatory sulfate reduction are constitutive.

Two low molecular mass hemoproteins with sulfite reductase activity (named P$_{590}$) have been purified and characterized from two sulfur reducers: *Ms. barkeri* and *Drm. acetoxidans* [87, 144, 145]. Both monomeric hemoproteins present visible spectra similar to that of the low-molecular-mass and low-spin assimilatory-type sulfite reductase of *D. vulgaris* H. The *Drm. acetoxidans* sulfite reductase has a molecular mass of 23.5 kDa and exhibits absorption maxima at 405, 545, and 587 nm [144]. The *Ms. barkeri* P$_{590}$ has a molecular mass of 23 kDa and its optical visible spectrum exhibits maxima at 395, 545 and 590 nm (Table 4) [145]. EPR spectra of the enzyme as isolated show that the siroheme is in a low-spin Fe(III) state (S = 1/2) with g-values at 2.40, 2.30, and 1.88 for the *Ms. barkeri* P$_{590}$ enzyme and g-values at 2.44, 2.33, and 1.81 for the *Drm. acetoxidans* enzyme [144].

Chemical analysis shows that both hemoproteins contain one siroheme and one [4Fe-4S] cluster per polypeptidic chain [144]. The specific sulfite reductase activity was 906 mU/mg protein for the *Drm. acetoxidans* P$_{590}$ enzyme and 2,790 mU/mg protein for the *Ms. barkeri* enzyme [87, 144]. The M*s. barkeri* P$_{590}$ enzyme has a higher specific sulfite reductase activity than that reported for the high-spin dSiR from SRB such as desulforubidin, desulfofuscidin, and desulfoviridin [25]. The two assimilatory-type sulfite reductases of sulfur reducers contain 5 labile sulfur atoms and 5 iron atoms; as it is the case for the *D. vulgaris* H enzyme, it was postulated that the extra sulfur atom could be the bridging ligand between the [4Fe-4S] center and the siroheme [87]. Both hemoproteins catalyze the direct six-electron reduction of sulfite to sulfide without the formation of free intermediates (thiosulfate and trithionate).

3   Enzymology of Hydrogen Sulfide Production from Elemental Sulfur

3.1   Eubacterial Sulfur Reductase

3.1.1   Sulfur Reductase in *Desulfovibrio* and *Desulfomicrobium* Species

The tetraheme cytochrome c_3 has the function of an elemental sulfur reductase in several *Desulfomicrobium* and *Desulfovibrio* species from which the sulfur reductase activity can be copurified with the tetrahemoprotein [26, 146, 147]. An exposed, low-potential heme of the tetraheme cytochrome c_3 from *Dsm. norvegicum* Norway 4 has been proposed to play an important mechanistic role. The polysulfide chains of colloidal S^0 are attacked by the reduced tetraheme cytochrome c_3, leading to a collapse of the micelles with the precipitation of S_8 molecules [148].

The sulfide produced by polysulfide reduction opens up the S_8 rings by a nucleophilic attack, leading to the production of new molecules of polysulfides, which are themselves quickly reduced to sulfide by *Dsm. norvegicum* tetraheme cytochrome c_3 [148]. Membranes isolated from *D. gigas* and *Dsm. norvegicum* contained hydrogenase and *c*-type cytochromes and catalyzed the dissimilatory sulfur reduction to sulfide. Sufficient hydrogenase and tetrahemic cytochrome c_3 must be linked with the *D. gigas* cytoplasmic membrane in the correct conformation to generate proton translocation sufficient for the chemiosmotic synthesis of ATP [149].

3.1.2 Polysulfide Reductase from *Wolinella succinogenes*

The mechanism of polysulfide respiration has been mainly investigated in the epsilon proteobacterial subclass *W. succinogenes* [27, 32]. The membrane fraction isolated from *W. succinogenes* cells grown with formate and either fumarate or polysulfide, as electron acceptor, catalyzes the polysulfide reduction by formate or H_2. The corresponding electron transport chain consists of 8-methyl-menaquinone, polysulfide reductase, and either hydrogenase or formate dehydrogenase. The isolated polysulfide reductase consists of the three subunits predicted by the *psrABC* operon, and contains a molybdenum ion coordinated by two molecules of MGD. A model of this membrane association with coupled proton pumping during sulfur respiration is given in Figure 5.

The PsrA subunit is the catalytic unit, PsrB is an iron-sulfur protein, and PsrC is an integral membrane protein that serves to anchor the other subunits on the membrane. Energy conservation via polysulfide respiration in Archaea and Bacteria appears to be similar. Membrane-bound respiratory chains produce a chemiosmotic potential, which is used by membrane-bound ATP synthases to form ATP [29, 32].

3.1.3 Polysulfide Reductase from *Desulfuromonas acetoxidans*

The final draft genome of *Drm. acetoxidans* codes for a "cytochromome" of 47 putative multiheme cytochromes *c* [150]. This strain contains several multiheme *c*-type cytochromes, the most abundant being the triheme cytochrome c_7. Polysulfides are formed in solution from the reaction of elemental sulfur with sulfide and are probably the *in vivo* substrate utilized by sulfur-reducing eubacteria.

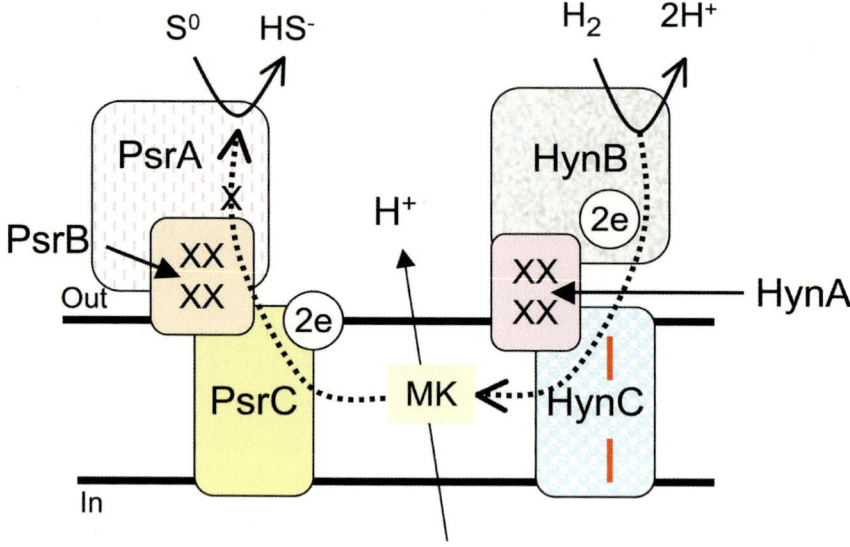

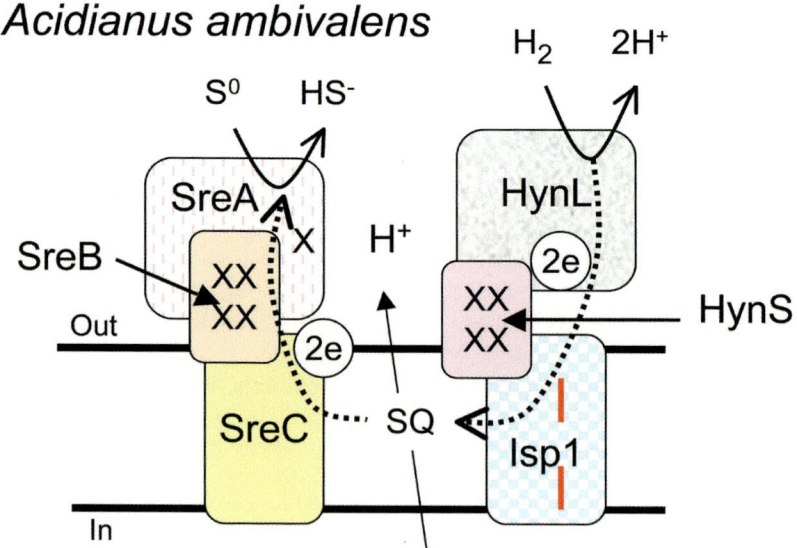

Figure 5 Model of sulfur-reducing complexes in membranes of *A. ambivalens* and *W. succinogenes*. PsrABC are sulfur-reducing subunits and HynABC are hydrogenase subunits of *W. succinogenes*. MK = menaquinone, SreABC are sulfur-reducing subunits and HynL, HynS, and Isp1 are hydrogenase subunits of *A. ambivalens*, SQ = sulfolobusquinone, X = [4Fe-4S] center, 2e = 2 electrons, and dashed lines indicates electron flow. The two vertical red lines in HynC and Isp1 indicate the presence of two b-type hemes. Out = periplasm and In = cytoplasm. Reproduced by permission from [25]; copyright 2012, Academic Press.

The *Drm. acetoxidans* triheme cytochrome c_7 is rapidly reduced by the *D. vulgaris* H [Fe] hydrogenase and it can completely reduce polysulfide with a very high specific activity of 20 µmoles of hydrogen consumed/min/mg protein [151]. This activity is twice as high as that reported for the purified tetraheme cytochrome c_3 with the highest specific SR activity from *Dsm. norvegicum* [26, 147]. These results indicate that cytochrome c_7 is probably the true terminal sulfur reductase in *Drm. acetoxidans* but the physiological electron donor for this triheme protein remains to be found.

3.1.4   Sulfur Oxidoreductase from *Sulfurospirillum deleyianum*

The *S. deleyianum* DSM 6946 sulfur oxidoreductase (SR) is a constitutive cytoplasmic enzyme. The *S. deleyianum* SR is energized by a [Ni-Fe] hydrogenase and is several times more active in crude extracts than in other sulfur-reducing eubacteria [27]. Sulfur reduction is enhanced by the presence of thiols and the SR contains at least one [4Fe-4S] center but no *b*- or *c*-type cytochromes [27].

3.2   Archaebacterial Sulfur Reductase

3.2.1   Membraneous Sulfur Reductase Complex from *Acidianus ambivalens*

The acidophilic *Acidianus (A.) ambivalens* DSM 3772 grows on elemental sulfur at 80 °C. A hydrogenase-sulfur reductase complex (SR) has been isolated from the membrane of *A. ambivalens* and the Sr subunits are similar to those found in *W. succinogenes* [25] (Figure 5). SR is a molybdoenzyme belonging to the DMSO reductase family [152]. The SR gene cluster consists of 5 subunits. The *sreA* gene produces a 110 kDa protein that has binding motifs for a [4Fe-4S] center. The *sreB* gene encodes for a protein rich in cysteines and could coordinate a [4Fe-4S] center. SreC is a hydrophobic protein that stabilizes the SR into the membrane and the role for SreD is unresolved at this time. The hydrogenase gene cluster consists of 12 genes and 3 of these encode for structural proteins. The HynL subunit contains nickel and the subunit HynS contains several [Fe-S] binding motifs, making this a [Ni-Fe] hydrogenase. The third structural gene, Isp1, is an integral membrane protein and serves to bind the 2 other subunits into the membrane. The quinone, presumed to be sulfolobusquinone, transfers electrons from hydrogenase to the SR.

3.2.2   Sulfur-Reducing Complex from *Pyrodictium abyssi*

Extensive studies on sulfur reduction have been performed with *Pyrococcus (P.) furiosus*, *A. ambivalens*, *Py. abyssi*, and *Py. brockii* as model organisms. The mechanism of sulfur reduction with H_2, in some members of the *Crenarchaeota* (e.g., *Pyrodictium* species and *A. ambivalens*) is similar to that of some eubacteria, such as *W. succinogenes*. This dissimilatory process involves two multi-subunits, membrane-bound enzymes: a nickel-iron-containing hydrogenase and a SR or PSR. These two enzymes together reduce sulfur to H_2S with H_2 as electron donor.

The composition of the described electron transfer chain shows participations of similar [NiFe] hydrogenases and similar PSR in the case of *A. ambivalens* and *W. succinogenes*, whereas the *Py. abyssi* SR seems to be different. Nine major subunits constitute the hydrogenase-sulfur multienzyme complex isolated from *Py. abyssi* DSM 6158 [153] and the subunits from *Py. abyssi* range from 24 to 82 kDa with 550 kDa for the entire complex. Analysis of this complex reveals the presence of a [NiFe] hydrogenase, a *c*-type heme, one or two *b*-type hemes, and an undetermined number of [Fe-S] centers.

3.2.3   Sulfur Reductase from *Pyrococcus furiosus*

P. furiosus has been studied as the model organism for the mechanism of the fermentation-based sulfur reduction. Two enzymes play key roles in the sulfur metabolism: a membrane-bound oxidoreductase complex (MBX) and a cytoplasmic coenzyme A-dependent NADPH sulfur oxidoreductase (NSR). MBX is encoded by an operon with 13 open reading frames and plays an essential role in mediating electron flow to sulfur [154]. NSR is a homodimeric flavoprotein (M_r 100,000) and reduces elemental sulfur to H_2S with NADPH as the electron donor [154]. This MBX-NSR complex sulfur-reduction system has only been reported so far in the *Thermococcales*.

4   Microbial Oxidation of Hydrogen Sulfide to Sulfate

The oxidative inorganic sulfur metabolism has been recently and extensively described both in eubacterial and archaebacterial microorganisms [28, 29, 49, 155]. Here, we briefly describe enzymes or multienzyme systems involved in sulfur compound oxidation in Bacteria and Archaea.

4.1   Archaebacterial Inorganic Sulfur Compound Oxidation

Aerobic dissimilatory sulfur oxidation is common in the order *Sulfolobales* of the *Crenarchaeota* which frequently thrive in continental solfataric fields [29, 155]. Mechanisms of archaeal inorganic sulfur compound oxidation were almost exclusively studied in *Acidianus* species with the thermoacidophilic *A. ambivalens* as the model organism [28, 29].

The initial step of *A. ambivalens* sulfur oxidation involves a cytoplasmic sulfur oxygenase reductase (SOR) (EC 1.13.11.55) catalyzing the oxygen-dependent sulfur disproportionation to form sulfide plus hydrogen sulfite [155] (reaction 14):

$$4\,S^0 \,+\, O_2 \,+\, 4\,H_2O \;\rightarrow\; 2\,H_2S \,+\, 2\,HSO_3^- \,+\, 2\,H^+ \tag{14}$$

Then thiosulfate is likely produced from sulfur and hydrogen sulfite in a non-enzymatic reaction.

SOR of *A. ambivalens* is a homo-oligomer composed of 24 identical monomers and the catalytic pocket of each subunit contains a low-potential mononuclear non-heme iron center and three conserved cysteinyl residues. The iron center is likely the site for both sulfur reduction and oxidation [156]. The three products of the *A. ambivalens* sulfur oxidation step (sulfite, thiosulfate, and sulfide) are presumably further oxidized to sulfate for energy conservation [29, 155].

Two sulfite oxidation pathways are present in *A. ambivalens*: (a) A membrane-bound sulfite:acceptor oxidoreductase as part of the Sox complex; (b) An alternative indirect soluble sulfite oxidation pathway coupled to substrate-level phosphorylation via APS reductase and APS:phosphate adenylyltransferase [155].

Two membrane-bound complexes are involved in *A. ambivalens* thiosulfate oxidation: (a) The terminal aa_3-type quinol oxidase which shuttles electrons from the caldariellaquinone pool to O_2 and consists of three subunits, which are encoded in a single operon in the *A. ambivalens* genome. (b) The membrane-bound tetrathionate-forming thiosulfate:quinone oxidoreductase which oxidizes thiosulfate to form tetrathionate with caldariellaquinone as electron acceptor [155, 157].

4.2   Eubacterial Inorganic Sulfur Compound Oxidation

The main enzymes or multienzyme complexes involved in sulfur compound oxidation are present both in CSB and phototrophic sulfur bacteria [49]. PSB and GSB use various combinations of sulfide, elemental sulfur, sulfite, and thiosulfate as electron donors in CO_2 fixation during anoxygenic photosynthetic growth [49]. Genetic and biochemical analyses show that the dissimilatory sulfur metabolism of the phototrophic organisms is very complex and still incompletely understood. We will only describe here the oxidative sulfur metabolism of anoxygenic phototrophic bacteria (GSB, PSB, and PNSB).

A variety of enzymes catalyzing inorganic sulfur oxidation reactions have been purified and biochemically and genetically characterized from PSB and GSB [49]. Complete genome sequence data are currently available for one strain of PSB and for ten strains of GSB. A number of genes potentially involved in the oxidative sulfur metabolism are present both in GSB and PSB: for example, genes for the sulfide:quinone oxidoreductase (SQR) and the sulfide-oxidizing enzyme flavocytochrome *c* [49]. On a molecular biochemical and genetic level, sulfur compound oxidation is best characterized in the GSB *Chlorobaculum tepidum* and in the PSB *Alc. vinosum*.

4.2.1   Oxidation of Sulfide

Enzymes that oxidize sulfide are the membrane-bound SQR and the periplasmic flavocytochrome *c* sulfide dehydrogenase (FccAB, EC 1.8.2.3) [49].

4.2.1.1   Sulfide:Quinone Oxidoreductase

SQR (EC 1.8.5.4) catalyzing sulfide oxidation with an isoprenoid quinone as the electron acceptor is present in both phototrophic and chemotrophic bacteria [49]. SQR plays an important role for the sulfide oxidation in PSB and FccAB appears less widespread. SQR is the only known sulfide-oxidizing enzyme that is found in all GSB strains. SQR is a member of the flavin disulfide reductases family.

SQR of *Rhodobacter capsulatus* is a membrane-bound flavoprotein with its active site located in the periplasm. The first X-ray structure for an SQR has been recently determined to 2.6 Å resolution in *A. ambivalens* [158]. This membrane-bound flavoprotein has two redox active sites: a covalently bound FAD and an adjacent pair of cysteine residues bridged by a trisulfide bridge between the two cysteine residues [158].

4.2.1.2   Flavocytochrome *c* Sulfide Dehydrogenase

Flavocytochrome *c* is usually a periplasmic enzyme consisting of a small FccA cytochrome *c* subunit (20 kDa) and a larger sulfide-binding FccB flavoprotein subunit (44 kDa). *In vitro*, flavocytochromes can catalyze electron transfer from sulfide to a variety of small *c*-type cytochromes (such as *Alc. vinosum* cytochrome c_{550}) that may then donate electrons to the photosynthetic reaction center [49]. The *in vivo* role of flavocytochrome *c* is unclear and if indeed the FccAB oxidizes H_2S *in vivo*, both PSB and GSB have alternative sulfide-oxidizing enzyme systems, such as SQR, that may be quantitatively more important.

4.2.2   Oxidation of Polysulfides

Polysulfides appear to be the primary product of sulfide oxidation in a number of PSB and GSB. The oxidation of sulfur deposits is the least understood step of sulfur metabolism. It is currently unknown how polysulfides are converted into sulfur globules and it could be a purely chemical, spontaneous process as longer polysulfides are in equilibrium with S^0 [49].

Many PSB and GSB can oxidize externally supplied solid elemental sulfur. Utilization of solid S^0 must include binding and/or activation of the sulfur as well as transport inside the cells. In the case of cyclo-octasulfur, this activation process could be an opening of the S_8 ring by nucleophilic reagents, leading to the formation of linear organic or inorganic polysulfanes. It may be speculated that "sulfur chains" rather than the more stable "sulfur rings" are the microbiologically preferred form of S^0 for most SOB. It has been proposed in *Alc. vinosum* that the stored sulfur has to be reductively activated to the oxidation state of sulfide in order to serve as a substrate for sulfite reductase operating in reverse performing the six-electron oxidation of sulfide to sulfite [159].

4.2.3   Oxidation of Stored Sulfur to Sulfite

The mechanism by which the periplasmically stored sulfur is made available to the cytoplasmic dissimilatory sulfite reductase is still unclear. The only gene region known so far to be essential for the oxidation of stored sulfur is the *dsr* operon. The reverse DsrAB of *Alc. vinosum* is encoded together with 13 other proteins in the *dsr* operon, *dsr ABEFHCMKLJOPNRS*. A model of the *Alc. vinosum* sulfur oxidation pathway has been proposed [160]. It is suggested that the sulfur is reductively activated, transported to and further oxidized in the cytoplasm, since the proteins encoded at the dsr locus are either membrane-bound or cytoplasmic and cannot act directly on the extracytoplasmic sulfur globules [160].

4.2.4   Oxidation of Sulfite to Sulfate

Two different pathways for sulfite oxidation are known in phototrophic and chemotrophic SOB: (a) Indirect, AMP-dependent sulfite oxidation via APS. Sulfite is oxidized by APS reductase in a cytoplasmic reaction that consumes sulfite and AMP and produces APS and reducing equivalents. This oxidative pathway occurs exclusively in members of the *Chromatiaceae* and in some GSB. (b) Direct sulfite oxidation by sulfite dehydrogenase (EC 1.8.2.1), typically a molybdenum-containing protein belonging to the sulfite oxidase family of molybdoenzymes [161].

4.2.5   Oxidation of Thiosulfate

In phototrophic and chemotrophic SOB that do not form sulfur deposits a periplasmic thiosulfate-oxidizing multienzyme complex (Sox complex) is responsible for formation of sulfate from thiosulfate. *Alc. vinosum* can pursue two different thiosulfate oxidation pathways, first the complete thiosulfate oxidation to sulfate, and second formation of tetrathionate.

4.2.5.1 Oxidation of Thiosulfate to Tetrathionate

Alc. vinosum tetrathionate-forming thiosulfate dehydrogenase (TsdA) (thio-sulfate:acceptor oxidoreductase EC 1.8.2.2) is a periplasmic, monomeric diheme 30 kDa cytochrome c_{554} with an isoelectric point of 4.2. UV-visible and EPR spectroscopies suggest methionine and cysteine as distal axial ligands of the two heme irons in TsdA.

4.2.5.2 Oxidation of Thiosulfate to Sulfate by the Sox Multienzyme System

Many GSB and PSB can oxidize thiosulfate completely to sulfate. The Sox complex, a periplasmic thiosulfate-oxidizing multienzyme complex was first found and char-acterized in *Paracoccus* (*Pc.*) *versutus* and *Pc. pantotrophus*. The Sox gene cluster of *Pc. pantotrophus* comprise 15 genes (*soxRSVWXYZABCDEFGH*). The so-called SoxAX cytochromes are heme-thiolate proteins playing a key role in bacterial thiosulfate oxidation [162]. They initiate the reaction cycle of a multienzyme com-plex in both photo- and chemotrophic SOB by catalyzing the attachment of sulfur substrates such as thiosulfate to a conserved cysteine present in a carrier protein.

Alc. vinosum possesses five *sox* genes in two independent loci (*soxBXA* and *soxYZ*) encoding proteins related to components of the *Pc. pantotrophus* Sox complex. Three sox-encoded proteins were purified and characterized from *Alc. vinosum*: the heterodimeric *c*-type cytochrome SoxXA, the monomeric SoxB containing a dimanganese center, and the heterodimeric thiosulfate-binding protein Sox YZ [160, 163].

In summary, Sox and Dsr proteins are absolutely essential in *Alc. vinosum* for the oxidation of thiosulfate and stored sulfur, respectively. Clusters of *dsr* and *sox* genes are also present in the only distantly related GSB as well as in the other sulfur-storing chemotrophic and phototrophic SOB. The mechanisms of thiosulfate oxidation via sulfur deposition and of the oxidation of deposited sulfur seem to be evolutionary highly conserved. Studies on *Alc. vinosum* can help to elucidate the sulfur oxidation pathway in other sulfur-storing bacteria. GSB oxidize sulfide and thiosulfate to sulfate, with extracellular sulfur globules as an intermediate and sulfur globule oxidation is strictly dependent on the dSiR system. In GSB, depending on the strain, sulfite is probably oxidized to sulfate by one or two mechanisms with different evolutionary origins, using either APS reductase or the polysulfide reductase-like complex 3 [164].

5 Conclusions

Anaerobic and aerobic microorganisms play a paramount role in biological cycling of inorganic sulfur compounds because they have developed enzyme systems to use inorganic sulfur compounds as electron donors or electron acceptors to facilitate growth.

Chemolithotrophic microorganisms, primarily SRB, shuttle electrons to sulfoxy compounds or S^0 with the accumulation of H_2S where sulfur is in the most reduced form. Aerobic SOB and CSB facilitate the removal of reduced sulfur compounds with the production of sulfate. Although there has been considerable advancement to understand the enzymology of enzymes for production and utilization of H_2S, further studies are needed to establish gene activities for these sulfur-metabolizing enzymes.

Such genomic studies should provide an insight into ancestry for horizontal gene transfer and to understand the energetic processes that drive the inorganic sulfur reactions. Since considerable similarity exists in the enzymology of H_2S production and H_2S utilization, future studies will be needed to determine which process evolved first.

Furthermore, investigations of the evolutionary links between the biological cycles of sulfur and nitrogen will be important to pursue since [4Fe-4S] clusters, iron porphyrins, and MGD cofactors are used in enzymes for both sulfite to hydrogen sulfide and nitrite to ammonia reduction [20].

Abbreviations and Definitions

A.	*Acidianus*
acetyl-CoA	acetyl-coenzyme A
ADP	adenosine 5'-diphosphate
Alc.	*Allochromatium*
AMP	adenosine 5'-monophosphate
APS	adenosine 5'-phosphosulfate
APSR	adenylylsulfate reductase
Ar.	*Archaeoglobus*
aSiR	assimilatory-type sulfite reductase
asr	anaerobic sulfite reduction
ATP	adenosine 5'-triphosphate
ATPS	ATP sulfurylase
CSB	colorless sulfur bacteria
D.	*Desulfovibrio*
Dba.	*Desulfobacter*
Dbu.	*Desulfobulbus*
DMSO	dimethylsulfoxide
Drm.	*Desulfuromonas*
Dsm.	*Desulfomicrobium*
dSiR	dissimilatory sulfite reductase
Dst.	*Desulfotomaculum*
D. vulgaris H	*Desulfovibrio vulgaris* Hildenborough
$\Delta G^{0'}$	standard free energy change
E.	*Escherichia*
$E^{0'}$	standard reduction potential

EPR	electron paramagnetic resonance
EXAFS	extended X-ray absorption fine structure
FAD	flavin adenine dinucleotide
GSB	green sulfur bacteria
MBX	membrane-bound oxidoreductase complex
MGD	molybdopterin guanine dinucleotide
MK	menaquinone
Ms.	*Methanosarcina*
NAD	nicotinamide adenine dinucleotide
NADH	nicotinamide adenine dinucleotide, reduced
NADPH	nicotinamide adenine dinucleotide phosphate, reduced
NSR	coenzyme A-dependent NADPH sulfur oxidoreductase
P.	*Pyrococcus*
Pc.	*Paracoccus*
P_i	inorganic phosphate
PNSB	purple non-sulfur bacteria
PP_i	inorganic pyrophosphate
PSB	purple sulfur bacteria
PSR	polysulfide reductase
Py	*Pyrodictium*
Pyb.	*Pyrobaculum*
S.	*Sulfurospirillum*
Sal.	*Salmonella*
SOB	sulfide-oxidizing bacteria
SOR	sulfur oxygenase reductase
SQR	sulfide:quinone reductase
SR	sulfur reductase
SRB	sulfate-reducing bacteria
SRP	sulfate-reducing prokaryotes
T.	*Thermodesulfobacterium*
W.	*Wolinella*

Acknowledgments Sequence data were produced by the US Department of Energy Joint Genome Institute http://www.jgi.doe.gov/ in collaboration with the user community.

References

1. R. Wang, *Physiol. Rev.* **2012**, *92*, 791–896.
2. L. L. Barton, G. D. Fauque, *Adv. Appl. Microbiol.* **2009**, *68*, 41–98.
3. R. Rabus, T. A. Hansen, F. Widdel, in *The Prokaryotes,* Vol. 2, *Ecophysiology and Biochemistry,* Eds M. Dworkin, S. Falkow, E. Rosenberg, K.-K. Scheifer, E. Stackebrandt, Springer, Berlin, 2006, pp. 659–768.

4. G. D. Fauque, in *Biotechnology Handbooks*, Vol. 8, *Sulfate-Reducing Bacteria*, Ed L.L. Barton, Plenum Press, New York, 1995, pp. 217–241.
5. J. Loubinoux, J.-P. Bronowicki, I. A. C. Pereira, J.-L. Mougenel, A. LeFaou, *FEMS Microbiol. Ecol.* **2002**, *40*, 107–112.
6. G. Muyzer, A. J. M. Stams, *Nature Rev. Microbiol.* **2008**, *6*, 441–454.
7. B. Ollivier, J.-L. Cayol, G. Fauque, in *Sulphate-Reducing Bacteria: Environmental and Engineered Systems*, Eds L. L. Barton, W. A. Hamilton, Cambridge University Press, Cambridge, UK, 2007, pp. 305–328.
8. Y. A. Shen, R. Buick, D. E. Canfield, *Nature* **2001**, *410*, 77–81.
9. Y. Shen, R. Buick, *Earth-Sci. Rev.* **2004**, *64*, 243–272.
10. G. Fauque, B. Ollivier, in *Microbial Diversity and Bioprospecting*, Ed A.T. Bull, ASM Press, Washington, D.C., 2004, pp. 169–176.
11. H. Castro, N. Williams, A. Ogram, *FEMS Microbiol. Ecol.* **2000**, *31*, 1–9.
12. V. M. Gumerov, A Mardanov, A Beletsky, M. Prokofeva, E. A. Bonch-Osmolovskaya, N. Ravin, K. Skyrabin, *J. Bacteriol.* **2011**, *193*, 2355–2356.
13. L. Jabari, H. Gannoun, J.-L. Cayol, M. Hamdi, B. Ollivier, G. Fauque, M.-L. Fardeau, *Int. J. Syst. Evol. Microbiol.* **2013**, *63*, 2082–2087.
14. H. Cypionka, *Annu. Rev. Microbiol.* **2000**, *54*, 827–848.
15. G. Fauque, J. LeGall, L. L. Barton, in *Variations in Autotrophic Life*, Eds J. M. Shively, L. L. Barton, Academic Press, London, 1991, pp. 271–337.
16. J. LeGall, G. Fauque, in *Biology of Anaerobic Microorganisms*, Ed A. J. B. Zehnder, Wiley, New York, 1988, pp. 587–639.
17. J. J. G. Moura, P. Gonzalez, I. Moura, G. Fauque, in *Sulphate-Reducing Bacteria: Environmental and Engineered Systems*, Eds L. L. Barton, W. A. Hamilton, Cambridge University Press, Cambridge, UK, 2007, pp. 241–264.
18. H. D. Peck, Jr., *Proc. Natl. Acad. Sci. USA* **1959**, *45*, 701–708.
19. D. R. Kremer, M. Veenhuis, G. Fauque, H. D. Peck, Jr., J. LeGall, J. Lampreia, J. J. Moura, T. A. Hansen, *Arch. Microbiol.* **1988**, *150*, 296–301.
20. J. Simon, P. M. H. Kroneck, *Adv. Microbial Physiol.* **2013**, *62*, 45–117.
21. G. Fauque, H. D. Peck, Jr., J. J. G. Moura, B. H. Huynh, Y. Berlier, D. V. DerVartanian, M. Teixeira, A. E. Przybila, P. A. Lespinat, I. Moura, J. LeGall, *FEMS Microbiol. Rev.* **1988**, *54*, 299–344.
22. J. J. G. Moura, I. Moura, M. Teixeira, A. V. Xavier, G. Fauque, J. LeGall, in *Metal Ions in Biological Systems*, Vol. 23, *Nickel and Its Role in Biology*, Ed H. Sigel, Marcel Dekker Inc., New York, 1988, pp. 285–314.
23. I. A. C. Pereira, A. V. Xavier, in *Encyclopedia of Inorganic Chemistry*, Ed R. B. King, Vol. 5, 2nd edn, Wiley, New York, 2005, pp. 3360–3376.
24. I. A. C. Pereira, in *Microbial Sulfur Metabolism*, Eds C. Dahl, C. G. Friedrich, Springer, Berlin, 2008, pp. 24–35.
25. G. D. Fauque, L. L. Barton, *Adv. Microbial Physiol.* **2012**, *60*, 1–90.
26. G. D. Fauque, *Meth. Enzymol.* **1994**, *243*, 353–367.
27. G. Fauque, O. Klimmek, A. Kröger, *Meth. Enzymol.* **1994**, *243*, 367–383.
28. A. Kletzin, T. Urich, F. Muller, T. M. Bandeiras, C. M. Gomez, *J. Bioenerg. Biomem.* **2004**, *36*, 77–91.
29. A. Kletzin, in *Archaea: Evolution, Physiology, and Molecular Biology*, Eds R. A. Garrett, H.-P. Klenk, Blackwell Publishing Ltd., Madden, MA, USA, 2007, pp. 261–274.
30. A. LeFaou, B. S. Rajagopal, L. Daniels, G. Fauque, *FEMS Microbiol. Rev.* **1990**, *75*, 351–382.
31. J. Boulègue, *Phosphorus Sulfur Silicon Relat. Elem.* **1978**, *5*, 127–128.
32. R. Hedderich, O. Klimmek, A. Kröger, R. Dirmeier, M. Keller, K. O. Stetter, *FEMS Microbiol. Rev.* **1999**, *22*, 353–381.
33. N. Pfennig, H. Biebl, *Arch. Microbiol.* **1976**, *110*, 3–12.
34. F. Widdel, N. Pfennig, in *The Prokaryotes*, Eds A. Balows, H. G. Trüper, M. Dworkin, W. Harder, K. H. Schleifer, 2nd edn, Vol. 4, New York, Springer, 1992, pp. 3379–3389.

35. A. Zöphel, M. C. Kennedy, H. Beinert, P. M. H. Kroneck, *Arch. Microbiol.* **1988**, *150*, 72–77.
36. A. Zöphel, M. C. Kennedy, H. Beinert, P. M. H. Kroneck, *Eur. J. Biochem.* **1991**, *195*, 849–856.
37. H. Biebl, N. Pfennig, *Arch. Microbiol.* **1977**, *112*, 115–117.
38. O. Ben Dhia Thabet, T. Wafa, K. Eltaief, J.-L. Cayol, M. Hamdi, G. Fauque, M.-L. Fardeau, *Curr. Microbiol.* **2011**, *62*, 486–491.
39. B. Ollivier, C. E. Hatchikian, G. Prensier, J. Guezennec, J.-L. Garcia, *Int. J. Syst. Bacteriol.* **1991**, *41*, 74–81.
40. R. Schauder, A. Kröger, *Arch. Microbiol.* **1993**, *159*, 491–497.
41. R. Thauer, K. Jungermann, K. Decker, *Bacteriol. Rev.* **1977**, *41*, 100–180.
42. P. Schönheit, T. Schäfer, *World J. Microbiol. Biotech.* **1995**, *11*, 26–57.
43. K. O. Stetter, G. Gaag, *Nature* **1983**, *305*, 309–311.
44. K. Finster, *J. Sulf. Chem.* **2008**, *29*, 281–292.
45. F. Bak, N. Pfennig, *Arch. Microbiol.* **1987**, *147*, 184–189.
46. K. Finster, W. Liesack, B. Thamdrup, *Appl. Environ. Microbiol.* **1998**, *64*, 119–125.
47. M. Kramer, H. Cypionka, *Arch. Microbiol.* **1989**, *151*, 232–237.
48. P. Philippot, M. Van Zuylen, K. Lepot, C. Fhomazzo, J. Farquhar, M. J. Van Kranendonk, *Science* **2007**, *317*, 1534–1537.
49. G. Muyzer, J. G. Kuenen, L. A. Robertson, in *The Prokaryotes – Prokaryotic Physiology and Biochemistry*, Eds E. Rosenberg, E. F. DeLong, S. Lory, E. Stackebrandt, F. Thompson, Springer-Verlag, Berlin, Heidelberg, 2013, pp. 555–588.
50. N.-U. Frigaard, C. Dahl, *Adv. Microbial Physiol.* **2009**, *54*, 103–200.
51. J. F. Imhoff, in *Sulfur Metabolism in Phototrophic Organisms*, Eds R. Hell, C. Dahl, D. Knaff, T. Leustek, Springer, The Netherlands, 2008, pp. 269–287.
52. D.Y. Sorokin, H. Banciu, L. A. Robertson, J. G. Kuenen, M. S. Muntyan, G. Muyzer, in *The Prokaryotes – Prokaryotic Physiology and Biochemistry*, Eds E. Rosenberg, E. F. DeLong, S. Lory, E. Stackebrandt, F. Thompson, Springer-Verlag, Berlin, Heidelberg, 2013, pp. 529–554.
53. M. S. Vandiver, S. H. Snyder, *J. Mol. Med.* **2012**, *90*, 255–263.
54. D. Julian, K. L April, S. Patel, J. R. Stein, S.E. Wohlgemuth, *J. Experiment. Biol.* **2005**, *208*, 4109–4122.
55. D. W. Kraus, J. E. Doeller, C.S. Powell, *J. Experiment. Biol.* **1996**, *199*, 1343–1352.
56. T. Bagarino, *Aquat. Toxicol.* **1992**, *24*, 21–62.
57. P. Nicholls, *Biochim. Biophys. Acta* **1975**, *396*, 24–35.
58. R. O. Beauchamp, Jr., J. S. Bus, J. A. Popp, C. J. Boreiko, D. A. Andjelkovich, *Crit. Rev. Toxicol.* **1984**, *13*, 25–97.
59. F. Widdel, in *Biology of Anaerobic Microorganisms*, Ed A. J. B. Zehnder, Wiley, New York, 1988, pp. 469–585.
60. S. M. Caffrey, G. Voordouw, *Antonie van Leeuwenhoek.* **2010**, *97*, 11–20.
61. J. M. Akagi, L. L. Campbell, *J. Bacteriol.* **1963**, *86*, 563–568.
62. H. Nakazawa, A. Arakaki, S. Narita-Yamada, I. Yashiro, K. Jinno, N. Aoki, A. Tsuruyama, Y. Okamura, S. Tanikawa, N. Fujita, H. Takeyama, *Genome Res.* **2009**, *19*, 1801–1808.
63. B. Suh, J. M. Akagi, *J. Bacteriol.* **1969**, *99*, 210–215.
64. J. P. Lee, H. D. Peck, Jr., *Biochem. Biophys. Res. Commun.* **1971**, *45*, 583–589.
65. J. M. Akagi, V. Adams, *J. Bacteriol.* **1973**, *116*, 392–396.
66. K. Kobayashi, S. Tachibana, M. Ishimoto, *J. Biochem.* **1969**, *65*, 155–157.
67. J.-P. Lee, J. LeGall, H. D. Peck, Jr., *J. Bacteriol.* **1973**, *115*, 529–542.
68. R. H. Haschke, L. L. Campbell, *J. Bacteriol.* **1971**, *106*, 603–607.
69. E. C. Hatchikian, J. G. Zeikus, *J. Bacteriol.* **1983**, *153*, 2111–1220.
70. W. Nakatsukasa, J. M. Akagi, *J. Bacteriol.* **1969**, *98*, 429–433.
71. E. C. Hatchikian, *Arch. Microbiol.* **1975**, *105*, 249–256.
72. R. M. Fitz, H. Cypionka, *Arch. Microbiol.* **1990**, *154*, 400–406.
73. M. Broco, M. Rousset, S. Oliveira, C. Rodrigues-Pousada, *FEBS Lett.* **2005**, *579*, 4803–4807.

74. H.D. Peck, Jr, J. LeGall, *Phil. Trans. R. Soc. B*. **1982**, *298*, 443–466.
75. K. Parey, U. Demmer, E. Warkentin, A. Wynen, U. Ermler, C. Dahl, *PLoS ONE* **2013**, 8: available on line, e74707. doi:10.1371/.
76. N. Sekulic, K. Dietrich, I. Paamann, S. Ort, M. Konrad, A. Lavie, *J. Mol. Biol.* **2007**, *367*, 488–500.
77. C. Dahl, H-G. Koch, O. Keuken, H. G. Trüper, *FEMS Microbiol. Lett.* **1990**, *67*, 27–32.
78. O. Y. Gavel, S. A. Bursakov, J. J. Calvete, G. N. George, J. J. Moura, I. Moura, *Biochemistry* **1998**, *37*, 16225–16232.
79. Y. Taguchi, M. Sugishima, K. Fukuyama, *Biochemistry* **2004**, *43*, 4111–4118.
80. O. Y. Gavel, A. V. Kladova, S. A. Bursakov, J. M. Dias, S. Texeira, V. L. Shnyrov, J. J. G. Moura, I. Moura, M. J. Romão, J. J. Trincão, *Acta Crystallogr. Sect . F, Struct. Biol. Cryst. Commun.* **2008**, *64*, 593–595.
81. Y.-L. Chiang, Y.-C. Hsieh, J-Y. Fang, E.-H. Liu, Y.-C. Huang, P. Chuankhayan, J. Jeyakanthan, M.-Y. Liu, S. I. Chan, C.-J. Chen, *J. Bacteriol.* **2009**, *191*, 7597–7608.
82. G. Fritz, T. Büchert, P. M. H. Kroneck, *J. Biol. Chem.* **2002**, *277*, 26066–26073.
83. B. Meyer, J. Kuever, *PLoSONE* **2008** 3, available online, e1514.doi10.1371/journal.pone. 0001514.
84. J. Lampreia, I. Moura, M. Teixeira, H. D. Peck, Jr., J. LeGall, B. H. Huynh, J. J. G. Moura, *Eur. J. Biochem.* **1990**, *188*, 653–664.
85. A. R. Ramos, K. L. Keller, J. D. Wall, *Front. Microbiol.* **2012**, *3*, 137, available online, doi:10.3389/fmicb.2012.00137.
86. G. M. Zane, H-c. B. Yen, J. D. Wall, *Appl. Environ. Microbiol.* **2010**, *76*, 5500–5509.
87. I. Moura, A. R. Lino, *Meth. Enzymol.* **1994**, *243*, 296–303.
88. M. J. Murphy, L. M. Siegel, *J. Biol. Chem.* **1973**, *248*, 6911–6919.
89. A. Schiffer, K. Parey, E. Warkentin, K. Diederichs, H. Huber, K. O. Stetter, P. M. Kroneck, U. Ermler, *J. Mol. Biol.* **2008**, *379*, 1063–1074.
90. J. Steuber, P. M. H. Kroneck, *Inorg. Chim. Acta* **1998**, *276*, 52–57.
91. G. Fritz, A. Schiffer, A. Behrens, T. Buchert, U. Ermler, P. M. H. Kroneck, in *Microbial Sulfur Metabolism*, Eds C. Dahl, C. G. Friedrich, Berlin, Springer-Verlag, 2008, pp. 13–23.
92. Y.-C. Hsieh, M.-L. Liu, V. C.-C. Wang, Y.-L. Chiang, E.-H. Liu, W. G. Wu, S. I. Shan, C.-J. Chen, *Mol. Microbiol.* **2010**, *78*, 1101–1116.
93. T. F. Oliveira, E. Franklin, J. P. Afonso, A. R. Khan, N. J. Oldham, I. A. C. Pereira, M. Archer, *Front. Microbiol.* **2011**, *2*, 71 doi:10.3389.
94. F. Grein, I. A. C. Pereira, C. Dahl. *J. Bacteriol.* **2010**, *192*, 6369–6377.
95. T. F. Oliveira, C. Vornhein, P. M. Matias, S. S. Venceslau, I. A. C. Pereira, M. Archer, *J. Struct. Biol.* **2008**, *164*, 236–239.
96. T. F. Oliveira, C. Vonrhein, P. M. Matias, S. S. Venceslau, I. A. C. Pereira, M. Archer, *J. Biol. Chem.* **2008**, *283*, 34141–34149.
97. R. R. Karkhoff-Schweizer, M. Bruschi, G. Voordouw, *Eur. J. Biochem.* **1993**, *211*, 501–507.
98. J. Ostrowski, J.-Y. Wu, D. C. Rueger, B. E. Miller, L. M. Siegel, N. M. Kredich, *J. Biol. Chem.* **1989**, *264*, 15726–15737.
99. E. T. Adman, L. Sieker, L. Jensen, *J. Biol. Chem.* **1976**, *248*, 3987–3996.
100. M. Bruschi, F. Guerlesquin, *FEMS Microbiol. Rev.* **1988**, *54*, 155–176.
101. D. H. George, L. T. Hunt, L. S. L. Yeh, W. C. Barker, *J. Mol. Evol.* **1985**, *22*, 117–143.
102. I. Moura, J. LeGall, A. R. Lino, H. D. Peck, Jr., G. Fauque, A. V. Xavier, D. V. DerVartanian, J. J. G. Moura, B. H. Huynh, *J. Am. Chem. Soc.* **1988**, *110*, 1075–1082.
103. J. Steuber, H. Cypionka, P. M. H. Kroneck, *Arch. Microbiol.* **1994**, *162*, 255–260.
104. J. Steuber, A. F. Arendsen, W. R. Hagen, P. M. H. Kroneck, *Eur. J. Biochem.* **1995**, *233*, 873–879.
105. B. M. Wolfe, S. Lui, J. A. Cowan, *Eur. J. Biochem.* **1994**, *223*, 79–89.
106. A. J. Pierik, M. G. Duyvis, J. M. L. M. van Helvoort, R. B. J. Wolbert, W. R. Hagen, *Eur. J. Biochem.* **1992**, *205*, 111–115.

107. R. H. Pires, S. S. Venceslao, F. Morais, M. Teixeira, A. V. Xavier, I. A. Pereira, *Biochemistry* **2006**, *45*, 249–262.

108. R. R. Karkhoff-Schweizer, D. P. W. Huber, G. Voordouw, *Appl. Env. Microbiol.* **1995**, *61*, 290–296.

109. N. Mizuno, G. Voordouw, K. Miki, A. Sarai, Y. Higuchi, *Structure*, **2003**, *11*, 1133–1140.

110. J. C. Mathews, R. Timkovich, M.-Y. Lin, J. LeGall, *Biochemistry* **1995**, *34*, 5248–5251.

111. P. A. Trudinger, *J. Bacteriol.* **1970**, *104*, 158–170.

112. Ø. Larsen, T. Lien, N. K. Birkeland, *Extremophiles*, **1999**, *3*, 63–70.

113. D. V. DerVartanian, *Meth. Enzymol.* **1994**, *243*, 270–276.

114. N . Klouche, O. Basso, J.-F. Lascourrèges, J.-L. Cayol, P. Thomas, G. Fauque, M.-L. Fardeau, M. Magot, *Int. J. Syst. Evol. Microbiol.* **2009**, *59*, 3100–3104.

115. Ø. Larsen, T. Lien, N. K. Birkeland, *FEMS Microbiol. Lett.* **2000**, *186*, 41–46.

116. B. R. Crane, L. M. Siegel, E. D. Getzoff, *Science*, **1995**, *270*, 59–67.

117. C. Dahl, N. M. Kredich, R. Deutzmann, H. G. Trüper, *J. Gen. Microbiol.* **1993**, *139*, 1817–1828.

118. L. Debussche, D. Thibaut, B. Cameron, J. Crouzet, F. Blanche, *J. Bacteriol.* **1990**, *172*, 6239–6244.

119. G. Fauque, A. Lino, M. Czechowski, L. Kang, D. V. DerVartanian, J. J. G. Moura, J. LeGall, I. Moura, *Biochim. Biophys. Acta* **1990**, *1040*, 112–118.

120. E. C. Hatchikian, *Meth. Enzymol.* **1994**, *243*, 276–295.

121. O. Haouari, M.-L. Fardeau, J.-L. Cayol, G. Fauque, C. Casiot, F. Elbaz-Poulichet, M. Hamdi, B. Ollivier, *Syst. Appl. Microbiol.* **2008**, *31*, 38–42.

122. A. M. Stolzenberg, S. H. Strauss, R. H. Holm, *J. Am. Chem. Soc.* **1981**, *103*, 4763–4778.

123. C. Dahl, H. G. Trüper, *Meth. Enzymol.* **2001**, *331*, 427–441.

124. B. B. Jørgensen, *Science*, **1990**, *249*, 152–154.

125. L. Stoffels, M. Krehenbrink, B. C. Berks, G. Unden, *J. Bacteriol.* **2012**, *194*, 475–485.

126. R. Starkey, *Soil Sci.* **1950**, *70*, 55–66.

127. S. E. Winter, P. Thiennimitr, M. G. Winter, B. P. Butler, D. L. Huseby, R. W. Crawford, J. M. Russell, C. L. Bevins, L. G. Adams, R. M. Tsolis, J. R. Roth, A. J. Bäumler, *Nature* **2010**, *467*, 426–429.

128. E. F. Johnson, B. Mukhopadhyay, *J. Biol. Chem.* **2005**, *280*, 38776–38786.

129. E. F. Johnson, B. Mukhopadhyay, *Appl. Env. Microbiol.* **2008**, *74*, 3591–3595.

130. H. Laue, M. Friedrich, J. Ruff, A. M. Cook, *J. Bacteriol.* **2001**, *183*, 1727–11733.

131. G. Harrison, C. Curle, E. J. Laishley, *Arch. Microbiol.* **1984**, *138*, 72–78.

132. V. L. Barbosa-Jefferson, F. J. Zhao, S. P. McGrath, N. Morgan, *Soil Biol. Biochem.* **1998**, *30*, 553–559.

133. E. L. Barrett, M. A. Clark, *Microbiol. Rev.* **1987**, *51*, 195–205.

134. M. Hinojosa-Leon, M. Dubourdieu, J. A. Sanchez-Crispin, M. Chippaux, *Biochem. Biophys. Res. Commun.* **1986**, *136*, 577–581.

135. J. L. Burns and T. J. DiChristina, *Appl. Environ. Microbiol.* **2009**, *75*, 5209–5217.

136. A. P. Hinsley, B. C. Berks, *Microbiology* **2002**, *148*, 3631–3638.

137. P. Hallenbeck, M. A. Clark, E. L. Barrett, *J. Bacteriol.* **1989**, *171*, 3008–3015.

138. C. J. Huang, E. L. Barrett, *J. Bacteriol.* **1991**, *173*, 1544–1553.

139. L. M. Siegel, P.S. Davis, *J. Biol. Chem.* **1974**, *249*, 1587–1598.

140. M. Kern, M. G. Klotz, J. Simon, *Mol. Microbiol.* **2011**, *82*, 1515–1530.

141. S. Shirodkar, S. Reed, M. Romine, D. Saffarini, *Environ. Microbiol.* **2011**, *158*, 287–293.

142. H. L. Drake, J. M. Akagi, *Biochem. Biophys. Res. Commun.* **1976**, *71*, 1214–1219.

143. B. H. Huynh, L. Kang, D. V. DerVartanian, H. D. Peck, Jr., J. LeGall, *J. Biol. Chem.* **1984**, *259*, 15373–15376.

144. I. Moura, A. R. Lino, J. J. G. Moura, A. V. Xavier, G. Fauque, H. D. Peck, Jr., J. LeGall, *Biochem. Biophys. Res. Commun.* **1986**, *141*, 1032–1041.

145. J. J. G. Moura, I. Moura, H. Santos, A. V. Xavier, M. Scandellari, J. LeGall, *Biochem. Biophys. Res. Commun.* **1982**, *108*, 1002–1009.

146. G. Fauque, Doctorat d'Etat Thesis in Physical Sciences, University of Technology of Compiègne, France, 1985, 222 pages.
147. G. Fauque, D. Hervé, J. LeGall, *Arch. Microbiol.* **1979**, *121*, 261–264.
148. R. Cammack, G. Fauque, J. J. G. Moura, J. LeGall, *Biochim. Biophys. Acta* **1984**, *784*, 68–74.
149. G. D. Fauque, L. L. Barton, J. LeGall, *Sulphur in Biology: Ciba Foundation Symposium* **1980**, *72*, 71–86.
150. A. S. Alves, C. M. Paquete, B. M. Fonseca, R. O. Louro, *Metallomics* **2011**, *3*, 349–353.
151. I. A. C. Pereira, I. Pacheco, M.-Y. Liu, J. LeGall, A. V. Xavier, M. Teixeira, *Eur. J. Biochem.* **1997**, *248*, 323–328.
152. S. Laska, F. Lottspeich, A. Kletzin, *Microbiology* **2003**, *149*, 2357–2371.
153. M. Keller, R. Dirmeier, *Meth. Enzymol.* **2001**, *331*, 442–451.
154. S. L. Bridger, S. M. Clarkson, K. Stirrett, M. B. Debarry, G. L. Lipscomb, G. J. Schut, J. Westpheling, R. A. Scott, M. W. W. Adams, *J. Bacteriol.* **2011**, *193*, 6498–6504.
155. A. Kletzin, in *Microbial Sulfur Metabolism,* Eds C. Dahl, C. G. Friedrich, Springer, Berlin, 2008, pp. 184–201.
156. T. Urich, C. M. Gomes, A. Kletzin, C. Frazao, *Science* **2006**, *311*, 996–1000.
157. W. Purschke, C. L. Schmidt, A. Petersen, G. Schafer, *J. Bacteriol.* **1997**, *179*, 1344–1353.
158. J. A. Brito, F. L. Sousa, M. Stelter, T. M. Bandeiras, C. Vonrhein, M. Teixeira, M. M. Pereira, M. Archer, *Biochemistry* **2009**, *48*, 5613–5622.
159. M. Schedel, M. Vanselow, H. G. Trüper, *Arch. Microbiol.* **1979**, *121*, 29–36
160. F. Grimm, B. Franz, C. Dahl, in *Microbial Sulfur Metabolism,* Eds C. Dahl, C. G. Friedrich, Springer, Berlin, 2008, pp. 101–116.
161. R. Hille, *Chem. Rev.* **1996**, *96*, 2757–2816
162. U. Kappler, M. J. Maher, Cell. *Mol. Life Sci.* **2013**, *70*, 977–992.
163. H. Sakurai, T. Ogawa, M. Shiga, K. Inoue, *Photosynth. Res.* **2010**, *104*, 163–176.
164. L. H. Gregersen, D. A Bryant, N.-U. Frigaard, *Front. Microbiol.* **2011**, available online, doi: 10.3389/fmicb.2011.00116.
165. J. P. Lee, C. Yi, J. LeGall, H. Peck, Jr, *J. Bacteriol.* **1973**, *115*, 453–455.
166. A. F. Arendsen, M. F. Verhagen, R. B. Wolbert, A. J. Pierik, A. J. Stams, M. S. Jetten, W. R. Hagen, *Biochemistry* **1993**, *32*, 10323–10330.

Chapter 11
Transformations of Dimethylsulfide

Ulrike Kappler and Hendrik Schäfer

Contents

ABSTRACT 280
1 INTRODUCTION 280
   1.1 Dimethylsulfide – Sources and Sinks in the Biosphere 280
   1.2 Overview of Microorganisms Capable of Transforming Dimethylsulfide
       and Known Dimethylsulfide-Transforming Reactions 283
       1.2.1 Dimethylsulfide Degradation
           by Aerobic Methylotrophic Bacteria 283
       1.2.2 Dimethylsulfide Degradation in Autotrophic Bacteria 283
       1.2.3 Anaerobic Microorganisms Utilizing Dimethylsulfide
           as a Carbon Source 283
       1.2.4 Dimethylsulfide Oxidation to Dimethylsulfoxide 284
2 ENZYMOLOGY OF DIMETHYLSULFIDE TRANSFORMATIONS 285
   2.1 Dimethylsulfoxide Reductases 286
       2.1.1 Basic Properties of Dimethylsulfoxide Reductases and Their Encoding
           Operons 286
       2.1.2 Kinetic and Structural Properties of Dimethylsulfoxide Reductases 292
       2.1.3 Dimethylsulfoxide Reductase-Related Enzymes and Dimethylsulfoxide
           Reductases in Unusual Biological Contexts 295
   2.2 Dimethylsulfide Dehydrogenases 298
       2.2.1 The *Rhodovulum sulfidophilum* Dimethylsulfide Dehydrogenase 298
       2.2.2 Dimethylsulfide Dehydrogenases: Environmental Distribution 299
   2.3 Phylogeny of Dimethylsulfoxide Reductases
       and Dimethylsulfide Dehydrogenases 300
   2.4 Dimethylsulfide Monooxygenases 303
       2.4.1 Properties of Dimethylsulfide Monooxygenases 303

U. Kappler (✉)
School of Chemistry and Molecular Biosciences, 76 Molecular Biosciences Building,
The University of Queensland, St. Lucia, QLD 4072, Australia
e-mail: u.kappler@uq.edu.au

H. Schäfer
School of Life Sciences, Gibbet Hill Campus, University of Warwick, Coventry, CV4 7AL, UK

© Springer Science+Business Media Dordrecht 2014
P.M.H. Kroneck, M.E. Sosa Torres (eds.), *The Metal-Driven Biogeochemistry
of Gaseous Compounds in the Environment*, Metal Ions in Life Sciences 14,
DOI 10.1007/978-94-017-9269-1_11

2.4.2   Distribution, Genomic Context, and Phylogeny of Dimethylsulfide
         Monooxygenases 304
3   GENERAL CONCLUSIONS 306
ABBREVIATIONS 307
ACKNOWLEDGMENTS 308
REFERENCES 308

Abstract Dimethylsulfide (DMS) is a naturally occurring chemical that is part of the biogeochemical sulfur cycle and has been implicated in climate-relevant atmospheric processes. In addition, DMS occurs in soil environments as well as in food stuff as a flavor compound and it can also be associated with disease states such as halitosis. A major environmental source of DMS is the marine algal osmoprotectant dimethyl-sulfoniopropionate (DMSP). A variety of bacterial enzyme systems lead either to the production of DMS from DMSP or dimethylsulfoxide (DMSO) or its oxidation to, e.g., DMSO. The interconversion of DMS and DMSO is catalyzed by molybdenum-containing metalloenzymes that have been very well studied, and recently another enzyme system, an NADH-dependent, flavin-containing monooxygenase, that produces formaldehyde and methanethiol from DMS has also been described.

DMS conversions are not limited to a specialized group of bacteria – evidence for DMS-based metabolism exists for heterotrophic, autotrophic and phototrophic bacteria and there is also evidence for the occurrence of this type of sulfur compound conversion in Archaea.

Keywords dimethylsulfide • dimethylsulfoxide • DMS dehydrogenase • DMS monooxygenase • DMSO reductase

Please cite as: *Met. Ions Life Sci.* 14 (2014) 279–313

1   Introduction

1.1   Dimethylsulfide – Sources and Sinks in the Biosphere

Dimethylsulfide (CH_3-S-CH_3, DMS) is a volatile organic sulfur compound. It is poorly soluble in water and has a low boiling point of 38 °C. Its characteristic odor may be unpleasant even at relatively low concentrations. Production of DMS can be the cause of malodorous emissions from a wide range of industries and operations, for example composting, brewing, animal husbandry, and paper milling, but also contributes to halitosis (bad breath). At low concentration, DMS has a sweet odor that is an important part of the aroma of a wide variety of foods and beverages ([1, 2] and references therein).

DMS has been the subject of much interest due to its prominent role in the biogeochemical cycle of sulfur. DMS is the largest biogenic source of sulfur to the atmosphere and it has been estimated that the annual flux of DMS from the oceans is approximately 21 Tg, representing about 80 % of the total global DMS flux to the atmosphere [3].

Sea-to-air transfer of DMS is a process that was recognized to provide a missing link in the global sulfur cycle that is responsible for sulfur transport from the oceans to the continents (Figure 1) [4]. Following its transfer into the atmosphere, DMS is oxidized by hydroxyl radicals and nitrate to mainly sulfate and methanesulfonic acid, which are important components of atmospheric aerosols [5, 6]. Atmospheric transport of these sulfur compounds and their deposition in terrestrial environments contribute to maintaining sulfur levels in soils which is important for plant productivity [7, 8].

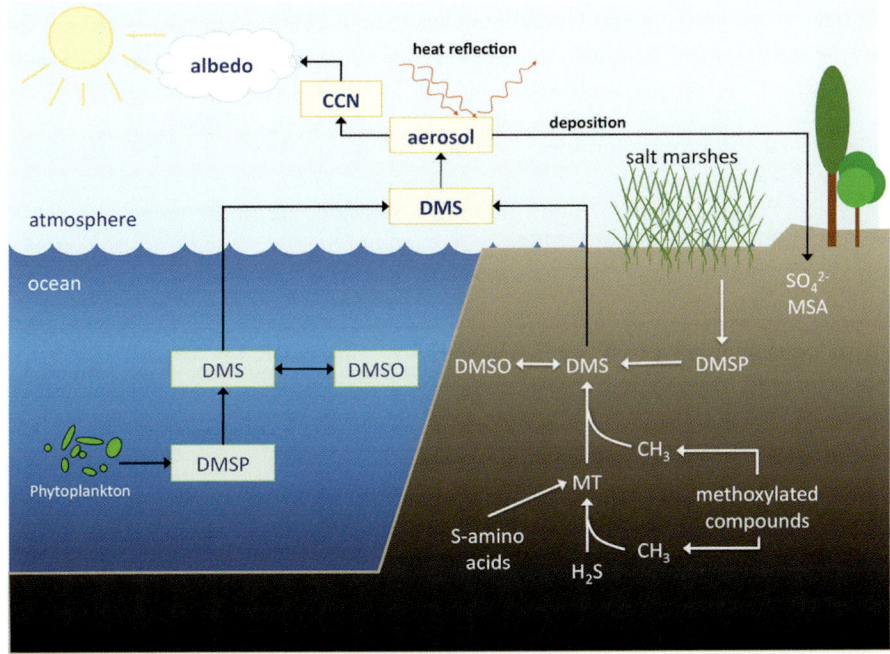

Figure 1   Transformations of dimethylsulfide in the biosphere. DMS, dimethylsulfide (CH_3-S-CH_3); DMSO, dimethylsulfoxide (CH_3-SO-CH_3); DMSP, dimethylsulfoniopropionate ((CH_3)$_2$S$^+$$CH_2$$CH_2COO^-$); MT, methanethiol ($CH_3$-SH); MSA, methanesulfonic acid (CH_3-SO_3H); CCN, cloud condensation nuclei. Reproduced with permission from [1]; copyright 2010, Oxford University Press.

In addition to its role in atmospheric sulfur transport, the sulfate aerosols derived from DMS in the atmosphere reflect heat radiation, and in doing so, reduce the radiative forcing of the atmosphere [9]. Furthermore, atmospheric sulfate aerosols can act as cloud condensation nuclei which may contribute to temperature regulation of the Earth through enhanced cloud cover and albedo formation. It was suggested that changes in albedo may reduce the incidence of sunlight on the surface of the ocean with potential consequences for primary productivity. This potential link between dimethylsulfoniopropionate (DMSP) production, DMS emission, atmospheric chemistry, and cloud formation was suggested to constitute a feedback loop which could affect climate via regulation of the amount of DMSP produced by phytoplankton and thence DMS emissions [9]. Recent modelling of the effect of DMS emissions into the atmosphere, however, suggests that the overall effect of DMS on climate may not be as big as previously thought, in comparison to other marine aerosol generation, e.g., from complex organic matter [10].

The main source of DMS emitted to the atmosphere from the oceans is DMSP, a sulfonium compound that is thought to act as a compatible solute, antioxidant, cryoprotectant and/or metabolic overflow mechanism [11–13] in diverse marine algae and phytoplankton [14] (Figure 1). Some 10^9 t of DMSP are thought to be produced in the oceans per annum. Upon release of DMSP into the water column, it is subject to microbial degradation that leads to about 30 % of the dissolved DMSP pool to be degraded to DMS by enzymes referred to as DMSP-lyases. Recent investigation of the genes and enzymes responsible for DMS production from DMSP have demonstrated that a wide range of different enzymes that belong to entirely different protein families are able to produce DMS in a DMSP-dependent manner (reviewed in [14]).

Typical concentrations of DMS in surface seawater are in the range of 1–7 nM [15]. Microbial utilization of DMS as carbon and/or energy source leads to degradation of the majority of DMS produced from DMSP. Degradation of DMS by methylotrophic microorganisms leads to assimilation into biomass and degradation of DMS to CO_2 and inorganic sulfur compounds including sulfate [16], thiosulfate [17], and tetrathionate [18]. DMS can also be utilized as an auxiliary energy source by heterotrophic bacteria [19, 20]. Utilization of DMS as a sulfur source has been documented in soil bacteria [21] and in a strain of *Marinobacter* in which DMS assimilation was dependent on light [22]. Only a fraction of the DMS produced from DMSP in the oceans is emitted into the atmosphere, while the remainder is subject to microbial degradation as outlined above [23].

While DMSP degradation is the main source of DMS in oxic marine environments, other pathways of DMS formation have been identified as sources of DMS in anoxic environments. The underlying processes include (i) methylation of sulfide and methanethiol (MT) [24], (ii) anaerobic degradation of sulfur containing amino acids [25, 26], and (iii) reduction of DMSO in anaerobic respiration [27].

Measurements of volatile organic sulfur compounds in anoxic freshwater sediments and overlying water columns detected MT and DMS concentrations in the nanomolar range (up to 76 nM for MT; up to 44 nM for DMS) [24]. It is thought that in these environments, DMS production is linked to the degradation of complex organic matter such as lignin, one of the most abundant pools of carbon in the biosphere, which generates methoxylated aromatic compounds that may be *o*-demethylated under anoxic conditions by anaerobic bacteria through methylation of sulfide [28–30].

Another process responsible for DMS production in anoxic environments is the respiratory reduction of DMSO [31], the enzymology of which is discussed in detail below.

1.2   Overview of Microorganisms Capable of Transforming Dimethylsulfide and Known Dimethylsulfide-Transforming Reactions

DMS can serve as a source of energy, carbon and/or sulfur for a wide variety of microorganisms. The key features of different groups of microorganisms that degrade DMS are discussed below and have also been reviewed recently in [1].

1.2.1   Dimethylsulfide Degradation by Aerobic Methylotrophic Bacteria

A range of methylotrophic bacteria are known to be able to grow on DMS as a sole carbon and energy source. Examples include strains of *Hyphomicrobium, Methylobacterium, Afipia, Arthrobacter, Methylophaga,* and others (e.g., [16, 17, 32–37]). Two different types of enzymes have been suggested to be responsible for DMS degradation in methylotrophs, DMS monooxygenases [16, 38] and putative methyltransferases [18]. Both of these enzymes generate methanethiol and formaldehyde as common intermediates. While MT is further degraded by methanethiol oxidases [39, 40], formaldehyde is either oxidized to formate and CO_2 or assimilated into biomass by the serine cycle or ribulose monophosphate cycle [41].

1.2.2   Dimethylsulfide Degradation in Autotrophic Bacteria

DMS-degrading isolates of autotrophic *Thiobacillus* species have also been reported [42–46] although the taxonomy of some of these isolates has not been confirmed by 16S rRNA gene sequencing. These strains grow aerobically [42, 44] and/or under denitrifying conditions [45, 46]. DMS degradation in *Thiobacillus thioparus* Tk-m was shown to proceed via the intermediates methanethiol and formaldehyde to the formation of CO_2 which was subsequently assimilated by the Calvin Benson Bassham cycle. Initial oxidation of DMS was strictly oxygen-dependent, but methanol was not detected as an intermediate [42]. Visscher and Taylor suggested, based on the observed inhibition by methyl t-butyl ether, that *Thiobacillus thioparus* T5 may utilize a DMS monooxygenase for DMS degradation, while *Thiobacillus* strain ASN-1 which was able to couple DMS oxidation to denitrification was assumed to degrade DMS using methyl transfer reactions [45].

1.2.3   Anaerobic Microorganisms Utilizing Dimethylsulfide as a Carbon Source

Both bacteria and Archaea able to grow on DMS anaerobically have been isolated [46–52]. In saltmarshes and anoxic sediments in marine and freshwater environments, sulfate-reducing bacteria and methanogenic Archaea are probably the main DMS-degrading populations [26, 31, 48, 53]. However, strains able to degrade DMS under anoxic conditions using nitrate as electron acceptor were also identified [52, 54, 55].

Enzymes of DMS metabolism have been identified in methanogens. In *Methanosarcina barkeri* DMS and methylmercaptopropionate (MMPA) were converted to methane with a corrinoid protein functioning as a coenzyme M methylase capable of DMS and MMPA degradation [56]. In *Methanosarcina acetivorans* three fused corrinoid/methyl transfer proteins have been implicated in methyl sulfide metabolism [57].

1.2.4 Dimethylsulfide Oxidation to Dimethylsulfoxide

Oxidation of DMS to DMSO has been reported for anoxygenic phototrophic purple sulfur bacteria [58] and phototrophic green sulfur bacteria [59] where this process provides electrons for carbon fixation [58]. DMS dehydrogenase was identified as the enzyme for DMS to DMSO oxidation in *Rhodovulum sulfidophilum* [60]. The enzyme and its encoding genes are discussed in further detail below.

Some heterotrophic bacteria were also shown to oxidize DMS to DMSO, including isolates of *Delftia acidovorans* [61] and *Sagittula stellata* [62]. While the underlying mechanism of DMS oxidation in these isolates remains to be identified, it was shown that DMS oxidation to DMSO provided an auxiliary source of energy for *S. stellata* when growing on fructose or succinate in the presence of DMS [19] allowing a higher growth yield to be attained. Similarly, it was suggested that a *Flavobacterium* strain was able to oxidize DMS to DMSO and might be capable of using this as an energy source [20]. The genetic and biochemical basis of DMS oxidation to DMSO in these strains has not been reported.

DMSO is also a product of co-oxidation of DMS by methanotrophic and ammonia-oxidizing bacteria [63, 64]. In the latter, the activity was shown to be due to ammonia monooxygenase (AMO), while it is assumed that methane monooxygenases may be responsible in methanotrophs, owing to the close evolutionary relationship between particulate methane monooxygenase and AMO [65] and the ability of the enzymes to co-oxidize a range of compounds [66].

Utilization of DMS, however, is not limited to energy-generating processes. Its use as a sulfur source has been reported for a wide range of bacteria including strains of *Marinobacter* [22], *Acinetobacter* [67], *Rhodococcus* [21], and *Pseudomonas putida* [68]. In *Marinobacter*, a flavin-containing enzyme appeared to be involved in the process which also required light [22]. In the other strains mentioned above, utilization of DMS as a sulfur source has been suggested to proceed via oxidation to DMSO by a multicomponent monooxygenase similar to phenol hydroxylase, followed by further oxidation to dimethylsulfone ($DMSO_2$) and methanesulfonic acid (MSA) [21, 67, 68].

2   Enzymology of Dimethylsulfide Transformations

As outlined above, DMS is a compound that forms naturally during the enzymatic breakdown of the algal osmolyte DMSP or the reduction of DMSO (Figure 1), which is a also a major industrially used chemical and is present in the environment as a result of natural degradation processes as well as waste disposal or pollution.

The DMS-consuming or -producing enzyme systems that will be discussed in detail below are the DMSO reductases [EC 1.8.5.3], DMS dehydrogenases [EC 1.8.2.4], and DMS monooxygenases [EC 1.14.13.131] (Figure 2). The first two of these enzymes belong to the mononuclear molybdenum enzymes, while the third is a FMN-dependent protein that also requires iron and magnesium as cofactors.

The reactions of DMSO reductase and DMS dehydrogenase are both reversible, and in combination allow a continuous cycle of DMS consumption and production (Figure 2).

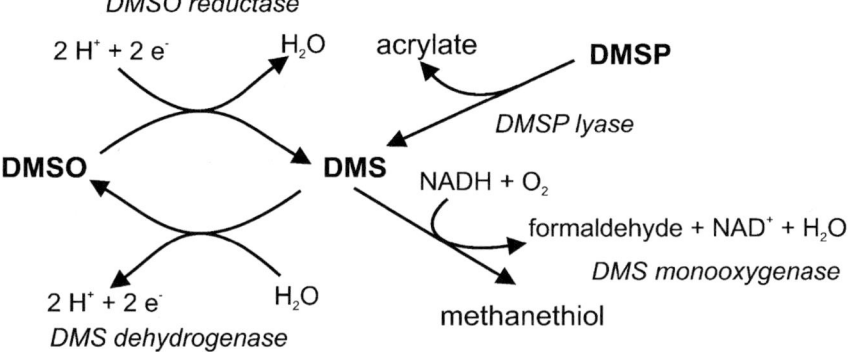

Figure 2 Enzymatic conversion of dimethylsulfide.

DMSP lyases which also produce DMS are a very diverse group of enzymes, where at least six different types of genes (*dddY*, *dddD*, *dddP*, *dddQ*, *dddL*, *dddW*) encoding proteins with different properties have been identified so far. These enzymes have been the subject of an excellent recent review [14] and will therefore not be covered here in detail. It should be noted, however, that the crystal structure of the DddQ DMSP lyase including details of the enzyme's mechanism has just been published [69].

A few other enzymes have been identified in the literature that might give rise to DMS as part of methylation reactions. These include the trimethylsulfonium-tetrahydrofolate N-methyltransferase [EC 2.1.1.19] that was enriched from a bacterium referred to as an 'unusual' species of *Pseudomonas* capable of using trimethylsulfonium as a source of cell carbon [70] and thioether S-methyltransferases [EC 2.1.1.96] that are involved in the metabolism of selenium and sulfur compounds in higher organisms [71, 72]. However, as these enzymes do not appear to play major roles in DMS turnover and details of their structure and function are mostly unknown they will not be covered in detail here.

2.1 Dimethylsulfoxide Reductases

2.1.1 Basic Properties of Dimethylsulfoxide Reductases and Their Encoding Operons

The production of DMS from DMSO is catalyzed by a particular class of metalloenzymes, the DMSO reductases [EC 1.8.5.3]. These enzymes are found in a variety of bacterial phyla and especially in members of the Proteobacteria where they are generally involved in mediating anaerobic respiration. The best-studied DMSO reductases are the enzymes from the γ-Proteobacterium *Escherichia coli* and the α-Proteobacteria *Rhodobacter capsulatus* and *Rhodobacter sphaeroides* [73–83]. DMSO reductases are mononuclear molybdenum enzymes, which means that the active site contains a single Mo metal center (Figure 3), and within that group of enzymes they belong to the DMSO reductase enzyme family.

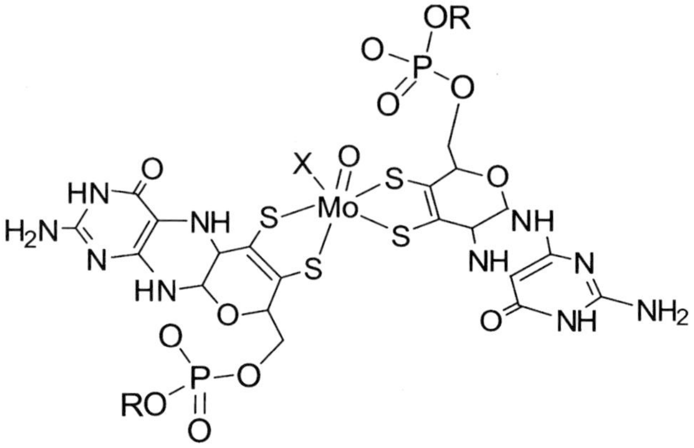

Figure 3 Schematic representation of the bis-MGD cofactor present in enzymes of the DMSO reductase enzyme family. Only the cofactors, the central molybdenum atom, and direct ligands to the molybdenum are shown, X = amino acid ligand to the Mo center.

The DMSO reductase enzyme family is very diverse and over 25 types of enzymes have been classified as belonging to this enzyme family [84–86] including several dehydrogenases such as formate dehydrogenase, ethylbenzene dehydrogenase, and DMS dehydrogenase.

As a result of this diversity of enzymes that make up this enzyme family other names have been suggested such as Complex Iron-Sulfur Molybdoenzyme family (CISM) and Mo/W Bis-PGD enzyme family [73, 84]. As a new name has not been settled on at this stage, we will use the older but widely adopted designation, DMSO reductase enzyme family.

The overall architecture of the catalytic subunits of DMSO reductase family enzymes is similar with four (and in some cases five) clearly identifiable domains as shown by available crystal structures [81, 82, 87–91]. However, there is great diversity of the general architecture of the enzymes which ranges from single subunit enzymes to membrane-bound multisubunit protein complexes with mostly three subunits. While the catalytic subunits of the various enzyme complexes are clearly related throughout the enzyme family, the other subunits and especially the membrane subunits show considerable diversity [73, 84].

Another key feature of DMSO reductase family enzymes is that all of them are of bacterial origin, and share a common active site assembly where a single Mo ion is coordinated by two molecules of a pyranopterin-dinucleotide cofactor (Figure 3). The pyranopterin ligands provide two dithiolene sulfur ligands to the Mo atom each, additional Mo ligands including amino acids, oxygen and sulfur atoms may also be present and will vary depending on the type of enzyme studied [84, 85]. The type of the dinucleotide attached to the pyranopterin cofactor may also vary, with guanosine and cytosine being the most commonly found derivatives [84, 85].

The reactions catalyzed by DMSO reductase family enzymes are very diverse and so far include both redox and non-redox reactions as well as hydroxylation reactions. The most common reactions are redox reactions that involve the transfer of oxygen or sulfur to or from a substrate, depending on whether the enzyme in question is a reductase or a dehydrogenase. The first of these two electron transfer reactions uses water as the oxygen donor or reaction product and is a reaction that is unique to Mo- or W-containing enzymes (eq. 1) [92].

$$R-O + 2\,e^- + 2\,H^+ \rightleftharpoons R + H_2O \tag{1}$$

For the sulfur transfer reactions HS^- or H_2S replace water in the reaction scheme (eq. 2)

$$R-S + 2\,e^- + 2\,H^+ \rightleftharpoons R + H_2S$$
$$R-S + 2\,e^- + H^+ \rightleftharpoons R + HS^- \tag{2}$$

A special case is the reaction of the formate dehydrogenase where catalysis leads to the formation of CO_2 and water from formate. This constitutes a formal hydrogen atom transfer as oxygen from water is not incorporated into the product. However, the reaction still involves a two electron transfer from the substrate to the enzyme [93].

In addition to the above reactions that are typical for mononuclear Mo and W enzymes, some recently identified enzymes of the DMSO reductase family, e.g., ethylbenzene dehydrogenase or C25 cholesterol dehydrogenase [94, 95], catalyze hydroxylation reactions (eq. 3):

$$R-H + H_2O \rightleftharpoons R-OH + 2\,e^- + 2\,H^+ \tag{3}$$

A few known members of the DMSO reductase family even catalyze non-redox reactions (acetylene hydratase, pyrogallol phloroglucinol transhydroxylase) [96–99]. Several recent reviews on enzymes of the DMSO reductase family exist and cover various aspects of the evolution, structural, spectroscopic, and kinetic properties of the enzymes in the enzyme family [73, 79, 84, 85, 93, 100, 101].

2.1.1.1   Dimethylsulfoxide Reductases – Basic Types and Properties

Within the DMSO reductase enzyme family there are two known types of DMSO reductases (DMSOR), the Dor- and the Dms-type. Both types of enzymes are located in the bacterial periplasm, i.e., in an extracytoplasmic compartment, and use quinol as the electron donor for their reaction. However, while the Dms-type enzymes are membrane-bound via additional subunits (DmsB, DmsC), the Dor-type enzymes are soluble and only transiently interact with the membrane-bound DorC-type cytochromes that act as electron donors for these systems (Figure 4).

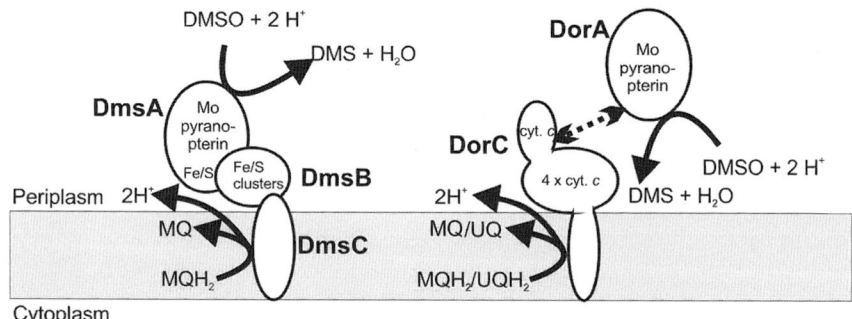

Figure 4 Cellular localization and reactions catalyzed by DMSO reductases of the Dor- and Dms-type. The enzymes shown are based on the structures of the *Rhodobacter* DorA DMSO reductase and the *Escherichia coli* DmsABC DMSO reductase. MQ/MQH$_2$, menaquinone/menaquinol; UQ/UQH$_2$, ubiquinone/ubiquinol; broken arrows indicate transient protein interactions.

Both types of enzyme systems are part of the bacterial respiratory chains where they function as alternative terminal reductases mediating anaerobic respiration and energy conservation in the absence of oxygen. However the primary sequences of these two types of enzymes only show 26 % amino acid (aa) identity (48 % aa similarity) which clearly indicates that they are not closely related (see Section 2.3 below).

2.1.1.2   Organization of Dimethylsulfoxide Reductase-Encoding Operons

The core operons encoding Dms-type DMSO reductases usually consist of three genes, *dmsABC*, where *dmsA* encodes the catalytic subunit of the enzyme that contains the Mo cofactor and an N-terminal iron-sulfur cluster that is usually designated as FS0. The *dmsB* gene encodes a ferredoxin-type electron transfer protein with four [4Fe-4S] clusters (FS1-4), and *dmsC* encodes a membrane subunit with 8 transmembrane helices (Figures 4 and 5).

For the maturation of the DmsABC protein complex a fourth protein, the DmsD molecular chaperone is required [102]. The DmsD protein is a so called 'REMP', a

Dms-type DMSO reductases

E. coli dms operon

H. influenzae dms gene cluster

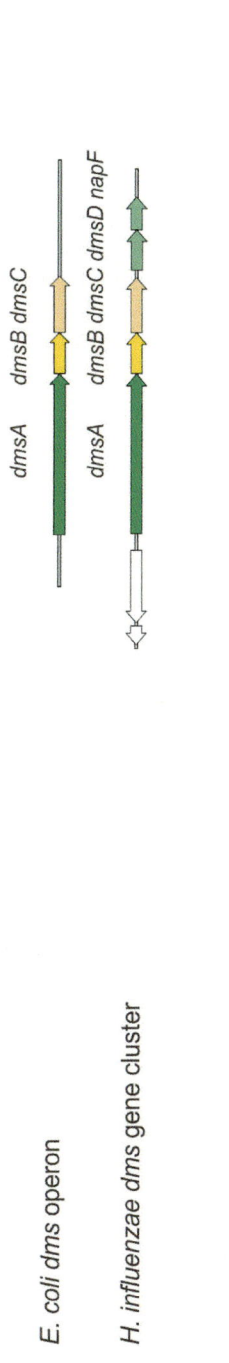

Dor-type DMSO reductases

R. sphaeroides dor gene region

R. capsulatus dor operon

E. coli tor operon

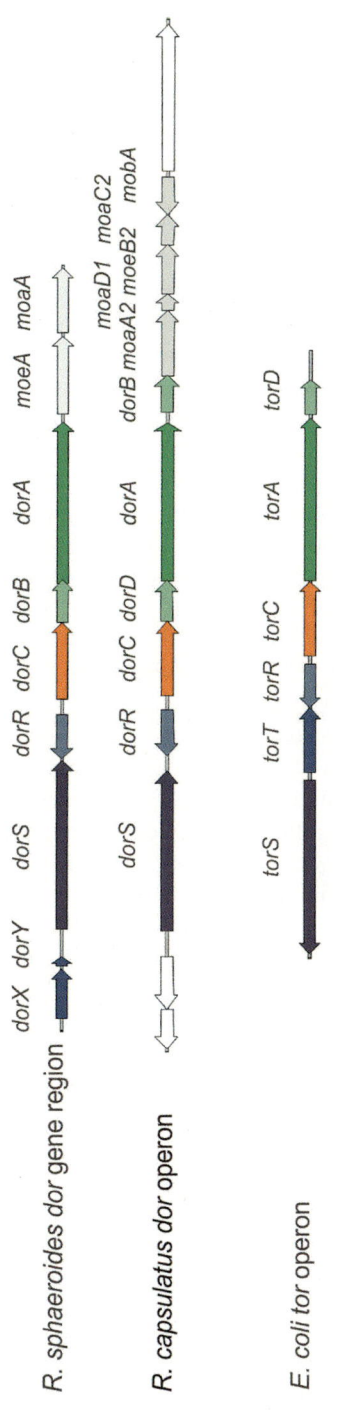

Figure 5 Structure of operons encoding DMSO reductases for the Dor- and Dms-type. The operons are aligned at the *dmsA*/*dorA* genes. Coloring scheme: Enzyme subunits and chaperones: *dorA*/*dmsA* genes, dark green; *dmsB* – yellow; *dmsC* – light orange; *dmsD*/*dorB*/*dorD* – light green. Regulatory genes: *dorR*, *dorX*, *dorY* – blue, *dorS* – dark blue. Molybdenum biosynthesis genes: light grey. Unrelated genes: white.

redox enzyme maturation protein, and is known to play a role in aiding cofactor insertion and export of DmsA to the periplasm [103–105]. The gene encoding this protein may be part of the *dmsABC* operon, as is the case in *Haemophilus influenzae* [106], however, it can also be encoded elsewhere on the chromosome (Figure 5). This is the case in *E. coli* where the DmsD protein required for the maturation of the DmsABC protein complex is part of an operon encoding a putative selenate reductase (*ynfE*) that is a paralogue of the DmsABC DMSO reductase [100, 107–109].

Additional genes encoding redox proteins such as ferredoxins may be present in the *dms* operons. This is the case for *Haemophilus influenzae*, where a gene encoding a NapF-like ferredoxin is present downstream of the *dmsABCD* genes. NapF proteins are normally associated with another type of DMSOR family molybdenum enzymes, the periplasmic nitrate reductases, and are thought to play a role in the insertion of the iron-sulfur cluster present in the NapA catalytic subunit [110].

The Dor-type DMSO reductases were first characterized from photohetero-trophic *Rhodobacter* species [80, 111–113], and the operons encoding these enzymes usually consist of three genes, *dorCDA*, where the first gene encodes a membrane-bound penta-heme cytochrome *c*, DorC, which acts as the electron donor to the catalytic subunit, DorA (Figures 4 and 5). The second gene encodes a DMSO reductase-specific chaperone protein that has been named either *dorB* (*R. sphaeroides*) or *dorD* (*R. capsulatus*), with the latter being the more common designation. The third gene, *dorA*, encodes the Mo cofactor containing catalytic subunit. DorA proteins differ from DmsA proteins in that they do not contain iron-sulfur clusters [114, 115].

The Dor-type DMSO reductases are closely related to the trimethylamine N-oxide (TMAO) reductases found in many bacteria including *E. coli*, and the operon structures of the TMAO reductases are very similar to those of the DMSO reductases although the order of the genes is usually *torCAD* [116].

A number of accessory genes located close to the *dorCDA* operons have been identified in the two *Rhodobacter* species in which these enzymes have been studied most. Upstream of the *dorCDA* operon two genes, *dorRS*, encoding a two-component regulatory system have been identified in both species. This two-component regulatory system is involved in regulating expression of the DMSO reductase structural genes [78, 114, 115].

In *Rhodobacter capsulatus* 37B4 an additional gene, *dorB*, encoding a NapD-like chaperone protein has been identified directly downstream of the *dorCDA* operon, and a similar gene appears to be present in *R. capsulatus* SB1003 [114, 115, 117].

Downstream of the *dor* operons in both *Rhodobacter* sp., two genes encoding molybdenum cofactor synthesis proteins, *moeA* and *moaA*, are found [115, 118]. The genome sequence of *R. capsulatus* SB1003 also shows that not only two but five genes encoding all major proteins involved in the production of the Mo-pyranopterin cofactor (acc. No YP_003578980-75) are found downstream of the *dor* operon [117, 119] (Figure 5). This is, however, not the case for *R. sphaeroides* 2.4.1 where only a *moeA* and a *moaA* gene are present [120, 121]. It is possible that more Mo-pyranopterin biosynthesis protein-encoding genes are also present in *R. capsulatus* 37B4 but in the absence of a genome sequence this cannot be verified at present.

Two additional proteins, DorXY, involved in regulating the activity of the *R. sphaeroides* DMSO reductase have been described [122]. The genes encoding these proteins are located upstream of the *dorSR* genes and appear to be conserved in the available genomes of *R. sphaeroides* strains, however, our analyses indicate that no homologues of DorY or DorX are found in the *R. capsulatus* SB1003 genome. DorY is thought to be directly involved in the activation of DMSOR expression [122].

2.1.1.3 Regulation of Dimethylsulfoxide Reductase Expression

Despite the very different structures of the operons encoding the enzymes, the expression of DMSOR is induced in the absence of oxygen in both *E. coli* and the *Rhodobacter* species.

In the *Rhodobacter* species expression of the *dorCDA* structural genes is directly controlled by the DorRS two-component system. DorR is an OmpR-type response regulator and has been shown to bind the *dorR-dorCDA* intergenic region where four binding sites (consensus pentanucleotide: 5′-TTC/AAC-3′ [123]) have been identified [123–125]. Deletion of the *dorR* gene leads to a loss of DMSO reductase activity [114, 115], and a similar effect was observed in *R. sphaeroides* for a *dorY* deletion strain [122]. While this indicates that DorR and, if present, DorY are the main regulators of *dorCDA* expression, the expression of this operon is closely linked to several other major regulatory circuits including light intensity, cellular redox state, availability of molybdate, and others [114, 119, 122, 126, 127].

In some cases specific binding sites for particular regulators have been identified, for example, it is known that fumarate nitrate reductase regulation (FNR) controls *dorCDA* expression by activating the *dorS* promoter, thus controlling the production of the complex sensor kinase that is part of the DorSR two component system [114, 126]. In other cases the links are not as obvious, e.g., reporter gene studies identified a reduced activity of the *dorCDA* promoter in the presence of high light intensities, an effect that may be linked to other work that observed a negative effect of the global RegA regulator on *dorCDA* expression [78, 122]. The RegAB two-component system is a global regulatory system that is highly prevalent in Proteobacteria and in *Rhodobacter* it is known to control processes as diverse as photosynthesis gene expression, nitrogen metabolism, and redox homeostasis [128–130]. The RegB sensor kinase can directly interact with molecules of the cellular quinone pool, while RegA has been shown to affect gene regulation not only by DNA binding, but also through direct interactions with other regulatory proteins [131–133].

In *E. coli* a dedicated regulator of the *dmsABC* operon such as the DorSR system appears to be absent and instead the expression of the *dmsABC* operon is controlled by two promoters, P1 and P2, located ~200 bp upstream of the *dmsA* gene that interact with a variety of transcription factors. P1 is controlled by FNR and a nitrate reductase regulator protein (NarL), while P2 is regulated by integration host factor (IHF) and the molybdate-responsive ModE regulator [134, 135]. In keeping with these regulatory circuits expression of the DmsA DMSO reductase is induced under anaerobic conditions but repressed in the presence of nitrate. Molybdate is required for full expression of the *dmsABC* operon, and defects in molybdate acquisition affect nitrate-dependent repression of *dmsABC* expression [134, 135].

While the molecular mechanisms of regulation are quite different for the Dor- and Dms-type DMSO reductases there appear to be some common themes in the regulation of DMSOR expression: In both systems anaerobiosis is a main factor controlling activity of the operons, and this effect appears to be mediated by FNR-type transcription factors. Anaerobiosis is an important regulatory element also in other DMSOR systems such as the enzymes from *Shewanella* and *Halobacterium* NRC-1 (see below). Both systems are responsive to the presence of molybdate in the growth medium through interactions with the ModE transcription factor in *E. coli* and the ModE homologue MopB in *R. capsulatus*.

In addition, there are then organism-specific regulatory patterns such as the integration of redox state and environmental light intensities in the phototrophic *Rhodobacter* species, while in *E. coli* competition with nitrate reduction appears to be a key aspect of DMSOR regulation. Overall both DMSOR systems are clearly regulated in response to several environmental parameters which involve integration of the action of a variety of different transcription factors.

2.1.2   Kinetic and Structural Properties of Dimethylsulfoxide Reductases

2.1.2.1   Kinetics of Dimethylsulfoxide Reduction

Kinetic parameters have been published for both Dor- and Dms-type enzymes using a variety of substrates, including the biologically relevant compounds DMSO, methionine sulfoxide (MetSO) and trimethylamine N-oxide (TMAO) (Table 1). Comprehensive lists of substrates tested and enzyme activities can be found in [136, 137].

The standard assay for DMSO/TMAO reductases is carried out under anoxic conditions and uses either methyl or benzyl viologen as the electron donor to the enzymes with dithionite being used as the reductant. The substrate added can be

Table 1 Kinetic parameters of Dor- and Dms-type DMSO reductases.

		DorA$_{RC}$	DorA$_{RS}$	DmsA$_{EC}$
DMSO	K_M (mM)	0.0097	0.007	0.18
	k_{cat} (s^{-1})	42.9	58	79.9
MetSO	K_M (mM)	n.r.	0.33	0.09
	k_{cat} (s^{-1})	n.r.	58	61.1
TMAO	K_M (mM)	0.193	68	20.2
	k_{cat} (s^{-1})	134.5	2,300	1,203

DorA$_{RC}$ – *Rhodobacter capsulatus* DorA DMSO reductase [142]; DorA$_{RS}$ – *Rhodobacter sphaeroides* DorA DMSO reductase [143], DmsA$_{EC}$ – *Escherichia coli* DmsABC DMSO reductase [137]

varied but is usually present in millimolar amounts [138]. The DMSOR reaction is reversible and can be assayed using the DMS dehydrogenase assay [139], but the K_M values for DMS as a substrate are very high, indicating that it is a poor natural substrate [140, 141].

Both characterized DorA proteins have very low K_M values (7–10 µM) for DMSO, with the K_M for MetSO being reported as 330 µM for the *R. sphaeroides* enzyme [142, 143] (Table 1). In contrast, the DmsABC enzyme had a lower K_M for methionine sulfoxide (MetSO) than for DMSO, although the turnover numbers for the two substrates were similar (61 and 78 s^{-1}) (Table 1).

For the N-oxide substrate TMAO, all three enzymes exhibited increased K_M values (Table 1), with the differences ranging from ~1 order of magnitude for the *R. capsulatus* enzyme to ~2–3 orders of magnitude for the other two enzymes. The turnover numbers were significantly higher for TMAO than for the S-oxide substrates, again with a ~2 order of magnitude increase for the *R. sphaeroides* and *E. coli* enzymes [143] (Table 1).

The reactivity of the Dor-type enzymes towards S-oxides as well as N-oxides has been shown to be modulated by two active site residues, Tyr114 and Trp116. Tyr114 is present in all enzymes with known reactivity towards S-oxides, but is absent in the TorA TMAO reductase. The nearby Trp116 influences substrate turnover, but not substrate affinity and is present in both S- and N-oxide reducing enzymes [142, 143].

Non-steady state kinetic parameters have been reported [140, 141] and the influence of redox potential on the DMSOR reaction has been studied by protein film voltammetry. These experiments revealed a potential-dependent switch in the activity of the *E. coli* DmsA protein that was postulated to have implications for the biological activity of the enzyme [144, 145]. Similar potential-dependent switches were later identified in other enzymes of the DMSOR enzyme family (e.g., periplasmic and membrane-bound nitrate reductases) and are now thought to reflect the structure of the Mo center of these enzymes as the proteins in question share little homology [146]. Early investigation of the Dor-type DMSOR from *Rhodobacter capsulatus* indicated that this enzyme might not possess this switch [147]. The protein film voltammetry experiments for the *Rhodobacter* enzyme also revealed the redox potentials of the Mo$^{(VI/V)}$ and Mo$^{(V/IV)}$ couples which were reported as a function of pH.

An interesting observation is that in the Dms-type DMSO reductase from *E. coli* the removal of a serine residue postulated to be equivalent to the Ser147 Mo center ligand of the *Rhodobacter* enzymes (see below) abolished activity of the enzyme and altered the redox properties of the Mo center [148].

2.1.2.2 Crystal Structures of Dimethylsulfoxide Reductases

At present only crystal structures of the Dor-type DMSO reductases are available. The first crystal structures of the enzymes from two different species of *Rhodobacter* were solved within the same year with additional structures being published in the following years [82, 149–151]. All structures agree very well regarding the overall architecture of the enzymes which comprises four distinct domains and a substrate access channel between domains II and III [82, 149–151] (Figure 6).

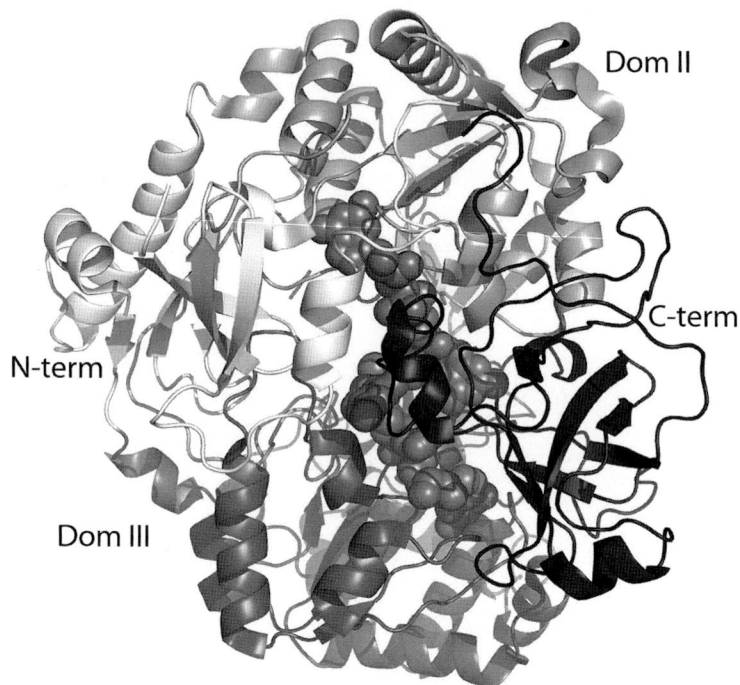

Figure 6 Crystal structure of the *Rhodobacter capsulatus* DMSO reductase (PDB 1EU1). The four domains of the protein are labeled (N-term = Dom I; C-term = Dom IV) and colored in deepening shades of grey, bis-MGD cofactor shown as space-filling model.

The substrate access channel in the DMSO reductases is quite wide which allows a variety of substrates to enter the active site, an observation that is consistent with the ability of the enzymes to reduce a variety of S- and N-oxide substrates. Solution of the crystal structure of the *R. capsulatus* enzyme in complex with DMSO showed that a polypeptide loop near the active site acts as a 'lid' for the active site during catalysis [149].

Regarding the structure of the catalytic Mo center (Figure 3), however, clear differences were observed in these initial structures. While the two pyranopterin-guanidine dinucleotide cofactors and the central Mo atom were clearly identified in both enzymes, the structures did not agree concerning the number and nature of the Mo ligands, where variously between 5 and 7 ligands were reported [82, 149–151]. As direct ligands to the Mo center either one or two oxo ligands were identified, and either one or both of the MGD cofactors were reported as being ligated to the Mo center. A variety of spectroscopic investigations followed the publication of the crystal structures, in attempts to resolve the issues caused by the changing ligand environments and trying to reconcile the observation with proposed mechanisms of catalysis (for more information see [84, 152]).

The most recent crystal structure of a Dor-type DMSO reductase [81], which had a resolution of 1.3 Å, resolved the issue of the apparently conflicting structures of the Mo center. Li et al. [81] showed that the Mo site was disordered in the crystals, and that at the high resolution of 1.3 Å it was possible to identify two distinct forms of the Mo center, a five-coordinate version similar to the structure reported by [83] and a six-coordinate form that now constitutes the accepted version of the DMSO reductase Mo center in the oxidized Mo(VI) state. The structure reported by Li et al. [81] also explained the seven-coordinate, dioxo Mo center published earlier [150] which likely arose from the presence of both five- and six-coordinate Mo centers in the crystals that could not be resolved at the lower resolution of the earlier structure. In the high-resolution structures, the four dithiolene ligands to the Mo center are always attached to the Mo ion, and the coordination sphere of the Mo consists of an oxo-methyl ligand (Ser147 in *R. capsulatus*) and an oxo ligand. In fact, it was shown by additional studies that the dissociation of one of the MGD cofactors observed in some of the earlier crystal structures was due to reversible damage caused by exposure of the enzyme to HEPES buffer under oxic conditions [153], and modifications of the enzyme by DMS and air have also been documented [140].

Based on the crystallographic, kinetic, and spectroscopic evidence the current view of the DMSO reductase reaction mechanism is that in the oxidized, Mo(VI) state the Mo center contains a serine ligand and a coordinated oxo group and four sulfur ligands provided by the dithiolene groups of the two MGD cofactors. On reduction of the center to the Mo(IV) state, the oxo group is lost, presumably to form water, leaving a five-coordinate Mo(IV) center. The substrate, DMSO then binds to the Mo(IV) center via its oxygen which remains as an oxo ligand on the reoxidized Mo center after release of the reaction product, DMS [84, 154].

To the best of our knowledge structural analyses of the effect of the substitutions of the two active site residues that control substrate specificity (see above), Tyr114, and Trp116 are not available at present [77, 142, 143, 155, 156].

A variety of spectroscopic techniques have been applied to the study of the Mo centers of both Dms and Dor-type DMSO reductases including electron paramagnetic resonance, resonance Raman, protein film voltammetry, various model compounds and, recently, density functional theory calculations. These studies have been reviewed in detail elsewhere [152, 157–161].

2.1.3 Dimethylsulfoxide Reductase-Related Enzymes and Dimethylsulfoxide Reductases in Unusual Biological Contexts

There is a wealth of sequence data indicating that both Dms- and Dor-type enzyme systems exist in many microorganisms (see Section 2.3), and there have been various reports of anaerobic, DMSO-based respiration in bacteria, but most of the enzyme systems have not been studied so far. This section briefly highlights some DMSO reductase systems that are of interest due to their occurrence in particular microorganisms and specific adaptations to the lifestyle of the source organisms. This highlights the functional diversity of the DMSOR enzymes that has not been explored in detail for most bacteria capable of DMSO reduction.

2.1.3.1 The *E. coli* Ynf Selenate Reductase

The *ynf* operon of *E. coli* has long been known to encode an enzyme related to the DmsA DMSO reductase, but the enzyme had also been shown to be unable to support growth of *E. coli* on DMSO in a DmsA-knockout background [108]. The operon encoding this enzyme is unusual as there are two genes (*ynfEF*) encoding catalytic subunits related to DmsA at the start of the *ynfEFGHI* operon. The *ynfGH* genes encode proteins related to DmsB and DmsC, respectively. The final gene in the gene cluster, *ynfI*, has been renamed and is now known as *DmsD* after it was recognized that it encodes a molecular chaperone that is required for the maturation of the DmsABC DMSO reductase and presumably also the Ynf selenate reductase [104, 105]. Interestingly, the *ynf* promoter is activated by FNR and repressed by NarL, both of which also control expression of the *dmsABC* promoter [107]. A recent study into selenate reduction in *Salmonella typhimurium* and *E. coli* K12 identified *ynfEF* as putative selenate reductases, which explains the inability of the *ynf* gene locus to enable DMSO-based anaerobic respiration in a *dmsABC* knock-out strain [162].

2.1.3.2 The *E. coli* TorZ S- and N-Oxide Reductase

Although it has been clearly shown that the DmsABC protein is the main DMSO reductase in *E. coli*, a 'Dor-type' DMSO reductase has been identified in *E. coli* [163]. This protein, TorZ, is encoded by a two gene operon, *torYZ* (also known as *yecK bisZ*), in which the second gene encodes a cytochrome that is related to the TorC/DorC cytochromes. Initially, TorZ was described as encoding an alternative "minor" biotin sulfoxide reductase and had been designated 'BisZ' [164], but it was later reclassified as a TMAO respiratory system, based on its structural similarity to the well characterized TorA TMAO reductase, and kinetic studies that showed that TorZ is capable of reducing both N- and S-oxides [163]. Sequence alignments clearly show that TorZ contains residues equivalent to both Y114 and W116, the residues that have been implicated in substrate specificity and the ability to reduce S-oxides in the *Rhodobacter* DMSO reductases. The exact function of TorZ in *E. coli* metabolism is unknown at present, but TorZ-like enzymes are present in a variety of Enterobacteriaceae and also Pasteurellaceae.

2.1.3.3 Extracellular Dimethylsulfoxide Reduction
in *Shewanella oneidensis*

The DMSO respiration system of *Shewanella oneidensis* is a special case as it is located in the outer membrane rather than in the periplasmic space (Figure 7). The core catalytic subunits of this complex, DmsA and DmsB, have high homology to the *E. coli* DMSO reductase, but homologues of the DmsC membrane anchor protein are absent from the *S. oneidensis* genome [165]. Instead, the two *S. oneidensis* gene clusters (SO_1427-SO_1432, *dmsEFABGH*; SO_4362-SO_4357) encoding DmsAB-related protein complexes encode homologues of proteins involved in metal reduction, MtrAB ('DmsEF' in the *Shewanella* DMSO reductase complex), that can mediate outer membrane attachment and electron transfer across the bacterial periplasm as well as a molecular chaperone ('DmsG') and an 'accessory protein' ('DmsH') of 155 aa. Within the second gene cluster, the *dmsGH* homologous genes are located upstream of the *mtrAB* related genes.

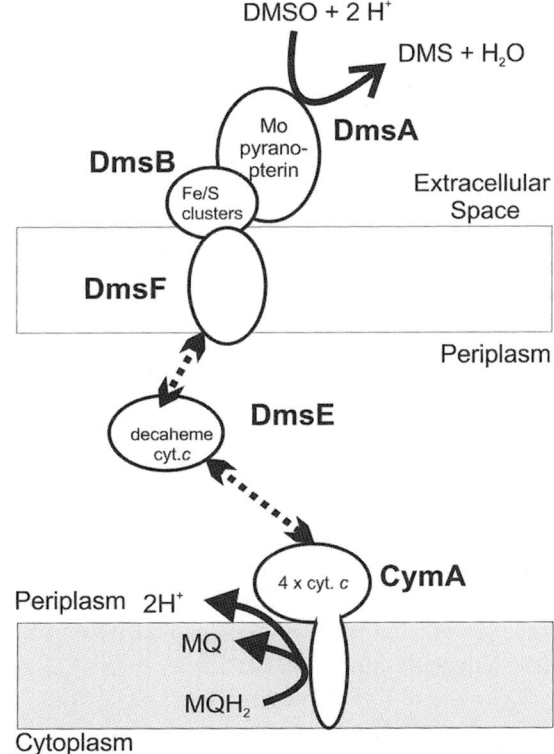

Figure 7 Structure and cellular organization of the *Shewanella oneidensis* outer membrane DMSO reductase in the bacterial cell envelope. MQ/MQH$_2$, menaquinone/menaquinol; cyt. *c*, cytochrome *c*; broken arrows indicate transient protein interactions.

Only the *dmsEFABGH* (SO_1427-SO_1432) operon shows significant induction during anaerobic growth and was shown by mutagenesis studies to be the main DMSO respiratory gene cluster in *S. oneidensis* [166]. Electron transfer to the DMSO reductase complex requires menaquinone as the electron donor, and proceeds via the CymA membrane-bound tetraheme cytochrome and the periplasmically located, decaheme DmsE cytochrome to the DmsABF complex in the outer membrane (Figure 7) [166–168]. The DmsE cytochrome is required for full activity of the complex, but results indicate that related, MtrA-like cytochromes such as MtrA, MtrD, SO_4360 or the CctA tetraheme cytochrome may be able to at least partially substitute for DmsE function [169].

Due to the outer membrane location of this and other enzyme systems (mediating, e.g., Fe(III) and Fe(III) citrate reduction) in *S. oneidensis*, the term 'extracellular respiration' has been coined to describe the function of these respiratory complexes [166, 170]. It has been hypothesized that the location of the DMSOR in the outer membrane facilitates the use of particulate or substrate-bound DMSO. Regulation of DMSO respiration in *S. oneidensis* appears to involve cyclic AMP (cAMP), cAMP receptor protein (Crp), and an ArcAB-related two-component regulatory system [171–173].

2.1.3.4   The *Halobacterium* NRC-1 Dimethylsulfoxide Respiratory System

An interesting observation is that Dms-type respiratory systems appear to also occur in Archaea, where a *dmsR-dmsEABCD* operon has been identified in *Halobacterium* NRC-1. In this operon the *dmsABCD* genes encode the subunits of the enzyme and the required molecular chaperone [174]. Based on sequence alignments these authors proposed that the Mo amino acid ligand of the DmsA protein should be an aspartate.

The function of the 82 aa DmsE protein is unknown, it is annotated as a hypothetical protein and related sequences appear to be confined to the Halobacteriaceae. Nevertheless, the *dmsE* gene is transcribed as part of the DMSO reductase operon in these Archaea, as shown by both transcript mapping and transcriptome analyses [174]. The *dmsR* gene encodes a regulator of the Bat family of transcriptional activators that is required for expression of the DMSO reductase operon. The *dmsEABCD* operon was shown to be induced during anaerobic growth and was required for growth on DMSO, but not on TMAO.

According to our analyses, enzymes related to the *Halobacterium* NRC-1 DMSO reductase are found exclusively in the euryarchaeotic lineage of the Halobacteriaceae, but close relatives are also present in the firmicutes and high GC Gram-positive bacteria (32 % aa identity; 50 % aa similarity), which might indicate an acquisition of the *dms* operon by the halophilic Archaea from bacterial lineages. For comparison – between the NRC_1 DmsA and the *E. coli* DmsA amino acid sequence similarities are 23 % identity/42 % similarity.

2.2   Dimethylsulfide Dehydrogenases

2.2.1   The *Rhodovulum sulfidophilum* Dimethylsulfide Dehydrogenase

Like the DMSO reductases, the DMS dehydrogenases belong to the DMSO reductase enzyme family, however, at present there is only one enzyme of this type that has been characterized. The *Rhodovulum sulfidophilum* DMS dehydrogenase is expressed during phototrophic growth of the bacteria in the presence of DMS which can serve as an electron donor to the photosynthetic apparatus via its electron acceptor, a cytochrome c_2 [139, 175–177]. DMS dehydrogenases essentially catalyze the reverse of the reaction of DMSO reductases (Figure 2), and like the DMSO reductases are capable of catalyzing both the forward and the reverse reaction [139].

The *R. sulfidophilum* DdhABC DMS dehydrogenase is a soluble, trimeric, periplasmic protein that consists of a catalytic Mo subunit (DdhA), that also contains an Fe/S cluster, a subunit with 4 Fe/S clusters (DdhB), and a subunit with a single heme *b* group (DdhC) [139, 175, 177]. The operon encoding DMS dehydrogenase, *ddhABDC*, also encodes a cytoplasmic protein, DdhD, that is likely to be a molecular chaperone involved in maturation of the enzyme, and upstream of this operon a gene (*ddhS*) encoding a signal kinase with homology to the DorS sensor kinase was identified [175], but no molecular studies of the regulation of the DMS dehydrogenase have been published so far, and a genome sequence is currently not available, although a sequencing project for *R. sulfidophilum* appears to exist (Source: www.ncbi.nlm.nih.gov/).

DMS dehydrogenase activity assays contain dichlorophenol-indophenol, phenazineethosulfate, and DMS as the substrate [139]. The purified DMS dehydrogenase had a maximal activity around pH 8, with pK_a values of 7.7 and 8.9 being reported, and K_M values for both DMS and cytochrome c_2 were in the low micromolar range with values of 52 ± 9 and 21 ± 2 µM, respectively [178]. Further analyses showed that DMS dehydrogenase uses a two-site ping-pong catalytic mechanism [178]. Cyclic voltammetry revealed a pH-dependent shift in the heme redox potential of approximately 20 mV/pH unit, and electron paramagnetic resonance was used to determine the redox potentials of the Mo, heme, and four of the Fe-S redox centers present in the enzyme at pH 8. The redox potentials of the two Mo couples were $55 +/- 10$ mV *versus* SHE for $Mo^{V/IV}$ and $123 +/- 13$ mV *versus* SHE for the $Mo^{VI/V}$ couple [179]. A comparison of the redox potentials of the various centers present in the enzyme showed that the potential range and variation was very similar to those observed in ethylbenzene dehydrogenase from *Aromatoleum aromaticum* and the NarG-type nitrate reductase [179].

2.2.2   Dimethylsulfide Dehydrogenases: Environmental Distribution

Analysis of DdhA-related sequences present in public databases (www.ncbi.nlm. nih.gov) reveals a tightly clustering group of sequences from the α-Proteobacteria *R. sulfidophilum, Sagittula stellata*, and *Citreicella* sp. SE45 that occurs together with a second group of sequences originating from the γ-Proteobacteria *Thiorhodococcus drewsii, Pseudomonas chloritidismutans*, and *Halomonas jeotgali*. These six species also contain genes encoding proteins with high homologies to DdhB and DdhC and can thus be regarded as likely being true DMS dehydrogenases, despite the fact that the enzyme from *P. chloritidismutans* is annotated as a putative chlorate reductase (ClrABC). Another group of sequences originating from members of the

Aquificales (mostly *Sulfurihydrogenibium* sp.), and some species of ε-Proteobacteria (e.g., *Sulfuricurvum*) is located on a separate branch close to the core DMS dehydrogenase group.

The finding of Ddh homologues in *Sagittula stellata* is interesting due to the reported ability of the organism to oxidize DMS to DMSO under aerobic conditions [62], providing an auxiliary energy source to the organism during heterotrophic growth [19]. However, a DMSDH enzyme assay of biomass grown on succinate and DMS was negative for the activity and SDS-PAGE analysis did not show obvious induction of Ddh subunits in the presence of DMS [19]. It is possible that a different enzyme system is responsible for oxidation of DMS to DMSO in this organism during aerobic heterotrophic growth and the role of the Ddh homologues and the conditions for their expression still need to be determined in *S. stellata*.

The DdhA-homologous sequences derived from the above organisms have ~40 % aa sequence identity to DdhA and the other two subunits are not as well conserved as in the first cluster of sequences. Despite this, these enzymes may still represent DMS dehydrogenases although the annotation of these sequences varies and mostly reflects their status as uncharacterized proteins. Experimental evidence will be required to determine the function of the encoded enzymes.

2.3   Phylogeny of Dimethylsulfoxide Reductases and Dimethylsulfide Dehydrogenases

As briefly set out above, within the DMSO reductase enzyme family at least three subgroups of enzymes have been identified using phylogenetic analyses of the amino acid sequences of the catalytic subunits. Type I enzymes are the formate dehydrogenase/Nap-type nitrate reductase enzymes, Type II enzymes are the functionally most diverse group and encompass DMS dehydrogenase, Nar-type nitrate reductase, ethylbenzene dehydrogenase, and related enzymes while the soluble DMSO/TMAO reductases (Dor-type) are classified as Type III enzymes [79, 175, 180].

An alternative way of classifying the enzymes in this enzyme family is to group them according to the nature of the amino acid ligand present at the Mo center, or based on the substrate/type of reaction catalyzed. The Mo centers of the DMSO reductase family enzymes fall into three broad categories using this classification method – Mo centers with a serine ligand (DMSO/TMAO reductases), Mo centers with Cys or SeCys ligands (formate dehydrogenases, Nap-type nitrate reductases), and with an Asp ligand (ethylbenzene dehydrogenases, Nar-type nitrate reductases) [84]. There is some evidence for the existence of a fourth group of enzymes, currently exemplified by the Aio arsenite oxidase, that completely lacks an amino acid ligand to the Mo center [87, 181].

With the exception of this arsenite oxidase, the groupings identified using the amino acid ligand to the Mo center correspond to the phylogenetic groups identified by McDevitt et al. [175] with the Type I enzymes having Cys or SeCys ligands, the Type II enzymes having an aspartate ligand to the Mo center and the Type III enzyme having a serine ligand.

The DMS dehydrogenases are part of the Type II enzyme group which is the group with the greatest diversity of reactions catalyzed by representative enzymes. Within this group the DMS dehydrogenases form a tightly clustered group of 6 sequences, all of which are of proteobacterial origin, and a second, closely related group of uncharacterized enzymes from ε-Proteobacteria and Aquificales may also be DMS dehydrogenases. The closest relatives of DMS dehydrogenases that have been characterized are the selenate reductases, and the group of enzymes comprising ethylbenzene and C25 cholesterol dehydrogenases (data not shown).

The DMSO reductases represent an interesting case with the Dor-type enzymes being representative of the Type III enzyme group, while the Dms-type enzymes are Type II enzymes and are distantly related to DMS dehydrogenase as indicated by their subunit structure and the presence of an Fe/S cluster in the catalytic subunit. Within the Type II enzymes, the DmsA enzymes form a deep branching, independent lineage [175, 180].

This raises an interesting question regarding the identity of the amino acid ligand to the Mo center in Dms-type DMSO reductases. The two available crystal structures of Type II enzymes (NarGHI nitrate reductase and ethylbenzene dehydrogenase) [89, 91, 182] as well as sequence alignments (Figure 8) indicate that the Type II enzymes contain an aspartate ligand to the Mo center, and some authors have suggested that, e.g., the DmsA protein from *Halobacterium* NRC-1 contains an aspartate as the Mo ligand [174].

However, there is also a study that indicates that at least the *E. coli* DmsA protein might contain a serine ligand to the Mo center. Trieber et al. identified Ser176 as a putative conserved Mo ligand in DmsA based on sequence alignments with Dor/Tor Type III proteins [148]. They created variants in this amino acid, all of which abolished the ability of *E. coli* to grow anaerobically with DMSO as the electron acceptor, and also abolished DMSO reductase enzyme activity *in vitro* [148]. Changes to the redox properties of the Mo center were also reported, and an S176H mutation led to a complete loss of the Mo electron paramagnetic resonance spectrum [148]. All of these results strongly indicate a role for DmsA-Ser176 in the function of the *E. coli* DMSO reductase.

Using sequence alignments containing a variety of sequences for Type II and Type III DMSOR family enzymes, we found that the alignment of DmsA sequences with the identified Mo ligands of Type II and Type III enzymes is very sensitive to the selection of sequences to be used in the alignment (Figure 8).

Alignments containing mostly sequences of Dor-type TMAO and DMSO reductases as well as sequences for a few Type II enzymes strongly resembled the alignment shown by [148] where *E. coli* DmsA-Ser176 clearly aligns with the known serine ligand of Dor/Tor-type enzymes. However, *E. coli* DmsA-Ser176 is located in a 'GDYS' sequence motif, and in this version of the alignment the aspartate residue located in this motif aligns with the known Asp Mo ligand of the Type II enzymes. The only exception to this is the *Halobacterium* DmsA sequence.

In contrast, if alignments are carried out with sequences of Type II enzymes only (DMS dehydrogenase, chlorate reductase, ethylbenzene dehydrogenase, selenate reductase, steroid C25 dehydrogenase, and perchlorate reductase) and sequences

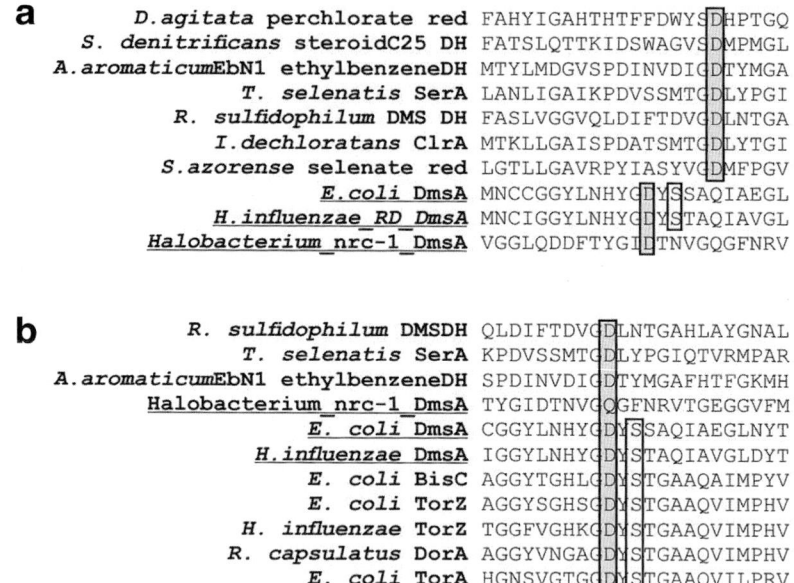

a
```
        D.agitata perchlorate red   FAHYIGAHTHTFFDWYSDHPTGQ
     S. denitrificans steroidC25 DH FATSLQTTKIDSWAGVSDMPMGL
  A.aromaticumEbN1 ethylbenzeneDH   MTYLMDGVSPDINVDIGDTYMGA
              T. selenatis SerA     LANLIGAIKPDVSSMTGDLYPGI
        R. sulfidophilum DMS DH     FASLVGGVQLDIFTDVGDLNTGA
           I.dechloratans ClrA      MTKLLGAISPDATSMTGDLYTGI
         S.azorense selenate red    LGTLLGAVRPYIASYVGDMFPGV
                  E.coli DmsA       MNCCGGYLNHYGDYSSAQIAEGL
             H.influenzae RD DmsA   MNCIGGYLNHYGDYSTAQIAVGL
        Halobacterium nrc-1 DmsA    VGGLQDDFTYGIDTNVGQGFNRV
```

b
```
        R. sulfidophilum DMSDH     QLDIFTDVGDLNTGAHLAYGNAL
              T. selenatis SerA    KPDVSSMTGDLYPGIQTVRMPAR
  A.aromaticumEbN1 ethylbenzeneDH  SPDINVDIGDTYMGAFHTFGKMH
        Halobacterium nrc-1 DmsA   TYGIDTNVGQGFNRVTGEGGVFM
                  E. coli DmsA     CGGYLNHYGDYSSAQIAEGLNYT
               H.influenzae DmsA   IGGYLNHYGDYSTAQIAVGLDYT
                  E. coli BisC     AGGYTGHLGDYSTGAAQAIMPYV
                  E. coli TorZ     AGGYSGHSGDYSTGAAQVIMPHV
              H. influenzae TorZ   TGGFVGHKGDYSTGAAQVIMPHV
              R. capsulatus DorA   AGGYVNGAGDYSTGAAQVIMPHV
                  E. coli TorA     HGNSVGTGGDYSTGAAQVILPRV
```

Figure 8 Alignments of amino acid sequences of Type II (**a, b**) and Type III (**b**) DMSO reductase family enzymes comparing the position of the putative amino acid ligand to the Mo center of DmsA DMSO reductases relative to other enzymes of the same type. Panel a: alignment of Type II DMSO reductase family enzyme, Panel b: alignment of Type II and Type III DMSO reductase family enzymes. The alignments only show the regions around the amino acid ligands to the Mo center present in these enzymes. The Mo-ligating amino acids are highlighted: grey – aspartate ligands typical of Type II enzymes; white – serine ligands typical of Type III enzymes. DH, dehydrogenase; red, reductase; SerA, selenite reductase; ClrA, chlorate reductase; DMSDH, DMS dehydrogenase; BisC, biotinsulfoxide reductase; DmsA, Dms-type DMSO reductase; DorA, Dor-type DMSO reductase; TorA, trimethylamine oxide reductase; TorZ, S- and N-oxide reductase.

for DmsA proteins from *E. coli, H. influenzae,* and *Halobacterium* NRC-1 (Figure 8), the aspartate Mo ligands of the non-DmsA sequences can be easily identified, but within the DmsA sequence group neither the DmsA-Ser176 residue nor the aspartate residue of the 'GDYS' motif align with the other aspartate residues. The DmsA aspartate residues of the 'GDYS' motif align with each other approximately 5 amino acids upstream of the conserved Type II enzyme aspartate residue, and it becomes obvious that an equivalent of the DmsA-Ser176 residue is not present in the *Halobacterium* NRC-1 DmsA sequence, confirming the predictions of [174].

So together this suggests that unlike the Dor-type DMSO reductases, DmsA-type DMSO reductases might indeed possess an aspartate ligand to the Mo center, and that the effects observed in the DmsA-Ser176 mutants of *E. coli* might be caused by the close proximity of this residue to the Mo center. Mutation of this residue could lead to perturbations of the protein backbone and the side chain packing which could explain the effects of such mutations on the properties of the DmsA Mo center reported by [148].

2.4 Dimethylsulfide Monooxygenases

2.4.1 Properties of Dimethylsulfide Monooxygenases

Hyphomicrobium species were amongst the earliest characterized DMS-degrading bacteria to be discovered. Several strains of DMS-degrading *Hyphomicrobium* species have been characterized physiologically (e.g., [16, 33, 37, 38, 183]). Some of these may also degrade other methylated sulfur compounds including dimethylsulfone and dimethylsulfoxide, which are initially reduced to DMS before a DMS monooxygenase (DMSMO) catalyzes its degradation to MT and formaldehyde [16, 38]. The activity of DMS monooxygenase was first identified in *Hyphomicrobium* S [16] based on NADH-dependent oxygen uptake in the presence of DMS, but the enzyme had not been purified and characterized until recently [184], and, similar to the situation with DMS dehydrogenase, there is currently only one characterized DMS monooxygenase.

The DMS monooxygenase from *Hyphomicrobium sulfonivorans* S1 was shown to be a flavin-dependent monooxygenase composed of two subunits (DmoAB) with molecular masses of 53 and 19 kDa, respectively. The large subunit DmoA is an FMN-dependent monooxygenase, while the small subunit is a NAD(P)H-dependent flavin oxidoreductase. Addition of FMN to the enzyme assay increased the enzyme activity approximately 12-fold. Activity assays with different combinations of reduced nicotinamides and flavin cofactors suggested that the small subunit is an NADH-dependent FMN reductase, although some activity was also shown with NADPH as electron donor. Use of alternative flavin cofactors resulted in negligible enzyme activities in line with the prediction that FMN is the flavin cofactor [184].

DMSMO converted DMS to MT and formaldehyde and had decreasing activity when alternative alkylsulfides with increasing length of the alkyl chain were used as substrates. Chelation experiments with EDTA, EGTA, and bathocuproine all decreased the enzyme activity. Reconstitution with a range of divalent metal cations showed the best restoration of enzyme activity (63 % compared to the native enzyme) when Fe^{2+} and Mg^{2+} were added in combination suggesting both are required for the activity of the enzyme, but further characterization of the metal content of DMSMO is still required [184].

DMS monooxygenase had a K_M of approximately 17 μM for DMS and a V_{max} of 1.25 μmol min^{-1} mg^{-1} protein giving a k_{cat} of 5.45 s^{-1}. This relatively low turnover number of DMSMO appears to be compensated for by a relatively high abundance of the enzyme which was estimated to be nearly 9 % of cell protein [184].

The large subunit was further characterized by peptide sequencing and gene cloning. PCR primers were designed based on peptide sequences and allowed identification of cloned genomic DNA fragments containing the gene encoding the DMSMO large subunit (*dmoA*). The genetic context of *dmoA* was shown to contain two genes encoding predicted FMN reductases, which may encode the small subunit of the enzyme (DmoB), but the *dmoB* gene has not been identified unequivocally to date due to lack of definitive mass spectrometry data and failure of

N-terminal sequencing for the DmoB subunit. Further genes in close proximity to *dmoA* were predicted to encode enzymes involved in the synthesis of the flavin cofactor as well as a sulfite oxidase and a further large subunit of a two subunit FMN-dependent monooxygenase that had high similarity to alkanesulfonate monooxygenase (SsuD) [184].

2.4.2 Distribution, Genomic Context, and Phylogeny of Dimethylsulfide Monooxygenases

In terms of its evolutionary relationship, the large subunit of DMS monooxygenase is a member of the luciferase superfamily. The most closely related homologues of DmoA are the enzymes of the NtaA/SnaA/SoxA(DszA) monooxygenase family which includes nitrilotriacetate monooxygenase (NtaA), pristinamycin IIA synthase subunit A (SnaA), dibenzothiophene sulfone monooxygenase (SoxA/DszA), and related enzymes [184].

Database searches with DmoA from *Hyphomicrobium sulfonivorans* identify a large number of homologous sequences of predicted FMN-dependent oxidoreductases with varying annotation, including some annotated as DMS monooxygenase large subunit genes. These originate from diverse bacteria including Proteobacteria and Actinobacteria, but the function of these homologues has yet to be determined. Potential homologues of DmoA with molecular masses of 53 kDa have been observed as dimethylsulfone-induced polypeptides in strains of *Arthrobacter methylotrophus* and *A. sulfonivorans* [38] and the presence of proteins with putative molecular masses of 53 kDa and 19 kDa was observed in $DMSO_2$-grown *Methylobacterium podarium* FM1 [32]. These proteins were absent when the strains were grown on methylamine, suggesting that these strains may potentially have similar DMS monooxygenases.

Phylogenetic analysis including representatives of characterized members of the NtaA/SnaA/SoxA(DszA) monooxygenase family and related FMN-dependent oxidoreductases shows that DmoA of *H. sulfonivorans* is closely related to proteins from *H. zavarzinii* and *Paracoccus* sp. TRP (Figure 9). These further cluster with DmoA homologues found in two *Pseudomonas* strains and a range of Actinobacteria including strains of *Rothia*, *Kocuria*, *Arthrobacter*, *Leifsonia*, *Cryocola*, *Clavibacter*, and *Pseudoclavibacter*. While not all strains of *Hyphomicrobium* and *Arthrobacter* may be able to degrade DMS, it is likely that the homologues found in these strains represent *dmoA* genes.

The most closely related homologue of DmoA is found in *Hyphomicrobium zavarzinii*, having an identity of 86 % at the amino acid level and a genomic neighbourhood very similar to that seen around *dmoA* in *Hyphomicrobium sulfonivorans*, encoding NADH-dependent flavin oxidoreductases, FMN-dependent alkanesulfonate monooxygenases, sulfite oxidase, and enzymes involved in flavin biosynthesis. In *Paracoccus* strains (J56, J4, TRP, J39, J55), the

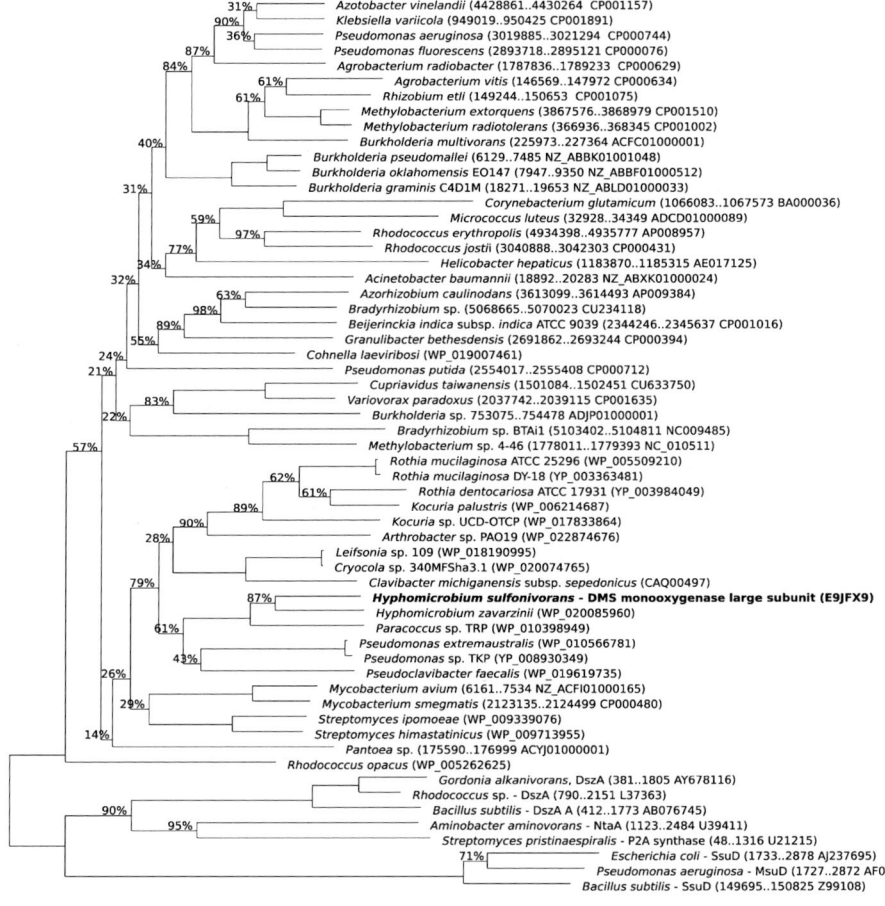

Figure 9 Distance tree showing the relationships of enzymes of the NtaA/SnaA/SoxA(DszA) family of enzymes including DmoA homologues deduced using the neighbor joining algorithm implemented in A using PAM correction. The tree is based on alignment positions present in the DmoA of *Hyphomicrobium sulfonivorans* (alignment positions 58 to 528). Bootstrap values are given for 100 replicates. SsuD, alkanesulfonate monooxygenase; MsuE, methanesulfonate monooxygenase; DszA, dibenzothiophene sulfone monooxygenase; P2A synthase, pristinamycin IIA synthase subunit A; NtaA, nitrilotriacetate monooxygenase.

genes encoding DmoA homologues were also associated with other alkanesulfonate monooxygenases and flavin oxidoreductases while *dmoA* homologues in *Rothia* strains are in close proximity to predicted sulfonate transporters. Overall, these findings suggest that some of the homologues of DmoA are likely to have a role in organosulfur compound degradation, but their substrate specificity and catalytic role remains to be investigated.

3   General Conclusions

The many roles of DMS and DMSO in microbial metabolism and physiology outlined above are reflected in the diversity of enzymes that are involved in the transformation of these organic sulfur compounds and the wide phylogenetic distribution of the ability to metabolize DMS within the bacterial and archaeal phyla.

There is still a huge discrepancy in the level of insight and understanding of the biochemistry, genetics, and structure-function relationships of DMS transforming enzymes, where DMSO reductases have been investigated in considerable detail, while there is still a lot to be learnt about the biochemistry of both DMS dehydrogenases and DMS monooxygenase enzymes, their structure-function relationships and, for the latter, the role of metals in catalysis. There is also a clear discrepancy in the distribution of DMS-converting enzymes and their relative abundance in the currently available genome sequences. While genes encoding DMSO reductase of either the Dor- or Dms-type are present in a large number of bacterial genomes, DMS dehydrogenases and DMS monooxygenases appear to have a much more limited distribution, and reasons for this apparent specialization are currently unknown, but the skewed distribution of these enzymes might at least partly be due to the relative overrepresentation of, e.g., Proteobacterial genomes in current databases, while DMS dehydrogenase and DMS monooxygenase genes are found in bacterial groups for which fewer genome sequences are available.

There is also clear potential for discovery of additional enzymes involved in DMS biotransformations, such as the methyltransferases in aerobic DMS-degrading methylotrophs and the NAD(P)H-dependent DMSO-converting enzymes reported to exist in $DMSO_2$ and DMSO-degrading *Hyphomicrobium* and *Arthrobacter* strains which are known to be expressed during aerobic heterotrophic growth of these bacteria.

DMS and related methylated sulfur compounds play important roles in the global biogeochemical cycle of sulfur and have relevance in many different areas due to their effect on atmospheric chemistry, atmospheric sulfur transport as well as being signalling compounds affecting the behavior of animals. They are important flavor compounds in foods and beverages, but may be produced in the human body due to disease or metabolic disorders. As industrially relevant chemicals as well as byproducts of sewage treatment processes, they can cause nuisance odors which need to be deodorized using biotechnological applications.

Considering these varied roles of DMS and related compounds it is clear that further insights into the molecular processes underlying DMS transformations will be required to advance our understanding of the interplay of DMS biotransformations and the global sulfur cycle.

Abbreviations

aa	amino acid
Aio	abbreviation for a type of arsenite (ai) oxidase (o)
AMO	ammonia monooxygenase
bp	base pair
cAMP	cyclic adenosine monophoshate
CCN	cloud condensation nuclei
Ddd	DMSP-dependent DMS; abbreviation used for DMSP lyase genes
DH	dehydrogenase
DMSMO	dimethylsulfide monooxygenase
DMS	dimethylsulfide
DMSO	dimethylsulfoxide
$DMSO_2$	dimethylsulfone
DMSOR	dimethylsulfoxide reductase
DMSP	dimethylsulfoniopropionate
DOR	dimethylsulfoxide reductase Dor-type
EDTA	ethylenediamine-N,N,N',N'-tetraacetic acid
EGTA	ethylene glycol-bis(2-aminoethylether)-N,N,N',N'-tetraacetic acid
FMN	flavin mononucleotide
FNR	fumarate nitrate reductase regulation
GC	% G + C content of DNA
HEPES	2-[4-(2-hydroxyethyl)piperazin-1-yl]ethanesulfonic acid
IHF	integration host factor
MetSO	methionine sulfoxide
MGD	molybdopteringuanine dinucleotide or pyranopterin guanine dinucleotide
MMPA	methylmercaptopropionate
MSA	methanesulfonic acid
MT	methanethiol
MQ	menaquinone
NADH	nicotinamide adenine dinucleotide (reduced)
NADPH	nicotinamide adenine dinucleotide phosphate (reduced)
NarL	nitrate reductase regulator protein
PCR	polymerase chain reaction
PGD	pyroanpterin guanindinucleotide
SDS-PAGE	sodium dodecyl sulfate-polyacrylamide gel electrophoresis
SHE	standard hydrogen electrode
Tg	Tera gram
TMAO	trimethylamine-N-oxide
UQ	ubiquinone

Acknowledgments   We would like to thank Dr. Megan Maher and Prof. Paul Bernhardt for their help in preparing this manuscript. HS is grateful to past and present coworkers and support from the UK-Natural Environment Research Council.

References

1. H. Schäfer, N. Myronova, R. Boden, *J. Exp. Botan.* **2010**, *61*, 315–334.
2. R. Bentley, T. G. Chasteen, *Chemosphere* **2004**, *55*, 291–317.
3. S. F. Watts, *Atmos. Environ.* **2000**, *34*, 761–779.
4. J. E. Lovelock, R. J. Maggs, R. A. Rasmussen, *Nature* **1972**, *237*, 452–453.
5. M. O. Andreae, *Marine Chem.* **1990**, *30*, 1–29.
6. D. P. Kelly, S. C. Baker, *FEMS Microbiol. Rev.* **1990**, *87*, 241–246.
7. F. J. Zhao, J. S. Knights, Z. Y. Hu, S. P. McGrath, *J. Environ. Qual.* **2003**, *32*, 33–39.
8. M. A. Kertesz, P. Mirleau, *J. Exp. Botan.* **2004**, *55*, 1939–1945.
9. R. J. Charlson, J. E. Lovelock, M. O. Andreae, S. G. Warren, *Nature* **1987**, *326*, 655–661.
10. P. K. Quinn, T. S. Bates, *Nature* **2011**, *480*, 51–56.
11. G. O. Kirst, C. Thiel, H. Wolff, J. Nothnagel, M. Wanzek, R. Ulmke, *Marine Chem.* **1991**, *35*, 381–388.
12. J. Stefels, *J. Sea Res.* **2000**, *43*, 183–197.
13. W. Sunda, D. J. Kieber, R. P. Kiene, S. Huntsman, *Nature* **2002**, *418*, 317–320.
14. A. R. J. Curson, J. D. Todd, M. J. Sullivan, A. W. B. Johnston, *Nat. Rev. Microbiol.* **2011**, *9*, 849–859.
15. A. Lana, T. G. Bell, R. Simo, S. M. Vallina, J. Ballabrera-Poy, A. J. Kettle, J. Dachs, L. Bopp, E. S. Saltzman, J. Stefels, J. E. Johnson, P. S. Liss, *Global Biogeochem. Cycles* **2011**, *25*, GB1004.
16. J. A. M. De Bont, J. P. van Dijken, W. Harder, *J. Gen. Microbiol.* **1981**, *127*, 315–323.
17. J. M. M. de Zwart, P. N. Nelisse, J. G. Kuenen, *FEMS Microbiol. Ecol.* **1996**, *20*, 261–270.
18. R. Boden, D. P. Kelly, J. C. Murrell, H. Schäfer, *Env. Microbiol.* **2010**, *12*, 2688–2699.
19. R. Boden, J. C. Murrell, H. Schäfer, *FEMS Microbiol. Lett.* **2011**, *322*, 188–193.
20. D. H. Green, D. M. Shenoy, M. C. Hart, A. D. Hatton, *Appl. Environ. Microbiol.* **2011**, *77*, 3137–3140.
21. T. Omori, Y. Saiki, K. Kasuga, T. Kodama, *Biosci. Biotechnol. Biochem.* **1995**, *59*, 1195–1198.
22. H. Fuse, O. Takimura, K. Murakami, Y. Yamoaka, T. Omori, *Appl. Environ. Microbiol.* **2000**, *66*, 5527–5532.
23. R. P. Kiene, T. S. Bates, *Nature* **1990**, *345*, 702–705.
24. B. P. Lomans, A. J. P. Smolders, L. M. Intven, A. Pol, H. J. M. Op Den Camp, C. Van Der Drift, *Appl. Environ. Microbiol.* **1997**, *63*, 4741–4747.
25. H. Kadota, Y. Ishida, *Annu. Rev. Microbiol.* **1972**, *26*, 127–138.
26. R. P. Kiene, D. G. Capone, *Microbiol. Ecol.* **1988**, *15*, 275–291.
27. S. H. Zinder, T. D. Brock, *Arch. Microbiol.* **1978**, *116*, 35–40.
28. F. Bak, K. Finster, F. Rothfuß, *Arch. Microbiol.* **1992**, *157*, 529–534.
29. J.-U. Kreft, B. Schink, *Arch. Microbiol.* **1993**, *159*, 308–315.
30. B. P. Lomans, P. Leijdekkers, J.-J. Wesselink, P. Bakkes, A. Pol, C. van der Drift, H. J. Op den Camp, *Appl. Environ. Microbiol.* **2001**, *67*, 4017–4023.
31. S. H. Zinder, T. D. Brock, *Nature* **1978**, *273*, 226–228.
32. V. Anesti, J. Vohra, S. Goonetilleka, I. R. McDonald, B. Sträubler, E. Stackebrandt, D. P. Kelly, A. P. Wood, *Env. Microbiol.* **2004**, *8*.
33. E. Borodina, D. P. Kelly, F. A. Rainey, N. L. Ward-Rainey, A. P. Wood, *Arch. Microbiol.* **2000**, *173*, 425–437.

34. H. G. Kim, N. V. Doronina, Y. A. Trotsenko, S. W. Kim, *Int. J. Syst. Evol. Microbiol.* **2007**, *57*, 2096–2101.
35. S. A. Moosvi, I. R. McDonald, D. A. Pearce, D. P. Kelly, A. P. Wood, *Sys. Appl. Microbiol.* **2005**, *28*, 541–554.
36. H. Schäfer, *Appl. Environ. Microbiol.* **2007**, *73*, 2580–2591.
37. G. M. H. Suylen, J. G. Kuenen, *Antonie van Leeuwenhoek* **1986**, *52*, 281–293.
38. E. Borodina, D. P. Kelly, P. Schumann, F. A. Rainey, N. L. Ward-Rainey, A. P. Wood, *Arch. Microbiol.* **2002**, *177*, 173–183.
39. W. D. Gould, T. Kanagawa, *J. Gen. Microbiol.* **1992**, *138*, 217–221.
40. G. M. H. Suylen, P. J. Large, J. P. Vandijken, J. G. Kuenen, *J. Gen. Microbiol.* **1987**, *133*, 2989–2997.
41. C. Anthony, *The Biochemistry of Methylotrophs*, Academic Press, London, 1982.
42. T. Kanagawa, D. P. Kelly, *FEMS Microbiol. Lett.* **1986**, *34*, 13–19.
43. S. Sivelä, V. Sundman, *Arch. Microbiol.* **1975**, *103*, 303–304.
44. N. A. Smith, D. P. Kelly, *J. Gen. Microbiol.* **1988**, *134*, 1407–1417.
45. P. T. Visscher, B. F. Taylor, *Appl. Environ. Microbiol.* **1993**, *59*, 3784–3789.
46. P. T. Visscher, B. F. Taylor, *Appl. Environ. Microbiol.* **1993**, *59*, 4083–4089.
47. K. Finster, Y. Tanimoto, F. Bak, *Arch. Microbiol.* **1992**, *157*, 425–430.
48. R. P. Kiene, R. S. Oremland, A. Catena, L. G. Miller, D. G. Capone, *Appl. Environ. Microbiol.* **1986**, *52*, 1037–1045.
49. B. P. Lomans, R. Maas, R. Luderer, H. J. Op den Camp, A. Pol, C. van der Drift, G. D. Vogels, *Appl. Environ. Microbiol.* **1999**, *65*, 3641–3650.
50. T. J. Lyimo, A. Pol, H. J. Op den Camp, H. R. Harhangi, G. D. Vogels, *Int. J. Syst. Evol. Microbiol.* **2000**, *50*, 171–178.
51. S. S. Ni, D. R. Boone, *Int. J. Syst. Bacteriol.* **1991**, *41*, 410–416.
52. Y. Tanimoto, F. Bak, *Appl. Environ. Microbiol.* **1994**, *60*, 2450–2455.
53. B. P. Lomans, H. J. Op den Camp, A. Pol, C. van der Drift, G. D. Vogels, *Appl. Environ. Microbiol.* **1999**, *65*, 2116–2121.
54. S. C. M. Haaijer, H. R. Harhangi, B. B. Meijerink, M. Strous, A. Pol, A. J. P. Smolders, K. Verwegen, M. S. M. Jetten, H. den Camp, *ISME J.* **2008**, *2*, 1231–1242.
55. P. T. Visscher, B. F. Taylor, R. P. Kiene, *FEMS Microbiol. Ecol.* **1995**, *18*, 145–153.
56. T. C. Tallant, J. A. Krzycki, *J. Bacteriol.* **1997**, *179*, 6902–6911.
57. E. Oelgeschlaeger, M. Rother, *Mol. Microbiol.* **2009**, *72*, 1260–1272.
58. J. Zeyer, P. Eicher, S. G. Wakeham, R. P. Schwarzenbach, *Appl. Environ. Microbiol.* **1987**, *53*, 2026–2032.
59. C. Vogt, A. Rabenstein, J. Rethmeier, U. Fischer, *Microbiology* **1997**, *143*, 767–773.
60. C. A. McDevitt, P. Hugenholtz, G. R. Hanson, A. G. McEwan, *Mol. Microbiol.* **2002**, *44*, 1575–1587.
61. L. Zhang, I. Kuniyoshi, M. Hirai, M. Shoda, *Biotechnol. Lett.* **1991**, *13*, 223–228.
62. J. M. González, F. Mayer, M. A. Moran, R. E. Hodson, W. B. Whitman, *Int. J. Syst. Bacteriol.* **1997**, *47*, 773–780.
63. D. Y. Sorokin, B. E. Jones, J. G. Kuenen, *Extremophiles* **2000**, *4*, 145–155.
64. L. Y. Juliette, M. R. Hyman, D. J. Arp, *Appl. Environ. Microbiol.* **1993**, *59*, 3718–3727.
65. A. J. Holmes, A. Costello, M. E. Lidstrom, J. C. Murrell, *FEMS Microbiol. Lett.* **1995**, *132*, 203–208.
66. H. B. Stirling, H. Dalton, *FEMS Microbiol. Lett.* **1979**, *5*, 315–318.
67. M. Horinouchi, K. Kasuga, H. Nojiri, H. Yamane, T. Omori, *FEMS Microbiol. Lett.* **1997**, *155*, 99–105.
68. T. Endoh, K. Kasuga, M. Horinouchi, T. Yoshida, H. Habe, H. Nojiri, T. Omori, *Appl. Microbiol. Biotechnol.* **2003**, *62*.
69. C.-Y. Li, T.-D. Wei, S.-H. Zhang, X.-L. Chen, X. Gao, P. Wang, B.-B. Xie, H.-N. Su, Q.-L. Qin, X.-Y. Zhang, J. Yu, H.-H. Zhang, B.-C. Zhou, G.-P. Yang, Y.-Z. Zhang, *Proc. Natl. Acad. Sci. USA* **2014**, in press.

70. C. Wagner, S. M. Lusty, H.-F. Kung, N. L. Rogers, *J. Biol. Chem.* **1967**, *242*, 1287–1293.
71. D. R. Warner, J. L. Hoffman, *Biochemistry* **1996**, *35*, 4480–4484.
72. N. M. Mozier, K. P. McConnell, J. L. Hoffman, *J. Biol. Chem.* **1988**, *263*, 4527–4531.
73. R. A. Rothery, G. J. Workun, J. H. Weiner, *Biochim. Biophys. Acta* **2008**, *1778*, 1897–1929.
74. D. Sambasivarao, H. A. Dawson, G. J. Zhang, G. Shaw, J. Hu, J. H. Weiner, *J. Biol. Chem.* **2001**, *276*, 20167–20174.
75. R. A. Rothery, C. A. Trieber, J. H. Weiner, *J. Biol. Chem.* **1999**, *274*, 13002–13009.
76. P. T. Bilous, S. T. Cole, W. F. Anderson, J. H. Weiner, *Mol. Microbiol.* **1998**, *2*, 785–795.
77. A. G. McEwan, J. P. Ridge, K. F. Aguey-Zinsou, P. V. Bernhardt, G. R. Hanson, *J. Inorg. Biochem.* **2003**, *96*, 54–54.
78. U. Kappler, W. M. Huston, A. G. McEwan, *Microbiology* **2002**, *148*, 605–614.
79. A. G. McEwan, J. P. Ridge, C. A. McDevitt, P. Hugenholtz, *Geomicrobiol. J.* **2002**, *19*, 3–21.
80. A. G. McEwan, G. R. Hanson, S. Bailey, *Biochem. Soc. Trans.* **1998**, *26*, 390–396.
81. H. K. Li, C. Temple, K. V. Rajagopalan, H. Schindelin, *J. Am. Chem. Soc.* **2000**, *122*, 7673–7680.
82. H. Schindelin, C. Kisker, J. Hilton, K. V. Rajagopalan, D. C. Rees, *Science* **1996**, *272*, 1615–1621.
83. F. Schneider, J. Loewe, R. Huber, H. Schindelin, C. Kisker, J. Knaeblein, *J. Mol. Biol.* **1996**, *263*, 53–69.
84. S. Grimaldi, B. Schoepp-Cothenet, P. Ceccaldi, B. Guigliarelli, A. Magalon, *Biochim. Biophys. Acta* **2013**, *1827*, 1048–1085.
85. A. Magalon, J. G. Fedor, A. Walburger, J. H. Weiner, *Coord. Chem. Rev.* **2011**, *255*, 1159–1178.
86. Y. Zhang, S. Rump, V. N. Gladyshev, *Coord. Chem. Rev.* **2011**, *255*, 1206–1217.
87. P. J. Ellis, T. Conrads, R. Hille, P. Kuhn, *Structure* **2001**, *9*, 125–132.
88. M. Jormakka, K. Yokoyama, T. Yano, M. Tamakoshi, S. Akimoto, T. Shimamura, P. Curmi, S. Iwata, *Nat. Struct. Mol. Biol.* **2008**, *15*, 730–737.
89. M. Jormakka, D. Richardson, B. Byrne, S. Iwata, *Structure* **2004**, *12*, 95–104.
90. M. Jormakka, S. Tornroth, B. Byrne, S. Iwata, *Science* **2002**, *295*, 1863–1868.
91. D. P. Kloer, C. Hagel, J. Heider, G. E. Schulz, *Structure* **2006**, *14*, 1377–1388.
92. J. J. R. Frausto da Silva, R. J. P. Williams, *The Biological Chemistry of the Elements – The Inorganic Chemistry of Life*, Oxford University Press, Oxford, 2001.
93. J. J. G. Moura, C. D. Brondino, J. Trincao, M. J. Romao, *J. Biol. Inorg. Chem.* **2004**, *9*, 791–799.
94. O. Kniemeyer, J. Heider, *J. Biol. Chem.* **2001**, *276*, 21381–21386.
95. J. Dermer, G. Fuchs, *J. Biol. Chem.* **2012**, *287*, 36905–36916.
96. W. Reichenbecher, A. Brune, B. Schink, *Biochim. Biophys. Acta* **1994**, *1204*, 217–224.
97. P. M. H. Kroneck, D. J. Abt, H. Niessen, B. Schink, *J. Inorg. Biochem.* **2001**, *86*, 300–300.
98. G. B. Seiffert, G. M. Ullmann, A. Messerschmidt, B. Schink, P. M. H. Kroneck, O. Einsle, *Proc. Natl. Acad. Sci. USA* **2007**, *104*, 3073–3077.
99. B. M. Martins, H. Dobbek, I. Çinkaya, W. Buckel, A. Messerschmidt, *Proc. Natl. Acad. Sci. USA* **2004**, *101*, 15645–15649.
100. S. M. McCrindle, U. Kappler, A. G. McEwan, *Adv. Microb. Phys.* **2005**, *50*, 147–198.
101. M. J. Romao, *Dalton Trans.* **2009**, 4053–4068.
102. R. J. Turner, A. L. Papish, F. Sargent, *Can. J. Microbiol.* **2004**, *50*, 225–238.
103. K. J. Sarfo, T. L. Winstone, A. L. Papish, J. M. Howell, H. Kadir, H. J. Vogel, R. J. Turner, *Biochem. Biophys. Res. Comm.* **2004**, *315*, 397–403.
104. A. L. Papish, C. L. Ladner, R. J. Turner, *J. Biol. Chem.* **2003**, *278*, 32501–32506.
105. N. Ray, J. Oates, R. J. Turner, C. Robinson, *FEBS Lett.* **2003**, *534*, 156–160.
106. R. D. Fleischmann, M. D. Adams, O. White, R. A. Clayton, E. F. Kirkness, A. R. Kerlavage, C. J. Bult, J. F. Tomb, B. A. Dougherty, J. M. Merrick, K. McKenney, G. Sutton, W. Fitzhugh, C. Fields, J. D. Gocayne, J. Scott, R. Shirley, L. I. Liu, A. Glodek, J. M. Kelley, J. F. Weidman, C. A. Phillips, T. Spriggs, E. Hedblom, M. D. Cotton, T. R. Utterback, M. C.

Hanna, D. T. Nguyen, D. M. Saudek, R. C. Brandon, L. D. Fine, J. L. Fritchman, J. L. Fuhrmann, N. S. M. Geoghagen, C. L. Gnehm, L. A. McDonald, K. V. Small, C. M. Fraser, H. O. Smith, J. C. Venter, *Science* **1995**, *269*, 496–512.

107. M. Xu, S. J. W. Busby, D. F. Browning, *J. Bacteriol.* **2009**, *191*, 3172–3176.
108. S. P. Lubitz, J. H. Weiner, *Arch. Biochem. Biophys.* **2003**, *418*, 205–216.
109. I. J. Oresnik, C. L. Ladner, R. J. Turner, *Mol. Microbiol.* **2001**, *40*, 323–331.
110. M. F. Olmo-Mira, M. Gavira, D. J. Richardson, F. Castillo, C. Moreno-Vivián, M. D. Roldán, *J. Biol. Chem.* **2004**, *279*, 49727–49735.
111. A. G. McEwan, S. J. Ferguson, J. B. Jackson, *Biochem. J.* **1991**, *274*, 305–308.
112. M. D. Moore, S. Kaplan, *J. Bacteriol.* **1989**, *171*, 4385–4394.
113. T. Satoh, F. N. Kurihara, *J. Biochem. (Tokyo)* **1987**, *102*, 191–197.
114. A. L. Shaw, S. Leimkuehler, W. Klipp, G. R. Hanson, A. G. McEwan, *Microbiology* **1999**, *145*, 1409–1420.
115. N. J. Mouncey, M. Choudhary, S. Kaplan, *J. Bacteriol.* **1997**, *179*, 7617–7624.
116. V. Mejean, C. Iobbi Nivol, M. Lepelletier, G. Giordano, M. Chippaux, M. C. Pascal, *Mol. Microbiol.* **1994**, *11*, 1169–1179.
117. H. Strnad, A. Lapidus, J. Paces, P. Ulbrich, C. Vlcek, V. Paces, R. Haselkorn, *J. Bacteriol.* **2010**, *192*, 3545–3546.
118. P. S. Solomon, A. L. Shaw, I. Lane, G. R. Hanson, T. Palmer, A. G. McEwan, *Microbiology* **1999**, *145*, 1421–1429.
119. P. S. Solomon, A. L. Shaw, M. D. Young, S. Leimkuehler, G. R. Hanson, W. Klipp, A. G. McEwan, *FEMS Microbiol. Lett.* **2000**, *190*, 203–208.
120. W. S. Kontur, W. S. Schackwitz, N. Ivanova, J. Martin, K. LaButti, S. Deshpande, H. N. Tice, C. Pennacchio, E. Sodergren, G. M. Weinstock, D. R. Noguera, T. J. Donohue, *J. Bacteriol.* **2012**, *194*, 7016–7017.
121. S. Zhou, E. Kvikstad, A. Kile, J. Severin, D. Forrest, R. Runnheim, C. Churas, J. W. Hickman, C. Mackenzie, M. Choudhary, T. Donohue, S. Kaplan, D. C. Schwartz, *Genome Res.* **2003**, *13*, 2142–2151.
122. N. J. Mouncey, S. Kaplan, *J. Bacteriol.* **1998**, *180*, 5612–5618.
123. I. Yamamoto, T. Ujiiye, Y. Ohshima, T. Satoh, *Plant Cell Physiol.* **2001**, *42*, 703–709.
124. I. Yamamoto, K. Takamatsu, Y. Ohshima, T. Ujiiye, T. Satoh, *Biochim. Biophys. Acta* **1999**, *1447*, 57–63.
125. T. Ujiiye, I. Yamamoto, T. Satoh, *Biochim. Biophys. Acta* **1997**, *1353*, 84–92.
126. N. J. Mouncey, S. Kaplan, *J. Bacteriol.* **1998**, *180*, 2924–2930.
127. J. H. Zeilstra-Ryalls, K. Gabbert, N. J. Mouncey, S. Kaplan, R. G. Kranz, *J. Bacteriol.* **1997**, *179*, 7264–7273.
128. S. Elsen, L. R. Swem, D. L. Swem, C. E. Bauer, *Microbiol. Mol. Biol. Rev.* **2004**, *68*, 263–279.
129. C. Bauer, S. Elsen, L. R. Swem, D. L. Swem, S. Masuda, *Philos. Trans. Roy. Soc. B* **2003**, *358*, 147–153.
130. S. Elsen, W. Dischert, A. Colbeau, C. E. Bauer, *J. Bacteriol.* **2000**, *182*, 2831–2837.
131. J. Gregor, T. Zeller, A. Balzer, K. Haberzettl, G. Klug, *J. Mol. Micobiol. Biotechnol.* **2007**, *13*, 126–139.
132. A. W. Dangel, F. R. Tabita, *Mol. Microbiol.* **2009**, *71*, 717–729.
133. J. Wu, C. E. Bauer, *mBio* **2010**, *1*, e00272–00210-e00272–00218.
134. P. M. McNicholas, R. C. Chiang, R. P. Gunsalus, *Mol. Microbiol.* **1998**, *27*, 197–208.
135. P. McNicholas, S. Rech, R. Gunsalus, *FASEB Journal* **1997**, *11*, A1378–A1378.
136. J. L. SimalaGrant, J. H. Weiner, *Eur. J. Biochem.* **1998**, *251*, 510–515.
137. J. L. SimalaGrant, J. H. Weiner, *Microbiology* **1996**, *142*, 3231–3239.
138. R. W. Jones, P. B. Garland, *Biochem. J.* **1977**, *164*, 199–211.
139. S. P. Hanlon, T. H. Toh, P. S. Solomon, R. A. Holt, A. G. McEwan, *Eur. J. Biochem.* **1996**, *239*, 391–396.

140. R. C. Bray, B. Adams, A. T. Smith, R. L. Richards, D. J. Lowe, S. Bailey, *Biochemistry* **2001**, *40*, 9810–9820.
141. B. Adams, A. T. Smith, S. Bailey, A. G. McEwan, R. C. Bray, *Biochemistry* **1999**, *38*, 8501–8511.
142. J. P. Ridge, K. F. Aguey-Zinsou, P. V. Bernhardt, I. M. Brereton, G. R. Hanson, A. G. McEwan, *Biochemistry* **2002**, *41*, 15762–15769.
143. K. E. Johnson, K. V. Rajagopalan, *J. Biol. Chem.* **2001**, *276*, 13178–13185.
144. K. Heffron, C. Leger, R. A. Rothery, J. H. Weiner, F. A. Armstrong, *Biochemistry* **2001**, *40*, 3117–3126.
145. K. Heffron, J. H. Weiner, R. A. Rothery, F. A. Armstrong, *J. Inorg. Biochem.* **1999**, *74*, 157–157.
146. C. Léger, P. Bertrand, *Chem. Rev.* **2008**, *108*, 2379–2438.
147. K. F. Aguey-Zinsou, P. V. Bernhardt, A. G. McEwan, J. P. Ridge, *J. Biol. Inorg. Chem.* **2002**, *7*, 879–883.
148. C. A. Trieber, R. A. Rothery, J. H. Weiner, *J. Biol. Chem.* **1996**, *271*, 27339–27345.
149. A. S. McAlpine, A. G. McEwan, S. Bailey, *J. Mol. Biol.* **1998**, *275*, 613–623.
150. A. S. McAlpine, A. G. McEwan, A. L. Shaw, S. Bailey, *J. Biol. Inorg. Chem.* **1997**, *2*, 690–701.
151. F. Schneider, J. Löwe, R. Huber, H. Schindelin, C. Kisker, J. Knäblein, *J. Mol. Biol.* **1996**, *263*, 53–69.
152. G. R. Hanson, I. Lane, in *Metals in Biology: Applications of High-Resolution Epr to Metalloenzymes*, Eds G. Hanson, L. Berliner, Springer, Vol. 29, 2010, pp. 169–199.
153. R. C. Bray, B. Adams, A. T. Smith, B. Bennett, S. Bailey, *Biochemistry* **2000**, *39*, 11258–11269.
154. S. Metz, W. Thiel, *Coord. Chem. Rev.* **2011**, *255*, 1085–1103.
155. N. Cobb, C. Hemann, G. A. Polsinelli, J. P. Ridge, A. G. McEwan, R. Hille, *J. Biol. Chem.* **2007**, *282*, 35519–35529.
156. J. P. Ridge, K. F. Aguey-Zinsou, P. V. Bernhardt, G. R. Hanson, A. G. McEwan, *FEBS Lett.* **2004**, *563*, 197–202.
157. M. J. Pushie, G. N. George, *Coord. Chem. Rev.* **2011**, *255*, 1055–1084.
158. G. N. George, K. J. Nelson, H. H. Harris, C. J. Doonan, K. V. Rajagopalan, *Inorg. Chem.* **2007**, *46*, 3097–3104.
159. N. Graham, C. J. Doonan, R. A. Rothery, N. Boroumand, J. H. Weiner, *Inorg. Chem.* **2007**, *46*, 2–4.
160. R. H. Holm, E. I. Solomon, A. Majumdar, A. Tenderholt, *Coord. Chem. Rev.* **2011**, *255*, 993–1015.
161. A. L. Tenderholt, J.-J. Wang, R. K. Szilagyi, R. H. Holm, K. O. Hodgson, B. Hedman, E. I. Solomon, *J. Am. Chem. Soc.* **2010**, *132*, 8359–8371.
162. D. Guymer, J. Maillard, F. Sargent, *Arch. Microbiol.* **2009**, *191*, 519–528.
163. S. Gon, J. C. Patte, V. Mejean, C. Iobbi-Nivol, *J. Bacteriol.* **2000**, *182*, 5779–5786.
164. A. del Campillo Campbell, A. Campbell, *J. Mol. Evol.* **1996**, *42*, 85–90.
165. J. F. Heidelberg, I. T. Paulsen, K. E. Nelson, E. J. Gaidos, W. C. Nelson, T. D. Read, J. A. Eisen, R. Seshadri, N. Ward, B. Methe, R. A. Clayton, T. Meyer, A. Tsapin, J. Scott, M. Beanan, L. Brinkac, S. Daugherty, R. T. DeBoy, R. J. Dodson, A. S. Durkin, D. H. Haft, J. F. Kolonay, R. Madupu, J. D. Peterson, L. A. Umayam, O. White, A. M. Wolf, J. Vamathevan, J. Weidman, M. Impraim, K. Lee, K. Berry, C. Lee, J. Mueller, H. Khouri, J. Gill, T. R. Utterback, L. A. McDonald, T. V. Feldblyum, H. O. Smith, J. C. Venter, K. H. Nealson, C. M. Fraser, *Nat. Biotech.* **2002**, *20*, 1118–1123.
166. J. A. Gralnick, H. Vali, D. P. Lies, D. K. Newman, *Proc. Natl. Acad. Sci. USA* **2006**, *103*, 4669–4674.
167. D. A. Saffarini, S. L. Blumerman, K. J. Mansoorabadi, *J. Bacteriol.* **2002**, *184*, 846–848.
168. C. Schwalb, S. K. Chapman, G. A. Reid, *Biochemistry* **2003**, *42*, 9491–9497.
169. D. Coursolle, J. A. Gralnick, *Mol. Microbiol.* **2010**, *77*, 995–1008.

170. J. A. Gralnick, D. K. Newman, *Mol. Microbiol.* **2007**, *65*, 1–11.
171. N. P. Shroff, M. A. Charania, D. A. Saffarini, *J. Bacteriol.* **2010**, *192*, 3227–3230.
172. D. A. Saffarini, R. Schultz, A. Beliaev, *J. Bacteriol.* **2003**, *185*, 3668–3671.
173. J. A. Gralnick, C. T. Brown, D. K. Newman, *Mol. Microbiol.* **2005**, *56*, 1347–1357.
174. J. A. Müller, S. DasSarma, *J. Bacteriol.* **2005**, *187*, 1659–1667.
175. C. A. McDevitt, P. Hugenholtz, G. R. Hanson, A. G. McEwan, *Mol. Microbiol.* **2002**, *44*, 1576–1587.
176. A. G. McEwan, T. H. Toh, P. S. Solomon, A. L. Shaw, S. P. Hanlon, M. E. Lidstrom, F. R. Tabita, in *Microbial Growth on C1 Compounds*, Kluwer Academic Publishers, Dordrecht, 1995, pp. 41–48.
177. S. P. Hanlon, R. A. Holt, G. R. Moore, A. G. McEwan, *Microbiology* **1994**, *140*, 1953–1958.
178. N. L. Creevey, A. G. McEwan, P. V. Bernhardt, *J. Biol. Inorg. Chem.* **2008**, *13*, 1231–1238.
179. N. L. Creevey, A. G. McEwan, G. R. Hanson, P. V. Bernhardt, *Biochemistry* **2008**, *47*, 3770–3776.
180. A. G. McEwan, U. Kappler, C. A. McDevitt, D. Zannoni, in *Respiration in Archaea and Bacteria*, Ed Govindjee, Kluwer Academic Publishers, Dordrecht, 2004, Vol. 1, pp. 175–202.
181. T. P. Warelow, M. Oke, B. Schoepp-Cothenet, J. U. Dahl, N. Bruselat, G. N. Sivalingam, S. Leimkuehler, K. Thalassinos, U. Kappler, J. H. Naismith, J. M. Santini, *PLoS ONE* **2013**, *8*, e72535.
182. M. Jormakka, B. Byrne, S. Iwata, *FEBS Lett.* **2003**, *545*, 25–30.
183. A. Pol, H. J. M. Op den Camp, S. G. M. Mees, M. A. S. H. Kersten, C. van der Drift, *Biodeg.* **1994**, *5*, 105–112.
184. R. Boden, E. Borodina, A. P. Wood, D. P. Kelly, J. C. Murrell, H. Schäfer, *J. Bacteriol.* **2011**, *193*, 1250–1258.

Index

A

A. sulfonivorans, 304
ABC transporter, 201, 202
Absorption spectra, 46, 47, 77, 134–136, 222
Acetaldehyde, 16, 18–20, 22, 30, 32, 33
Acetate, 18–20, 22, 42, 57, 127
Acetobacterium woodii, 103
Acetogens, 42, 57, 105
Acetoin, 19
Acetonitrile, 28, 32
Acetyl-CoA synthase(s) (ACS), 39, 42, 43,
    49, 50, 52, 54, 56–62, 64, 73–75, 79,
    83–85, 126
Acetyl-coenzyme A (acetyl-CoA), 20, 39, 42,
    49, 54, 56, 57, 60, 62, 73, 74, 85, 126
Acetylene (C$_2$H$_2$), 16, 165, 170
    acidity constant, 17
    bioavailability, 17, 18
    fermentation, 20
    properties, 17
    reduction, 16, 17
Acetylene hydratase (AH), 16, 18–32, 287
    active site, 21, 24, 27, 28
    AH(Mo), 21, 22
    AH(W), 22
    coordination geometry, 23
    X-ray structure, 22
Acetylides, 17
Achromobacter cycloclastes, 184, 190, 195,
    198, 203
Acidianus, 241, 244, 248, 265, 266
    ambivalens, 241, 264–268
Acidilobus, 241
Acidiphilium, 248
Acidithiobacillaceae, 247

Acidithiobacillus, 248
Acidity constants (pK_a)
    acetylene, 17
    methane, 140
Acidobacteria, 224
Acireductone dioxygenase, 126
Acinetobacter, 284
Actinobacteria, 41, 221, 224, 304
Active site of
    acetylene hydratase, 21, 24, 27, 28
    carbon monoxide dehydrogenase, 76
    cytochrome *c* nitrite reductase, 226
Adenosine 5'-diphosphate (5'-ADP), 12,
    110, 245
    MgADP, 152, 153, 157, 159, 160, 167
Adenosine 5'-monophosphate (5'-AMP),
    12, 251, 269
    cyclic, 298
Adenosine 5'-phosphosulfate (APS), 242, 245,
    250–253, 267, 269, 270
    reductase (APSR), 242, 245, 251, 253, 267,
    269, 270
Adenosine 5'-triphosphate (5'-ATP), 12,
    20, 43, 103, 104, 149–152, 157–160,
    166, 171, 200, 218, 242, 243, 245,
    250–253, 263
    hydrolysis, 151, 159, 171
    MgATP, 151
    sulfurylase (ATPS), 243, 245, 250–253
    synthase, 43, 243, 263
    synthesis, 43, 103, 104
S-Adenosylmethionine, 132
Adenylylsulfate pyrophosphorylase, 251
Aerobes, 37–65, 72, 73, 239
Aerosols, 216, 281

© Springer Science+Business Media Dordrecht 2014   315
P.M.H. Kroneck, M.E. Sosa Torres (eds.), *The Metal-Driven Biogeochemistry
of Gaseous Compounds in the Environment*, Metal Ions in Life Sciences 14,
DOI 10.1007/978-94-017-9269-1

Afipia, 283
Agriculture, 184
Albedo formation, 281
Alcaligenes faecalis, 184
Alcohols, *see also* individual names, 28, 30,
  31, 240
Aldehyde, 12
  acet-, 16, 18–20, 22, 30, 32, 33
Algae, 109, 110, 239, 282
  marine, 282
  mats, 102, 109, 110, 127
  photosynthesis, 109
Alkalilimnicola, 248
Alkenes, 17, 148
Alkyl thioethers, 136
Alkynes, *see also* individual names, 17, 27
Allochromatium vinosum, 247
Allophane, 12
Alphaproteobacteria, 245, 247
Aluminosilicate, 4
Aluminum, 3–5, 11
Amicyanin, 193
Amino acids, *see also* individual names, 7, 8,
  11, 12, 16, 20, 22, 24–28, 44, 78, 113,
  114, 132, 179, 185, 193, 205, 212, 226,
  227, 240, 251, 255, 257, 286–288, 298,
  300–302, 304
  sequences, 20, 113, 300, 302
Ammonia, *see also* Ammonium, 16, 126,
  147–173, 179, 180, 211–233, 271, 284
  in the environment, 214
  production, 212, 214, 218, 231, 232
  turnover, 217
Ammonia monooxygenase (AMO), 16, 180,
  213, 217, 219, 284, 293
Ammonifex, 241
Ammonification pathway, 180
Ammonium (NH$_4^+$), 180, 212
  aerosols, 216
  chloride, 215
  nitrate, 216
  oxidation, 180, 213, 214, 217
  sulfate, 216
Amoebacter, 247
Anaerobes, 37–65, 230, 239, 241, 270, 283, 284
Anaerobic
  archaea, 42
  bacteria, 16, 42, 44, 49, 282
  deltaproteobacteria, 19
  fermentation, 18
  oxidation of methane (AOM),
    127, 128, 140
  respiration, 221, 282, 286, 288, 296

sludge, 240
sulfate reducers, 43
Anaerobic ammonium oxidation (anammox),
  180, 213, 214, 217
Anammoxosome, 213, 219
Anatase (TiO$_2$), 92
Ancalochloris, 247
Andesite, 3
Anesthetic, 181
Anoxic sediments, 19, 282, 283
Antarctica, 183
Anthropogenic N$_2$O release, 183
Anticancer drug, 216
APSR. *See* Adenosine 5'-phosphosulfate
  reductase
Aquaspirillum, 248
Aquifex, 482
Aquificales, 300, 301
Arcanobacterium haemolyticum, 221
Archaea, *see also* individual names, 38, 42, 43,
  54, 57, 73, 101, 102, 127, 221, 240–249,
  255, 256, 259, 263, 266, 280, 283, 284,
  298, 306
  anaerobic, 42
  methanogenic, 43, 57, 73, 244, 284
  methanotrophic, 132
  sulfur reducing, 240–242, 244, 255, 256,
    259, 260, 265, 266, 306
Archaeoglobus (*Ar.*), 240, 252, 258, 260
  fulgidus, 241, 252, 258, 259
  profundus, 256, 258–260
  veneficus, 260
Arcobacter, 248
Aromatoleum aromaticum, 299
Arsenite oxidase, 300
Arthrobacter, 283, 304, 306
  methylotrophus, 283
Ascorbate, 188, 195, 205
Assembly of Cu$_Z$, 189, 202–203
Assimilatory nitrite reductase, 213
Assimilatory sulfate reduction, 239, 240, 242,
  245, 250–252, 254, 260, 262
Assimilatory sulfite reductase, 221, 222, 254,
  256, 261, 262
Asthma, 216
Atmosphere
  aerosols, 281
  concentration of N$_2$O, 18
  dimethylsulfide flux, 281
  Early Earth, 1–13, 18, 101
  Earth, 16, 17, 32
  primordial, 40
  Titan, 18

ATP. *See* Adenosine 5'-triphosphate
Autotrophic
   bacteria, 247, 283
   methanogens, 57
Azane, 212
Azide, 22
Azotobacter
   chroococcum, 150
   vinelandii, 150, 151, 154, 155, 166, 167,
      169, 170, 218

B
Bacillus (B.)
   azotoformans, 221
   bataviensis, 221
   selenatireducens, 221
   subtilis, 191
   vireti, 221
Bacteria(l), *see also* individual names, 73, 103
   anaerobic, 16, 42, 44, 49, 282
   alphaproteo-, 41, 286, 299
   autotrophic, 283
   betaproteo-, 41, 247
   CODH/ACS, 58, 59
   colorless sulfur, 239, 246–249, 267, 271
   cyano-, 102, 105, 109, 110, 240, 246
   deltaproteo-, 19, 242
   denitrification, 183–184, 217
   dimethylsulfide-degrading, 303
   enteric, 220, 229, 261
   epsilonproteo-, 245, 247
   flavo-, 284
   gammaproteo-, 41, 245–247, 249, 299
   Gram-negative, 18, 183, 200, 221, 232
   green sulfur, 239, 246, 247, 267–270, 284
   heterotrophic, 282, 284
   myco-, 41
   nitrate-reducing, 127
   pathogens, 261
   phototrophic, 246, 267, 268, 299
   phototrophic green sulfur, 245–247, 267,
      268, 270, 284
   phototrophic purple sulfur, 246, 247, 267,
      268, 284
   phototrophic sulfur, 246, 247, 267
   purple non-sulfur, 246, 247, 267
   purple sulfur, 239, 246, 247, 267–270
   respiration, 288
   sulfate-reducing, 18, 72, 105, 222,
      240–243, 249–251, 253, 254, 257, 260,
      262, 271, 284

   sulfide-oxidizing, 239, 242, 245–249
   sulfite-oxidizing, 245–249
   sulfur-reducing eu-, 242, 244, 263, 265
   thiosulfate-oxidizing, 245–249
Bactericidal agent, 41
Basalt, 3
Bathocuproine, 303
Beggiatoa, 248
Bentonite, 4
Berthierine, 5
Betaproteobacteria, 247
Bilophila wadsworthia, 260
Biogeochemical cycle of
   sulfur, 240, 280, 306
   nitrogen, 179, 180, 212, 214, 232
Biomass, 42, 102, 207, 214, 217, 282,
   283, 300
Biomimetic model, 141, 193
Biotin sulfoxide reductase, 296
Biotite, 5
2,2'-Bipyridine, 215
Bis-molybdopterin guanine dinucleotide
   (bis-MGD, MGD), 20, 22–25, 32,
   261, 263, 271, 286, 294, 295
Blastochloris, 247
Blue copper proteins, 193
Bonds
   π, 17
   σ, 17, 39
   C–C, 73
   C≡C, 16–18, 26, 28
   C–H, 44, 49, 140
   C=O, 39
   C≡O, 38, 172
   C–S, 48, 137–140
   H. *See* Hydrogen bonds
   Mo=S, 44
   Ni–Fe, 51, 63
   N–N bond, 181
   N–O bond, 181, 225, 227
   N,N triple bond, 147–172
   O–C–O bond, 39
   phosphodiester, 7
   triple, 3, 16–18, 26, 28, 38, 147–172
Bradyrhizobium japonicum, 44
Brewing, 280
Bromoethanesulfonate, 137
Bromopropanesulfonate, 136
Brucite, 12
Butane, 18
Butene, 18
n-Butyl isocyanate, 54, 55, 79

C

Cacodylate, 22
Cadmium, 250
    sulfide, 93
Calcium, Ca^{2+}, 3, 11, 215, 223, 226, 227
    hydroxide (Ca(OH)$_2$), 215
Caldisphaera, 241
Caldivirga, 240, 241
Calvin cycle, 109, 110, 179
Calvin-Benson-Bassham cycle, 41
Campylobacter
    concisus, 218
    curvus, 218
    fetus, 218
    jejuni, 218
Candidatus
    desulforudis, 241
    Kuenenia stuttgartiensis, 219
Carbamates, 27
Carbohydrates, 221
Carbon
    ^{13}C, 79
    cycling, 242
    fixation, 41, 101, 284
Carbon dioxide (CO$_2$), 2, 3, 18, 38–43, 46,
        48–52, 55–57, 63, 71–95, 101, 103, 107,
        109, 110, 115, 117, 119, 126, 134, 136,
        140, 171–183, 246, 267, 282, 283, 287
    fixation, 41, 72, 101, 109, 179, 246, 267
    reduction, 52, 55, 56, 63, 72, 80, 83–88, 92,
        93, 95, 101, 140
    CO$_2$/CO interconversions, 72
    CO$_2$/CO redox couple, 93
Carbon monoxide (CO), 37–64, 38–40, 71–95,
        102, 111, 115, 117, 119, 121, 126, 131,
        134, 141, 149, 150, 157, 164, 171, 172,
        215, 222, 239, 243, 249, 250
    ^{13}CO, 115
    CO$_2$/CO interconversions, 72,
    CO$_2$/CO redox couple, 93
    biological cycle, 40, 41
    distribution, 56–58
    in the biosphere, 40–43
    oxidation, 40, 43, 46–48, 50, 55, 56, 63, 73,
        77, 79, 83–91, 93, 102, 284
    reduction, 40, 42, 92, 150, 171, 172
Carbon monoxide dehydrogenase(s) (CODH),
        37–64, 71–95, 126, 243
    3D-structures, 75
    ACDS/CODH, 79
    active site, 76
    Ag,Mo-CODH, 47
    catalytic cycle, 85
    Class IV, 86–91

CODH/ACS, 39, 42, 43, 50, 52, 54–59,
        72–75, 79, 83–85
CODH I, 85, 90
CODH II, 79
CODH III, 73
CODH IV, 75
CODH V, 75
Cu,Mo-CODHs, 43–49, 55, 62, 63
Ni-CODHs, 71–95
NiFe-containing, 74, 75
sources, 40
Carbonates, 2
Carboxydothermus (*C.*) *hydrogenoformans*,
        43, 49, 50, 59, 72, 73, 102
Catalase, 230
Catalysis by
    molybdenum nitrogenase, 150
    vanadium nitrogenase, 149, 166–172
Catalysts in the early Earth, 4–6
Catalytic mechanism of
    methyl-coenzyme M reductase, 137–140
    molybdenum nitrogenase, 149, 150, 154,
        157–166, 169
Catalytic properties of Cu$_Z$, 198–200
Cation
    exchange capacity, 5
    radical, 137, 227
CdS nanoparticles, 93
Cellulose, 127
Chamosite, 5
Channels, 41, 46, 52–53, 58, 59, 79, 85, 114,
        115, 130–132, 203–206, 227, 228, 231,
        249, 255, 257, 294
Chaperone(s), 25, 202
    metallo-, 127
Charcoal, 41
Charge transfer, 49, 188, 189, 195
Chemolithoautotrophs, 247
Chlamydomonas reinhardtii, 109
Chlorate reductase, 299, 301, 302
Chlorine oxide radicals, 182
Chlorobaculum tepidum, 247, 268
Chlorobiaceae, 246
Chlorobium, 247
Chloroflexaceae, 246
Chlorofluorocarbons, 182, 183
Chloroherpeton, 247
Cholesterol dehydrogenase, 287, 301
Choline, 19
Chromatiaceae, 246, 269
Chromatiales, 246
Chromosome(s), 222, 255, 290
Chronoamperometric measurements,
        81, 82, 88

Circular dichroism, 22
  magnetic, 78, 133
Cisplatin, 216
Citrate(s), 150, 156, 157, 169, 218
Citreicella sp., 299
Citrobacter, 261
Clathrochloris, 247
Clavibacter, 304
Clay minerals, 4–13
Climate, 128, 182, 183, 216, 282
Clinochlore, 5
Clostridium
  nigrificans, 251
  pasteurianum, 117, 150, 260
Cloud formation, 282
Cluster(s)
  [4Cu:2S], 178, 197
  [4Fe] cubane, 22
  [2Fe2S], 45, 46, 48
  [3Fe-4S], 73, 76, 77, 257
  [4Fe-4S], 21–23, 26, 27, 73, 74, 76, 220,
    253–258, 261, 262, 271
  [Fe$_2$S$_2$], 167
  [Fe$_3$S$_4$], 63
  [Fe$_4$S$_3$], 156
  [Fe$_4$S$_4$], 74, 148, 150–154, 158, 166–168
  [Fe$_8$S$_7$], 150, 156, 168
  [MoFe$_3$S$_3$], 156
  [Ni3Fe-4S], 73, 76
  [Ni4Fe-4S], 73, 76
  A-, 74, 85
  B-, 74, 76, 77
  C-, 73–77, 79, 85, 87
  D-, 74–76, 78, 81, 82, 84, 85, 93
  ferredoxin-type, 21, 22, 253, 255
  H-, 115–117, 120
  iron sulfur, 21, 23, 53, 75, 80–82, 111,
    113–117, 121, 126, 150, 180, 202, 229,
    255, 257, 261–263, 286, 288, 290
  Ni,Fe, 39
  P-, 150, 154-156, 159, 160, 167-169
CO. *See* Carbon monoxide
CO$_2$. *See* Carbon dioxide
CoA. *See* Coenzyme A
Cobalt, Co^{2+}, 52
Coal, 41, 126, 140
CoBSH. *See* Coenzyme B
Coenzyme A (CoA), 59, 60, 73, 126, 266
Coenzyme B (CoBSH), 130–132, 136, 137,
  139, 140
Coenzyme F$_{420}$, 113, 260
Coenzyme F$_{430}$, 129–132
  structure, 129, 130
  properties, 129, 130
  reactivity, 129, 130

Coenzyme M methylase, 284
Cofactor(s), *see also* individual names, 20–26,
  28, 32, 43–46, 62, 119, 129, 133–135,
  137, 139, 140, 150, 157, 166, 167, 169,
  201, 218, 220, 231, 253, 255, 258,
  261, 271, 285–288, 290, 294, 295,
  303, 304
  F$_{430}$, 129–133, 135, 137, 139, 140
  F$_{430}$-Ni(II)/F$_{430}$-Ni(I), 129
  flavin, 303
  iron-molybdenum, 150, 154–157, 160, 161,
    163–165, 168, 169
  nickel, 129
  pyranopterin, 43, 46, 287, 290
Colorless sulfur bacteria (CSB), 239, 246–249,
  267, 271
Combustion, 17, 38, 40, 100, 102, 128,
  183, 216
Copper, 38, 43–49, 55, 59, 62–64, 180,
  184–187, 189–195, 198–202, 204,
  206, 213, 217, 250
  $^{63/65}$Cu, 190
  ^{63}Cu, 193
  ^{65}Cu, 193
  Cu,Mo-CODH(s), 43–49, 55, 62, 63
  Cu$_A$, 185–195, 199, 201–206
  Cu$_A$ in *P. stutzeri* nitrous oxide
    reductase, 194
  Cu$_Z$, 187–189, 192–195, 197–200
  Cu$_Z$*, 197
  clusters. *See* Clusters
  trafficking, 202
Copper(I), 17, 45, 48, 49, 63, 192, 201–203
  Cu$^+$/Cu^{2+}, 191, 193, 195
Copper(II), 133, 192, 203, 215
  Cu$^+$/Cu^{2+}, 191, 193, 195
  tetraammine, 215
Corrensite, 5
Chromium, Cr(III), 215
Crenarchaeota, 244, 248, 265, 266
Crust, 3, 127
Cryocola, 304
Crystal structures of, *see also* Structures *and*
    X-ray structures
  adenosine 5'-phosphosulfate reductase, 253
  Cu,Mo-CODHs, 43–45
  dimethylsulfoxide reductases, 293–295
  iron hydrogenase, 119
  iron-iron hydrogenase, 119
  MoFe protein, 155
  Rhodobacter capsulatus DMSO
    reductase, 291
CSB. *See* Colorless sulfur bacteria
Cubane cluster, 22, 76
Cupredoxin, 185, 186, 189, 191, 193, 202, 204

Cyanate (NCO⁻), 78, 79, 85, 87–90
    iso-, 54, 55, 79
    thio-, 85, 89–91
Cyanide (CN⁻), 18, 78, 79, 87, 88, 90, 249
    ¹³CN, 79
    -inhibited CODH, 54, 88
Cyanobacteria, 102, 105, 109, 110, 240, 246
Cycles of
    carbon monoxide, 40, 41
    carbon, 242
    dihydrogen, 101–103, 126
    nitrogen, 148, 180, 220
    ribulose monophosphate, 283
    serine, 283
    sulfur, 239, 240, 245, 246, 281, 306
    tricarboxylic acid, 41
Cyclic voltammetry, 81, 82, 84, 90, 94, 299
Cysteine
    seleno-, 114, 300
Cytochrome(s), 41, 104, 110, 117, 185, 187,
    189–191, 193, 200, 201, 211–232, 242,
    243, 246, 249, 261–263, 265, 268, 270,
    290, 296–299
    high-molecular weight, 242
    octaheme, 242, 243
Cytochrome b, 261
Cytochrome bo₃ oxidase, 191
Cytochrome(s) c, 185, 187, 200, 201, 214, 217,
    219, 220, 222, 227, 230–232, 242, 263,
    265, 268, 270, 288
    c₂, 299
    c₅₅₀, 268
    c₅₅₂, 187, 220, 222
    c₅₅₄, 270
    c₇, 263, 265
    cd₁, 187, 213, 217
    biosynthesis, 220
    flavo-, 267, 268
    multiheme, 211–232, 242, 263
    octaheme, 231, 243, 261–263
    oxidase, 189–191, 193, 249
    tetrathionate reductase, 231
Cytochrome c nitrite reductase (NrfA),
    180, 213, 217, 218, 220–232
    cd₁, 187, 213
    active site, 226
    biochemistry, 221–227
    in stress defense, 230
    structure, 221–227
Cytoplasmic membrane, 43, 45, 47, 103,
    200, 201, 227, 232, 263
Cytosine, 44, 45

D
D. baculatus, 256
D. desulfuricans, 222, 245, 251, 252, 255, 260
D. gigas, 252–258, 263
D. thermophilus, 258, 259
D. vulgaris, 218, 221, 222, 227, 228, 241,
    250, 253–259, 261–262, 265
    Hildenborough, 222, 250, 253–259, 261,
    262, 265
Deep sea, 19
Dehydrogenases
    cholesterol, 287, 301
    dimethylsulfide, 284–286, 293,
        298–303, 306
    ethylbenzene, 286, 287, 299–301
    formate, 103, 243, 263, 286, 287, 300
    glutamate, 179
    hydrazine, 180
    hydroxylamine, 180
    methylenetetrahydromethanopterin, 117
    steroid C25, 301
    xanthine, 44
Delftia acidovorans, 284
Deltaproteobacteria, 19, 242
Denitrification, 17, 149, 179, 180, 183–184,
    187, 200, 212–214, 217, 220, 283
Density functional theory (DFT), 27–28, 30,
    32, 52, 60, 134, 139, 225, 227, 295
Desulfobacter, 241, 258
    vibrioformis, 258, 259
Desulfobacterium autotrophicum, 241
Desulfobulbus, 240, 241, 258
    rhabdoformis, 258, 259
Desulfocapsa sulfoexigens, 245
Desulfococcus, 241, 255
Desulfocurvus, 241, 258
Desulfofuscidin, 255, 256, 259, 262
Desulfofustis, 241, 258
Desulfohalobium, 241, 243, 258
Desulfolobus, 241
Desulfoluna, 241
Desulfomicrobium (*Dsm.*), 241, 243, 244,
    258, 262, 263
    norvegicum, 241, 256, 258, 263, 265
Desulfomonile, 255
Desulfonatronovibrio, 241
Desulfonema, 255
Desulforegula, 255
Desulforubidin, 255, 256, 258, 262
Desulfosarcina, 258
Desulfosporomusa, 240
Desulfosporosinus, 240, 241, 243, 257

Desulfotomaculum (*Ds.*), 72, 241, 243, 250, 251, 257, 258
  carboxydovorans, 72
  nigrificans, 251, 256
Desulfovibrio, 116, 218, 221, 227, 239–245, 250, 252, 255, 257–262
  sulfodismutans, 245
  vulgaris, 218, 221, 222, 227, 228, 241, 250, 253–259, 261–262, 265
Desulfovirgula, 240
Desulfoviridin, 255, 256, 260, 262
Desulfurella, 241
Desulfurispora, 240
Desulfurococcales, 244
Desulfuromonas (*Drm.*) *acetoxidans*, 241, 256, 262–265
Desulfurylation, 239
DFT. *See* Density functional theory
Diabase, 3
Diazene (N_2H_2), 164, 166
Diazotrophs, 149
Digestive tracts, 240
Dihydrogen (H_2), 2, 18, 39, 43, 46–49, 57, 63, 64, 73, 91, 92, 99–121, 126, 127, 134, 135, 141, 149, 157, 160, 163–165, 170, 171, 203, 214, 216, 240, 244, 263, 265, 266
  cleavage, 100
  cycles, 100–111, 126
  energy, 100
  formation, 170
  oxidation, 48, 49, 103, 109, 114
  oxygenic metabolisms, 104
  photosynthetic, 102, 105, 109-111
  production, 100, 101, 105, 107–109, 113, 114, 117, 121
  solar-H_2 economy, 105–111
  solar production, 105, 108, 109, 114, 121
Dimethylsulfide (DMS), 279–306
  -degrading bacteria, 303
  flux to the atmosphere, 281
  in surface seawater, 282
  metabolism, 284
  monooxygenase, 283, 285, 303, 304, 306
  oxidation to dimethylsulfoxide, 284
  production, 282
  sources, 280–282
  transformations, 285–305
Dimethylsulfide degradation in
  anaerobic methylotrophic bacteria, 283
  autotrophic bacteria, 283
Dimethylsulfide dehydrogenases, 284–286, 293, 298–303, 306

environmental distribution, 299–300
  phylogeny, 300–302
Dimethylsulfide monooxygenases, 283, 285, 303–306
  phylogeny, 304, 305
Dimethylsulfone ($DMSO_2$), 115, 120, 284, 304, 306
Dimethylsulfoniopropionate (DMSP), 281, 282, 285
  degradation, 282
  lyase, 285
Dimethylsulfoxide (DMSO), 20, 22, 23, 26, 32, 43, 265, 281, 282, 284–302, 304, 306
  respiration, 297, 298
Dimethylsulfoxide reductase(s) (DMSOR), 20, 22, 23, 26, 32, 285–302, 306
  Dms-type, 288, 292, 301
  Dor-type, 290, 293, 295, 302
  phylogeny, 300–302
  properties, 286–295
  regulation of expression, 290–292
$DMSO_2$. *See* Dimethylsulfone
Dinitrogen (N_2), 2, 18, 22, 38, 42, 105, 109, 147–172, 179, 180, 184, 199, 205, 206, 212, 213, 216, 217, 220
  reduction, 42, 150, 165–166, 169, 170, 172
Dinitrogen monoxide. *See* Nitrous oxide (N_2O)
Dioxygen, O_2, 2–5, 18, 23, 25, 26, 28, 30, 38, 41, 48, 63, 76, 100, 102, 103, 105, 107–113, 116–120, 126, 132, 134, 135, 156, 171, 179–182, 184, 189, 190, 192, 198, 205, 206, 214, 215, 225, 232, 247, 249, 257, 267, 283, 287, 288, 291, 295, 303
  O_2/H_2O, 103, 108
  reduction, 103
  tolerance, 111–114
Disproportionation of
  inorganic sulfur intermediates, 245
  thiosulfate, 239, 245
Dissimilatory sulfate, 237–271
  reduction, 239, 242, 245, 250–252, 254, 260, 262
Dissimilatory sulfite reductase (Dsr, DSR), 242, 243, 250, 254, 255, 257–261, 269, 270
Dissimilatory thiosulfate reduction, 261
Dithiocarbamates (dtc), 27
Dithiolate(s) (dtl), 27, 120
Dithiolene, 23, 27, 28, 287, 295
Dithiomethylamine, 115, 120

Dithionite, 16, 20–22, 77, 152, 156–158, 188,
    195, 197, 199, 205, 222, 259, 292
DMSP. *See* Dimethylsulfoniopropionate
DMSO. *See* Dimethylsulfoxide
DNA, 7, 10, 11, 19, 222, 257, 291, 303
    B-, 10
Dolerite, 183
Ds. variabilis, 256
Dst. thermocisternum, 256–259

E
$E^{0'}$, *see also* Redox potentials, 40, 43, 51,
    244, 251, 260
Early Earth, 4–6, 12, 13, 40
    atmosphere, 1–13, 18, 101
Early life catalysts, 1–13
Earth, 1–13, 17–18, 40, 100, 101, 105, 108,
    126, 127, 148, 149, 211, 212, 231,
    245, 281
    atmosphere, 16, 17, 32
Ectothiorhodosinus, 247
Ectothiorhodospira, 247
Ectothiorhodospiraceae, 246
EDTA. *See* Ethylenediamine-*N,N,N'*,
    N'-tetraacetic acid
EGTA. *See* Ethylene glycol-bis
    (2-aminoethylether)-*N,N,N'*,
    N'-tetraacetic acid
Electrocatalysis, 81–84, 94, 95
Electron density, 39, 54, 198, 204
Electron nuclear double resonance (ENDOR),
    46, 79, 133, 136, 162–164, 193
    ^{19}F-, 132, 137
Electron paramagnetic resonance (EPR), 21,
    22, 46, 50, 51, 60, 63, 77, 79, 87, 111,
    113, 117, 119, 132–134, 136, 140, 152,
    156, 157, 163, 164, 166, 169, 187, 188,
    190, 192, 193, 195, 220, 252, 255, 259,
    262, 270, 299, 301
Electron spin echo envelope modulation
    (ESEEM), 133, 162, 164
Electron transfer or transport, 26, 46, 48, 50,
    52, 53, 58–60, 72, 74, 84, 85, 104, 111,
    114, 150, 154, 157, 159, 165, 166, 185,
    187, 190–194, 200, 201, 204–206, 224,
    227–230, 239, 242, 243, 255, 257, 263,
    266, 268, 287, 288, 297
    gated, 205
Energy
    conversion, 73
    metabolism, 73, 214, 244
Enterobacteriaceae, 220, 296

Enzymes, *see also* individual names
    Class I, 73
    Class II, 57, 73, 74
    Class III, 57, 72, 73
    Class IV, 82–84
    copper, 180, 185, 186, 217
    [Fe], 115–119
    [FeFe], 117, 120, 121
    involved in ammonia turnover, 217
    nickel, 120, 126
    [NiFe], 109, 113, 114, 121
    [NiFeSe], 114
Enzymology of
    dimethylsulfide transformations, 285–305
    hydrogen sulfide production, 250–262, 271
EPR. *See* Electron paramagnetic resonance
Epsilonproteobacteria, 245, 247
Escherichia (E.) coli, 25, 79, 92, 105, 106, 112,
    114, 115, 117, 191, 218, 221, 222, 229,
    230, 257, 262, 286, 288, 290–293,
    296–298, 301, 302
ESEEM. *See* Electron spin echo envelope
    modulation
Estuarine muds, 240
Ethane, 17, 18, 150
Ethanolamine, 19
Ethylbenzene dehydrogenase, 286, 287,
    299–301
Ethylene (C_2H_4), 16, 17, 19, 28, 31, 32,
    150, 171
Ethylene glycol-bis(2-aminoethylether)-*N,N,*
    N',N'-tetraacetic acid (EGTA), 303
Ethylenediamine, 215
Ethylenediamine-*N,N,N',N'*-tetraacetic
    acid (EDTA), 303
Ethyne, 16
Eubacterial sulfur reductase, 242–244,
    259, 262–265
Eukaryotic organisms, 214
Euryarchaeota, 244
Evolution, 6, 10, 62, 64, 91, 92, 95, 102, 110,
    113, 120, 121, 170, 179, 230, 287
Extended absorption fine structure (EXAFS),
    133, 167–169, 191, 193, 252

F
F_{430} cofactor. *See* Cofactors
FAD. *See* Flavin adenine dinucleotide
Fe-only nitrogenase, 149
[Fe] hydrogenase(s), 100, 117–119, 121,
    243, 265
    crystal structure, 119

[FeFe] hydrogenase(s), 100, 102, 107, 109, 110, 114–117, 120, 121
  biosynthesis, 117
  crystal structure, 119
  structure and function, 115–117
FeMoco. *See* Iron-molybdenum cofactor
Fermentation, 18, 20, 221, 245
  acetylene, 20
  carbohydrates, 221
Ferredoxin(s), 43, 46, 58, 76, 92, 109, 110, 158, 243, 258, 260, 261, 290
  -type cluster, 21, 22, 253, 255
Fertilizer, 183, 184, 215
Firmicutes, 41, 218, 221, 240, 248, 298
Fischer-Tropsch process, 91–92, 171
Flavin
  biosynthesis, 304
  cofactor, 303
  disulfide reductase, 268
Flavin adenine dinucleotide (FAD), 43, 45–48, 58, 64, 253, 268
Flavin mononucleotide (FMN), 285, 303, 304
  -dependent alkanesulfonate monooxygenases, 304
Flavobacterium, 284
Flavocytochrome *c*, 267, 268
Flavorubredoxins, 230
FMN. *See* Flavin mononucleotide
FNR. *See* Fumarate nitrate reductase
Formaldehyde, 12, 117, 280, 283, 303
Formate, 38, 40, 42, 106, 126, 127, 222, 243, 263, 283, 287
  dehydrogenase, 103, 243, 263, 286, 287, 300
  reductase, 23, 32
Fossil fuels, 40, 100, 102, 128
Fourier transform infrared (FTIR), 111, 113, 117
Freshwater, 246, 249, 283, 284
  sediments, 19, 260, 282
FTIR. *See* Fourier transform infrared
Fuel conversion, 102
Fumarate, 221, 228, 263
Fumarate nitrate reductase (FNR), 110, 291, 296
  -type transcription factors, 292
Fungi, 239

G
ΔG values, 19, 20, 27, 30, 103, 127, 128, 245, 251
Gabbro, 3
Gammaproteobacteria, 245–247, 249
Gas channel, 52, 58, 59, 79, 114
Gas chromatography, 101

Gas chromatography-mass spectrometry (GC-MS), 165, 171, 298
Gene expression, 250, 291
Genome, 49, 64, 221, 240, 244, 251, 257, 261, 263, 267, 290, 291, 297, 299, 306
Geobacter, 218
Geothermal springs, 102
Gibbs free energy. *See* ΔG
Global
  climate, 182, 216
  dihydrogen cycle, 101–103, 126
  nitrogen cycle, 148, 180, 220
  sulfur cycle, 240, 281, 306
Glutamate dehydrogenase, 179
Glutamine synthase, 179
Glyceraldehyde, 6, 110
Glycerol, 19
Gram-negative bacteria, 18, 183, 200, 221, 232
Granite, 3
Green sulfur bacteria (GSB), 239, 246, 247, 267–270, 284
Greenhouse
  effect, 2, 182, 183
  gas, 102, 125–142, 177–207
GSB. *See* Green sulfur bacteria

H
H. pseudoflava, 44, 46
H^+, 11, 73, 101, 103, 107
H^+/H_2, 103, 107
Haber-Bosch process, 184, 216
Haemophilus influenzae, 290, 302
Half-life of
  CO, 40
  N_2O, 182
Halitosis, 280
Halobacterium, 292, 298, 301, 302
Halochromatium, 247
Halomonas jeotgali, 299
Halorhodospira halophila, 247
Halothane, 181
Halothiobacillus, 247, 248
Hectorite, 4
Heliobacteria, 246
Helium, 2
Heme(s), 40, 180, 185, 191, 194, 201, 214, 217–220, 222–231, 256, 259, 261, 263, 266, 270, 299
  Fe(II), 222
  Fe(III), 222, 259
  octa-, 231, 242, 243, 261–263
  siro-, 220, 231, 254–259, 261, 262

Hemoglobin(s), 85, 230, 249
HEPES. *See* 2-[4-(2-Hydroxyethyl)piperazin-
     1-yl]ethanesulfonic acid
Hexacyanoferrate ([Fe(CN)$_6$]$^{3-}$), 21
Highest occupied molecular orbital (HOMO),
     39, 65, 225, 232
HOMO. *See* Highest occupied molecular
     orbital
Homochirality, 6, 7, 10–12
Homocitrate, 150, 156, 157, 169, 218
Hot springs, 127, 240, 246
Hydration, 19, 31, 182
Hydrazine (N$_2$H$_4$), 164–166, 170, 180, 213,
     217, 219
   dehydrogenase, 180
   hydrolase, 180
Hydrocarbons, 18, 91, 149, 150, 171,
     172, 250
Hydrocorphin, 129
Hydrogen bond(s), 10, 23, 24, 26–28, 30,
     45, 46, 54–56, 63, 157, 194, 205,
     206, 215, 225, 257
Hydrogen cyanide, *see also* Cyanide, 18, 249
Hydrogen peroxide, 126, 218, 230
Hydrogen sulfide (H$_2$S), 216, 220, 221, 231,
     237–272, 287
   effects on gene expression, 250
   metabolism, 240–249
   oxidation, 266–270
   production, 250–266, 271
   properties, 249
   toxicity, 249–251
Hydrogenase(s),16, 47, 73, 91, 92, 100–121,
     126, 160, 243, 260, 263–266
   [Fe], 100, 117–119, 121, 243, 265
   [FeFe], 100, 102, 107, 109, 110, 114–117,
     120, 121
   mechanism, 119, 120
   mimics, 108, 120
   [NiFe], 91, 100–103, 105, 109–116, 243,
     265, 266
   [NiFeSe] hydrogenases, 114
   nickel-free, 115–119
   photoelectrolysis, 107–109
Hydrogenivirga, 248
Hydrogenobacter, 248
Hydrogenovibrio, 248
Hydrolase
   hydrazine, 180
Hydrolysis of MgATP, 151
Hydrophobic channels, 53
Hydrothermal vent, 102, 127, 249
Hydroxo ligand, 25, 28

Hydroxide ion (OH$^-$), 5, 25, 28, 39, 77,
     79, 113, 115
Hydroxyapatite, 12
Hydroxylamine, 180, 213, 217, 218, 221,
     226, 227, 230–232, 261
   dehydrogenase, 180
2-[4-(2-Hydroxyethyl)piperazin-1-yl]
     ethanesulfonic acid (HEPES),
     22, 33, 295, 307
Hydroxylases
   molybdenum, 43–46, 49, 64
Hyperfine sublevel correlation spectroscopy
     (HYSCORE), 133, 136, 142, 162,
     164, 173
Hyperthermus, 241, 244
Hyphomicrobium sp., 303
   sulfonivorans, 303–305
HYSCORE. *See* Hyperfine sublevel
     correlation spectroscopy

I

Ice age, 182
ICP-MS. *See* Inductively coupled plasma-mass
     spectrometry
IDS. *See* Indigodisulfonate
Igneous rock, 3, 4
Ignicoccus, 244
Illite, 5, 11
Indigodisulfonate (IDS), 154, 156, 169, 173
Inductively coupled plasma-mass spectrometry
     (ICP-MS), 20, 22, 33
Infra-red, 111, 119, 121, 182
Inhibition of
   CO oxidation, 86, 90
   CO$_2$ reduction, 86, 87
   CODH, 89–91
Inhibitor(s), 16, 25, 47, 55, 71–95, 119, 136,
     139, 161, 197, 198, 252
Interstellar gas clouds, 18
IR. *See* Infrared
Iron, Fe, 3–5, 16, 21, 22, 24, 25, 38, 50, 53,
     62, 73, 76, 77, 79, 100–102, 110–112,
     115–119, 121, 129, 149, 150, 154–158,
     162–169, 173, 214, 219, 220, 222–225,
     227, 243, 250, 255–257, 262, 267,
     271, 285
   ^{57}Fe, 162
   oxides, 3
Iron(II), Fe^{2+}, 51, 63, 77, 119, 183, 222,
     225, 227, 303
Iron(III), Fe^{3+}, 225, 259, 262, 298
   Fe$^{3+/2+}$, 5

Iron-iron hydrogenase, 100, 102, 107, 109,
    110, 114–117, 120, 121
Iron-molybdenum cofactor, 150, 154–157,
    160, 161, 163–165, 168, 169
Iron sulfur clusters. *See* Cluster(s)
Isobacteriochlorin, 220
Isochromatium, 247

K
Kaolinite, 5, 11, 12
Klebsiella pneumoniae, 150
Kocuria, 304
Kyoto Protocol, 183

L
Lakes, 18, 127, 246
Lamprobacter, 247
Lamprocystis, 247
Last universal common ancestor
    (LUCA), 64
Laughing gas, *see also* Nitrous oxide, 181
Lead, 250
Leifsonia, 304
Lignin, 127, 282
Lipids, 11, 179
Liquid natural gas, 128, 140, 141
Lithium nitride (Li$_3$N), 215
Lowest unoccupied molecular orbital (LUMO),
    39, 40, 182, 225
LUCA. *See* Last universal common ancestor
LUMO. *See* Lowest unoccupied
    molecular orbital

M
M. thermophila, 57
Magnesium, Mg^{2+}, 3–5, 152, 251, 285, 303
    MgADP, 152, 153, 157, 159, 160, 167
    MgATP, 151
Magnetic circular dichroism (MCD), 78, 133
Magnetospirillum, 248
MALDI-TOF. *See* Matrix-assisted laser
    desorption/ionization time of flight
    analysis
Manganese, Mn^{2+}, 109, 250, 251
Mantle, 3
Marcus theory, 227
Marichromatium, 247
Marine
    algae, 282
    sediments, 19, 40, 127, 241, 249, 260, 283

Marinobacter, 282, 284
    hydrocarbonoclasticus, 184, 187, 189,
        195, 198, 199
Mars, 2, 4
Marshes, 127, 283
Mass spectrometry, 119, 132, 303
    gas chromatography-, 165, 171, 298
Matrix-assisted laser desorption/
        ionization time of flight analysis
        (MALDI-TOF), 20
MCR. *See* Methyl-coenzyme M reductase
Mechanism of
    CO oxidation, 48, 56
    dioxygen tolerance, 111–114
    methyl-coenzyme M reductase, 125–142
    N$_2$ reduction, 150, 172
    Ni-Fe CODH, 78
    ping-pong, 299
    reversible carbon dioxide reduction, 55, 56
Menaquinol
    8-methyl-, 263
    oxidase, 201
Mercaptoheptanoylthreonine phosphate
        (CoBSH), 130–132, 136, 137, 139, 140
Mercury, 2
Metabolism of
    dihydrogen, 100
    dimethylsulfide, 284
    energy, 73, 214, 244
    hydrogen sulfide, 240–249
    methane, 128
    methylsulfide, 284
    oxygenic dihydrogen, 104
    selenium, 285
    sulfur, 246, 249, 266–268, 285
    tetrathionate, 260
    thiosulfate, 260
Metallochaperones, 127
Metamorphic rock, 4
Meteorites, 12
Methane, CH$_4$, 16, 18, 43, 73, 101, 117,
        125–142, 171, 180, 183, 284
    acidity constant, 140
    generation, 127, 128, 139, 140
    metabolism, 128
    monooxygenase, 16, 127, 180, 284
    oxidation, 40, 127, 128, 140–142, 180
    utilization, 127, 128
Methanesulfonic acid (MSA), 281, 284
Methanethiol, 281–283, 303
Methanobacterium thermoautotrophicum, 57
    Marburg, 244
Methanobacterium, 241, 244

Methanocaldococcus jannaschii, 118, 132, 260
Methanococcoides burtonii, 260
Methanococcus, 241, 244
    thermolithotrophicus, 244
    voltae, 132
Methanoculleus thermophilus, 132
Methanogenesis, 127–129, 139, 140
    reverse, 127, 128, 140
Methanogens, 42, 57, 73, 101, 105, 117, 127,
        129, 132, 140, 244, 284
    acetoclastic, 73
Methanol, 39, 180, 244
Methanopyrus, 244
    kandleri, 130, 132, 260
Methanosarcina, 244
    acetivorans, 284
    barkeri (Ms.), 54, 79, 130, 132, 241, 244,
        256, 261–262, 284
    frisia, 57
Methanothermobacter (M.) marburgensis,
        130, 132
Methanothermus, 244
Methionine
    S-adenosyl-, 132
    sulfoxide (MetSO), 292, 293
    synthase, 139
Methyl
    amine, 115, 120
    iodide, 133, 136, 139
    -Ni(III), 130, 132, 134–137
    radical, 136, 137, 139
    -SCoM methyltransferase, 139
    sulfide metabolism, 284
    transfer, 283, 284
Methyl-coenzyme M reductase (MCR),
        125–142
    catalytic mechanism, 137–140
    MCR I, 132
    MCR II, 132
    nickel center, 133–137
    properties, 130–132
    reactivity, 130–132
    structure, 128–132
Methyl-viologen (MV$^+$), 84, 195, 197,
        199, 261
    radical, 84
Methylation of sulfide, 282
Methylenetetrahydromethanopterin
        dehydrogenase, 117
N-Methylformamide (NMF), 157, 168
Methylmercaptopropionate (MMPA), 284
Methylobacterium, 283
    podarium, 304

Methylophaga, 283
Methyltransferase(s), 42, 283, 285, 306
8-Methyl-menaquinone, 263
Michaelis-Menten constants (K_M), 20, 47, 50,
        84, 292, 293, 299, 303
Microbes, 40–42, 64, 100, 102, 103,
        105, 109, 111, 114, 127, 129, 149,
        212, 221
Microbial
    H$_2$ metabolism, 100
    mats, 102, 127, 240, 246
    oxidation of hydrogen sulfide, 266–270
Microorganism(s), *see also* individual names,
        41–43, 72, 92, 105, 114, 150, 212, 230,
        232, 239–241, 245, 246, 249, 250, 252,
        259, 260, 266, 270, 271, 282–284,
        295, 296
    methylotrophic, 282, 283
Midpoint potential, 51, 154, 166, 193
Mineral(s), *see also* individual names, 3–6, 11,
        12, 183
    clay, 4–13
MMPA. *See* Methylmercaptopropionate
Molecular dynamics, 52
Molecular oxygen, *see also* Dioxygen,
        41, 100, 180, 184
Molybdate, 19–21, 291, 292
Molybdenum
    ^{95}Mo, 163
    Cu,Mo-CODHs, 43–49, 55, 62, 63
    Fe,Mo cofactor, 150, 154–157, 160, 161,
        163–165, 168, 169
    hydroxylases, 43–46, 49, 64
    MoFe clusters. *See* Clusters
    MoFe protein, 150–152, 154–165,
        169, 218
    MoFe$_7$S$_9$C-homocitrate, 150
    Mo=S bond, 44
Molybdenum(IV), 46, 48, 295
Molybdenum(V), 22, 46, 49
    Mo$^{(V/IV)}$ couple, 293, 299
Molybdenum(VI), 46, 48, 63, 295
    Mo$^{(VI/V)}$ couple, 293, 299
Molybdenum nitrogenase, 64, 149–167,
        169–172
    catalytic mechanism, 149, 150, 154,
        157–166, 169
    properties, 149–157
Molybdopterin
    bis-guanine dinucleotide. *See*
        Bis-molybdopterin guanine
        dinucleotide
    cytosine dinucleotide, 44, 45

Monooxygenase(s)
  ammonia. *See* Ammonia-
  dimethylsulfide, 283, 285, 303–306
  flavin mononucleotide-dependent, 304
  methane, 16, 127, 180, 284
Montmorillonite, 4–6, 11, 12
Moorella (M.) thermoacetica, 50, 58, 73, 78, 79
Mössbauer spectroscopy, 63, 77, 156, 169, 262
MSA. *See* Methanesulfonic acid
Multiheme cytochromes *c*, 211–232, 242, 263
Muscovite, 5
Mutagenesis, 20, 25–27
Mycobacteria, 41
Mycobacterium tuberculosis, 41, 231

N
NAD. *See* Nicotinamide adenine dinucleotide
NADH. *See* Nicotinamide adenine
      dinucleotide reduced
NapA, 213, 228, 290
NapC, 218, 227, 228
Naphthoquinone-8, 261
NarG. *See* Soluble nitrate reductase
Natural gas, 100, 128, 140, 141, 240, 249
Nautilia profundicola, 218
NCO⁻, 78, 79, 85, 87–90
Neisseria gonorrhoeae, 218
Neurotransmission, 41
Neutron scattering analysis, 119
Nickel, 3, 50, 51, 53–56, 58, 59, 61–63, 76–79,
      101, 111, 112, 114, 115, 120, 125–142,
      150, 265
  ^{63}Ni-F$_{430}$, 129
  Ni,Fe clusters. *See* Clusters
  Ni-CODHs, 71–95
  transporters, 127
Nickel(0), Ni^0, 51, 60, 61, 63, 77, 78
Nickel(I), Ni$^+$, 60, 129, 132–137, 139, 141
  Ni(I)-F$_{430M}$, 133
Nickel(II), Ni^{2+}, 51, 59–61, 63, 77, 78, 111,
      129, 130, 132–137, 139, 140
  Ni(II)-F$_{430M}$, 133
  Ni(II)/Ni(I), 133, 135
Nickel(III), Ni^{3+}, 60, 91, 111, 113, 129, 130,
      132–137, 139
[NiFe] hydrogenase(s), 91, 100–103, 105,
      109–116, 243, 265, 266
  crystal structure, 119
  function, 111–114
[NiFeSe] hydrogenases, 114
Nicotinamide adenine dinucleotide
      (NAD), 23

Nicotinamide adenine dinucleotide reduced
      (NADH), 108, 243, 261, 303
  -dependent flavin oxidoreductases, 304
NifH, 151, 166, 167, 218
Nitrate, 216
  ammonification, 180, 212, 214
  reduction, 42, 127, 128, 179, 184,
      213–215, 292
Nitrate reductase, 16, 23, 25, 32, 179, 213,
      293, 299–301
  Nap-type, 300
  Nar-type, 300
  soluble (NarG), 25, 299
Nitrate reduction, 42, 127, 128, 179, 184,
      213–215, 247, 292
Nitric acid (HNO$_3$), 215, 216
Nitric oxide (NO), 179–181, 183, 184, 212,
      213, 215–221, 226, 227, 230–232, 249
Nitric oxide reductase, 183, 184, 213
Nitrification, 180, 183, 212, 217
  de-. *See* Denitrification
Nitrite
  oxidase, 180
  denitrification, 213
Nitrite reductase, 180, 187, 213, 217–231
  cytochrome *c* (NrfA), 180, 213, 217, 218,
      220–232
Nitrogen, *see also* Dinitrogen
  ^{15}N, 164, 193
  fixation, 16, 148–150, 180, 213, 214
  oxides, *see also* individual names, 182, 216
Nitrogenase(s), 16, 64, 105, 147–172, 180, 213,
      214, 217, 218
  Fe-only, 149
  molybdenum, 64, 149–167, 169–172
  superoxide dismutase-dependent, 149
  vanadium, 149, 150, 157, 166–172
Nitrosomonas europaea, 219, 231
Nitrospirae, 240
Nitrous oxide (N$_2$O), 16, 179–189,
      191–195, 197–207, 212, 216, 220,
      221, 226, 232
  atmospheric chemistry, 18, 181–184
  activation, 203–206
  emissions, 182, 183
  environmental effects, 181–184
  properties, 181, 182
Nitrous oxide reductase (N$_2$OR), 16,
      177–207, 213
  three-dimensional structures, 189, 190
NMF. *See* *N*-Methylformamide
NMR. *See* Nuclear magnetic resonance
Nocardia rhodochrous, 18

nos operon, 185, 200, 201
NosZ protein, 185, 187, 200–202
NO$_x$, *see also* individual names, 182, 216
Nrf-dependent ammonification, 213
NrfA, 180, 213, 217, 218, 220–232
NrfH, 218, 222, 227–230
Nuclear magnetic resonance (NMR), 119, 129,
     132, 202
    proton, 119
Nucleic acids, *see also* DNA *and* RNA, 6, 7,
     11, 12, 179
Nucleophilic attack, 29–32, 39, 48, 49, 55,
     60, 77, 79, 136, 139, 263

O
Ocean, 2, 3, 5, 127, 214, 280–282
Oil wells, 127
Oligopeptides, 11
Oligotropha carboxidovorans, 44, 46, 47, 72
Operons
    encoding DMSO reductases, 289
    nos, 185, 200, 201
Organic matter, 40, 214, 249, 282
Origin of life, 1, 6, 10, 101
Ostwald process, 215, 216
Oxidases
    arsenite, 300
    cytochrome, 189–191, 193, 194
    cytochrome *bo*$_3$, 191
    menaquinol, 201
    nitrite, 180
    sulfite, 269, 304
    xanthine, 44
Oxidation of
    carbon monoxide. *See* Carbon monoxide
    hydrogen sulfide to sulfate, 245, 266–270
    polysulfides, 268–269
    sulfide, 268
    sulfite, 269
    thiosulfate, 269, 270
Oxidative stress, 75
Oxides, 3, 195, 293, 296
Oxidoreductases
    hydroxylamine, 213, 231, 232
    quinolone 2-, 49
    quinone-interacting, 242
    sulfide:quinone (SQR), 267, 268
    sulfite, 245
Oxygen. *See* Dioxygen
Oxy-sulfur reductases, 260, 261
Ozone (O$_3$), 108, 182, 183
    depletion, 182, 183

P
P-582-type sulfite reductase, 255–258
Pa. denitrificans, 184, 189-191, 193, 195, 198
Paper mill industry, 249, 280
Paracoccus (Pc.), 248, 270, 304
    denitrificans, 184, 189-191, 193, 195, 198
    pantotrophus, 270
    versutus, 270
Particulate matter, 216
Pasteurellaceae, 296
PDB. *See* Protein Data Bank
Pelobacter acetylenicus, 16, 18–32
Pelodictyon, 247
Perchlorate reductase, 301
Petroleum refinery, 249
PFE. *See* Protein film electrochemistry
PGE. *See* Pyrolytic graphite 'edge' electrode
Phaeovibrio, 247
1,10-Phenanthroline, 85
Phosphodiester bonds, 7
Photosynthesis, 3, 72, 102, 109, 110, 179, 291
Photosynthetic
    algal mats, 102, 109, 110
    dihydrogen, 102, 105, 109–111
Photosystem II, 109, 110
Phototrophic bacteria. *See* Bacteria *and*
     individual names
Phyllosilicate, 4, 5
Phylogeny of
    dimethylsulfide dehydrogenases, 300–302
    dimethylsulfide monooxygenases,
     304–305
    dimethylsulfoxide reductases, 300–302
Physiology of *D. vulgaris*, 221, 250
Phytoplankton, 282
Ping-pong catalytic mechanism, 299
Planctomycetes, 213, 219, 223
Planets, 2, 4, 17, 18, 101, 149, 240
Plant(s), 3, 40, 72, 180, 239, 246, 249, 281
PNSB. *See* Purple non-sulfur bacteria
Polar ice cores, 182
Pollution, 285
Polyethylene glycol (PEG), 19, 22
Polysulfide reductase (Psr), 229, 244, 263–266
Polysulfides, 136, 229, 239, 244, 246, 263, 265,
     268, 269
Polythionates, 239, 242
Ponds, 127, 183
Potassium, K$^+$, 3, 89
Primordial
    atmosphere, 40
    energy, 15–33
    solar nabula, 2

Production of
   ammonia, 126, 211-232
   ammonium, 180
   carbon monoxide, 126, 250
Prokaryotes, 214
   sulfate-reducing (SRP), 240–242,
      253–255
Propane, 18, 150
1,2-Propanediol, 19
Propyne, 28, 31, 32
Prosthecochloris, 247
Protein Data Bank (PDB), 10, 21, 22, 58, 75,
      76, 112, 116, 118, 130, 131, 149, 153,
      155, 158, 167, 168, 186, 192, 197, 198,
      204, 223, 224, 226, 294
Protein film electrochemistry (PFE), 75, 77–82,
      85, 87, 95, 113
Protein film voltammetry, 293, 295
Protein(s), *see also* individual names
   blue copper, 193
   Fe, 150–154, 157–160, 166–167
   MoFe, 150–152, 154–165,
      169, 218
   NosZ, 185, 187, 200–202
   TATA binding, 7
   VFe, 166–170
Proteobacteria, 223, 245, 247, 248, 286,
      291, 304
Proteus, 261
Proton coupled electron transfer, 60
PSB. *See* Purple sulfur bacteria
Pseudaminobacter, 248
Pseudoclavibacter, 304
Pseudomonas (P.)
   chloritidismutans, 299
   nautica, 184, 189
   perfectomarina, 184
   putida, 284
   stutzeri, 184, 185, 187, 189, 190-195, 198,
      199, 201–205
Psr. *See* Polysulfide reductase
Pterin(s), 43, 46, 117, 287, 290
Purines, 11
Purple non-sulfur bacteria (PNSB), 246,
      247, 267
Purple sulfur bacteria (PSB), 239, 246,
      247, 267–270
Pyranopterin cofactor, 43, 46, 287, 290
Pyridine, 215
Pyrimidines, 11
Pyrobaculum (Pyb.), 241, 244
   aerophilum, 185
   islandicum, 244, 258, 259

Pyrococcus, 244
   furiosus, 241, 265, 266
Pyrodictium (Py.), 244, 266
   abyssi, 241, 252, 265, 266
   brockii, 265
Pyrolytic graphite 'edge' electrode (PGE),
      81–84, 94, 95
Pyrophyllite, 5

Q
Quinolone 2-oxidoreductase, 49
Quinone(s), 45, 47, 104, 106, 219, 229, 230,
      265, 268, 291
   -interacting membrane oxidoreductase, 242

R
Radical
   cation, 137, 227
   chlorine oxide, 182
   hydroxyl, 40, 281
   methyl, 136, 137, 139
   MV^+, 84
   $^{\bullet}$SCoM, 137, 139
   thiyl, 136, 137, 139, 140
   transfer, 227
Ralstonia (R.) eutropha, 103, 104, 109,
      113–115
Rate constants, 47, 63, 85, 141, 250
   k_{cat}, 47, 50, 82, 136, 292, 303
Reaction mechanism (for), 25, 27–32, 48–49,
      55, 59–62, 129, 141, 150, 172, 295
   acetylene hydratase, 27–32
   acetyl-coenzyme A synthase, 59–62
Redox potential(s), *see also* $E^{0'}$, 21, 51, 76–78,
      84, 103, 107, 129, 133, 154, 200, 253,
      293, 299
   midpoint, 51, 154, 166, 193
Reductases
   adenosine 5'-phosphosulfate, 242, 245,
      251, 253, 267, 269, 270
   assimilatory sulfite, 221, 222, 254, 256,
      261, 262
   biotin sulfoxide, 296
   chlorate, 299, 301, 302
   cytochrome *c* nitrite (NrfA), 180, 213, 217,
      218, 220–232
   dimethylsulfide, 286–295, 300–302
   dimethylsulfoxide, 20, 22, 23, 26, 32,
      285–302, 306
   dissimilatory sulfite, 242, 243, 250, 254,
      255, 257–261, 269, 270

Reductases (*cont.*)
  eubacterial sulfur, 242–244, 259, 262–265
  formate, 23, 32
  fumarate nitrate, 110, 291, 296
  methyl-coenzyme M, 125–142
  nitrate, 16, 23, 25, 32, 179, 213, 293,
    299–301
  nitric oxide, 183, 184, 213
  nitrite, 180, 187, 213, 217–231
  nitrous oxide, 16, 177–207, 213
  oxy-sulfur, 260, 261
  perchlorate, 301
  polysulfide, 229, 244, 263–266
  quinolone 2-oxido-, 49
  selenate, 290, 296, 301
  sulfide:quinone oxido-, 267, 268
  sulfite, 221, 222, 226, 231, 242, 251,
    254–262, 269
  tetrathionate, 231, 260, 261, 267
  oxygenase-, 267
  thiosulfate, 251, 253, 260, 261
  tetrathionate, 231, 260, 261, 267
  TorZ N-oxide, 296, 302
  TorZ S-oxide, 296, 302
  trimethylamine N-oxide, 290, 292, 293,
    296, 300, 301
  Ynf selenate, 296
Reduction of
  acetylene, 16, 17
  carbon dioxide. *See* Carbon dioxide
    reduction
  carbon monoxide, 40, 42, 92, 150, 171, 172
  dinitrogen, 42, 165–166, 169, 170
  dioxygen, 103
  flavin disulfide, 268
  nitrate, 42, 127, 128, 179, 184, 213–215,
    220, 292
  sulfate, 239, 240, 242, 245, 250–252, 254,
    260, 262
  sulfur, 237–272
Respiration (or Respiratory)
  anaerobic, 221, 282, 26, 288, 296
  chain, 41, 228, 230, 263, 288
  mechanism, 41, 110, 184, 200, 214, 221,
    228–230, 240, 242, 243, 260, 263, 282,
    288, 295–298
  cellular, 110
  nitrate reduction, 213, 214, 220
  pathway, 184
Reverse
  methanogenesis, 127, 128, 140
  tricarboxylic acid cycle, 41
Rhabdochromatium, 247

Rhodobaca, 247
Rhodobacter (R.)
  capsulatus, 247, 268, 286, 290–295
  sphaeroides, 286, 290–293
Rhodobium, 247
Rhodococcus, 284
  A1, 18
Rhodomicrobium, 247
Rhodopila, 247
Rhodoplanes, 247
Rhodopseudomonas palustris, 247
Rhodospirillum rubrum, 43, 49, 50, 72, 73,
    78, 247
Rhodothallasium, 247
Rhodovibrio, 247
Rhodovivax, 247
Rhodovulum, 247
  sulfidophilum, 284, 298, 299
Ribulose monophosphate cycle, 283
RNA, 7, 10–12
Rock, 2–4, 127
  igneous, 3, 4
  metamorphic, 4
Roseospira, 247
Roseospirillum, 247
Rothia, 304, 305
Ruthenium
  dyes, 107
  photosensitizer, 92
Rubrivivax, 247
Ruegeria pomeroyi, 41

S
S^0 state, 43, 239, 241–246, 249, 262–266, 269
Sagittula stellata, 284, 299, 300
Salmonella, 102, 241, 261
  enterica Serovar Typhimurium, 261
  typhimurium, 296
Saltmarshes, 283
Saponite, 4, 5
Saturn, 18
SDS-PAGE. *See* Sodium dodecyl sulfate
    polyacrylamide gel electrophoresis
Seawater, 282
Sedimentary rock(s), 2, 4
Sediments, 5, 18, 19, 40, 127, 214, 241, 242,
    246, 249, 260, 282, 283
  anoxic, 19, 282, 283
  aquatic, 127
  freshwater, 19, 260, 282
  marine, 19, 40, 127, 241, 249, 260, 283
Selenate reductase, 290, 296, 301

Selenium, 285
Selenocysteine (SeCys), 114, 300
Semiconductor nanoparticles, 93
Sepiolite, 5
Sequence alignments, 296, 298, 301
Serine cycle, 283
Shewanella, 228, 241, 292, 297
   oneidensis, 218, 221, 222, 228, 231,
     261, 297, 298
Siderite, 183
Silicates, 4
Silicibacter pomeroyi, 41
Silicon, 3–5
Silver, Ag$^+$, 17
   diammine, 215
Single-crystal spectroscopy, 207
Singly occupied molecular orbital (SOMO), 46
Siroheme, 220, 231, 254–259, 261, 262
Site-directed mutagenesis, 20, 25–27
Smectite, 4, 5
Sodium, Na$^+$, 3, 11, 22, 133
   azide, 22
   cacodylate, 22
   dithionite, 16, 188, 195, 197, 199,
     205, 222
Sodium dodecyl sulfate polyacrylamide
   gel electrophoresis (SDS-PAGE),
   20, 300
Soil, 4, 6, 40, 102, 127, 128, 183, 184, 221, 232,
   240, 246, 249, 260, 261, 281, 282
   microbes, 40, 41, 102
   sulfur levels, 281
Solar
   dihydrogen economy, 105–111
   energy, 105, 108, 109
   fuel, 95, 102, 109, 111
   -H$_2$ economy, 105–111
   H$_2$ production, 105, 108, 109, 114, 121
   system, 2, 101
Soluble nitrate reductase (NarG), 25, 299
SOMO. *See* Singly occupied molecular orbital
Sphaerotilus, 248
SQR. *See* Sulfide:quinone oxidoreductase
SRB. *See* Sulfate-reducing bacteria
SRP. *See* Sulfate-reducing prokaryotes
Staphylothermus, 241
Stappia stellulata, 41
Starkeya spp, 247, 248
Steroid C25 dehydrogenase, 301
Stetteria, 241, 244
Stratosphere, 183
Streptomyces, 113
   thermoautotrophicus, 149

Structures of, *see also* Crystal structures
   and X-ray structures
   bacterial CODH/ACS, 58, 59
   cluster C from CODH, 50–52
   coenzyme F$_{430}$, 129, 130
   Cu,Mo-CODHs, 44–46
   cytochrome c nitrite reductase, 221–227
   Fe protein, 166, 167
   hydrogenases, 43–64, 111–117
   methyl-coenzyme M reductase, 128–132
   [NiFe] hydrogenases, 111–114, 119
   nitrous oxide reductase, 189, 190
Stygiolobus, 241, 244, 248
Sudoite, 5
Sugars, 6, 7, 10, 12, 19, 240
Sulfate, 42, 43, 110, 127, 128, 216, 221, 226,
   237–272, 281, 282
   adenylyltransferase, 251
   dissimilatory, 237–271
   reduction, 18, 43, 72, 105, 114, 127, 128,
     222, 237–272, 284
   respiration, 240, 242
   -reducing archaea, 240–242, 259–260
   -reducing bacteria (SRB), 18, 72, 105, 222,
     240–243, 249–251, 253, 254, 257, 260,
     262, 271, 284
   -reducing prokaryotes (SRP), 240–242,
     253–255
Sulfide (S^{2-}), 43, 47, 77, 79, 85, 89–91, 156,
   157, 164, 195, 202, 203, 205, 239, 240,
   245, 249–251, 254, 255, 257, 260–263,
   267–270, 282, 284
   hydrogen. *See* Hydrogen sulfide
   methylation, 282
   oxidation, 239, 247, 249, 268
   -oxidizing bacteria, 239, 242, 245–249
   poly-. *See* Polysulfides
Sulfide:quinone oxidoreductase (SQR),
   267, 268
   X-ray structure, 268
Sulfite (SO$_3^{2-}$), 218, 220, 221, 224, 226, 231,
   243, 245, 246, 253–257, 259–262, 267,
   269–271
   oxidase, 269, 304
   oxidoreductase, 245
   reductase, 221, 222, 226, 231, 242, 251,
     254–262, 269, 270
   reduction, 239, 251, 254, 258, 260, 261
   -disproportionating bacteria, 245
   -oxidizing bacteria, 245–249
Sulfobacillus, 248
Sulfolobaceae, 248
Sulfolobales, 244, 266

Sulfolobus solfataricus, 252
Sulfospirillum carboxydovorans, 43
Sulfonates, 136, 137
Sulfur
    ^{33}S, 134
    bacteria (colorless), 239, 246–249,
        267, 271
    bacteria (green), 239, 246, 247,
        267–270, 284
    cycle, 239, 240, 245, 246, 281, 306
    soil levels, 281
    metabolism, 246, 249, 266–268, 285
    oxygenase reductase, 267
    phototrophic bacteria. *See* Bacteria
    reducing archaea, 240–242, 244, 255, 259,
        260, 265, 266, 306
    reduction, 237–272
    -oxidizing bacteria, 239, 245, 248, 252,
        253, 269–271
    -reducing eubacteria, 242, 244, 263, 265
Sulfuric acid, 38, 129, 216
Sulfuricella denitrificans, 248
Sulfuricurvum, 248, 300
Sulfurihydrogenibium sp., 248, 300
Sulfurimonas spp., 247, 248
Sulfurisphaera ohwakuensis, 248
Sulfuritalea, 248
Sulfurospirillum deleyianum, 221–224, 241,
        244, 265
Sulfurovum, 247, 248
Sun, 2, 3, 105
Superoxide, 126
Superoxide dismutase-dependent
        nitrogenase, 149
Swamps, 127
Synechocystis, 109
Syngas, 91, 141
Synthesis of ATP, 43, 103, 104
Syntrophobacter, 241

T
T. mobile, 256, 259
Taenia solium, 7
Talc, 5
Tat pathway, 201, 202
TATA binding protein, 7
Tetrahydrofolate (H$_4$F), 42, 57, 65, 285
Tetrapyrrole, 8, 126, 129, 132–135, 257
Tetrathionate, 218, 260, 261, 267, 269,
        270, 282
    reductase, 231, 260, 261, 267
Therapeutic agent(s), 41, 216
Thermocrinis, 248
Thermithiobacillus, 248

Thermochromatium, 247
Thermocladium, 240, 244
Thermococcales, 244, 266
Thermococcus, 241, 244
Thermodesulfatator, 240, 241
Thermodesulfobacterium, 240, 241, 259
    commune, 241, 256, 259
Thermodesulfobium narugense, 240
Thermodesulfobium, 241
Thermodesulfovibrio, 240, 241, 259
    hydrogeniphilus, 259
    yellowstoni, 259
Thermodiscus, 244
Thermoplasma, 244
Thermoplasmatales, 244
Thermoproteales, 240, 244
Thermoproteus, 244
Thermothrix, 248
Thermus thermophilus, 191, 229, 252
Thioalkalicoccus, 247
Thioalkalimicrobium, 247, 248
Thioalkalivibrio, 218, 231, 247, 248
    nitratireducens, 218, 231
    paradoxus, 218, 231
Thiobaca, 247
Thiobacillus, 247, 248, 283
    thioparus, 283
Thiocapsa roseopersicina, 247
Thioclava, 247, 248
    pacifica, 247
Thiocyanate, 85, 89–91
Thiocystis, 247
Thiodictyon, 247
Thioethers, 136
Thiofaba, 248
Thioflavicoccus, 247
Thiohalomonas, 247, 248
Thiohalophilus, 248
Thiohalorhabdus, 248
Thiohalospira, 247, 248
Thiolamprovum, 247
Thiolates, 27, 120, 281–283, 303
Thiomargarita, 248
Thiomicrospira, 247, 248
    denitrificans, 185
Thiomonas, 248
Thiopedia, 247
Thiophaeococcus, 247
Thioploca, 248
Thioprofundum, 248
Thiorhodococcus, 247, 299
    drewsii, 299
Thiorhodospira, 247
Thiorhodovibrio, 247
Thiospira, 248

Thiospirillum, 247

Thiosulfate, 43, 218, 239, 243, 245, 246, 249,
     251, 257, 260–262, 267, 270, 282
  disproportionation, 239, 245
  reductases, 251, 253, 260, 261
  oxidation, 269, 270
  -oxidizing bacteria, 245–249
Thiothrix, 247, 248
Thiovirga, 248
Thiovulum, 247, 248
Thiyl radical, 136, 137, 139, 140
Thorneley–Lowe model, 158–166
Three-dimensional structures of, *see also*
     Crystal structures, Structures *and*
     X-ray structures
  Cu$_A$, 191, 192
  nitrous oxide reductase, 189, 190
Titan, 18
  atmosphere, 18
Titanium
  dioxide (TiO$_2$), 92–95, 107–108
  Ti(III) citrate, 16, 20, 51, 133, 137, 154
TMAO. *See* Trimethylamine N-oxide
TorZ N-oxide reductase, 296, 302
TorZ S-oxide reductase, 296, 302
Toxic gas, 37–64, 237–271
Toxicity, 239
  of hydrogen sulfide, 249–251
Trace gas, 16, 17, 41, 182
Transformation of dinitrogen to ammonia,
     147–172
Transcription factors
  fumarate nitrate reductase-type, 292
Trimethylamine N-oxide (TMAO), 290, 292,
     293, 296, 298
Trimethylamine N-oxide reductase, 290, 292,
     293, 296, 300, 301
Triple bond, 3, 16–18, 26, 28, 38, 147–172
Trithionate, 251, 257, 261, 262
Troposphere, 40
Tungstate, 19–21
Tungsten, W, 20–30, 32
  W(IV), 21
  W(V), 21
  W(VI), 21

U
Ubiquinones, 41, 288
United States of America, 19
Urease(s), 64, 126
Utilization of methane, 127, 128
UV-visible, 130, 133, 136, 188, 270

V
Vanadium nitrogenase, 149, 150, 157, 166–172
Venus, 2
Vermiculite, 5, 12
Verrucomicrobia, 224
VFe protein, 166–170
Vinyl alcohol, 28, 30, 31
Volcanoe(s), 2
  emission, 18, 40
Voltammetry, 81, 82, 84, 90, 94, 293, 295, 299
Vulcanisaeta, 240, 241

W
Waste disposal, 285
Water(s), 2–5, 16, 18, 19, 22, 24–32, 38–40,
     48, 51–53, 55, 56, 62, 63, 91, 92, 100,
     105, 107–109, 114, 115, 127, 179,
     181, 182, 184, 197, 198, 203, 205,
     206, 214, 215, 221, 225, 227, 230,
     240–242, 246, 249, 257, 260, 280–282,
     287, 295
  oxidation, 108, 109
  treatment, 249
  vapor (H$_2$O$_v$), 2, 39
Wolinella (*W.*) *succinogenes*, 185, 187, 201,
     218, 221, 222, 224, 226, 227, 229,
     230, 241, 261, 263–266
Wood-Ljungdahl pathway, 42, 56
Würm ice age, 182

X
X-ray absorption spectroscopy (XAS), 46, 47,
     77, 134–136, 167, 169
X-ray crystallography, 77, 111, 129, 136,
     203, 222
X-ray diffraction, 198
X-ray structure of, *see also* Crystal structures
     and Structures
  SQR, 268
  acetylene hydratase, 22
Xanthine dehydrogenase, 44
Xanthine oxidase, 44
XAS. *See* X-ray absorption spectroscopy

Y
Ynf selenate reductase, 296

Z
Zinc, Zn^{2+}, 12, 58, 59, 62, 215, 250–252

Printed by Printforce, the Netherlands